BIOSTRATIGRAPHIC AND GEOLOGICAL SIGNIFICANCE OF PLANKTONIC FORAMINIFERA

BIOSTRATIGRAPHIC AND GEOLOGICAL SIGNIFICANCE OF PLANKTONIC FORAMINIFERA

UPDATED SECOND EDITION

MARCELLE K. BOUDAGHER-FADEL

UCLPRESS

First published in 2015 by
UCL Press
University College London
Gower Street
London WC1E 6BT

Available to download free: www.ucl.ac.uk/ucl-press

First edition 2012
Second edition 2013

A CIP catalogue record for this book is available
from The British Library.

ISBN: 978-1-910634-24-0 (Hbk.)
ISBN: 978-1-910634-25-7 (Pbk.)
ISBN: 978-1-910634-26-4 (PDF)
ISBN: 978-1-910634-27-1 (epub)
ISBN: 978-1-910634-28-8 (mobi)
DOI: 10.14324/111.9781910634257

Acknowledgments

Following the great success of making the second edition of *Biostratigraphic and Geological Significance of Planktonic Foraminifera* available freely online (over 3000 copies were downloaded in less than two years), I am delighted that an updated second edition will now be published by UCL Press. This edition contains a few minor corrections, an index and some colour figures (in the online PDF). However, as well as being freely available online, print on-demand hard-copies will also be available, for those who prefer their books as physical entities. The charts mentioned in the book are available in the online PDF (http://dx.doi.org/10.14324/111.9781910634257).

The creation of this revised second edition has been enabled by my excellent colleagues at UCL Press, Jaimee Biggins and Lara Speicher.

I should repeat here the acknowledgements from the first two editions of this book. The second edition was created and enabled by my colleague and friend Prof. David Price. While, in the writing of the first edition, I was helped by numerous other friends and colleagues. Specifically, I wrote for the first edition that:

I am indebted to former colleagues Prof. Alan Lord and the late Prof. Fred Banner, who enabled me to establish my research career at UCL. In that context, I would also like to express my gratitude to my colleagues in the Department of Earth Sciences at UCL and especially to those involved with the Micropalaeontology Unit collection. I am particularly grateful for the assistance of Mr James Davy with material preparations over the years and to the Natural History Museum London, especially for Prof. Norman Macleod and colleagues, for giving me access to their excellent collection. I would like to thank Dr Kate Darling for providing the photographs of living planktonic foraminifera used in this book. My work has also been enriched by working with the members of the South East Asia Consortium Group, Royal Holloway, and I have benefited greatly from working with industrial colleagues, including those from Petrobras, Chevron and Corelab.

However, notwithstanding all of the above, the book would not have been possible without the help and support of my colleague Prof. David Price. Prof. Price's advice throughout the writing of this book, and our valuable and stimulating discussions gave me insights into, and new understanding of, the relationship between the small floating planktonic foraminifera and large-scale, global geological processes. I would also like to thank him for helping me delve into the wider processes involved in evolution and for his unstinting encouragement throughout the project. Finally I would like to thank my family, and especially my sons Nicholas and Michael, for their support and consideration throughout the writing of this book.

There are many photographs and illustrations in this book. Most are original, but some are reproduced from standard sources. I have tried to contact or reference all potential copyright holders. If I have overlooked any or been inaccurate in any acknowledgement, I apologise unreservedly and I will ensure that suitable corrections are made in subsequent editions.

Needless to say, despite all the help and assistance that I have had, there will undoubtedly still be errors and omissions in this book. For these I must take full responsibility.

Marcelle K. BouDagher-Fadel
London, 2015

Contents

Chapter 1

An introduction to planktonic foraminifera

1.1 The biological classification of the foraminifera

Foraminifera are marine, free-living, amoeboid protozoa (in Greek, proto = first and zoa = animals). They are single-celled eukaryotes (organisms the cytoplasm of which is organized into a complex structure with internal membranes and contains a nucleus, mitochondria, chloroplasts, and Golgi bodies, see Fig. 1.1), and they exhibit animal-like (cf. plant-like) behaviour. Usually, they secrete an elaborate, solid carbonate skeleton (or test) that contains the bulk of the cell, but some forms accrete and cement tests made of sedimentary particles. The foraminiferal test is divided into a series of chambers, which increase in number during growth. In life, they exhibit extra-skeletal pseudopodia (temporary organic projections) and web-like filaments that can be granular, branched and fused (rhizopodia), or pencil-shaped and pointed (filopodia). The pseudopodia emerge from the cell body (see Plate 1.1 below) and enable bidirectional cytoplasmic flow that transports nutrients to the body of the cell (Baldauf, 2008). Foraminifera first appeared in the Cambrian with a benthic mode of life and, over the course of the Phanerozoic, invaded most marginal to fully marine environments. They diversified to exploit a wide variety of niches, including, from the Late Triassic or Jurassic, the planktonic realm. These planktonic forms are the focus of this book.

Both living and fossil foraminifera come in a wide variety of shapes. They occupy different micro-habitats and exploit a diversity of trophic mechanisms. Today, they are extremely abundant in most marine environments from near-shore to the deep sea, and from near surface to the ocean floor. Some even live in brackish habitats.

The complexity and specific characteristics of the structure of foraminiferal tests (and their evolution over deep-time) are the basis of their geological usefulness. After the first appearance of benthic forms in the Cambrian, foraminifera became abundant, and by the late Palaeozoic, they exhibited a relatively large range of complicated test architectures. Their continued evolution and diversification throughout the Mesozoic and Cenozoic, and the fact that they still play a vital role in the marine ecosystem today, means that foraminifera are of outstanding value in zonal stratigraphy, paleoenvironmental, paleobiological, paleoceanographic, and paleoclimatic interpretation and analysis.

Fossil and living foraminifera have been known and studied for centuries. They were first mentioned in Herodotus (in the fifth century BC), who noted that the limestone of the Egyptian pyramids contained the larger benthic foraminifera *Nummulites*. Their name is derived from a hybrid of Latin and Greek terms meaning "bearing pores or holes," as the surfaces of most foraminiferal tests are covered with microscopic perforations, normally visible at about 40x magnification. Among the earliest, workers who described and drew foraminiferal tests were Anthony van Leeuwnhock in 1600 and Robert Hooke in 1665, but the accurate description of foraminiferal architecture was not given until the nineteenth century (see Brady, 1884; Carpenter *et al.*, 1862; see Fig. 1.2).

The systematic taxonomy of the foraminifera is still undergoing active revision. The first attempts to classify foraminifera placed them in the Mollusca, within the genus *Nautilus*. In 1781, Spengler was among the first to note that foraminiferal chambers are, in fact, divided by septa. In 1826, d'Orbigny, having made the same observation, named the group foraminifera. In 1835, foraminifera were recognized by Dujardin as protozoa, and shortly afterwards, d'Orbigny produced the first classification of foraminifera, which was based on test morphology. The taxonomic understanding of foraminifera has advanced considerably over the past two decades, and recent studies of molecular systematics on living forms are revealing their very early divergence from other protoctistan lineages (Wray *et al.*, 1995). In this book, we follow Lee's (1990) elevation of the Order Foraminiferida to Class Foraminifera, and the concomitant elevating of the previously recognized suborders to ordinal level. Throughout this book, therefore, the suffix "-oidea" is used in the systematics to denote superfamilies, rather than the older suffix "-acea", following the recommendation of the International Commission on Zoological Nomenclature (see the International Code of Zoological Nomenclature 1999, p. 32, Article 29.2). Modern workers normally use the structure and composition of the test wall as a basis of primary classification, and this approach will be followed here.

Despite the diversity and usefulness of the foraminifera, the phylogenetic relationship of foraminifera to other eukaryotes remains unclear. According to early genetic work on the origin of the foraminifera by Wray *et al.* (1995), the phylogenetic analysis of verified foraminiferal DNA sequences indicates that the foraminiferal taxa are a divergent "alveolate" lineage, within the major eukaryotic radiation. Their findings cast doubt upon the assumption that foraminifera are derived from an amoeba-like ancestor, and they suggested that foraminifera were derived from a heterokaryotic flagellated marine protist. For these authors, the phylogenetic placement of the foraminifera lineage is a problem, as the precise branching order of the foraminifera and the "alveolates" remained uncertain. Following the work of Wray *et al.* (1995), many scientists have tried to trace the origin of the foraminifera using a variety of methods, but molecular data from foraminifera have generated conflicting conclusions. Molecular phylogenetic trees have assigned most of the characterized eukaryotes to one of the eight major groups. Archibald *et al.* (2003) indicated that cercozoan and foraminiferan polyubiquitin genes (76 amino acid proteins) contain a shared derived character, a unique insertion, which implies that foraminifera and cercozoa share a common ancestor. They proposed a cercozoan–foraminiferan supergroup to unite these two large and diverse eukaryotic groups. However, in other recent molecular phylogenetic studies, the foraminifera are assigned to the Rhizaria, which are largely amoeboid unicellular forms with root-like filose or reticulosed pseudopodia (Archibald, 2008; Cavalier-Smith, 2002; Nikolaev *et al.*, 2004). The cercozoa and foraminifera groups are included within this supergroup (see Fig. 1.3). Additional protein data, and further molecular studies on rhizarian, cercozoan, and foraminiferan forms, are necessary in order to provide a more conclusive insight into the evolution and origins of these pseudopodial groups.

Figure 1.1. An equatorial section of a foraminiferal test, *Spirillina vivipara* Ehrenberg. D, ingested diatoms; N, nuclei Pr, proloculus, P, large phagosomal vacuole (after Alexander, 1985).

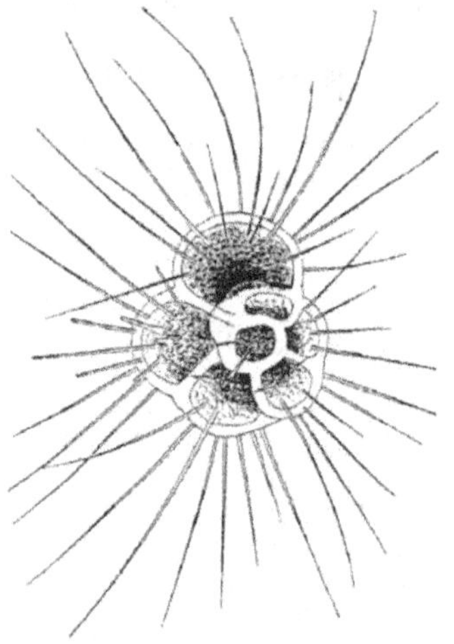

Figure 1.2. A living species of *Globorotalia* (probably belonging to the *G. cultrata* (d'Orbigny) group) drawn by J.J. Wild from off New Guinea, during the HMS Challenger Expedition (from Brady, 1884). Although Brady supposed Wild's drawing to represent "spines," no known *Globorotalia* has spines. We believe them to be pseudopodia. This figure is the first representation of the extrathalamous cytoplasm giving rise to pseudopodia on the surface and near the periphery of a globorotaliid foraminifera.

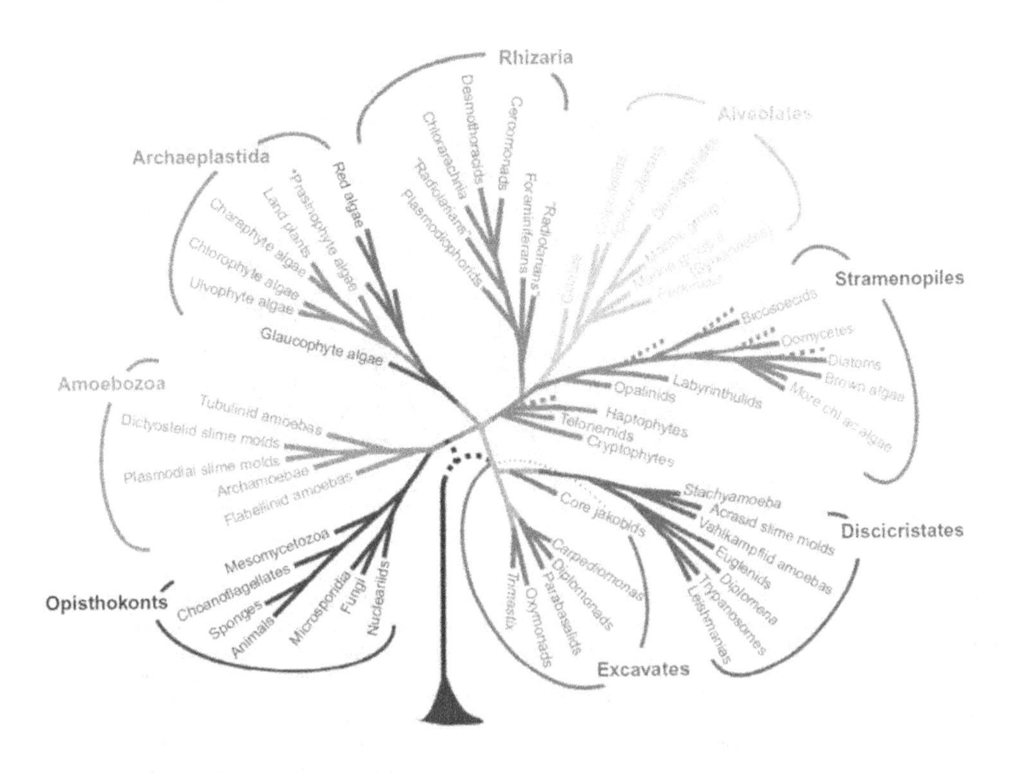

Figure 1.3. A consensus phylogeny of eukaryotes from Baldauf (2008).

1.2 Planktonic foraminifera

Foraminifera are separated into two types following their life strategy, namely, the benthic and the planktonic foraminifera. The benthic forms occur at all depths in the marine realm. They vary in size from less than 100 µm in diameter to a maximum breadth of many centimetres. Benthic foraminiferal tests may be agglutinated (quartz or other inorganic particles being stuck together by calcitic or organic cements), or may be primarily secreted and composed of calcite, aragonite, or (rarely) silica. They include many species that live attached to a substrate or that live freely and include organic-walled and agglutinated small foraminifera that dominate the deep-sea benthic microfauna, as well as a major group of foraminifera with complicated internal structures, the so-called larger benthic foraminifera (BouDagher-Fadel, 2008), that include major reef-forming species. However, the other type of foraminifera, which is just as successful as their benthic ancestors, namely, the planktonic foraminifera, is the subject of our study, and the remainder of this book will be focused on them.

Planktonic foraminifera have tests that are made of relatively globular chambers (that provide buoyancy) composed of secreted calcite or aragonite. They float freely in the upper water of the world's oceans, with species not exceeding 600 µm in diameter. They have a global occurrence and occupy a broad latitudinal and temperature zone. The majority of planktonic foraminifera float in the surface or near-surface waters of the open ocean as part of the marine zooplankton. The depth at which a given species lives is determined in part by the relative mass of its test, with deeper dwelling forms usually having more ornamented and hence more massive tests. Upon death, the tests sink to the ocean floor and on occasion can form what is known as a foraminiferal ooze. On today's ocean floors, *Globigerina* oozes (named after the important foraminiferal genus *Globigerina* that dominates the death assemblage ooze) may attain great thickness and cover large areas of the ocean floor that lie above the calcium carbonate compensation depth, the depth below which all $CaCO_3$ dissolves. Today, planktonic and larger benthic calcareous foraminifera are among the main calcifying protists, contributing almost 25% of the present-day carbonate production in the oceans (Langer, 2008). Planktonic foraminifera occur, therefore, in many types of marine sediment, which on lithification yield carbonates or limestones. These rocks become hardened and denser on lithification, and their constituent microfossils often can only be studied in thin section. They can be dated by the presence of a few key planktonic foraminiferal taxa, which provide excellent biostratigraphic markers, and are sometimes the only forms that can be used to date carbonate successions (see Fig. 1.4 and subsequent chapters).

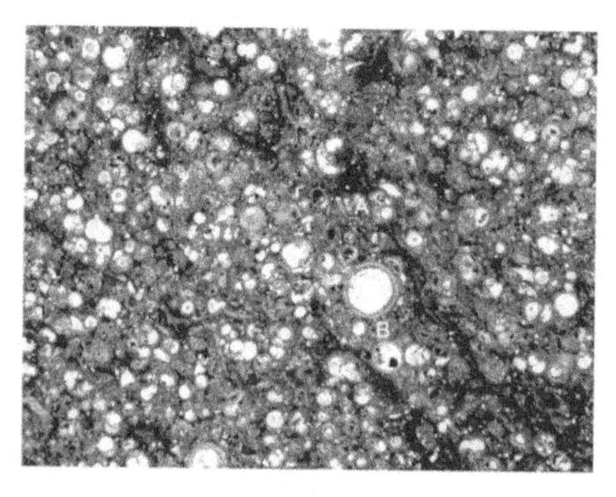

Figure 1.4. A photomicrograph showing a carbonate thin section of a micritic packstone (x10) composed of a Miocene planktonic foraminiferal assemblage in which species of *Globigerina* and *Globigerinoides* are abundant. (A) *Globoquadrina dehiscens* (Chapman, Parr, and Collins) and (B) *Orbulina universa* d'Orbigny (B) are also present. Image from the UCL Collection.

Planktonic foraminifera show high diversity and adaptability, both in their morphology and biology. Planktonic foraminifera have undergone significant evolution since their first development from benthic forms in the Late Triassic or Jurassic (see Chapter 3). They consist of a large number of identified and stratigraphically defined species, and exhibit a rich and complex phylogenetic history. Foraminiferal tests of fossil and living forms have been systematically described (at generic and suprageneric levels) by Loeblich and Tappan (1964, 1988). What is known about living foraminifera has been reviewed by Lee and Anderson (1991) and their colleagues, while the biology of modern planktonic foraminifera has been presented by Hemleben *et al.* (1989) and Sen Gupta *et al.* (1997). More recently, their proteins and molecular biology have been analyzed in greater detail, and this will be discussed further in Chapter 2. Fossilized forms, however, are known, of course, only by their tests. Morphological criteria such as the globular nature of the test and other features (discussed below and in more detail in subsequent chapters) have been used to determine that fossil forms were indeed planktonic, while the other microfossils associated with them have helped to ascertain their deep-water marine habitat and, in some cases, to constrain their age determination.

Because of the abundance of planktonic foraminiferal tests in most marine sediments, and because of the regularity of their structures and their taxonomic diversity, they provide continuous evidence of evolutionary changes (see Fig. 1.5) from which detailed phylogenetic relationships can be established. Their well-defined biostratigraphic ranges and phylogenetic relationships have been found to be useful in both academic studies of global evolution and by the hydrocarbon industry for correlation in sedimentary sequences. In particular, the petroleum exploration industry finds the planktonic foraminifera to be of great utility, because they are easy to extract from both outcrop and subsurface samples, and enable biostratigraphic dating to be carried out in new exploration areas very quickly. Examples of their industrial use come from publications sponsored, for example, by Exxon (Stainforth *et al.*, 1975a, b), the Royal Dutch Shell Group (Postuma, 1971), and British Petroleum (Blow, 1979). Postuma (1971) presented illustrations of Albian and younger forms, while Stainforth *et al.* (1975a, b) and Blow (1979) deal solely with Cenozoic taxa in the West, and Subbotina (1953) with those of the former Soviet Union. The planktonic foraminifera that are found in sediments of Middle Cretaceous and younger age have, for over 50 years, been used for worldwide biostratigraphic correlation (e.g., Bolli, 1957).

Taxonomic research is fundamental to maximizing the usefulness of planktonic foraminifera in stratigraphical studies, as precise zonal stratigraphy depends upon precise discrimination of genera and species. Planktonic foraminifera are classified taxonomically using criteria based on the characteristics of their external calcareous test. Identification is based on general morphology as well as the ultrastructural and microstructural features of the test (Hemleben *et al.*, 1989) as seen by transmission electron microscopy (TEM) (Bé *et al.*, 1966) and scanning electron microscopy (SEM) (Cifelli, 1982; Lipps, 1966; Scott, 1974). The features of the planktonic foraminiferal tests of importance in classification at the generic and specific levels usually deal with their chamber arrangements, the nature of sutures, the wall structures, and the nature of external ornamentation, perforations, apertures, and accessory structures. Some of the classifications put emphasis upon aperture position and external apertural modifications, while others distinguish the different families on the basis of fine details of wall structure and wall surface, including whether they are smooth, pitted (possessing distinct external pore-funnels with externally enlarged outlet of the pores), or spinose (possessing spines).

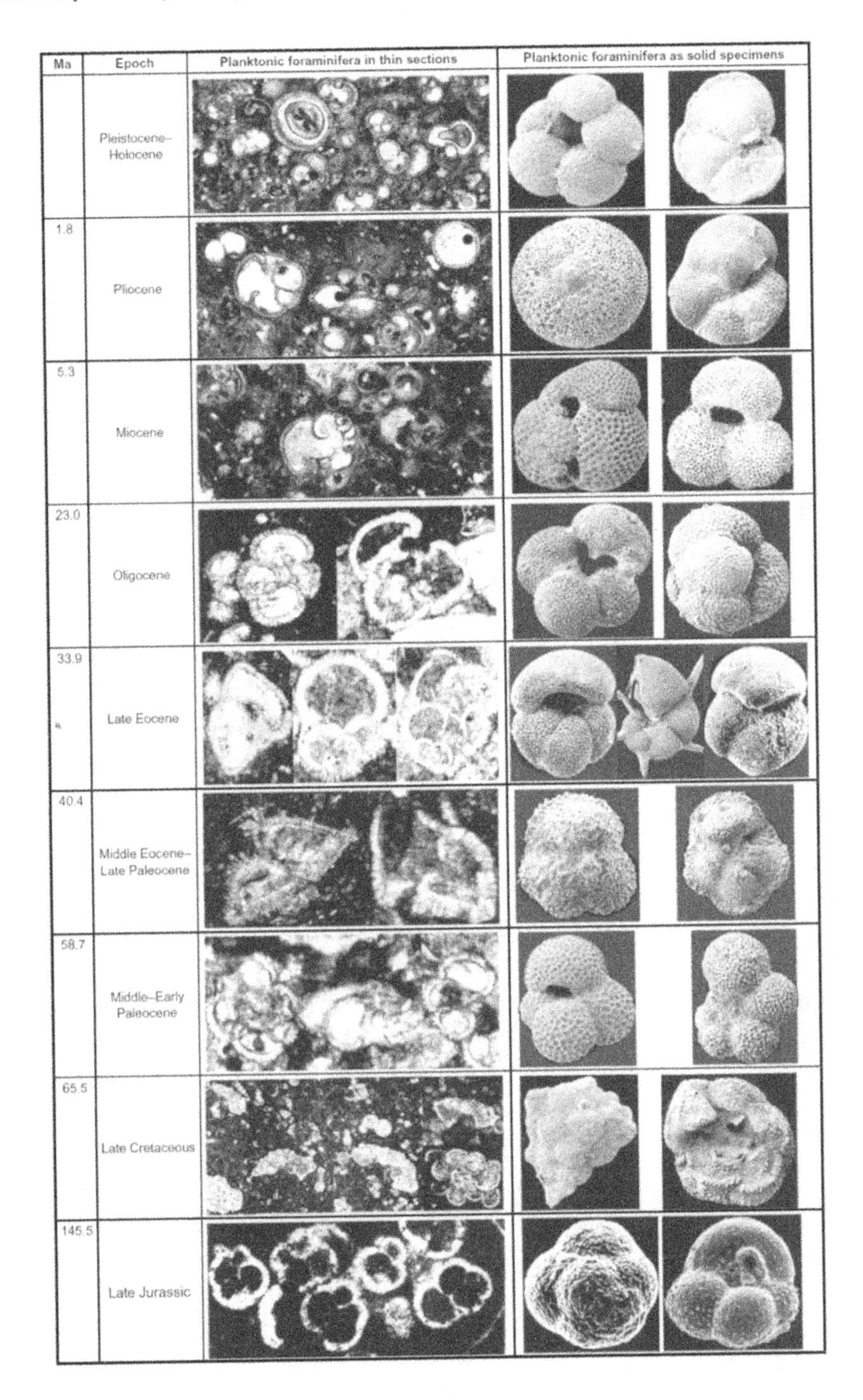

Figure 1.5. The succession of the main planktonic foraminiferal assemblages through geological time. Images from UCL Collection, except were indicated. Species of *Conoglobigerina* and *Globuligerina* are found in the Late Jurassic, from Wernli and Görög (2000) and BouDagher-Fadel, *et al.* (1997) ; *Globotruncana* and *Racemiguembelina* in the Late Cretaceous; *Subbotina* and *Praemurica* in the Early and Middle Paleocene; *Acarinina* and *Morozovella* in the Late Paleocene and Middle Eocene; *Turborotalia*, from Blow (1979), and *Hantkenina* in the Late Eocene; *Dentoglobigerina* and *Catapsydrax* in the Oligocene; *Globigerinoides* in the Miocene; *Orbulina* and *Globorotalia* in the Pliocene; and *Neogloboquadrina* and *Globorotalia* in Pleistocene and Holocene.

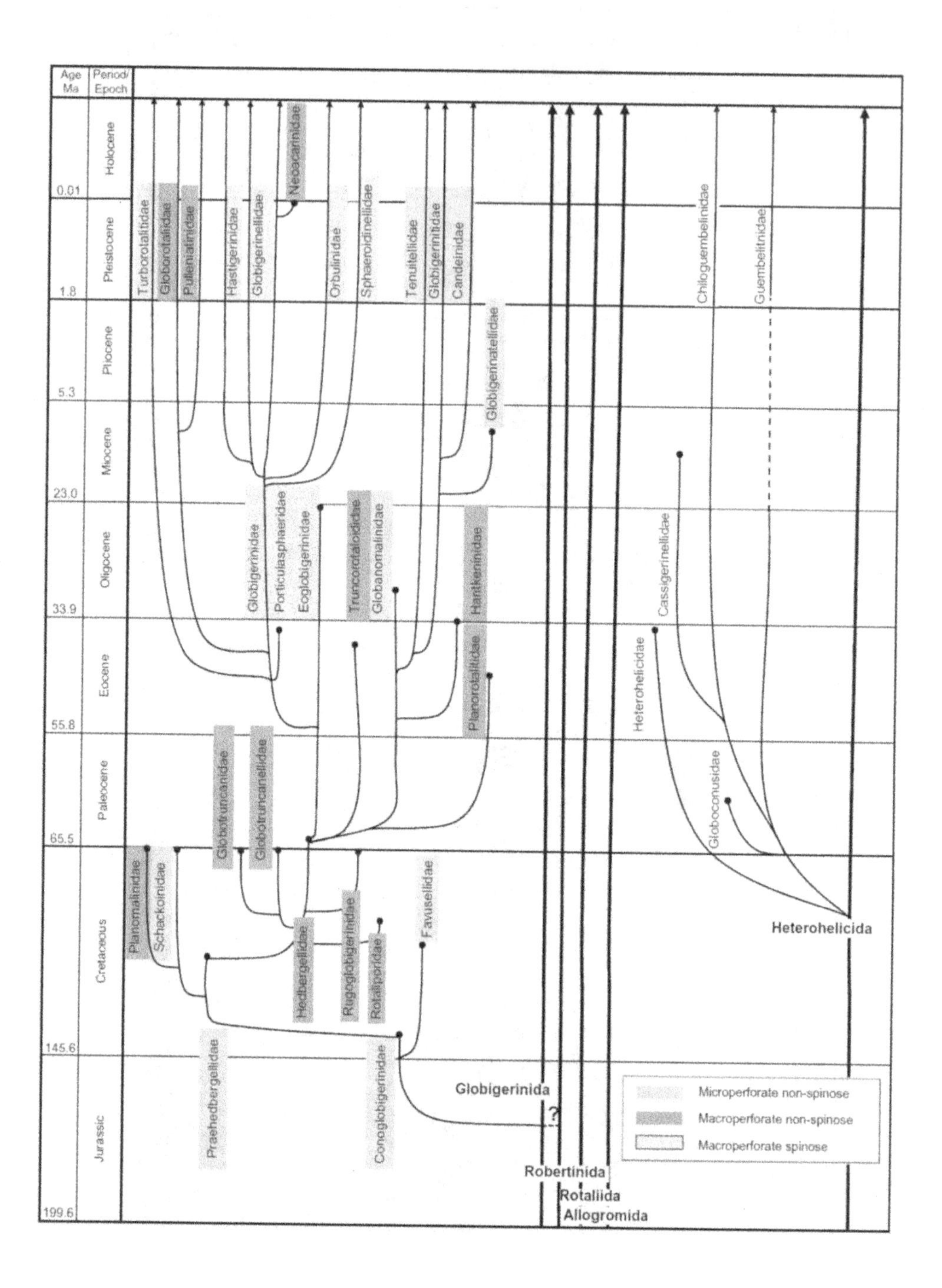

Figure 1.6. The evolution of the planktonic foraminiferal families (thin lines) from their benthic ancestors.

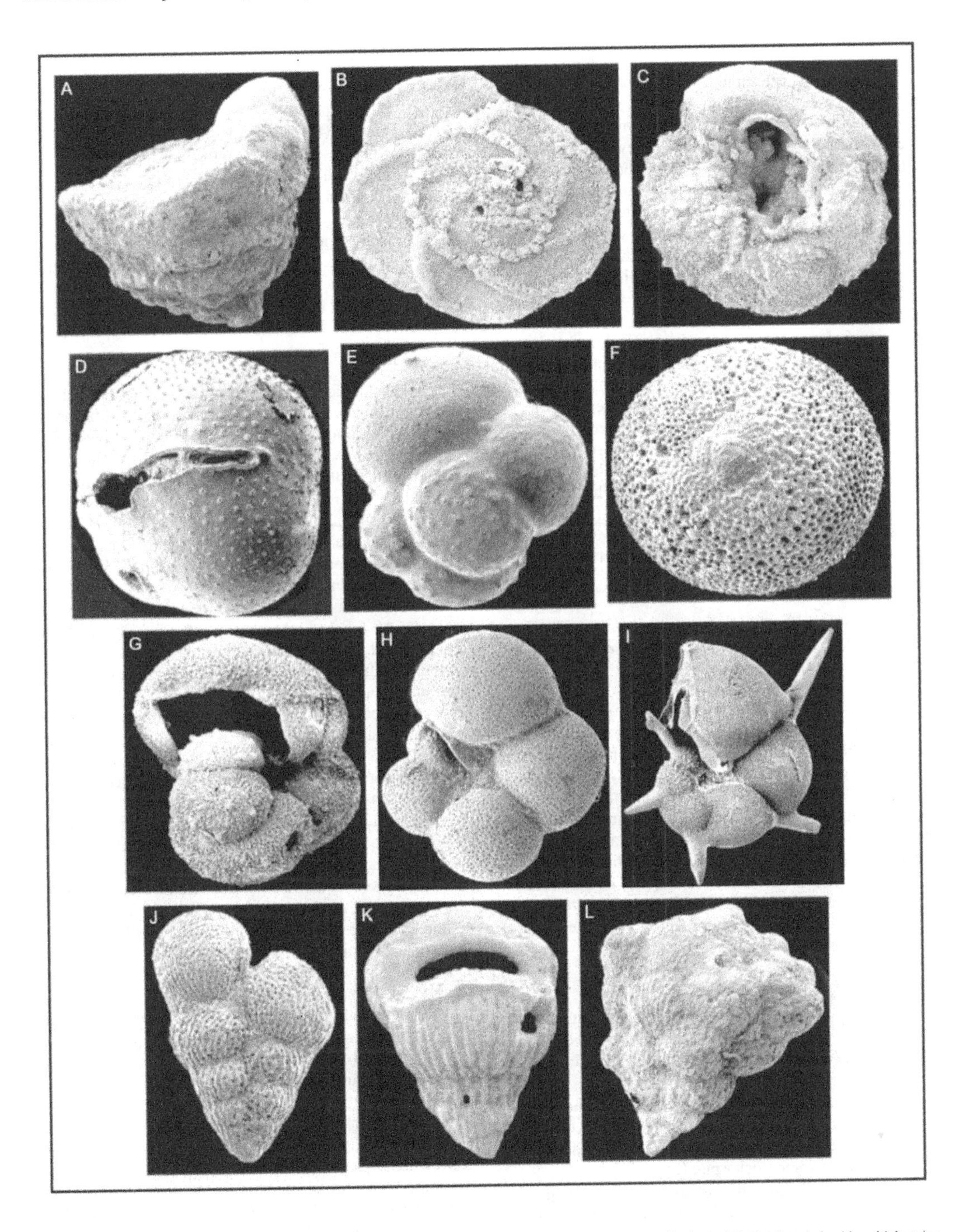

Figure 1.7. Examples of different styles of coiling in planktonic foraminifera (images from UCL Collection), (A) Trochospiral with a high spire, *Contusotruncana*; (B, C) Trochospiral with low spire, *Globotruncana*; (D) Trochospiral with the adult test coated in a thick, smooth cortex of calcite, *Sphaeroidinella*; (E) Inflated enrolled biserial form coiled into a tight, involute trochospire, *Cassigerinella*; (F) Streptospiral test with the last globular chamber completely embracing the umbilical side, *Orbulina* ; (G) Planispiral, biumbilicate test, *Hastigerina*;(H) Trochospiral with a compressed smooth test, *Turbeogloborotalia*; (I) Planispiral, biumbilicate test with tubulospines, *Hantkenina*;(J) Biserial test, *Heterohelix*; (K) Biserial test strongly increasing in thickness, *Pseudotextularia*; (L) Multiserial test, *Racemiguembelina*.

This morphological approach to taxonomy has led to the identification of many families of planktonic foraminifera that have evolved, and (on many occasions) gone into extinction, since their initial development from benthic ancestors in the Late Triassic or Jurassic. The entire phylogenetic lineage of the planktonic foraminifera is shown in Fig. 1.6. In subsequent chapters, the process by which this evolutionary tree has been established will be explained, but first in this chapter, the different morphological characteristics, upon which the taxonomy of planktonic foraminifera is based, will be discussed with reference to exemplar forms belonging to the specific families of foraminifera named in Fig. 1.6. These foraminifera will be discussed in a systematic way in subsequent chapters, where we will combine all criteria of structure, sculpture, and morphological features to present an overview of the taxonomy of the different families, genera, and species at different stages of the stratigraphic column. An excellent glossary of the terminology used in the description of foraminiferal morphology has recently been electronically published by Hottinger (2006), and this should be referred to as necessary for exact definitions of some of the terms introduced below.

1.2.1 The morphology, sculpture, and structure of the test of planktonic foraminifera

Foraminiferal tests rarely consist of only one chamber; usually, as the organism grows, it adds successively additional, progressively larger chambers to produce a test of varying complexity. The intrinsic buoyancy of the planktonic foraminifera is provided by the generally globular nature of their chambers. Some living planktonic foraminifera add a new chamber every day and grow at a rate that sees them increase their diameter by about 25% per day (Anderson and Faber, 1984; Bé *et al.*, 1982; Caron *et al.*, 1981; Erez, 1983; Hemleben *et al.*, 1989).

Planktonic foraminifera have different patterns of chamber disposition (see Fig. 1.7):

- Trochospiral growth has the chambers coiling along the growth axis while also diverging away from the axis. The test has dissimilar evolute spiral and involute umbilical sides (Fig. 1.7A–C, H).
- Involute trochospiral growth has the chambers biserial or triserial in early stages, later becoming enrolled biserial, but with biseries coiled into a tight, involute trochospire (Fig. 1.7E).
- Planispiral growth has the chambers coiling along the growth axis but showing no divergence away from the axis. The test is biumbilicate, with both the spiral and umbilical sides of the test being identical and symmetrical relative to the plane of bilateral symmetry (Fig. 1.7G, I).
- Streptospiral growth has the chambers coiling in successively changing planes, or with the last globular chamber completely embracing the umbilical side (Fig. 1.7F).
- Uniserial, biserial, triserial multiserial, etc., patterns of growth have (after an initial planispiral or trochospiral stage) chambers arranged in one, two, three, or more rows in a regularly superposed sequence. The biserial form is planar (Fig. 1.7J, K), but multiserial forms can be three dimensions forming a conical test (Fig. 1.7L).

The planktonic foraminifera have a simple test with no internal structures and are, therefore, quite distinct from the larger benthic foraminifera. These latter can develop canal systems within the walls (Fig. 1.8A), plugs and pillars within the septa and umbilici, and internal toothplates that modify the routes of exit and ingress of the cytoplasm through the aperture (Fig. 1.8B, Ca; see BouDagher-Fadel, 2008). As can be seen in thin section, planktonic foraminifera do not develop plugs, pillars, or canal systems (see Fig. 1.8Cb–F).

The aperture of the planktonic foraminifera is the main opening of the last chamber cavity into the ambient environment. It can open completely in the umbilicus, umbilical/intraumbilical aperture (Fig. 1.9H), extend from the umbilicus toward the periphery of the test, intra–extraumbilical (Fig. 1.9K), or open completely over the periphery, unconnected to the umbilicus, extraumbilical (Fig. 1.9L). It may also be modified exteriorly by the development of an apertural tooth (inward projection(s) of the inner portion of the chamber wall into the aperture, see Fig. 1.9D), lips (Fig. 1.9F), a tegillum or tegilla (Fig. 1.9B, C), or a porticus or portici (Fig. 1.9A, E, I). However, the umbilical plates, such as the tegillum and portici, are similar to those of the benthic Rotaliina (e.g., *Haynesina*, *Rosalina*), which partially enclose the umbilical digestive cytoplasm with similar skeletal material (Alexander, 1985). The latter seems to be advantageous for extrathalamous digestion of disaggregated particles. As will be discussed in Chapter 4, in the Cretaceous, advanced forms of the Hedbergellidae evolved these analogous structures partially to enclose their umbilici (e.g., *Ticinella*, *Rotalipora*; see Fig. 4.9). By analogy with the benthic forms, therefore, this could have enabled a similar partial enclosure of an umbilical digestive 'reservoir' of cytoplasm to facilitate the effective ingestion of absorbed particles. Even predatory, carnivorous Cenozoic globigerinids partially digested disaggregated prey in extrathalamous digestion vacuoles (Spindler *et al.*, 1984). In contrast to the Hedbergellidae, however, other Cretaceous forms, such as the Praehedbergellidae and the Schackoinidae, never developed extended portici.

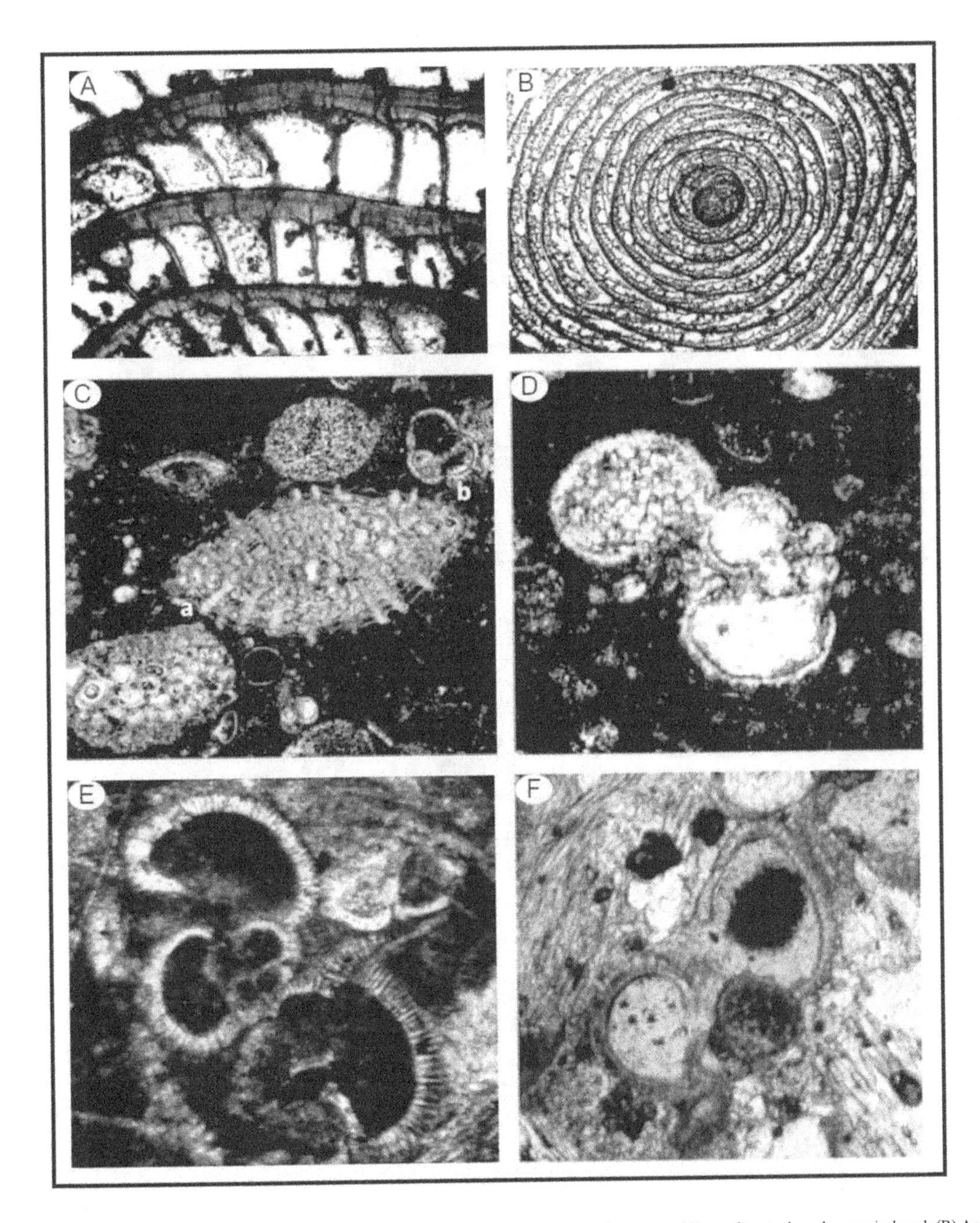

Figure 1.8. A, B, Ca; Examples of internally complicated benthic foraminiferal tests. (A) Enlargement of *Nummulites* to show the marginal cord; (B) An enlargement of *Loftusia* to show the interiors of the chambers partially filled by networks of irregular projections; (Ca) Three layered miogypsinids. Cb, D–F; Examples of planktonic foraminifera in thin section showing internally simple tests lacking the additional skeletal structures characteristic of benthic taxa. (Cb) *Globigerina*; (D) *Favusella*; (E) *Dentoglobigerina*; (F) *Globorotalia*. All images from the UCL Collection.

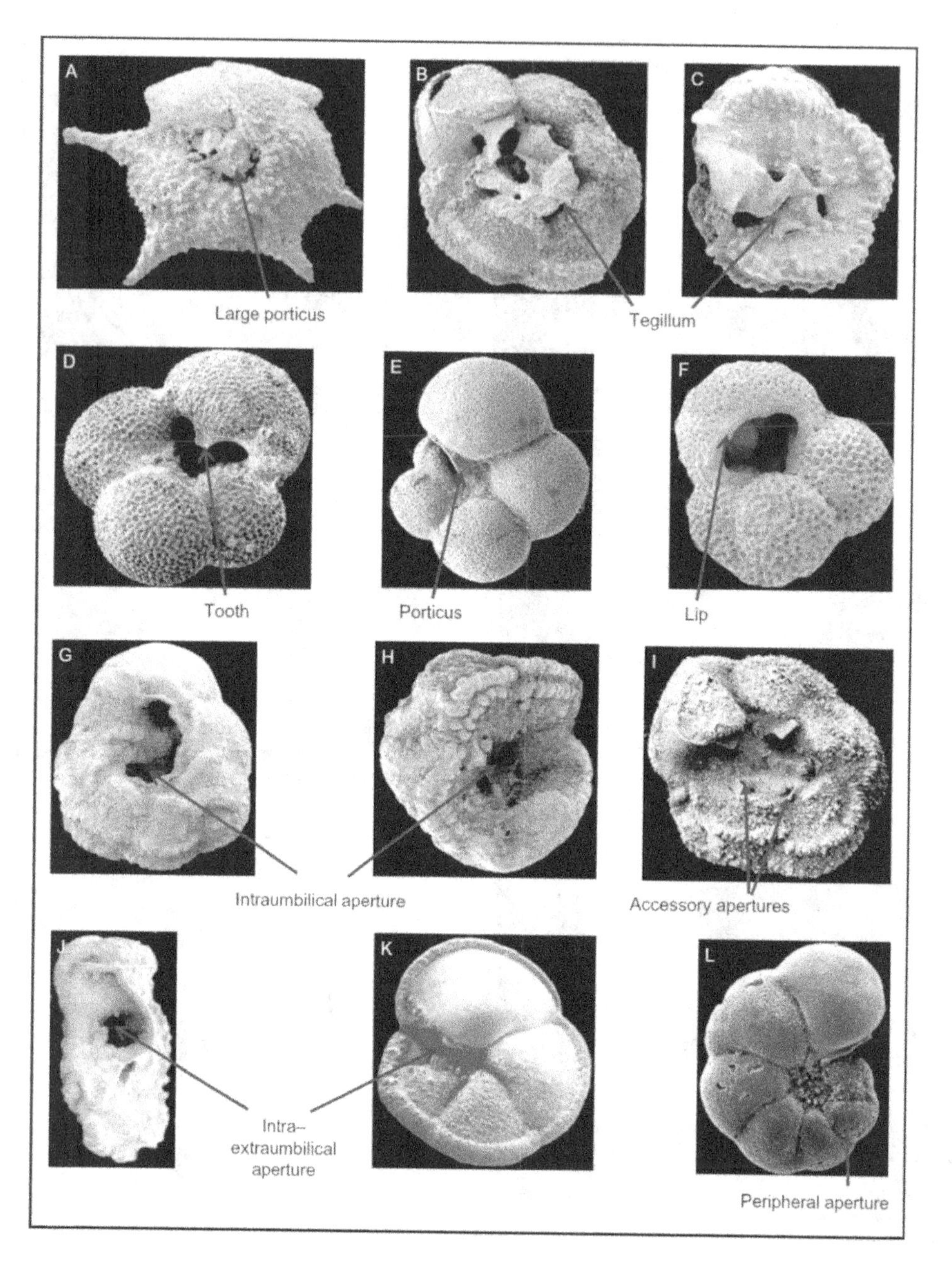

Figure 1.9. Examples of apertural variations in planktonic foraminifera. (A) *Radotruncana* with a primary umbilical aperture bordered by a large porticus; (B) *Globotruncana* with an intraumbilical aperture covered by a tegillum; (C, J) *Abathomphalus* with an intra–extraumbilical aperture covered by a tegillum; (D) *Dentoglobigerina* with an axiointraumbilical aperture with a tooth-like, sub-triangular, symmetrical porticus projecting into the umbilicus; (E) *Turbeogloborotalia* with an intra–extraumbilival aperture bordered by a porticus; (F) *Guembelitrioides* with an intraumbilical aperture bordered by a lip; (G) *Globotruncana* showing an intraumbilical aperture with a broken tegillum; (I, H) *Contusotruncana* showing primary umbilical aperture with short portici fusing in places to form accessory apertures; (K) *Globorotalia* with an intra–extraumbilical aperture; (L) *Pseudohastigerina* with extraumbilical, peripheral aperture bordered by a lip. All images from the UCL Collection.

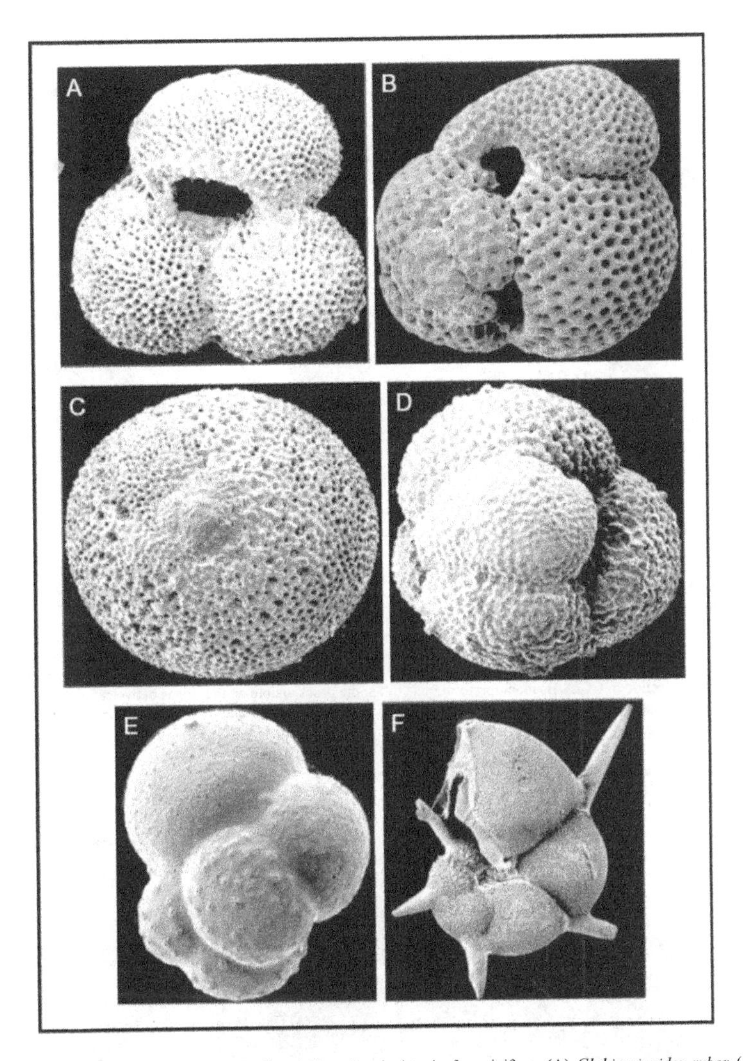

Figure 1.10. Examples of perforation types and apertural variations in planktonic foraminifera. (A) *Globigerinoides ruber* (d'Orbigny) showing a macroperforate test with thin secondary calcitic crusts surrounding the spine bases; (B) *Globigerinoides sacculifer* (Brady) showing a surface with regular pore pits and a supplementary sutural aperture; (C) *Orbulina suturalis* Brönnimann showing a densely perforated test with areal apertures; (D) *Catapsydrax dissimilis* (Cushman and Bermudez) showing distinct perforation pits and an aperture covered by a single umbilical bulla; (E) *Cassigerinella chipolensis* (Cushman and Ponton) showing a microperforate test with scattered pustules concentrated near the umbilicus; (F) *Hantkenina alabamensis* Cushman showing a high aperture, with a porticus which broadens laterally, and a smooth but densely perforated surface. All images from the UCL Collection.

In the Cenozoic, chamber growth developed so that some forms exhibited supplementary apertures (such as in *Globigerinoides*, Fig. 1.10B) or areal apertures (as in *Orbulina*, Fig. 1.10C). Occasionally, the aperture of a planktonic foraminifera is covered by a skeletal structure, a bulla (Plate 4.1, Fig. 18; Fig. 1.10D), or sometimes only a trace of it can be seen. Bullae extend over the umbilicus of the ultimate whorl and cover the primary, main, or supplementary apertures. They may also have marginal accessory apertures. These occurred in the Mesozoic and Cenozoic (see Chapters 4 and 5). However, the bullae of the Mesozoic differ from those of the Cenozoic species in often possessing short discontinuous ridges formed from pseudomuricae (see below), while those of the Cenozoic are perforate, not muricate (see below), and from one to four accessory, infralaminal apertures (e.g., Catapsydrax, Fig. 1.10D). In life, bullae would have allowed contact with the exterior through the accessory, infralaminal apertures at their margins. They too may have covered the umbilicus to conceal a mass of extrathalamous digestive cytoplasm by analogy with what is seen in some living benthic rotaliines (Alexander and Banner, 1984). It is possible that with the extraumbilical extension of the aperture and narrowing of the umbilicus, bullae became unnecessary.

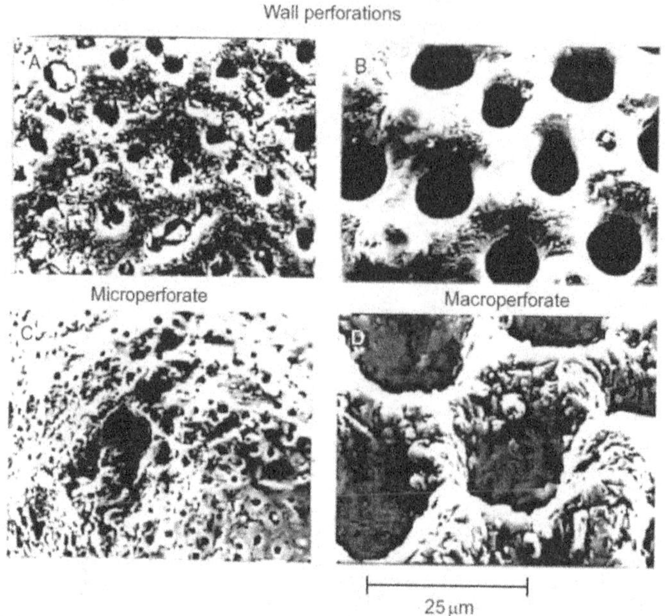

Figure 1.11. Enlargement of planktonic foraminiferal walls to show various types of perforations. Fine (A) or dense (C) micro-perforations in *Cassigerinella,* from Banner's Collection in UCL; (B) macroperforations with spine bases visible on the ridges between the perforations as in *Globigerinoides subquadratus*; (D) macroperforations with distinct cancelations as in *Globigerinoides immaturus*, from Kennett and Srinivasan (1983).

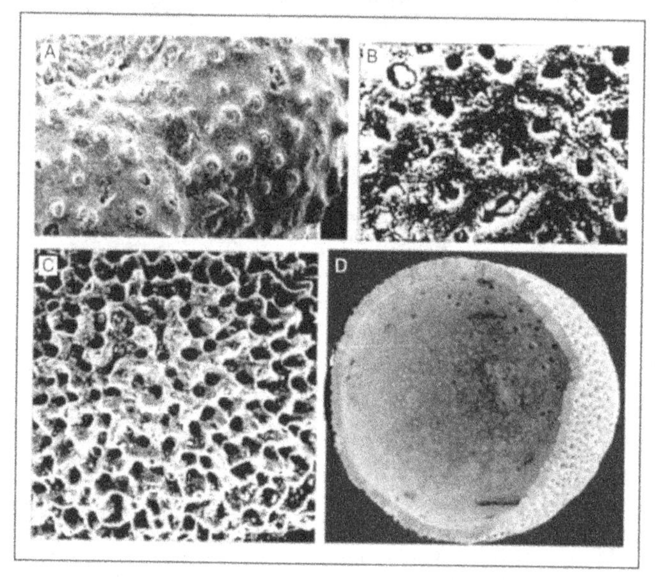

Figure 1.12. Enlargement of wall surfaces to show different types of walls. (A) Pseudomuricae, some apparently hollow (perforation cones), in *Globuligerina*; (B) Irregular microperforation, in *Blefuscuiana*; (C) Macroperforate test with widely scattered spine bases, in *Globigerina bulloides*; (D) Enlargement of a broken *Orbulina* showing small microperforations and larger pores or areal apertures. All images from the UCL Collection.

The secreted calcitic or aragonitic tests of the planktonic foraminifera are always perforated by multitudinous small holes (Fig. 1.7). These perforations did not, however, form open connections between the cell within the test to the surrounding sea water, rather they enable the internal cytoplasm to make biochemical contact with a cytoplasmic sheet on the outer surface of the test, the extrathalamous cytoplasm (see Plate 1.1 below). They are closed off internally by the inner organic lining, allowing gas and solute exchange between intrathalamous and extrathalamous cytoplasm. Only small molecules and nutritional salts can penetrate via the perforation. They are also important for the nutrition of any algal endosymbionts hosted by the foram.

These perforations vary in size, with diameters between 0.28 and 2.5 μm in the microperforate forms (see Figs. 1.10E and 1.11A), and between 2.5 and 10 μm in the macroperforate forms (see Fig. 1.11B). They are abundant in the walls of each chamber of the test of, for example, all genera of the Globigerinida (e.g., Fig. 1.10A). However, microperforations are usually irregularly dispersed, while macroperforations are dense (see Fig. 1.11). Perforations must be distinguished from the much larger pores/areal apertures, which occur in the final chambers of some Cenozoic taxa, and which allow direct ingress and egress of cytoplasmic strands (pseudopodia) to and from the test limits. Pores occur commonly in the walls of the Miocene genera such as *Praeorbulina*, *Orbulina* (see Fig. 1.12D), and *Globigerinatella*, where they replace the primary aperture of ancestral forms (see Chapter 6; Plate 6.11, Fig. 1). Pore diameter appears to be directly related to environmental temperature (Bé, 1968), and test porosity is relatively uniform for species coexisting in the same latitudinal belts (Hemleben *et al.*, 1989).

A variety of surface ornamentations also characterize planktonic foraminifera species, and these include perforation cones, pseudomuricae, muricae, short ridges or costellae, favose reticulations, pustules, muricocarinae or keels, spines, or smooth surfaces bearing a thin veneer of highly uniform calcite.

Perforation cones are conical-like structures (see Figs. 1.12A and 1.18A; and BouDagher-Fadel *et al.*, 1997) with axial vents, which were built by extrathalamous cytoplasm over the perforations of many taxa of the Praehedbergellidae and the Heterohelicidae (*Guembelitria*, Plate 4.19, Figs. 1–3; *Pseudotextularia*, Plate 4.18, Figs. 22–28). Perforation cones are not "pore cones" (as they have been called by many authors, especially referring to their occurrence in the Heterohelicoidea), because they have nothing to do with pores (cf. perforations). The perforation cones on the surface of the test of Praehedbergellidae make their tests heavier and thicker toward the aperture; they become much lighter on the later chambers and may be absent on the last. This was achieved by depositing a new surface lamella of calcite over the earlier test as each new chamber was formed. The penultimate chamber had just one extra lamella of calcite, but earlier chambers have more and more lamellae, thickening the test and strengthening the cones. As a consequence, this gives rise to the strongest, most prominent cones nearest to the aperture.

The perforations have been shown in many benthic rotalines to act as links between the cytoplasm that is inside each chamber (intrathalamous) and that part which forms a layer outside the test (extrathalamous). It is from the extrathalamous cytoplasm that the pseudopodia arise. They are referred to as "granular", as they carry granules that include excreted particles being carried away from the test and captured organic particles being carried into the test for ingestion. When the captured particles reach the extrathalamous cytoplasm covering the test, they are dragged over the test surface to the umbilicus and aperture for ingestion and digestion. The pseudopodia and the rest of the extrathalamous cytoplasm must also be the means by which the protozoan gathers oxygen and excretes waste gases. Ions pass through the perforations into and out of the intrathalamous cytoplasm (Angell, 1967; Banner and Williams, 1973; BouDagher-Fadel *et al.*, 1997; Sheehan and Banner, 1972). Microperforate tests were adequate to allow gas and cation exchange when planktonic foraminifera were living in well oxygenated surface waters of shelf seas and open oceans, but the increased porosity that was created by the of macroperforate tests enabled efficient gas and cation exchange in less well oxygenated subsurface waters. The macroperforate Hedbergellidae and their descendants, therefore, are likely to have been more capable of populating deeper, cooler, and less oxygenated oceanic waters than were their microperforate ancestors.

In the Mesozoic, many Globigerinida have walls with surfaces that possess mound-like structures located between the surface perforations. In the Late Cretaceous, these structures were conical and pointed and were often as tall or taller than they were wide (Fig. 1.13E); these features were named muricae by Blow (1979). Muricae first appeared on species of the globigerinid genus *Hedbergella* in the Late Aptian. Large perforations were also developed by these forms, enabling them to occupy deeper water niches, prior to evolution of muricae. Phylogenetically, the muricae appear gradually first near the aperture and then evolve to be near the periphery and sutures. Muricae can also fuse meridionally across the test surface at right angles to the periphery, to form a costellae, as in *Rugoglobigerina* (Fig. 1.13D, G) and related genera. Muricae were present on the tests of the Globigerinoidea up to the close of the Eocene (e.g., *Morozovella*, Fig. 1.14A, D; see Chapter 5). The muricae developed in forms belonging to the superfamily Globigerinoidea independently and at different stages. The muricae of Cretaceous and Paleogene forms are morphologically indistinguishable, yet because of the major End Cretaceous extinction event they must have developed independently one from another (see Chapters 4 and 5). In both cases, however, they presumably were covered by extrathalamous extensions of the cytoplasm and were capable of carrying and transporting vacuoles, which would have given them a competitive advantage, and might explain why they were able to spread faster than the smooth praehedbergellids in the Cretaceous and globanomalids, which died out in the early part of the Middle Eocene. The smooth, nonmuricate planktonic foraminifera would have had no skeletal control on the distribution of sites for pseudopodial extensions into the surrounding seawater. The pseudopodia would either have extended irregularly or would have occupied a volume approximating to that of a sphere. If the same volume is compressed into a disc, the surface area for prey capture or collection must be very greatly increased. Therefore, species which could extrude the pseudopodia via muricae with greater structure would have had a considerable advantage in the collection of available suspended particulate nutrients.

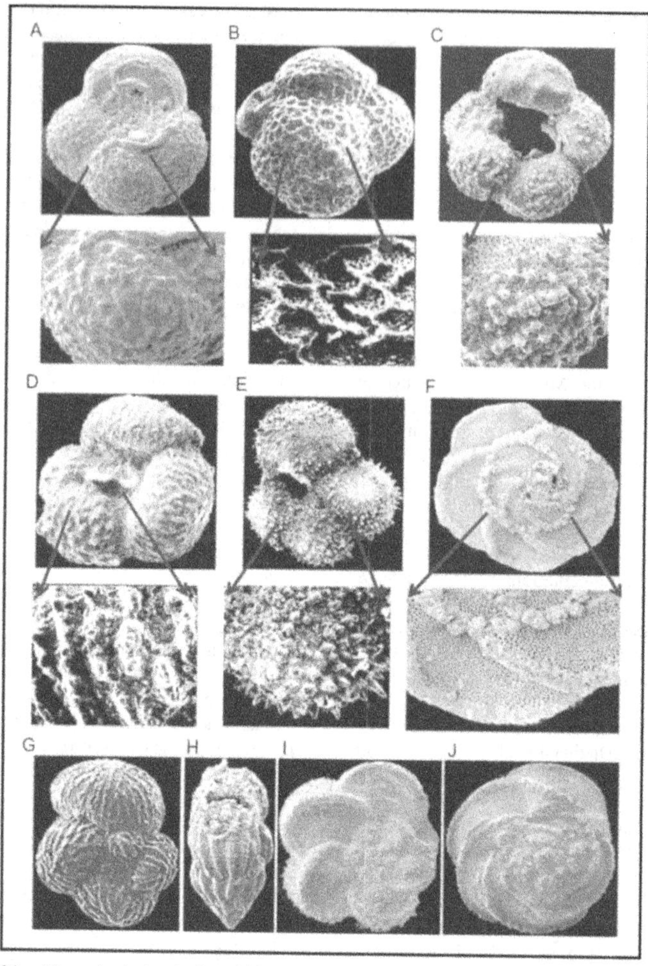

Figure 1.13. Specimens typifying Mesozoic planktonic foraminiferal forms. (A) Microperforate with pseudomuricae, *Globuligerina;* (B) Favose reticulations, *Favusella;* (C) Muricae fusing to form short ridges or costellae, *Archaeoglobigerina;* (D, G) Well developed costellae, *Rugoglobigerina;* (E) Pointed muricae strongest at periphery of test, *Globotruncanella;* (F-J) A keel along the sutures and on the periphery, (F, I) *Globotruncana* and (J) *Contusotruncana;* (H) Muricae fusing to form irregularly arranged longitudinal costellae on the surface of the test, *Pseudoguembelina.* All images from the UCL Collection.

In the Globotruncanidae of the Late Cretaceous, the muricae fuse together, along the periphery of the test, to form a muricocarina or keel (e.g., in *Globotruncana*, Fig. 1.13F). This also occurred in the planispiral *Planomalina* (Plate 4.15, Figs. 23 and 24). Some globotruncanids possessed two muricocarinae, which can be close together (appressed) or separated with imperforate peripheral band in between them (see Fig. 1.15). The keel can extend into the cameral sutures of the dorsal side and can extend into the ventral sutures of each chamber on the umbilical side. All the Globotruncanidae are extinct, so it is impossible to observe the function of these structures. However, it has been noted that the carinae of some predatory, biconvex benthic Rotaliina (e.g., *Amphistegina*), form the foundation of the "take-off" of food-gathering pseudopodia (Banner, 1978). We can surmise that the double muricocarinae of *Globotruncana* (Fig. 1.16A) were also associated with nutrition. If fans of pseudopodia radiated from the extrathalamous cytoplasmic layer, which covered the test, and arose from each muricocarina, then they would drag organic particles to the muricacarinae where they would become disaggregated. These disaggregated particles would then be channelled along the imperforate band to the terminal face and then to the umbilicus. This would mean that no food particles would impede the function of the perforations on the dorsal and ventral faces of the chambers but would ensure that simple cytoplasmic flow would direct the nutrient particles straight to the umbilical aperture for their ingestion.

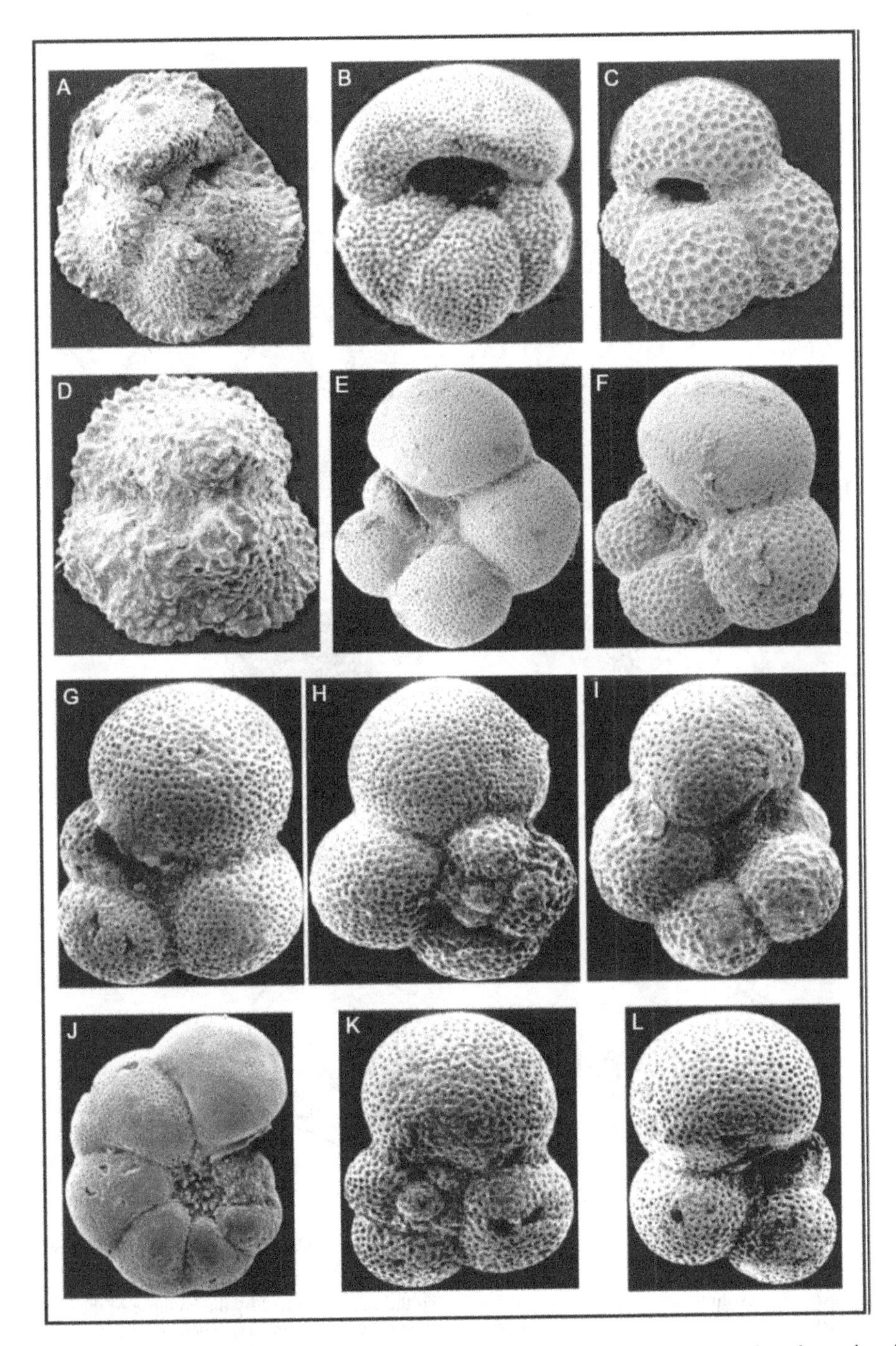

Figure 1.14. Specimens typifying Paleogene planktonic foraminiferal tests. (A) *Morozovella acuta* (Toulmin); (B) *Turborotalia pseudoampliapertura* (Blow and Banner), from Blow (1979); (C) *Subbotina triangularis* (White); (D) *Morozovella aequa* (Cushman and Renz); (E) *Turbeogloborotalia compressa* (Plummer); (F) *Parasubbotina pseudobulloides* (Plummer); (G, H, I) *Parasubbotina variant* (Subbotina); (J) *Pseudohastigerina micra* (Cole); (K, L) *Subbotina triloculinoides* (Plummer). Images A and C-L from the UCL Collection.

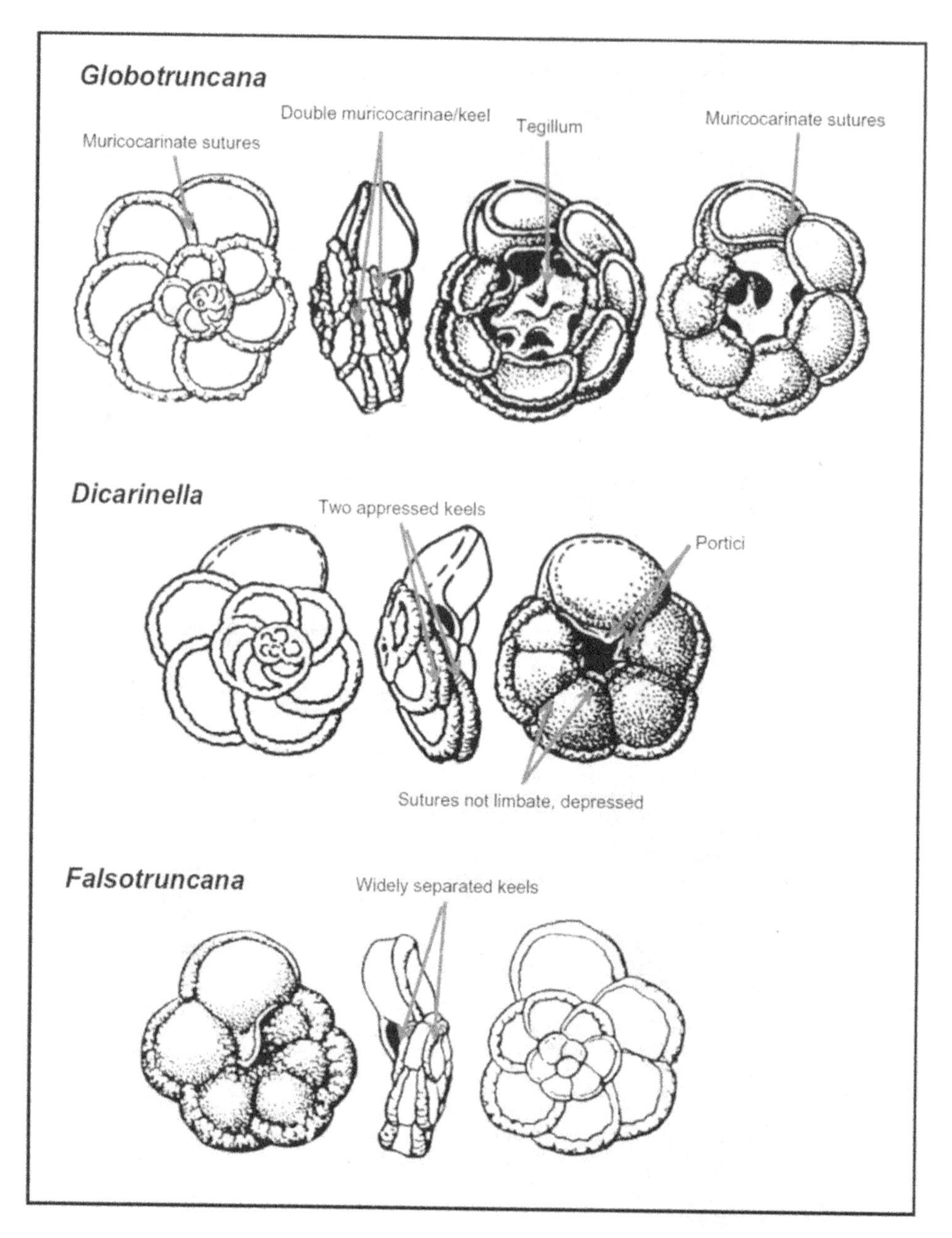

Figure 1.15. Double peripheral keels developed by Globotruncanidae. *Globotruncana* with double keels with an intermediate imperforate band; *Dicarinella* with closely spaced, appressed keels; and *Falsotruncana* with widely separated keels with an imperforate band in between them.

In the Paleogene, this evolutionary trend was repeated, with muricae again fusing at the periphery to form keeled *Morozovella* (see Figs. 1.14A, D and 1.16B), while in the Neogene, the compressed test of the globorotaliids is similarly reinforced by peripheral thickening of the chamber to give a keel (Fig. 1.16C). However, in the globorotaliids, muricae are absent and the test is smooth, so the keel is formed by the collapse of the chamber wall along the test periphery, where it serves an architectural function of strengthening the test (Hemleben *et al.*, 1989). In the early stages of the globorotaliids, the keel area is still covered by perforations (see Fig. 1.16C), while in older chambers, the keel is strengthened by additional calcite layers (Hemleben *et al.*, 1989).

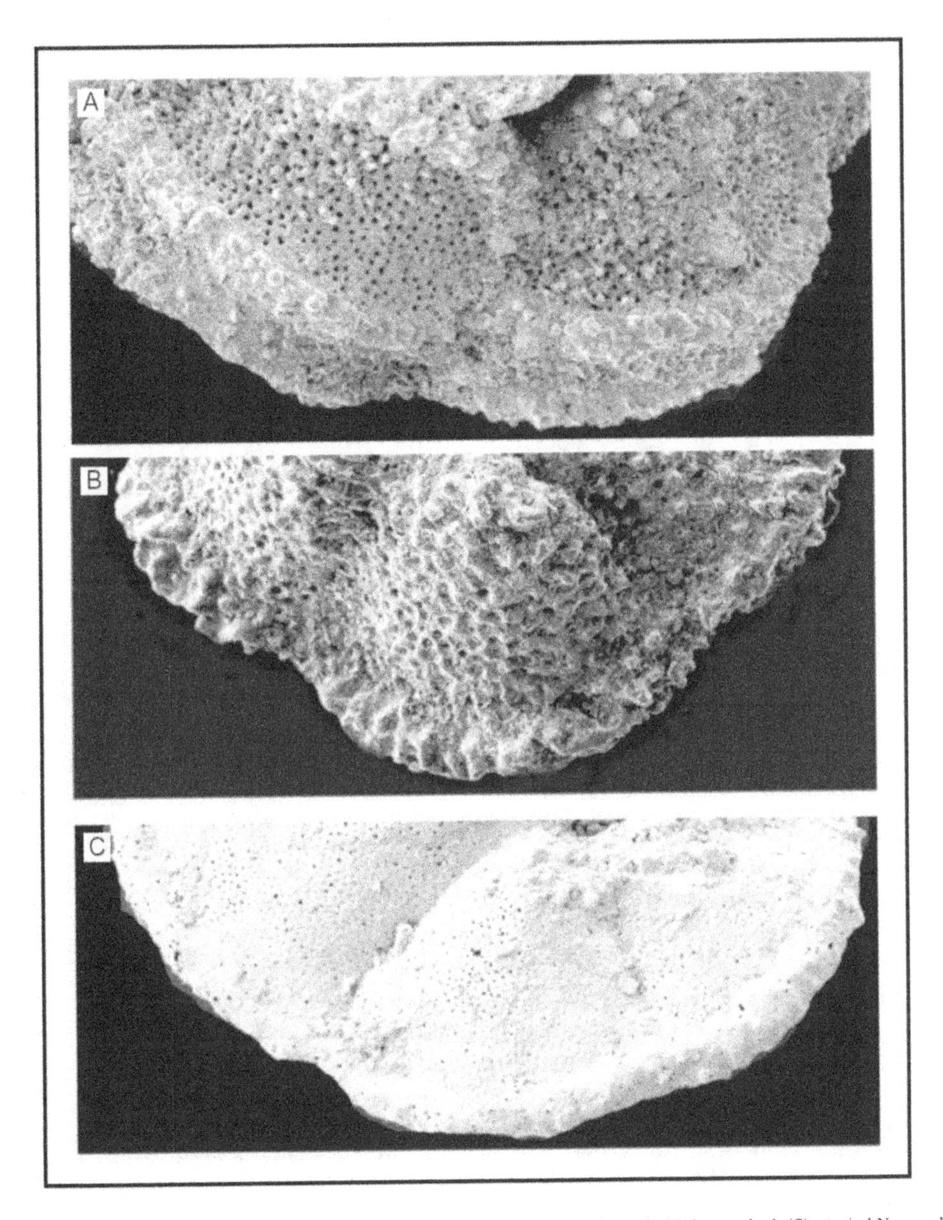

Figure 1.16. Enlargement of three different tests to show: (A) a typical Cretaceous keels; (B) a typical Paleogene keel; (C) a typical Neogene keel. All images from the UCL Collection.

The surface of the test in most nonspinose Neogene species bears small rounded calcitic knobs which have been called pustules (emended by Blow, 1979). They look similar to the perforation cones of the Praehedbergellidae or the muricae of the Hedbergellidae and their related forms and are also concentrated on the test surface in the vicinity of the aperture(s) and the umbilicus as exemplified in *Turborotalia cerroazulensis* (Fig. 1.17A) and *Globorotalia margaritae* (Fig. 1.17B). Here, their function seems to be the disaggregation, by rasping, of food particles prior to their ingestion through the apertures (Alexander and Banner, 1984) or digestion in the umbilical cytoplasmic "reservoir" (Alexander, 1985). They also may have served as anchor points for masses of rhizopodia (Hemleben, 1975; Hemleben *et al.*, 1989). In the Paleogene, microperforate taxa can also possess pustules (e.g., *Globastica/Globoconusa* of the Danian, and the Globigerinitidae of the Late Palaeogene to Holocene; Plate 5.3, Figs. 1–5; Plate 6.1, Figs. 8–13; Chapters 5 and 6).

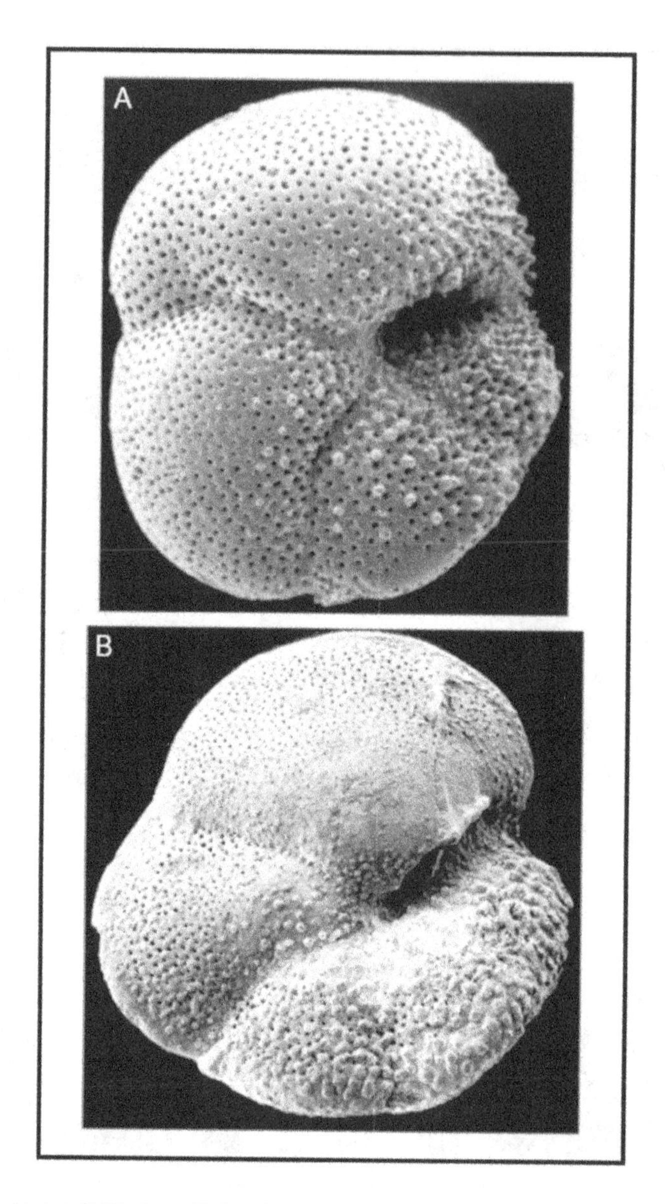

Figure 1.17. An enlargement of the test of (A) the Eocene *Turborotalia cerroazulensis* (Cole) and (B) the Pliocene *Globorotalia margaritae* Bolli, (x125). They show pustules concentrated on the test surface in the vicinity of the aperture and the umbilicus. Both images from the UCL Collection.

The Favuselloidea of the Jurassic and Early Cretaceous possessed neither pointed muricae nor blunt, scattered pustules. Instead, many of their tests possessed narrow, blunt or pointed, scattered projections which were more elevated and regularly spaced than pustules. BouDagher-Fadel *et al.* (1997) called these structures pseudomuricae. Pseudomuricae are often hollow (Samson *et al.*, 1992; see Chapters 3 and 4; Figs. 1.13A and 1.18A) as muricae can be (Blow, 1979, plate 208), but never fuse peripherally into muricocarinae. However, in contrast, they can fuse into discontinuous ridges (e.g., as on *Globuligerina*, see Fig. 1.13A), which are at random angles to each other, and which eventually themselves fuse together into favose reticulations (Fig. 1.13B; see Chapter 4), which appear to be solid (Fig. 1.18E). Muricae never fuse to form reticulations but occur between the macroperforations of the taxa which have them. The ridges of *Favusella* may still retain pseudomuricae at the junctions between the ridges of the reticulum (Fig. 1.13B). Pseudomuricae can enclose patches of microperforations, unlike anything seen in macroperforate forms (Fig. 1.18A).

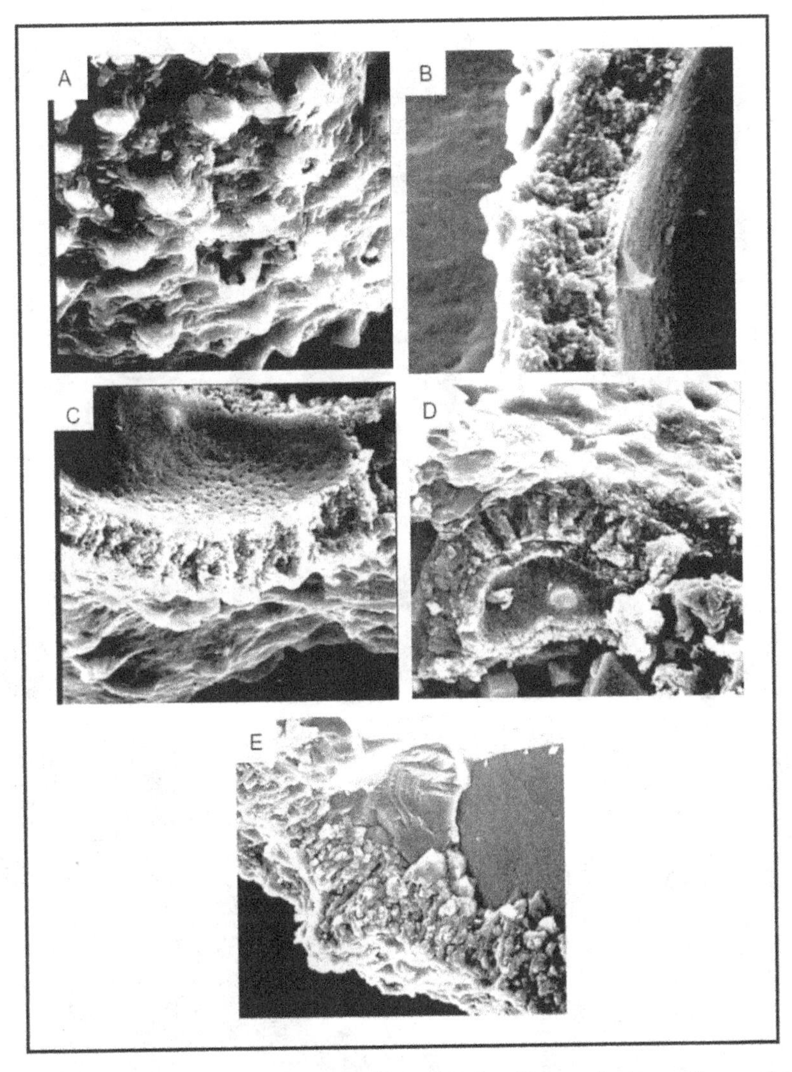

Figure 1.18. Enlargements of *Globuligerina oxfordiana* (Grigelis) from the Oxfordian of Moscow showing a well preserved aragonitic wall. (A) Enlargement of the wall surface to show pseudomuricae, some apparently hollow forming perforation cones (x2200); (B–D) Dissection of later chambers showing the inner lining becoming perforate and the canaliculated microperforate aragonitic wall (x2000); (E) *Favusella washitensis* (Carsey), dissection of recrystallized wall, showing canaliculated wall but solid reticulation (x867). Images from BouDagher-Fadel *et al.* (1997).

True spines really only came to full development in the Eocene, yet by the end of the Paleogene spinose taxa had replaced muricate forms entirely (Fig. 1.19). Spines are very long, needle-shaped (aciculate) structures, circular or triangular in cross section (Hemleben *et al.*, 1989, pp. 209–210), which have sub-cylindrical bases embedded in the chamber wall. Their morphology includes five general types (Hemleben *et al.*, 1989; Saito *et al.*, 1976):

- the *Globigerina*-type, round in cross section throughout (Fig. 1.19D);
- the *Globigerinoides*- and *Orbulina*-type, with round spines, but which may become triradiate distally (Fig. 1.19B);
- the *Orcadia*-type triangular spines throughout (Fig. 1.19A);
- the *Globigerinella*-type with triangular thin section at the base becoming triradiate (Plate 6.4, Fig. 16);
- the *Hastigerina*-type which are clearly triradiate from their base (Fig. 1.19H).

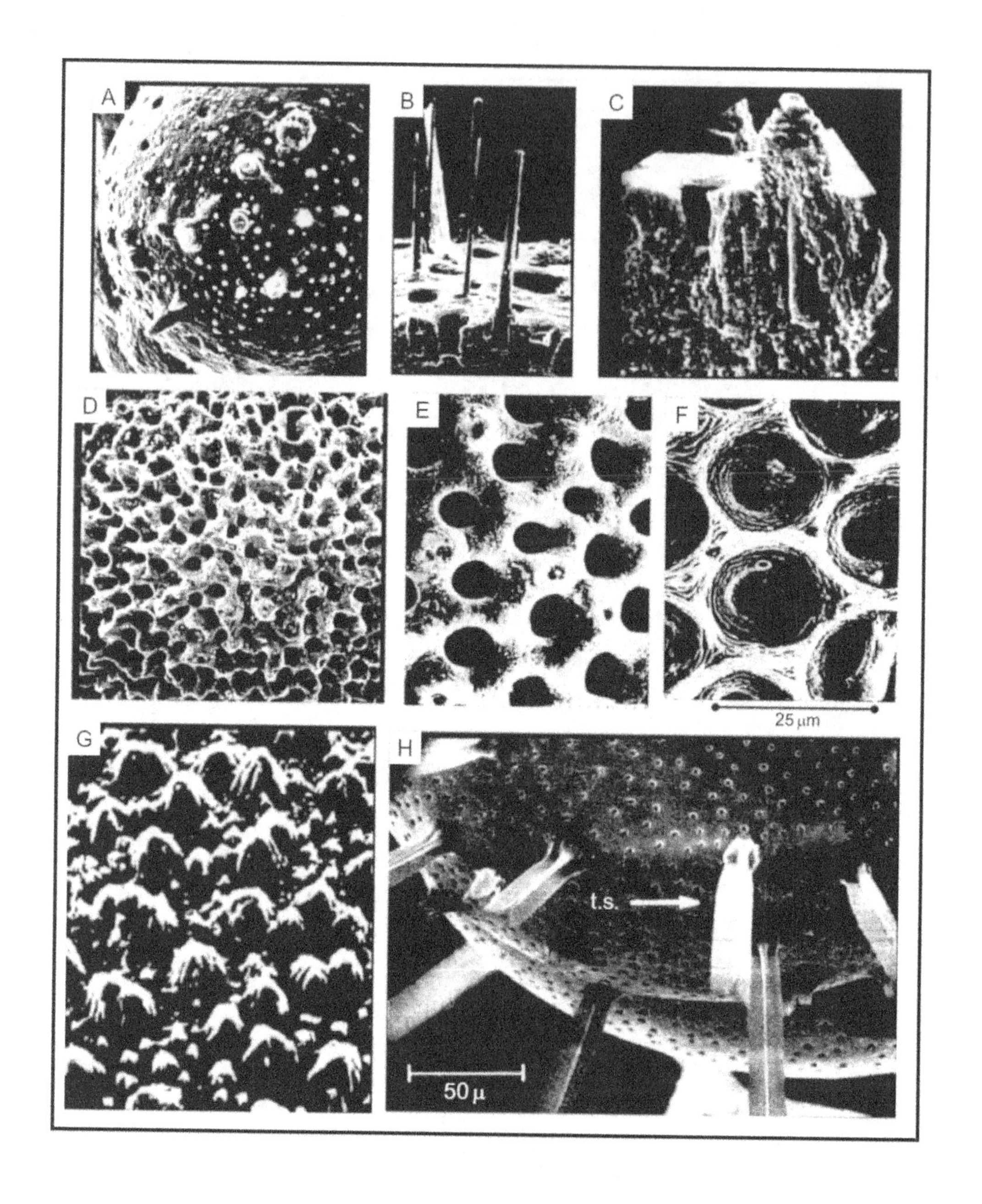

Figure 1.19. (A) Detail of the surface of *Orcadia riedeli* (Rögl and Bolli) showing a macroperforate test with widely scattered spine bases, from BouDagher-Fadel *et al.* (1997), x890; (B, C) Enlargement of the surface of *Orbulina universa* d'Orbigny, from Bé *et al.* (1973): (B) Showing pores and originally round spines become triradiate distally, (C) Enlargement of the wall surface showing a spine originating from a primary organic membrane, which divides the inner and outer calcite lamellae, x700; (D) Enlargement of the test of *Globigerina bulloides* d'Orbigny showing a macroperforate, nonmuricate surface, but with abundant and crowded spine bases from BouDagher-Fadel *et al.* (1997), x450; (E) Enlargement of a surface showing a wall without perforation pits; (F) Enlargement of a surface showing perforation pits, (E-F) from Kennett and Srinivasan (1983); (G) Enlargement of *Globigerina bulloides* d'Orbigny showing a macroperforate surface with abundant spine bases, from Banner's Collection in UCL, x450; (H) Enlargement of the perforate wall surface of *Hastigerina pelagica* (d'Orbigny), from Bé (1969), showing triradiate spines (t.s.) from their bases, x1675.

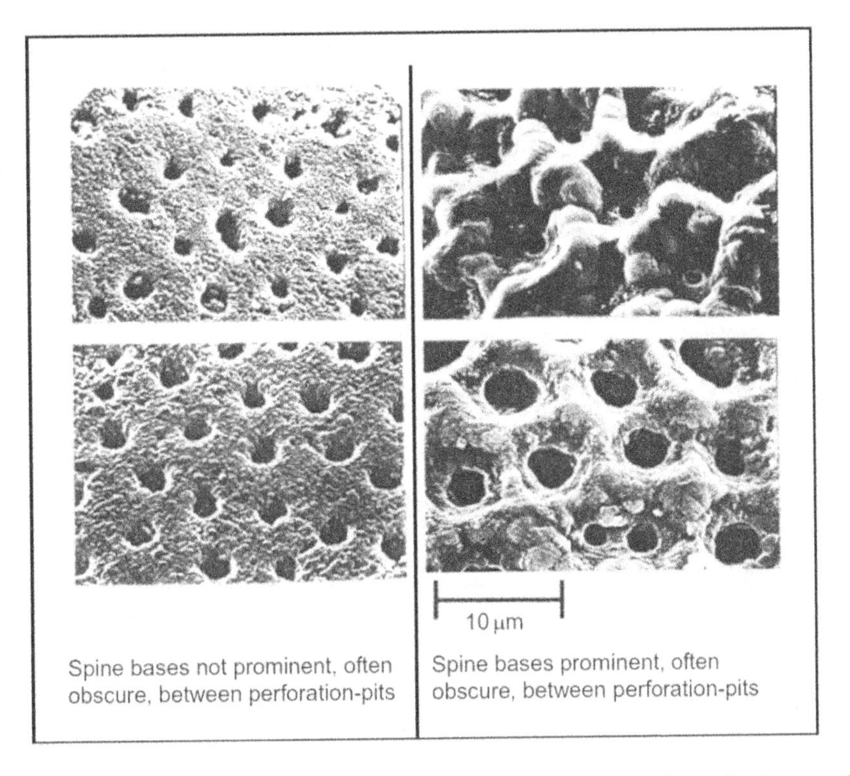

Figure 1.20. Two types of wall showing spine bases, from Kennett and Srinivasan (1983). The spine bases are often prominent between perforation pits as in *Clavatorella* (right), or rarely prominent, and often obscure, as in *Globorotalia* (left).

Clearly, spines could have a greater variety of function than the muricae of their ancestral taxa. Spines might have a dense or sparse distribution. In Fig. 1.19, one can note the different distribution of the spine bases found on extinct *Globigerina* (Fig. 1.19D) and *Orcadia* (Fig. 1.19A) from the recent seas. The spine bases, which become embedded in the ridges (Fig. 1.12C) between the macroperforations of post-Paleogene taxa, have been well studied in SEM by Bé (1969), Bé *et al.* (1973), Bé and Hemleben (1970), and Bé (1980). The spine bases are often prominent between perforation pits as in *Clavatorella*, or rarely prominent, and often obscure, as in *Globorotalia* (see Fig. 1.20).

It is known that the spines of living Globigerinidae are covered by extrathalamous extensions of the cytoplasm, capable not only of carrying vacuoles and transporting symbionts but also of performing the feeding and ion exchange functions (Adshead, 1980). They can ingest food particles from as far as possible from the host test and reject waste matter from similar distances (Hemleben *et al.*, 1989). This is important particularly in turbulent shallow water where they are found. Spinose globigerinid species occupy shallow (mixed layer and intermediate depths, 0–100 m) planktonic habitats (Fairbanks *et al.*, 1982), and harbour algal symbionts within their cytoplasm (Bé, 1982; Boltovskoy and Wright, 1976; Hemleben *et al.*, 1989; Lipps, 1979; Murray, 1991). The presence of spines has been suggested to being linked to maintaining control over the depth of the habitat and to help the organism resist sinking through the water column (Bé, 1982; Hutchinson 1967; Lipps 1979). However, the presence or absence of spines cannot be the sole determining factor here, as there are a number of nonspinose planktonic foraminiferal species that occupy shallow–intermediate planktonic habitats too (e.g., see Figs. 5.22 and 6.21). An alternative hypothesis for the function of spines is that they may provide the organism with a metabolically inexpensive means of greatly increasing the area of its pseudopodial network, and in so doing, increase the opportunity for that network to capture food particles (Murray, 1991). Neither the acquisition of spines nor the ability to harbour algal symbionts can be regarded as an adaption to a specific depth habitat. Spines were developed by the globigerinid lineage well after the transition to a shallow–intermediate depth habitat had been made by their ancestral species (Macleod, 2001). Spines, however, also may have been an adaptation to assist in dispersing the symbiotic algae in such a way as to optimize diffusion of CO_2 and oxygen.

1.2.2 The crystallographic structure and chemical composition of the test walls of planktonic foraminifera

All planktonic foraminifera are perforate and, therefore, known or assumed to construct bilamellar shells (Sen Gupta, 1999), where the chamber wall is formed primarily by two mineralized layers (outer and inner lamellae) on either side of a primary organic sheet, the median layer. The only exception is the monolemellar genus *Hastigerina* (Hemleben *et al.*, 1989; Spindler *et al.*, 1979). Their hyaline test is made of calcium carbonate crystals. The crystals form microgranules (as shown by Banner and Williams, 1973), which are themselves aligned into rows parallel to the perforations and perpendicular to the surface of the test wall. These crystals are so arranged that their optic c-axes are in alignment; when the aligned optic axes are themselves perpendicular to the test surface, the test is truly hyaline. Optically, the crystal structure is not discernible and must be determined by other means. These basic crystallographic criteria, coupled with the underlying understanding that the tests are calcitic, are fundamental to the definition of the foraminiferal order Globigerinida as given by Loeblich and Tappan (1988). Moreover, throughout their book, Hemleben *et al.* (1989) either assume or state that all Globigerinida are calcitic. Analysis of the crystallography and chemical composition of the tests of the early planktonic foraminifera is still in its infancy. Gorbachik and Kusnetsova (1986) and BouDagher-Fadel *et al.* (1997) suggested that the early Favuselloidea, as we describe in ensuing chapters, were aragonitic, while the tests of the Globigerinoidea were calcitic. However, there is still some uncertainty in this claim. The instability of aragonite is well known. Aragonitic fossils are highly susceptible to dissolution, and their recovery is unlikely to occur unless there were high sedimentation rates or rapid sealing of sediments soon after deposition (cf. the preservation of fossil didemnid ascidian spicules (Varol and Houghton, 1996, p. 136)). This could explain the relative rarity of occurrence of Jurassic favusellid taxa. It is well known that many benthic foraminiferal families are consistently calcitic, while others are consistently aragonitic. This shows that the preferred mineralogy is genetically controlled, not influenced by the environment they inhabit. If Loeblich and Tappan's scheme (1964, 1988) was strictly followed, aragonitic planktonic foraminifera would have to be assigned to a completely separate suborder/superfamily (as are indeed the aragonitic benthic foraminifera). Given the subject is still in flux, in the meantime, it is far better to emend the definition of the Globigerinida (as, indeed, we do in this book) to include both aragonitic and calcitic genera.

1.2.3 Morphology, stratified biocenoses, and the death assemblages of planktonic foraminifera

As we have seen, the planktonic mode of life for the foraminifera was enabled by the development of globular, buoyant chambers sometime in the Late Triassic or Jurassic. Initially, simple globular forms would have occupied the surface waters, but later as ornamentation of the test developed, more massive forms could have achieved neutral buoyancy at greater depth beneath the ocean surface.

In the fossil assemblage, there should be an accumulation of all the species of planktonic foraminifera which lived in the overlying water column. The depth-restricted living assemblages would be mixed together as dead tests on the sea floor. Therefore, ignoring calcareous dissolution, the living assemblage (biocenosis) of the most superficial water should always be represented in the death assemblage (thanatocenosis) on the sea bed; the biocenoses of slightly deeper, less oxygen saturated water would be present as well if such waters existed. Therefore, if our inference on the relationship between morphology and habitat is correct, we would find assemblages of the subglobular-chambered taxa (associated with surface-water habitats) in all normal marine samples, but these would on occasion also be accompanied by species with heavily ornamented or elongated or flattened chambers (characteristics of deep-dwelling forms). In contrast, however, the latter species, with elongated or flattened chambers, should never be found in isolation without forms with subglobular chambers. This is, in fact, what is found. Many sediments yield samples that only contain foraminifera with the sub-globular chambers. For example, in Tunisia, many Hauterivian-Early Aptian samples have yielded *Gorbachikella* spp. but no other planktonic foraminifera; these species have the most subglobular chambers of any Praehedbergellidae. In the central North Sea area, the Barremian–Aptian succession contains (often abundant) *Praehedbergella* and *Blefuscuiana* but only very rare *Lilliputianella* (Banner *et al.*, 1993). There are no known cases of the flattened or elongated *Lilliputianella*, *Schackoina*, *Leupoldina*, or *Claviblowiella* occurring in microfossil assemblages without the subglobular chambered taxa. The chamber shape of the Praehedbergellidae may, therefore, be taken to characterize different levels in the upper waters of the Early Cretaceous ocean.

An outstanding exception to the analysis made above is presented by the case of *Wondersella*, which occurs in thin beds at the very top of the Late Aptian Shuaiba Formation from offshore United Arab Emirates. This genus has the most flattened and extended adult chambers of any known praehedbergellid, but it occurs commonly and uniquely in samples from these beds (Banner and Strank, 1987, 1988). However, these sediments are clearly those of an anoxic–dysoxic sedimentary basin. The micritic sediment has microlayering but no signs of bioturbation and contains no benthic microfossils; the bottom waters were anoxic. It is very probable, therefore, that even the upper waters in this basin were themselves suboxic and that normal surface-water forms found this basin intolerable, and that *Wondersella*, a planktonic foraminifera with flattening and radial extension of the adult chambers, was the only form with a morphology compatible with living in such an oxygen-depleted environment.

This discussion on the diversity of the morphology of planktonic foraminifera in death assemblages has so far only considered the effects of oxygen saturation in the wave swept uppermost waters, the less-saturated waters below the wave-troughs, and the strongly dysoxic waters of very deep waters or environmentally exceptional basins. It has not considered the effects of other variables such as salinity, temperature, illumination, food supply, and so on. These are unlikely to have had marked effect on test shape and structure. Studies of living Globigerinoidea suggest that oceanic variations in salinity do not limit their biogeographic distribution (Bijma *et al.*, 1990), and the same was probably true for the Praehedbergellidae. With the exception of illumination, the effects of the other variables are difficult to determine. Temperature may be estimated from oxygen isotope ratios (e.g., Shackleton, 1967; Shackleton *et al.*, 1985 and many other papers), and differences in ambient temperature may change test porosity (i.e., the number of perforations in any unit area of the test surface) but do not cause changes in the diameter of the perforations (Bijma *et al.*, 1990).

There is still more work to be done to fully interpret the effect of paleoenvironment on the details of the morphology of fossil foraminifera and the composition of their death assemblages.

1.3 The evolution and paleontological history of the planktonic foraminifera

1.3.1 Evolution of the planktonic foraminifera

Unlike larger benthic foraminifera, the planktonic foraminifera are not biofacies bound, and regionally constrained. They float freely in the oceans, with variations in their morphology allowing them to exploit fully the wide range of niches in the planktonic realm. Genes are the fundamental units of life and determine the genotypical properties of any life form. However, environment, ontogeny, and conditions during growth determine the phenotypical character of a species, determining, for example, the shape of the test. The interaction between the gene in the embryont and the selection of the external features that the foraminifera develop during growth is an example of this selection process discussed by Gould (2002) in his book "The Structure of Evolutionary Theory". This process acts on features that emerge from complex gene interaction during ontogeny and not from individual genes. Many characters in the evolution of foraminifera are gradual and linked to the timing of phases in growth. They could have been started by DNA mutation at one stage of foraminiferal life and amplified as a consequence of having to cope with stress and adverse environments.

The calcareous test of a planktonic foraminifer must have had a function related to the separation of different parts of the cytoplasm of the protozoan for different but related functions. Everything in the test must have provided, at one stage or another, a competitive advantage. The test can have had little or no defensive value, because larger predators that ingest plankton would not have been deterred by such a small calcareous object. Therefore, it would seem that through evolution, the features of the test were selectively amplified to provide the foraminifera themselves a competitive advantage to exploit different niches in the oceans. Thus, as we have seen the development of globular chambers enabled a planktonic mode of life to be established in the first place, while the development of keels and other more massive ornamentation may have led to developing neutral buoyancy at some depth below the surface waters.

This view is supported by the observation that the evolutionary histories of many morphological features of the planktonic foraminifera have shown adaptive convergent trends. Identical shapes or structures appear again and again within the same lineage or in parallel lineages from different stocks. Thus from example:

- the trend from high to low trochospiral coiling, and the repeated development of planispiral forms with tubulospines (as developed in the Cretaceous Hedbergellidae and the Paleogene Globanomalidae),
- the trend from globose to compressed keeled (as seen in the Globotruncanidae in the Cretaceous and the Globorotaliidae in the Neogene),
- the development of apertural and umbilical coverplates by coalescence of umbilical portici and teeth in the Cretaceous, to the development of intense streptospiral coiling, with the last globular chamber completely embracing the umbilical side in the Neogene.

These trends are seen to develop independently in Mesozoic, Paleogene, and Neogene families, and in all likelihood develop because each specific ecological niche has an optimal shape or processes that are required for its most effective exploitation.

One driving force for the evolution of planktonic foraminifera would have been climate change and the resulting changes in oceanic conditions. Planktonic foraminifera have latitudinal distributions that correlate with temperature (Fraile *et al.*, 2008). Their phylogenetic evolutions are closely associated with global and regional changes in climate and oceanography during their history (see Fig. 1.5). Large-scale changes in the climate caused stress to the planktonic assemblages. Evolutionary convergence, that produces morphologically very similar tests, implies that similar habitats became occupied. Biostratigraphic studies have demonstrated that feeding mechanisms and reproductive strategies are key traits that affect survival rates (Twitchett, 2006). As in the benthic realms (see BouDagher-Fadel, 2008), small, non-specialized planktonic forms fare better than large and advanced forms during times of stress. After significant extinction events, the surviving planktonic foraminifera were usually dominated by small globular forms. After the End Cretaceous event, for example, where a large impact probably caused massive disruption to the climate system, the post-impact environment was one of low productivity. This low productivity state continued for many hundreds of thousands of years (the Early Danian) and was characterized by the development of small, stunted forms (the "Lilliput effect") and a low biomass ecosystem. During the post-impact recovery, these survival forms adapted to their new environment and eventually new, larger forms evolved and established themselves in a variety of specialized niches. Many of these new forms exhibited morphological similarities to earlier Mesozoic forms that had previously occupied the same niches (see Chapter 5).

An example of such convergence is seen in the evolution of digitate species, which evolved repeatedly within the planktonic foraminifera throughout the Cretaceous and Cenozoic (Fig. 1.21). In the Late Aptian, the Schackoinidae evolved near-trapezoidal chambers with multiple spines, as in *Schackoina* (Fig. 1.21A). In the Early Eocene (Late Ypresian, P9) forms with hollow tubulospines gave rise to *Hantkenina* (Fig. 1.21B), and again in the Miocene, planispiral forms (e.g., *Protentella*, Fig. 1.21C) or trochospiral forms (e.g., *Beella*, Fig. 1.21D) with high, radially elongate chambers made their appearance.

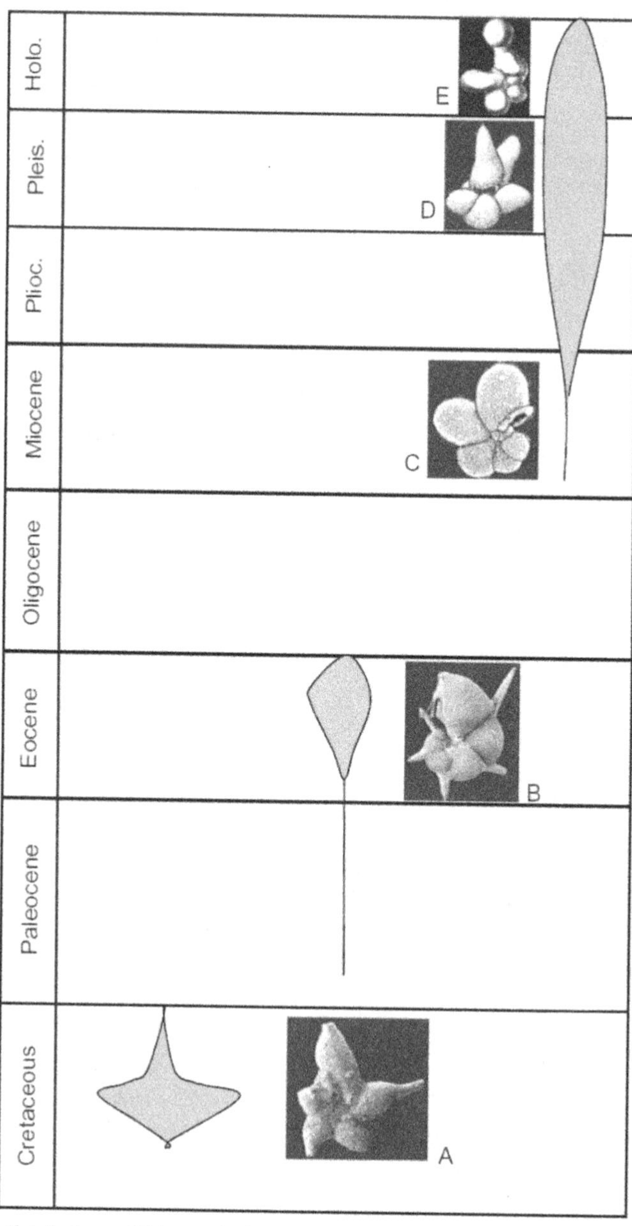

Figure 1.21. The convergence of similar shapes of digitate species through the stratigraphic column. (A) *Schackoina*, (B) *Hantkenina*, (C) *Protentella*, (D) *Beella*, (E) *Hastigerinella*. Images A-B from the UCL Collection, (C) from Kennett and Srinivasan (1983), and D-E from the Banner Collection in UCL.

In this convergent adaption of chamber elongation, the surface area of the chambers increased and therefore the potential proportion of extrathalamous cytoplasm also increased. This enhanced the ability of these forms to survive in dysoxic seawater (BouDagher-Fadel *et al.*, 1997; Coxall *et al.*, 2007). If we consider solely the concentration of dissolved oxygen in the seawater, then, if the water is saturated or supersaturated in oxygen, a relatively small area of extrathalamous cytoplasm would be needed to maintain effective gas exchange. However, if the seawater becomes dysoxic, then relatively more extrathalamous cytoplasm, capable of gas exchange, is needed for the foram to survive. A sphere has a minimum surface area to volume ratio; therefore, a spherical chamber would have the greatest possible ratio of intrathalamous to extrathalamous cytoplasm. This can be modified by producing vacuolated cytoplasm, as in the clavate chambers of *Hastigerinella* (Fig. 1.21E), or by extending the extrathalamous cytoplasm into thin spines, as in the *Schackoina* (Fig. 1.21A) and *Hantkenina* (Fig. 1.21B). Trochospiral forms had no vacuolated cytoplasm. Therefore, the relative amount of extrathalamous cytoplasm in the Praehedbergellidae and Globigerinidae had to be increased in other ways; one simple way of doing this was to modify the chamber shape from that of a sphere to something approaching a trapezoid, which can maximum surface area relative to volume, as in the Schackoinidae (Plate 4.15, Fig. 6).

Another driver of the evolution of the planktonic foraminifera (see Chapter 5) was linked to the development of photosymbiosis, which led to the abrupt evolution of the morozovellids and acarinids in the Paleogene. However, Quillévéré *et al.* (2001) demonstrated that photosymbiosis did not trigger an immediate species-level radiation in this group. Instead, the acarininids remained a low-diversity taxon restricted to high latitudes for nearly 1.8 million years before radiating ecologically and taxonomically. Their eventual radiation is tied to an expansion of their geographic range into tropical environments that enabled them to fully exploit their photosymbiotic ecology. This suggests that even in the Darwinian world, the evolution of a new mode of life (i.e., photosymbiosis), which would be expected in principle to give a competitive advantage, is no guarantee of an immediate increase in diversity and abundance. Chance and opportunity seem also to play a role. As we will see in the complex story of the phylogenetic development of the planktonic foraminifera, there have been many evolutionary branches; some have prospered, others dwindled, and still others were abruptly cut short.

1.3.2 Paleontological history of the planktonic foraminifera

As a consequence of their ability to evolve rapidly and fill a range of ecological niches, planktonic foraminifera are very valuable indicators of major global and regional geological events, including impacts, tectonics, climate and sea-level changes (see Fig. 1.5). The story of the evolution and periodic extinction of the planktonic foraminifera will be told in detail in Chapters 3 to 6, but below is a brief introductory summary of their paleontological evolution (see Fig. 1.6).

As will be discussed in Chapter 3, the planktonic mode of life evolved in the Mesozoic, and became established with the appearance of the Conoglobigerinidae in the Late Bajocian. Early Conoglobigerinidae were only meroplanktonic (i.e., they were planktonic for only a part of their life cycle) and provincial, only occurring in central and northern Tethys. The appearance in the Bathonian of *Globuligerina*, however, saw the first truly holoplanktonic and hence more globally widespread form.

Conoglobuligerina persisted into the Early Cretaceous (see Chapter 4) and gave rise to the favusellids in the Berriasian. The Early Valanginian saw a shift from the relatively cool climate of the end of the Jurassic and earlier Berriasian to the "greenhouse" world that continued for the rest of the Cretaceous. This change coincided with the disappearance of the last of the conoglobigerinids and the appearance of the small globular *Gorbachikella* species, the ancestor stock of the Cretaceous Praehedbergellidae. A short-lived anoxic event in Western Tethys during the Late Hauterivian interval (at ~127.5 Ma), called the "Faraoni Event," coincided with the spreading and diversification of the small globular opportunistic *Gorbachickella* and the subsequent explosion of the praehedbergellid lineages in the Barremian. However, in the Late Barremian, another major oceanic anoxic event triggered the extinction of some species of the subspherical *Blefuscuiana*. The survivors were species of *Blowiella*, *Lilliputianella*, and *Leupoldina* that had a competitive shape, suitable for adaptation in increasingly dysoxic seawater. The dramatic rises in temperature recorded in the Early to Middle Aptian coincided with a significant turnover of planktonic foraminifera worldwide, the highest in the Cretaceous period. The Late Aptian saw a significant number of planktonic foraminiferal extinctions, but these were on the whole compensated for by the establishment of a large number of new genera at the Aptian–Albian boundary. The Cenomanian–Turonian boundary coincided with the onset of yet another major anoxic event, the "Bonarelli Event". It is marked by the extinctions of the rotaliporids. Planktonic foraminifera flourished in the Late Cretaceous, but the cosmopolitan assemblages of the Maastrichtian were brought to an end by one of the greatest mass extinctions of all time.

Only a few Cretaceous species survived the K-P mass extinction, but from these few survivors, the entire subsequent Paleogene stock was derived. As will be discussed in Chapter 5, despite these simple beginning, there is still controversy over the taxonomic and phylogenetic classification of the resulting Paleogene forms. Overall, species diversity increased gradually from the Paleocene into the Lutetian during a time of globally warm conditions, but with cooling from the Bartonian to the Chattian in the Late Oligocene, there was a notable decrease in species diversity. Throughout the Paleogene, extinction rates remained fairly constant; however, extinction events were more notable at the family level. The eoglobigerinids dwindled gradually until they disappeared at the end of the Oligocene, which also saw the final extinction of the Guembelitriidae. This stratigraphic boundary is not marked by any major discontinuity, and the planktonic foraminifera of the Neogene show a gradual, continuous development from those forms of the seven families of Paleogene planktonic foraminifera that survived uninterrupted into the Miocene.

As is outlined in Chapter 6, in the Early Miocene, the planktonic foraminifera were most abundant and diverse in the tropics and subtropics, but following the Mid-Miocene Climatic Optimum, global cooling was established and many species adapted their development to populate temperate and sub-polar oceans. The Early Zanclean saw the highest species diversity of the Neogene. The Middle and Late Pliocene saw a continuation in the reduction of global temperatures and the final closure of the Central American seaway, which profoundly changed oceanic circulation and drove a significant number of species extinctions. There were, however, few extinctions or new species at the end of the Pleistocene, and most modern, living species originated in the Pliocene and Pleistocene

1.4 Conclusion

From this introductory chapter, we have seen that fossil planktonic foraminifera are extremely abundant in most marine sediments and occur worldwide within broad latitudinal temperature belts. When this wide geographical range is combined with frequently short specific time ranges, they make excellent tools for biostratigraphy. Their study is, therefore, essential for the exploitation of economically vital deposits of oil and gas, as planktonic foraminifera are central to our ability to date, correlate, and analyze the sedimentary basins that are key to the economic well-being of the world. A detailed understanding of the taxonomy of planktonic foraminifera is essential, therefore, for any applied biostratigraphic analysis.

However, we have also seen that planktonic foraminifera are biologically complex and highly versatile. They are both generically robust (having survived in one form or another from the Jurassic) but can be highly ecologically specialized and hence specifically vulnerable to climatic and temperature changes. Their study provides, as a result, considerable insight into the evolutionary process as well as into the major geological mechanisms associated with extinction and recovery. Planktonic foraminifera also continue to occupy the present-day oceans. The study of their worldwide distribution and current ecological functions enables important information to be inferred on environmental factors that control the oceanographic realm.

In the following Chapter 2, therefore, we discuss the biology of living forms and we will address some of the recent advances being enabled by molecular studies of living planktonic foraminifera. Then, each of the four subsequent chapters will outline the paleobiological and the geological significance of the planktonic foraminifera through geological time. Specifically, the taxonomy, phylogenetic evolution, paleoecology, and biogeography of the planktonic foraminifera will be outlined and discussed, relative to the biostratigraphical time scale (as defined by Gradstein *et al.*, 2004) of the Middle and Late Phanerozic. In addition to describing and discussing the morphology of whole specimens, the chapters will also contain examples of carbonate thin sections, the study of which enable the identification of planktonic foraminifera in cemented carbonates, and hence expand their usefulness in geological investigations and hydrocarbon exploration still further.

Finally, many of the plate figures presented are new and/or are type figures from the UCL micropaleontological collection, and the Natural History Museum, London (referred to as NHM), while other type figures are from the U.S. National Museum of Natural History, Washington, DC (referred to as USNM) and the Geological Survey of Canada (referred to as GSC). In addition, some figures also reproduced from standard sources. I have tried to contact or reference all potential copyright holders. If I have overlooked any or been inaccurate in any acknowledgement, I apologise unreservedly and I will ensure that suitable corrections are made in subsequent editions.

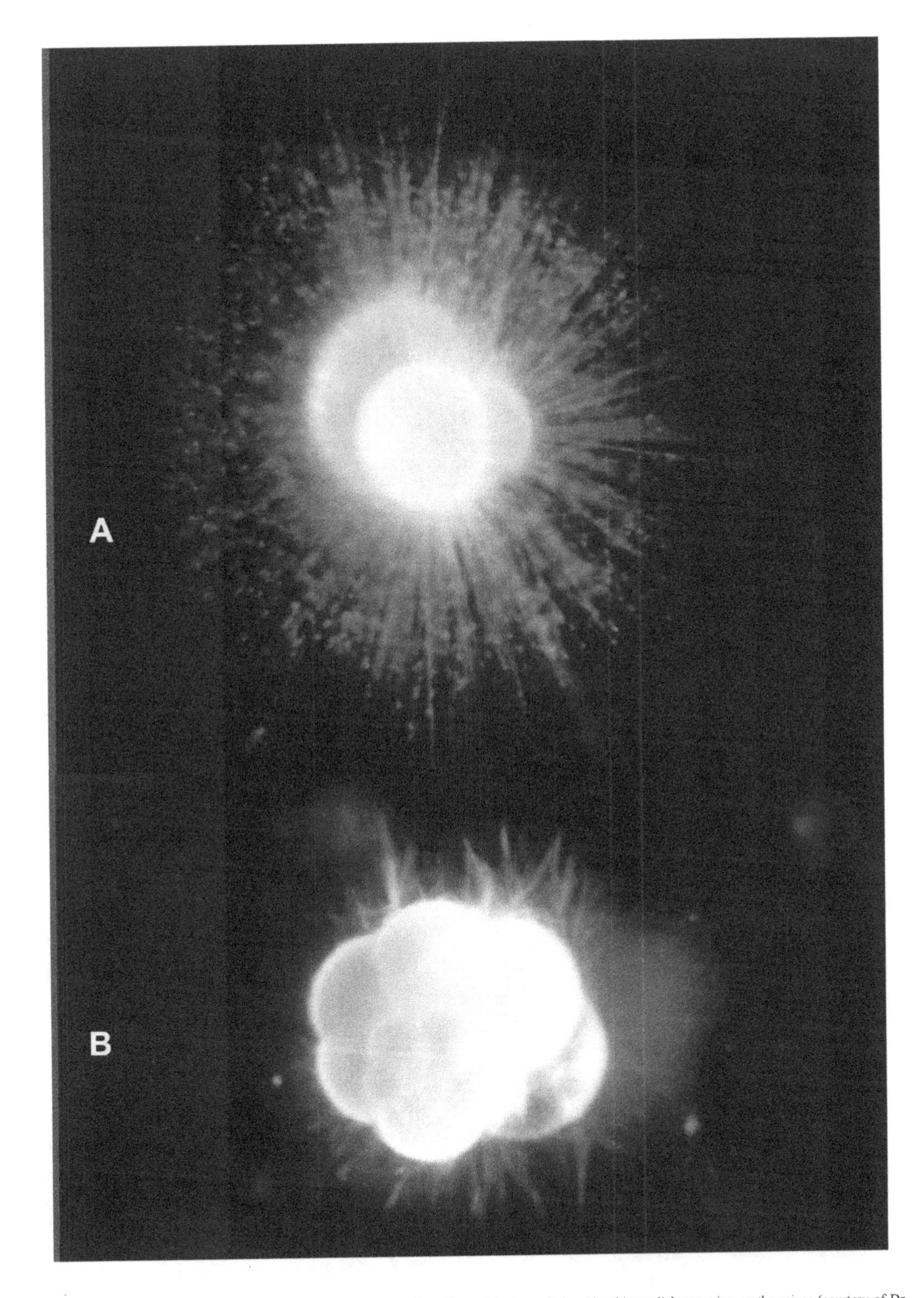

Plate 1.1. (A) *Globigerinoides sacculifer* (Brady), a spinose species with symbionts carried out by rhizopodial streaming on the spines (courtesy of Dr. Kate Darling); (B) *Neogloboquadrina dutertrei* (d'Orbigny), a nonspinose species (courtesy of Dr. Kate Darling).

Chapter 2

The biological and molecular characteristics of living planktonic foraminifera

2.1 The biological characteristics of modern planktonic foraminifera

Living planktonic foraminifera show a significant diversity and are highly adaptable, both in their morphology and biology. They exhibit characteristic cytoplasmic pseudopodal features that stream nutrients into, and waste out of the main body of the test. Many planktonic species harbour photosynthesizing symbionts in the pseudopodal structures, while others merely sequester chloroplasts on a temporary basis. In their open ocean habitat, planktonic species have developed a number of mechanisms for coping with the difficulties of reproduction. In this section, we briefly review the nature of the planktonic foraminiferal cytoplasm and their mechanism for test growth, the nature of their symbionts, and their reproductive strategy.

2.1.1 Cytoplasm and test growth

The vast bulk of the cytoplasm of the single cell of a living planktonic foraminifera is enclosed within a hard test, and fills the internal space to match its shape. Much of the inner space of the test is filled by vacuolated cytoplasm, and the nucleus is typically located in one of the inner chambers (Hemleben *et al.*, 1989). The cytoplasm possesses organelles, known as fibrillar bodies (Lee *et al.*, 1965), which in the deep evolutionary past may have originated as intracellular symbiotic bacteria (West, 1995). There are three zones of intergrading cytoplasm: the compact cytoplasm inside the test, the frothy or reticulate cytoplasm (usually observed in the final chamber or at the aperture), and the external cytoplasm (extrathalamous cytoplasm) comprising alveolate masses or reticulate to fibrose strands of pseudopodia emanating from it. Three types of pseudopodia may be observed, rhizopodia (bifurcating and anastomosing), filopodia (long, thin, and straight), and reticulopodia (net-like). The sticky rhizopodia (pseudopodia) stream outward into the surrounding environment forming a radial net that is used for feeding (Hemleben *et al.*, 1989). They may radiate from the test to form an outer envelope or may extend along the spines in spinose species (see Plate 1.1).

Test (or shell) growth and the events leading to the formation and calcification of new chambers in foraminifera have been studied since 1854, when Schultze investigated chamber formation in *Peneroplis*. However, it is only since the pioneering work of Bé *et al.* (1977), that a more detailed understanding of the biomineralization process has become available. The formation of a new chamber is accompanied by the withdrawing of feeding rhizopodia, an increase of cytoplasmic streaming inside the test, and the development of a translucent bulge of cytoplasm from the aperture. This is followed by the radiation of thick fan-like rhizopodia from the bulge to define the margins of what will become the next chamber. The cytoplasmic bulge gradually extends out to fill the rhizopodal envelope, where upon it forms the outline of the new chamber. The establishment of the anlage, the organic structure that is primarily responsible for the calcification of the wall, then begins on the surface of the cytoplasmic bulge. Calcite crystallization and growth subsequently occurs, and the whole process is complete in approximately 6 hours (Hemleben *et al.*, 1989). The wall of the chamber can be further thickened by lamellar accretion when subsequent chambers are added (Bé *et al.*, 1979). Chamber formation is generally similar in spinose and nonspinose taxa; however, spines are developed in the spinose taxa after the chamber is largely complete (Hemleben *et al.*, 1986). For a more detailed discussion of the complexities and details of biomineralization in nature as a whole, see Mann (2001).

The chemical and isotopic composition of planktonic foraminiferal tests are becoming increasingly important in the study of paleoclimatology and paleoceanography (e.g., Heuser *et al.*, 2005; Shackleton and Opdyke, 1973; Wade *et al.*, 2008). Thus, for example, the carbon and oxygen isotopic ratios, as well as the ratios of several trace elements to calcium content (e.g., Mg/Ca ratios), of the calcitic tests are being used as indicators of the following:

- environmental conditions during test growth (Pearson *et al.*, 1993, 1997; Spezzaferri and Pearson, 2009);
- the temperature of the water column, seasonal effects, light intensity variations, and ocean water density stratification (Emelyanov, 2005; Haarmann *et al.*, 2011); and
- the acquisition of photosymbionts in different phylogenetic lineages over time (Wade *et al.*, 2008).

For a detailed discussion on the use of stable isotopes in the reconstruction of the evolutionary history of planktonic foraminifera, see Pearson (1998), and for a summary of the information that can now be inferred from the chemical signature of planktonic foraminiferal tests, see Rohling and Cooke (1999).

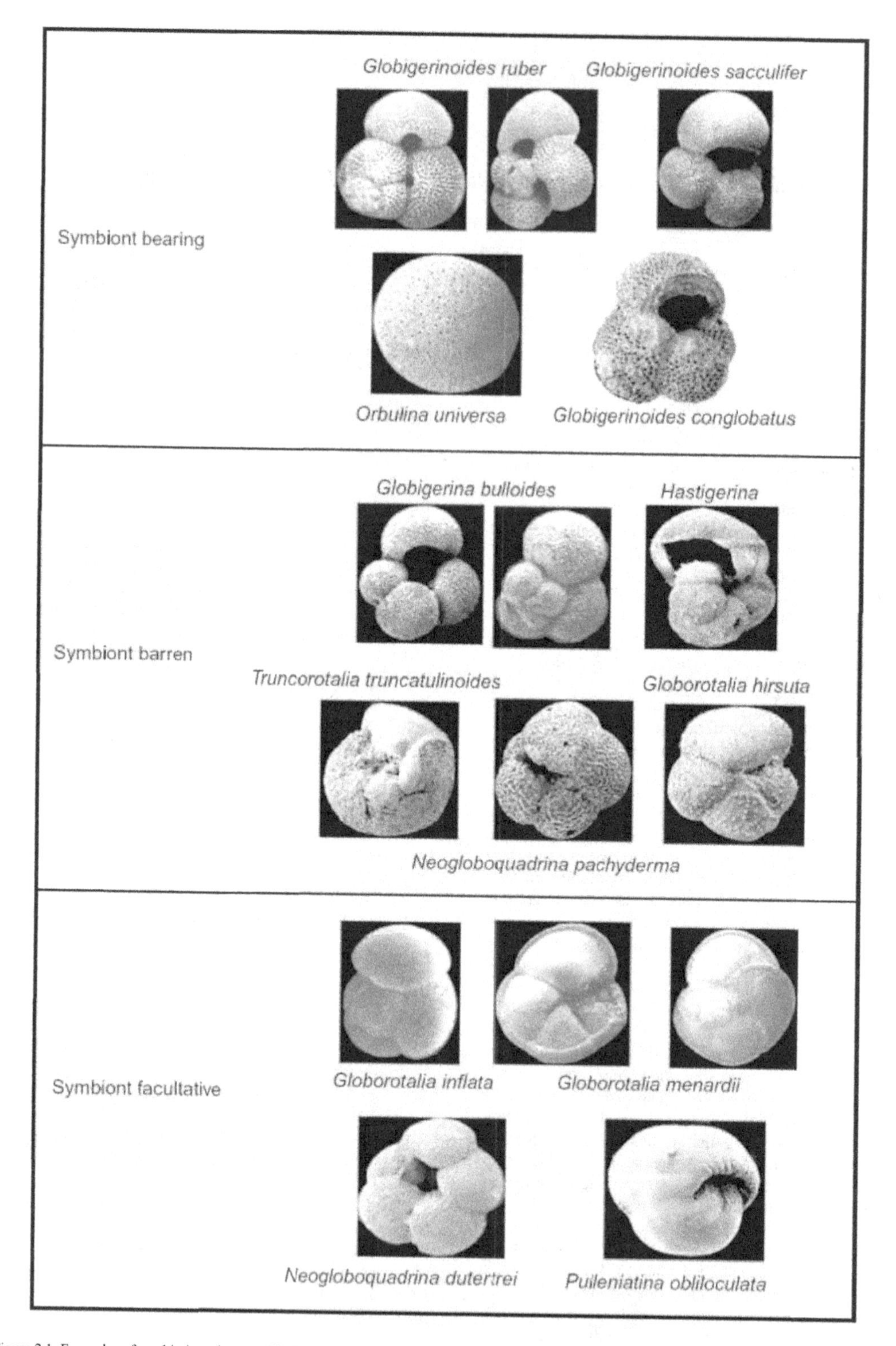

Figure 2.1. Examples of symbiotic and nonsymbiotic planktonic foraminifera. All images from the UCL Collection,

2.1.2 Symbionts

Symbiosis is particularly prevalent in tropical larger benthic foraminifera and in the planktonic foraminifera. However, unlike the larger benthic foraminifera, which harbour different types of symbionts (including diatoms, dinoflagellates, red algae, and chlorophytes; see BouDagher-Fadel, 2008), planktonic foraminifera have a symbiotic relationship only with two types of algal symbiont, either (but not both) dinoflagellates or chrysophytes (Bé, 1982; Hemleben *et al.*, 1989). These symbionts play an important role in foraminiferal reproduction, calcification, and growth (Bé, 1982; Bé *et al.*, 1982; Caron *et al.*, 1981; Duguay, 1983; ter Kuile, 1991). The symbionts benefit the foraminiferal hosts by providing a source of energy via photosynthesis (Falkowski *et al.*, 1993; Hallock, 1981) and by possibly removing host metabolites (Hallock, 1999). The endosymbionts (e.g., dinoflagellates) are advantageous for planktonic foraminifera, as they allow them to thrive in environments that are oligotrophic (low in nutrients) but which have abundant sun light (Hallock, 1981). Photosymbiosis may have contributed to species diversification (Norris, 1991), and its development in the geological past was a key innovation in the evolutionary development of the planktonic foraminifera (Margulis and Fester, 1991).

Foraminifera do not inherit their photosymbionts but acquire them throughout their life cycle from the ambient sea water (Bijma *et al.*, 1990; Hemleben *et al.*, 1989). About one-quarter of extant tropical to subtropical surface-dwelling, spinose species of planktonic foraminifera (such as *Globigerinoides ruber*, *Gdes conglobatus*, *Gdes sacculifer*, and *Orbulina universa*) harbour dinoflagellate symbionts (Faber *et al.*, 1985; Hemleben *et al.*, 1989; Fig. 2.1), all of which possibly belong to a single species, *Gymnodinium béii* (Gast and Caron, 1996; Lee and Anderson, 1991). The spines of spinose planktonic foraminifera may aid flotation but certainly allow them to capture prey items (e.g., zooplanton and phytoplankton) and to carry symbionts (in Plate 1.1, the algal symbionts can be clearly seen among the radially arranged spines). In some species (e.g., *Globigerinoides sacculifer*), dinoflagellate symbionts are transported out to the distal parts of rhizopodia in the morning and are returned back into the test at night. For symbiotic spinose forms, the symbiosis is thought to be obligative (i.e., survival outside the relationship is impossible; Hemleben *et al.*, 1989). However, a few spinose foraminifera (e.g., *Globigerina bulloides* and *Hastigerina pelagica*) are unusual in being symbiont barren, though the latter houses commensals (Spindler and Hemleben, 1980). The planispiral spinose *Globigerinella* possesses two different photosymbionts, both of them chrysophytes.

In contrast to the carnivorous spinose forms, the nonspinose planktonic foraminifera are herbivorous, like benthic forms. Many nonspinose planktonic foraminifera (e.g., *Globigerinita glutinata*, *Neogloboquadrina dutertrei*, *Pulleniatina obliquiloculata*, *Globorotalia inflata*, and *Gt. menardii*) harbours facultative chrysophytes (see Fig. 2.1). These nonobligative symbionts are housed either on a nonpermanent basis, photosynthesizing within perialgal vacuoles, or they are sometimes digested (Hemleben *et al.*, 1988). Other nonspinose taxa (e.g., *Neogloboquadrina pachyderma*, *Truncorotalia truncatulinoides*, *Gt. hirsuta*) are symbiont barren (Hemleben *et al.*, 1989; Fig. 2.1).

The dominant factors controlling the size and distribution of both symbiotic and asymbiotic planktonic foraminiferal species are light and nutrient density. Nutrient flux decreases offshore, while light, needed for symbiont photosynthesis, increases offshore as water turbidity lessens, but naturally decreases with increasing water depth. Asymbiotic foraminifera, which survive by grazing, dominate coastal fauna. Species that benefit from symbiont photosynthesis dominate the offshore fauna (Ortiz *et al.*, 1995). They tend to have relatively large size tests, which facilitate the support of the symbionts and which benefit from enhanced photosynthetic activity (Norris, 1996; Spero and DeNiro, 1987). They have a more cosmopolitan distribution than asymbiotic foraminifera and a greater ability to withstand periods of nutrient stress (Norris, 1996).

Isotopic methods have been used to determine possible symbiotic associations in fossil planktonic foraminifera (e.g., D'Hondt and Zachos, 1993; D'Hondt *et al.*, 1994; Norris, 1996), as the vital processes often leave a characteristic isotopic signature. Correlation between test size and characteristic stable isotopic ratios indicated that photosymbiosis existed in some Late Maastrichtian planktonic foraminifera (Bornemann and Norris, 2007; D'Hondt and Zachos, 1998). It has been suggested that photosymbiosis as a life strategy was given a competitive advantage by the development at this time of oligotrophic conditions associated with increased water-mass stratification (Abramovich and Keller, 2003). All photosymbiotic forms, however, went extinct at the End Cretaceous catastrophe. In the fossil record, however, photosymbiosis reappeared in the acarinids and morozovellids in the Paleogene. D'Hondt *et al.* (1994) hypothesized that their reappearance was closely linked to habitat and not to test morphology; *Acarinina* and *Morozovella* occupied the same habitat as modern photosymbiotic taxa, but they differ greatly in morphology from living forms.

2.1.3 Reproduction

In contrast to the variety of reproductive strategies seen in the benthic foraminifera (BouDagher-Fadel, 2008), sexual reproduction is the only strategy that has ever been recorded for the planktonic foraminifera (Goldstein, 1999; Hemleben *et al.*, 1989). It is not possible to make detailed observations of the physiological and morphological changes during the entire life cycle of planktonic foraminifera, as they have not as yet not produced viable offspring in the laboratory (Lee and Anderson, 1991). However, test size distributions, coupled with abundance changes and more general laboratory observations, have been used by Hemleben *et al.* (1989) and Lee and Anderson (1991) to infer the life cycle of *Globigerinoides sacculifer* (Plate 6.3, Figs. 9–10) and *Hastigerina pelagica* (Plate 6.8, Fig. 11).

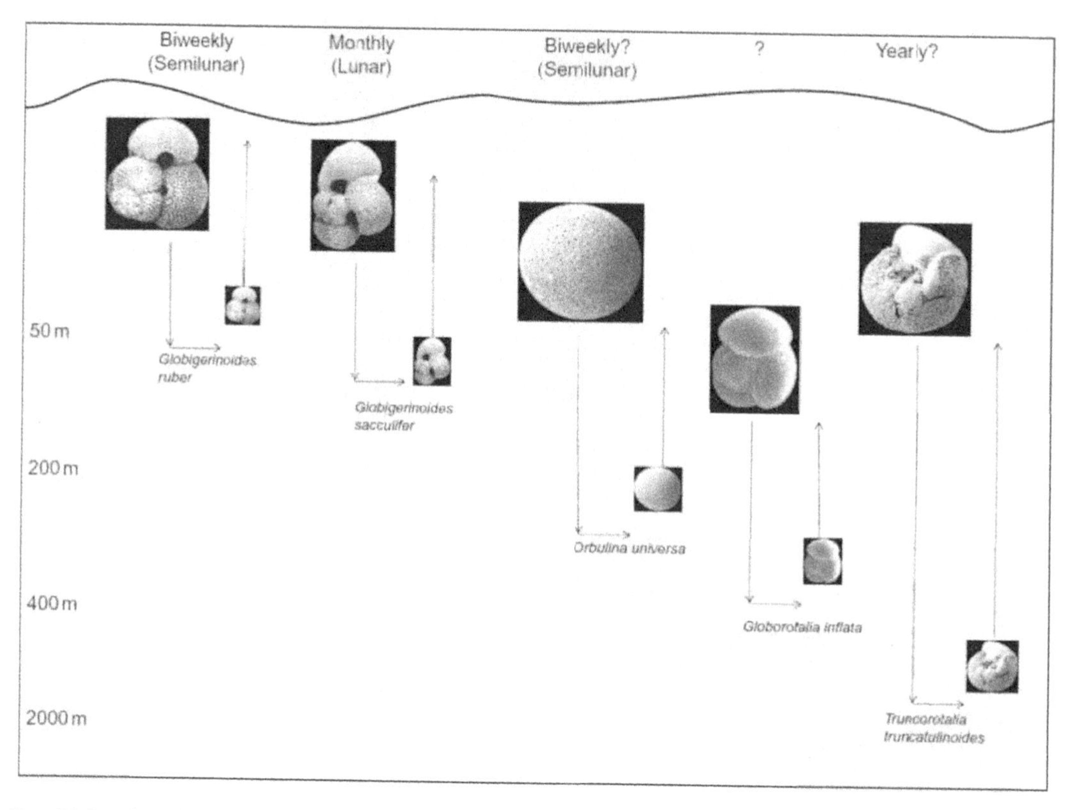

Figure 2.2. Reproduction cycles of some planktonic foraminifera from the subtropical and tropical realms. Cycle patterns span a range from near-surface-dwelling forms that reproduce at shorter time intervals to those that reproduce at depth and rise to the surface (adapted from Hemleben *et al.*, 1989). All images from the UCL Collection.

Hemleben *et al.* (1988) recorded that the planktonic foraminifera have developed a number of mechanisms for coping with the difficulties of reproduction in an open ocean environment, in order to maximize the chances that compatible gametes of the correct species will meet. Throughout the year, most species will migrate throughout the water column, a strategy that is thought to maximize the access to available food sources. Approximately a day preceding gamete release, mature individuals of spinose planktonic foraminifera (e.g., *Globigerinoides sacculifer*) start to sink down below the photic zone, while digesting their symbionts, and accumulate in the thermocline to release their gametes. During this time, they produce additional calcification of final chambers just before shedding their spines. Then nuclear division occurs, filling the cytoplasm with daughter nuclei, and the gametes are released from an expanding bulge of the cytoplasm (Lee and Anderson, 1991); at least 10^5 gametes are released from each parent cell (Bé and Anderson, 1976). Several days later, juvenile individuals appear again in the productive surface waters and the cycle repeats itself (Erez *et al.*, 1991; Taniguchit and Bé, 1985). The sinking of planktonic foraminifera into deeper water in their reproductive cycles may have many advantages, such as accessing a stable breeding environment where food source for juveniles is available, and enhancing the survival of the gametes and resulting juveniles by placing them where they can avoid predators.

Gamete release in planktonic foraminifera is mainly synchronized by a lunar, semilunar, or diurnal cycle (Hemleben *et al.*, 1989), although food availability may also play some role in the timing of reproduction (Hemleben *et al.*, 1989). Gametes are released in their hundreds of thousands and though not proven conclusively, evidence suggests that the primary reproductive strategy may be dioecious (i.e., each individual has reproductive units that are either simply male or simply female), with gametes from different parents fusing to form the new juvenile (Hemleben *et al.*, 1989). The general reproductive strategies and habitation-horizons of some modern planktonic foraminifera, as recorded by Hemleben *et al.* (1989), are shown in Fig. 2.2.

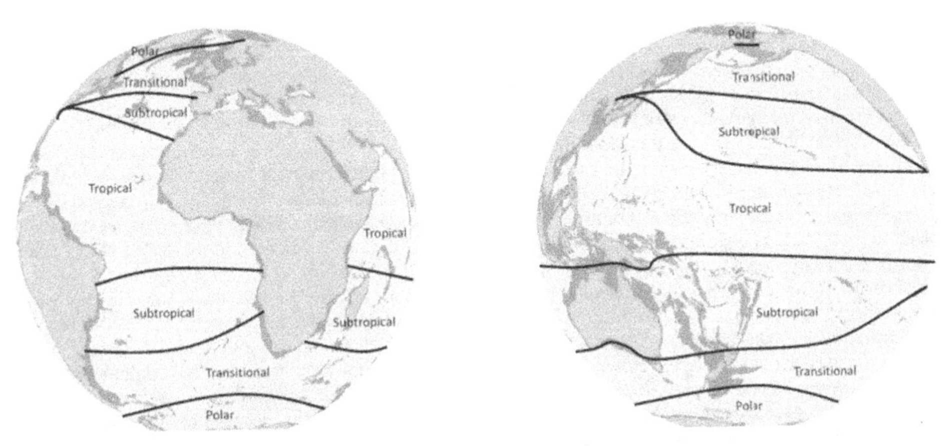

Figure 2.3. World map showing the four major planktonic foraminiferal faunal provinces: tropical, subtropical, temperate (or transitional), and polar.

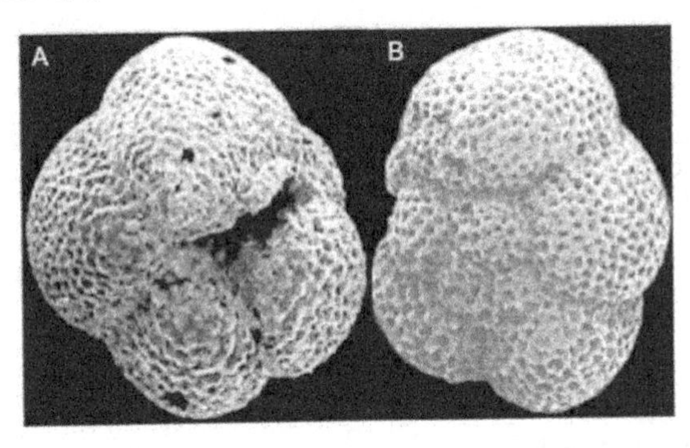

Figure 2.4. *Neogloboquadrina pachyderma* (Ehrenberg), left coiling from South Africa, with a secondary calcite crust, A, umbilical view; B, spiral view, x84. Both images from the UCL Collection.

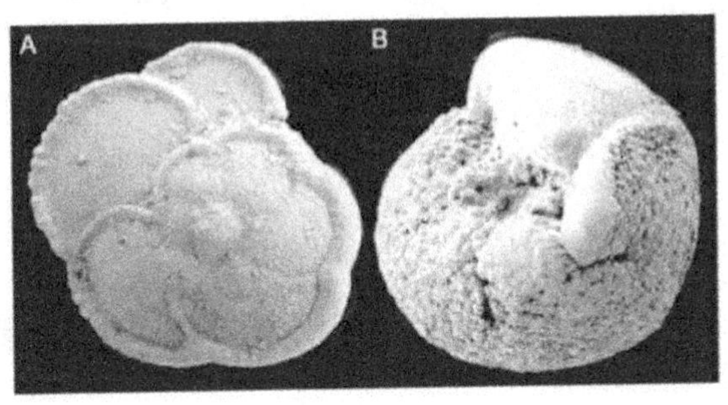

Figure 2.5. *Truncorotalia truncatulinoides* (d'Orbigny), right coiling from the Tasman Sea, A, spiral view; B, umbilical view, x65. Both images from the UCL Collection.

2.2 Biogeography and ecology of the modern planktonic foraminifera

Living planktonic foraminifera have expanded to fill a wide variety of niches within the global ocean, including tropical waters as well as subtropical and polar water masses. They live mostly in the photic zone but descend to waters as deep as several thousand meters. In comparison with their highly diverse benthic counterparts, there are only about 90 extant species of planktonic foraminifera (see Charts 6.1–6.3 online). Their species diversity peaks in the subtropics and decreases steeply toward the poles (see Fig. 2.3). They are most abundant in euphotic, near-surface waters, in water depths between 10 and 50 m. The surface waters of the sea are usually saturated or supersaturated in oxygen, which is dissolved from the air that is mixed into the oceans by wave action. It is in these most superficial layers of seawaters that symbiotic or asymbiotic planktonic foraminifera with subglobular chambers are found (e.g., *Globigerina* and *Hastigerina*). At greater depths, the oceanic oxygen levels drop, and forms with a greater surface area to volume ratio have an advantage (as they enable oxygen to diffuse into the cell more effectively than the more spherical forms, e.g., *Beella* and *Hastigerinopsis*, see Chapter 6).

The distribution of each species is controlled by various factors, such as temperature, light intensity, and prey or nutrient availability, and this can be used as a proxy for the reconstruction of paleoenvironments. For example, the geographic distribution of *Neogloboquadrina pachyderma* (Fig. 2.4) is mainly controlled by seawater temperature. However, two morphotypes of *Globigerinoides ruber* (*Gdes ruber* s.s. and *Gdes ruber* s.l.) collected from the South China Sea showed different vertical distributions, with *Gdes ruber* s.s. predominant in surface waters and *Gdes ruber* s.l. in a deeper ones. This difference in habitat can be attributed to a difference in light intensity and food availability, with the shallower-living morphotype being more dependent upon symbionts (Kuroyanagi and Kawahata, 2004).

The direction of coiling (dextral or sinistral) of some planktonic foraminifera has been used for some time as a proxy for paleoclimate. For example, variants of *Neogloboquadrina pachyderma* have long been used as tools to infer changes in polar ocean temperatures over geological time (Peck *et al.*, 2008). It had been thought that the sense of coiling within *N. pachyderma* was temperature dependent, such that during periods of relatively colder oceanic temperatures, its test coils sinisterly (Fig. 2.4), while during periods of relative warmth, assemblages with more than 50% dextral coiling are found. However, more recent research (see the following section) now shows that these two coiling variants are in fact genetically distinct species (Darling and Wade, 2008; Darling *et al.*, 2006; de Vargas *et al.*, 2001), with *N. pachyderma* having the sinistral coiling (Fig. 2.4) and *Paragloborotalia incompta* being dextral. Similar behavior is exhibited by *Truncorotalia* (Fig. 2.5; Fig. 6.13). *Truncorotalia* species have always been deep dwelling, with left coiling groups found in subpolar regions, and right coiling forms (Fig. 2.5) found in relatively warmer waters. These changes in coiling direction correlate with isotopic shifts associated with temperature changes (Feldman, 2003). However, the coiling variants of *T. truncatulinoides* have also now been shown to be genetically different (de Vargas *et al.*, 2001) and may even be distinct species.

2.3 The molecular biological studies of the planktonic foraminifera

Relatively recent molecular biological studies based on the analysis of DNA sequences have led to an extensive increase in our understanding of the evolutionary relationships between the species of planktonic foraminifera living in the oceans today, the origin of major phylogenetic lineages, and the likelihood that stable morphospecies are in fact species clusters. Furthermore, as seen above, genetic studies have also given new insights into their biogeography and ecological diversity.

The focus of research in this field has been on the ribosomal ribonucleic acid (rRNA) of the planktonic foraminifera. rRNA is the RNA component of the ribosome, which is the enzyme that is the location of protein synthesis in all living cells. The rRNA forms two subunits, the large subunit (LSU) and small subunit (SSU). Planktonic foraminifera show an unusually high level of genetic diversity in their SSU rRNA (Aurahs *et al.*, 2009; Darling *et al.*, 1996; Pawlowski *et al.*, 1996; Wade *et al.*, 1996), and many contain more than one genetically distinct unit that can be used to differentiate between different species. These sequences of protein-coding genes provide an important alternative source of phylogenetic information to the morphological data upon which fossil analysis depends (Flakowski *et al.*, 2005). It has been found that some different types of "morphospecies" can indeed be highly divergent genotypes, having slight morphological differences and displaying distinct ecological adaptations. They can, therefore, be separated taxonomically into clusters of cryptic or sibling species, which may have diverged many millions of years ago (Darling *et al.*, 1999, 2007; de Vargas *et al.*, 2002; Huber *et al.*, 1997; Wade and Darling, 2002). Not only do data on rRNA give us insight into the genetics of modern planktonic foraminifera, but they can also be used to confirm the validity (or otherwise) of the phylogeny of these forms inferred from the fossil, morphological record.

2.3.1 Molecular and genetic insights into the origin of planktonic foraminifera

As introduced above, recent studies on planktonic foraminifera have focused on partial sequences of the SSU rRNA gene (Aurahs *et al.*, 2009; Darling *et al.*, 1996; Pawlowski *et al.*, 1996; Wade *et al.*, 1996) and the LSU rRNA gene (Pawlowski *et al.*, 1994). Both units can easily be obtained from single cells captured in environmental samples (Flakowski *et al.*, 2005). SSU rRNA sequences can be amplified using the "universal" eukaryote primers of White *et al.* (1990), originally designed for use on fungi. As noted in Chapter 1, molecular analyses have shown that the Order Foraminifera are a mono-phyletic group within the eukaryotic phylogeny (Archibald *et al.*, 2003; Berney and Pawlowski, 2003; Darling *et al.*, 1996; Flakowski *et al.*, 2005; Keeling, 2001; Longet *et al.*, 2003; Pawlowski *et al.*, 1996; Wade *et al.*, 1996), and they seem to form one of the earliest diverging eukaryote lineages in the "tree of life."

Molecular studies have provided information about subtle features of the evolutionary history of planktonic forms (Aze *et al.*, 2011). For example, as referred to above, prior to genetic studies, it had been thought that the morphologically similar *Neogloboquadrina pachyderma* and *"N." incompta* were a single species, but with coiling direction being an ecophenotypic response to temperature (Ericson, 1959; Ottens, 1992). Genetic studies, however, have revealed substantial divergence between the two forms, and the coiling direction in *N. pachyderma* is a consistent genetic trait, heritable through time (Darling *et al.*, 2004, 2006). Indeed, detailed morphological study reveals them even to belong to different genera. The species *incompta* belongs to the genus *Paragloborotalia*, because of its wall structure (pustules on surface, with no ridge growth) and the shape of its aperture (umbilical–extraumbilical throughout adult growth), while *Neogloboquadrina pachyderma* has an anterointraumbilical aperture and a build-up of a calcitic crust on its test (see Fig. 2.4 and Chapter 6).

Molecular systematics and genetic analyses have also been used to determine the phylogenetic evolutionary lineages of the planktonic forms and their relationship to benthic form. Fossil evidence suggested that the earliest planktonic foraminifera evolved from benthic lineages, originating in the Late Triassic and the Jurassic (see Chapter 3). Their evolution occurred initially by adopting a meroplanktonic mode of life, in which benthic forms (living mainly attached to the sediment) adopted the planktonic mode for a part of their life cycle (perhaps the reproductive stage). Such transitions from the benthic to the planktonic mode of life are well documented (Banner *et al.*, 1985; BouDagher-Fadel *et al.*, 1997) in many modern benthic rotaline foraminifera, which are usually resident on firm substrates (stones or algae, etc.) during which time the adherent side of their tests become flattened. When the planktonic mode of life is adopted, the chambers lose their adherence and become more fully globular (see Fig. 2.6). The development of a planktonic state, before reproduction, enables a species to disperse its gametes more freely throughout the seawater and so to spread its progeny more widely. The holoplanktonic (fully planktonic) mode of life did not appear in the fossil record before the Middle Jurassic, with the development of *Globuligerina*. A second development of holoplanktonism occurred later in the Cretaceous, giving rise to the triserial guembelitriid lineage. Genetic studies of the only living representative of this guembelitriid lineage (*Gallitellia vivans*) show that this lineage is quite distinct from the globigerinid planktonic foraminifera and that genetically the guembelitriids are more closely related to the benthic rotaliids than they are to the globigerinids (see Chapter 6). The genetically inferred divergence time of *Gallitellia vivans* (Ujiié *et al.*, 2008) is, however, estimated as being no older than 18 Ma (i.e., the Early Miocene). This suggests, therefore, that the origin of this living guembelitriid is in fact independent from the other Cretaceous and Paleogene Heterohelicida (see Fig. 1.6), and that *Gallitellia vivans* might have acquired the planktonic mode of life by a relatively recent transition from the benthos to the planktonic mode somewhere between the Miocene and the Pliocene.

This multiple occurrence of the evolutionary transition from benthic to planktonic is further corroborated by an increasing amount of evidence from molecular studies, which suggest that even living globigerinid foraminifera are polyphyletic in origin. Most analyses of living globigerinid planktonic foraminifera show three separate clusters in the molecular tree (Fig. 2.7), which equate to the morphologically defined microperforate, nonspinose forms (e.g., *Globigerinita uvula*; Plate 6.1, Fig. 15); macroperforate, nonspinose forms (e.g., *Globoratalia menardii*, *Neogloboquadrina dutertrei*; Plate 6.7, Figs. 14–15); and spinose forms (e.g., *Globigerinoides ruber*; Plate 6.3, Figs. 7–8). Such distinct genetic clustering may indicate that the planktonic way of life of these living Globigerinida evolved from three (or more) independent preexisting benthic lines (de Vargas *et al.*, 1997).

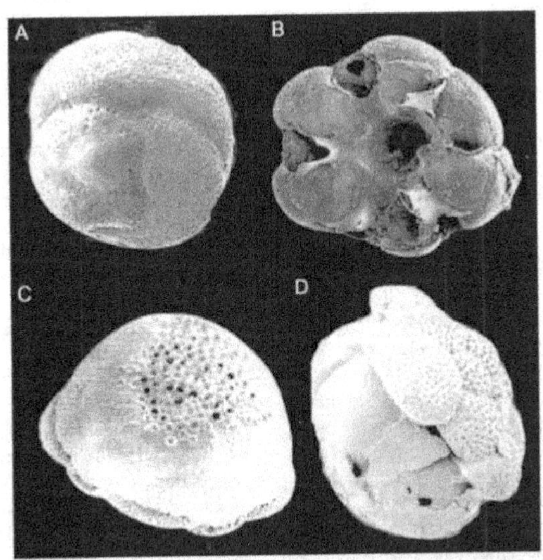

Figure 2.6. The gamonts of some species of the normally benthic Discorbidae and Cymbaloporidae have a planktonic stage just before and during reproduction. They independently evolved flotation structures in subglobular chambers at the last stage of growth. (A) *Neoconorbina concinna* Brady, Holocene, Kenya, spiral view, x85; (B) *Cymbaloporetta* sp., Holocene, Kenya, x60; (C) *Cymbaloporetta cifelli* Banner, Pereira, and Desai, Holocene, Kenya, x40; (D) *Cymbaloporetta* sp. Holocene, offshore Lebanon, x40. All images from the UCL Collection.

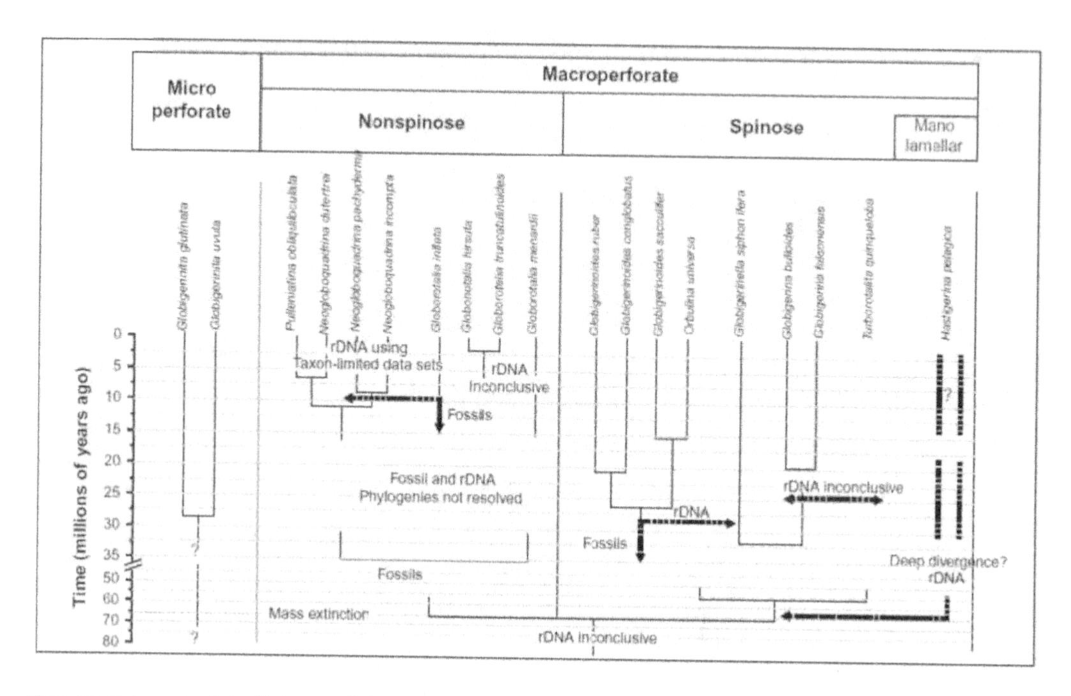

Figure 2.7. Phylogenetic relationship between planktonic foraminifera as inferred from genetic data. Incongruences between genetic and fossil data are highlighted; fossil evidence that is contradictory to molecular phylogenies but poorly resolved is also indicated (after Aurahs *et al.*, 2009).

The placement in Fig. 2.7 of the spinose Globigerinidae and the nonspinose Globorotaliidae in separate lineages contrasts with the traditional paleontological view of their common origin. Evidence that the globorotaliids diverged from a globigerinid ancestor in the Eocene (Fig. 1.6) is well documented in the fossil record (see Chapter 5). However, the molecular data suggest that some living globorotaliids are far more closely related to some benthic taxa than they are to living globigerinids (Aurahs *et al.*, 2009; Darling *et al.*, 1997, 1999, 2000, 2006; de Vargas *et al.*, 1997; Stewart *et al.*, 2001). However, this differentiation between planktonic and benthic affiliations may be a false dichotomy. In laboratory experiments, Hilbrecht and Thierstein (1996) were able to observe forms with compressed tests, conventionally classified as planktonic forms, *Globigerinella siphonifera* and *Globorotalia menardii* undertaking benthic activities through their life cycle. *Glla siphonifera* was able to create well-organized burrows and to excavate sediment in a circular pile. *Gt. menardii* and *Glla siphonifera* have specific crawling and burrowing orientations. This behavior is in contrast to that of planktonic foraminifera with globular chambers and buoyant tests, such as *Globigerinoides ruber*, *Globigerinoides sacculifer*, *Hastigerina pelagica*, and *Orbulina universa*, which when observed did not exhibit any benthic activity and in fact died if placed on a sediment substrate (Hilbrecht and Thierstein, 1996).

The monophylogeny of the *Neogloboquadrina* and its relationships to *Globorotalia* is strongly supported in the fossil record (see Chapter 6). However, Darling *et al.* (1999) found that the only two globorotaliids in their sample of analyzed species, *N. dutertrei* and *Gt. menardii*, do not cluster together, but rather branch off deeply within the benthic group and are genetically widely separated from each other (Fig. 2.7). According to Darling *et al.*, the nonspinose *Neogloboquadrina dutertrei* (Plate 6.7, Figs. 14–15) is genetically more closely related to benthic species, from which they infer either that it has only recently evolved from a benthic habitat or that it exhibits a much slower rate of genetic evolution than other planktonic species. Similarly, the origin of the extant family Hastigerinidae (which possesses monolamellar shells) remains genetically ill-defined (Aurahs *et al.*, 2009), while the microperforate *Tinophodella glutinata*, the only representative of the Globigerinitidae so far studied, appears to be genetically closer to the globorotaliids and the benthic rotaliids than it is to the globigerinids (Aurahs *et al.*, 2009). In contrast to the inconclusive genetic data in the nonspinose species, the relationships between the globular spinose forms (with globular buoyant tests), such as *Globigerina bulloides–Globigerina falconensis*, *Globigerinoides ruber–Globigerinoides conglobatus*, and *Globigerinoides sacculifer–Orbulina universa*, are in agreement with the fossil record (see Fig. 2.7; Aurahs *et al.*, 2009).

In the fossil record, the relationship between the microperforate, the spinose macroperforate, and the nonspinose macroperforate foraminifera is inferred from their morphology (overviewed in Chapter 1 and discussed further in Chapter 6). Figure 1.6 shows continuous evolutionary lineages of planktonic forms over the past

200 Ma. However, the recent genetic and observational evidence discussed above suggests a more complex picture in which the interchange between the planktonic and benthic realms may be partially reversible when, for example, compressed planktonic forms that are specialized in the exploitation of deeper water columns (e.g., *Globorotalia* and *Globigerinella*) can also exploit the benthic realm. This might explain the observed molecular branching of *Globorotalia* deeply within the benthic group and separately from the spinose globular planktonic foraminifera, and hence provides evidence of a polyphyletic origin for the planktonic foraminifera. Similarly, the multiserial planktonic foraminifera have almost certainly experienced repeated transitions between the benthic and the planktonic realms, suggesting that they too can exploit both niches. Indeed, recently, Darling *et al.*(2009) demonstrated that the extant biserial planktonic *Streptochilus globigerus* is genetically identical (and so belongs to the same biological species) as the benthic form *Bolivina variabilis*. This "benthic" or "pseudobenthic" behaviour in planktonic foraminifera, especially in the compressed or uncoiled species, may have been widespread among fossil planktonic foraminifera; however, no direct evidence of such behavioural complexity is or can be retained in the fossil record. As a result, a simple and apparently continuous phylogenetic tree (e.g., Fig. 1.6) is the inevitable consequence of morphologically based analyses; however, the interpretation of such trees must always be tempered by the suspicion that the behaviour of many now extinct species may have been more complex when they were alive than is evident from their fossilized forms.

2.3.2 Biogeography, cryptic speciation, and molecular biology

The global open ocean is not a uniform ecosystem, but one comprised of regionally distinct climate zones. Global circulation patterns and climate zones define basic physical boundaries for the planktonic foraminifera that inhabit the ocean. The ocean environment is also a complex ecosystem, and within the faunal provinces, other factors such as salinity, prey abundance or nutrient levels, turbidity, illumination, chemistry, and thermocline gradient may also affect diversity, abundance, and distribution of the planktonic foraminifera. In studying the global distributions of modern planktonic foraminifera, broad planktonic provinces can be designated as tropical, subtropical, temperate, and polar (see Fig. 2.3). Specific assemblages of morphologically adapted species are associated with each of these regions. The fact that planktonic foraminiferal species adhere to these faunal provinces, coupled with their widespread global latitudinal distribution, has made them extremely useful in the study of both modern and ancient marine ecosystems (Hemleben *et al.*, 1989). Discrete assemblages are also found in transitory provinces associated with regional upwelling (Hemleben *et al.*, 1989; Lipps, 1979). Indeed, individual species may be found across several zones, such as tropical–subtropical, subtropical–temperate, and even across all zones, as cosmopolitan forms (see Charts 6.1–6.3 online), suggesting that gene flow is common globally, with genetic intermixing between populations occupying several climatic zones.

Superficially, land-mass barriers and water-mass fronts do not generally seem to form insuperable divisions. Morphologically, identical assemblages of warm-water species, such as *Orbulina universa*, are found in geographically distinct bioprovinces, such as the Caribbean, the Coral Seas, and the Mediterranean. Furthermore, data from the SSU rRNA genes of specimens of *Globigerina bulloides*, *Turboratalita quinqueloba*, and *Paragloborotalia incompta*, collected from the Arctic and Antarctic regions, show that species from these two isolated regions possess identical rRNA genotypes (Darling and Wade, 2008). Thus, although these species are only found in the high latitudes and are absent from tropical regions, it appears that either the water-mass that separates them is not a major barrier to gene flow or that these assemblages have been separated for too short a period of time for them to develop distinct genetic identities. In contrast, however, genetically distinct forms of *Neogloboquadrina pachyderma* are found in the Arctic and Antarctic waters, while *Globorotaloides hexagonus* is found only in the tropics of the Indo-Pacific suggesting that for this form at least the Southern Africa Cape currently presents a barrier to their passage into the Atlantic (Darling and Wade, 2008).

However, increasingly more detailed studies that combine molecular, ecological, and morphological evidence are now revealing multiple cases of "cryptic speciation" among planktonic foraminifera (Darling and Wade, 2008; Ujiié and Lipps, 2009), which challenge the morphospecies concept and the paleoceanographic interpretations based up on them. So, for example, *Truncorotalia truncatulinoides* has been found to be a complex of four genetic species adapted to particular oceanic conditions, two occurring in the subtropics, one in the sub-Antarctic convergence zone, and one in Antarctic waters (de Vargas *et al.*, 2001). Likewise, it seems that *Orbulina universa* comprises three cryptic species (Darling and Wade, 2008; de Vargas *et al.*, 1999), also distributed according to oceanic provinces and particularly to chlorophyll concentration at the sea surface. *Globigerinella siphonifera* comprises over five types, which can be divided into at least two sibling species that can be distinguished by isotopic variations, shell porosity, and the species of their photosymbionts (Darling and Wade, 2008; Darling *et al.*, 1997, 1999; Huber *et al.*, 1997). Within the realm of planktonic foraminifera, therefore, there indeed seems to be an abundance of cryptic species. Seears *et al.* (2012) proposed that sea surface, primary productivity is the main factor driving the segregation of planktonic foraminifera, with variations in symbiotic associations possibly playing a role in the specific ecological adaptations observed. They suggested that ecological partitioning could be contributing to the high levels of cryptic genetic diversity observed within the planktonic foraminifera and support the view that ecological processes may play a key role in the diversification of marine pelagic organisms. The newly discovered genotypes show non-random distributions, suggestive of distinct ecology ecotypes (Darling and Wade, 2008; de

Vargas *et al.*, 2002). This widespread development of cryptic speciation among the planktonic foraminifera hints, therefore, at the processes that underpin Darwinian evolution, namely, that genetic divergence and eventual speciation is driven when communities become isolated, colonize different niches, or adopt different strategies for survival. This process has been exemplified par excellence by the planktonic foraminifera as they evolved over geological time to produce the wide diversity of species seen today and in the fossil record.

In conclusion, therefore, in this chapter, we have reviewed evidence linking molecular biology to species diversification and shown how molecular studies help constrain the phylogenetic evolution of recent forms. When studying fossil forms, however, genetic data are not available and only information drawn from test morphology can be used to develop phylogenetic relationships. We have explained, however, that such fossil-based studies may miss some of the true complexity of the modes of life that the foraminifera exploited when extant. So bearing this caveat in mind, we use in the following chapters the morphological approach to the definition of the taxonomy and the phylogenetic evolution of the fossil record to define the biostratigraphic and environmental significance of the Mesozoic and Cenozoic planktonic foraminifera. In Chapter 3, we review the earliest planktonic radiation, which possibly occurred in the Late Triassic, but which certainly was established in the Jurassic. In Chapters 4 and 5, we extend the analysis of the development of the planktonic foraminifera through the Cretaceous and the Paleogene, and finally in Chapter 6, we show how the morphologically based studies of the Neogene forms relate to the developing genetic understanding that has been described in this chapter. We will see that more molecular data, covering all known planktonic species, are needed to resolve the many issues that remain outstanding, and only then will it be possible to combine the insights from genetics with morphological and fossil data to provide a complete (or as complete as will ever be possible) understanding of the evolutionary unfolding of planktonic foraminifera over geological time.

Chapter 3

The Mesozoic planktonic foraminifera: The Late Triassic–Jurassic

3.1 Introduction

The first foraminifera with hard parts appeared in the Cambrian. These were the Allogromida, which are unilocular, simple agglutinated forms. Textularida, the agglutinated foraminifera with a test of quartz, or other inorganic particles, stuck together by calcitic or organic cements, evolved from Allogromida late in the Cambrian. The members of the Textularida remained the dominant group in the early Paleozoic. However, the Rotaliida, with tests made of primarily secreted calcite or rarely aragonite, made their first appearance in the Triassic. These benthic forms (or the related Robertinida) gave rise to forms with a meroplanktonic mode of life (see Chapter 1 and 3.3 in this chapter) in the Late Triassic and again in the Jurassic, as represented by the Favuselloidea. This superfamily includes the short-lived Sphaerogerinidae of the Late Triassic and the Conoglobigerinidae of the Jurassic (see Fig.3.1), which became widespread in the Middle Jurassic, and subsequently colonized the oceans by adopting a (holo-)planktonic mode of life.

The appearance of foraminifera with a planktonic mode of life in the Mesozoic was arguably the most important event in the entire evolutionary history of the Cenozoic, as subsequently their development had a significant impact on the whole marine biosphere. Since 1958, when Grigelis first described *Globuligerina oxfordiana* (Plate 3.4, Figs. 1–6; Plate 3.5, Figs. 9–12; Plate 3.6, Fig. 7) from the Upper Jurassic of Lithuania, our knowledge of early planktonic foraminifera has changed significantly. All species of Late Triassic and Jurassic planktonic foraminifera are placed in order Globigerinida and are members of the superfamily Favuselloidea. Early Jurassic favusellids were originally described from Eastern Europe (e.g., Kasimova and Aliyeva, 1984; Kuznetsova and Gorbachik, 1980), but their occurrence in sediments dating from the Bajocian have now been reported from across the greater part of Europe and North Africa (Görög, 1994; Wernli and Görög, 2007). BouDagher-Fadel *et al.* (1997) produced as comprehensive a study of their evolutionary history as was possible at that time, but since 1997, further discoveries have been made, including the occurrence of the short-ranged Sphaerogerinidae from the Late Triassic (Korchagin *et al.*, 2003). However, despite these recent discoveries, the Triassic–Jurassic taxa are still poorly understood. Most forms are represented by only a few specimens, and their intraspecific variations are still imperfectly defined. Below, the morphologies and taxonomies of the Triassic–Jurassic Favuselloidea are presented. This is followed by a discussion of their biostratigraphic and phylogenetic significance, and finally, their wider geological and paleoenvironmental role is reviewed.

3.2 Morphology and taxonomy of the Triassic–Jurassic planktonic foraminifera

The Triassic–Jurassic Favuselloidea are aragonitic, microperforate, pseudomuricate, with an umbilical aperture and sub-globular adult chambers. They show a marked dimorphism, a feature not known in true (holo-)planktonic foraminifera. The taxonomy of the Triassic–Jurassic genera can be summarized as follows:

CLASS FORAMINIFERA Lee, 1990

ORDER GLOBIGERINIDA LANKASTER, 1885

Tests are planispiral or trochospiral, at least in the early stage, and microperforate or macroperforate, smooth, muricate, or with spines. Apertures are terminal, umbilical, intra–extraumbilical, or peripheral. Walls are calcitic, but the walls of early may be aragonitic. Late Triassic (Rhaetian) to Holocene.

Superfamily FAVUSELLOIDEA Longoria, 1974 emend. Banner and Desai, 1988

For the members of this superfamily, the test is trochospiral with two or more whorls of sub-globular chambers. The test surface is covered by microperforations and pseudomuricae. They may also have perforated cones or may possess a favose surface structure of fused pseudomuricae, forming an anastomosing reticulation. The spiral side is evolute, while the umbilical side is involute with a small umbilicus. The aperture is intraumbilical, often covered by a bulla, or intra–extraumbilical and develops a high or low arch, which may be strongly asymmetric. The walls are believed to be composed primarily of aragonite. They range from the Late Triassic (Rhaetian) to Cretaceous (Early Cenomanian).

Family Conoglobigerinidae BouDagher-Fadel, Banner and Whittaker, 1997

This family consists of those genera of the Favuselloidea that do not possess a favose surface structure and have an intraumbilical or intra–extraumbilical aperture. The intraumbilical aperture may be an interiomarginal low arch or have a loop-shaped opening, and may also be covered by a bulla. Jurassic (Late Bajocian) to Cretaceous (Early Valanginian).

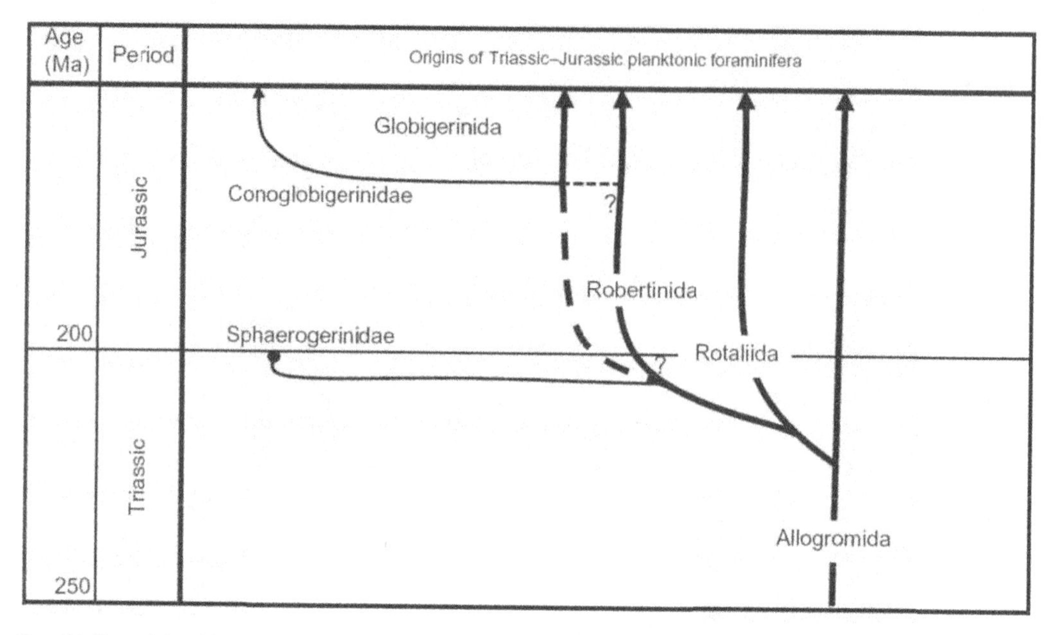

Figure 3.1. The evolution of the early planktonic foraminiferal families (thin lines) from the benthic family Oberhauserellidae.

- *Compactogerina* Simmons *et al.*, in BouDagher-Fadel et al., 1997 (Type species: *Globuligerina stellapolaris* Grigelis, 1977). This genus has an intra–extraumbilical aperture possessing a narrow lip that partly covers the umbilicus. The pseudomuricae fuse into short disconnected ridges, mostly parallel to the growth direction. Late Jurassic (Oxfordian to Tithonian; Plate 3.5, Figs. 13–20).
- *Conoglobigerina* Morozova, 1961, emend. BouDagher-Fadel *et al.*, 1997 (Synonym: *Woletzina* Fuchs, 1973) (Type species: *Conoglobigerina dagestanica* Morozova, 1961). This conoglobigerinid has an intraumbilical aperture, which is an interiomarginal low arch covered occasionally by a bulla. The surface of the test is covered with pseudomuricae that occasionally fuse into very short, discontinuous, blunt bridges. Middle Jurassic (Late Bajocian) to Early Cretaceous (Early Valanginian; Plate 3.2, Figs. 2–20; Plate 3.3, Figs. 1–3, 5–7, and 11–21; Plate 3.6, Figs. 1–4 and 7–9).
- *Globuligerina* Bignot and Guyader, 1971, emend. BouDagher-Fadel *et al.*, 1997 (Type species: *G. oxfordiana* Grigelis, 1958). This form has an intraumbilical aperture that has a loop-shaped bulimine, with a distinct rim, and its posterior margin is set forward into the umbilicus. A bulla covers the primary aperture and the umbilicus. If it is broken, its trace always remains. The surface of the test is covered with pseudomuricae that always fuse into very short, discontinuous ridges. Jurassic (Late Bathonian to Early Kimmeridgian; Plate 3.3, Figs. 5–7; Plate 3.4, Figs. 1–6; Plate 3.5, Figs. 1–12; Plate 3.6, Figs. 1–4 and 8–9).
- *Haeuslerina* Simmons *et al.*, in BouDagher-Fadel *et al.*, 1997 (Type species: *Globigerina helvetojurassica* Hauesler, 1881). This genus has an aperture that is intraumbilical–extraumbilical in extent, reaching almost to the periphery of the penultimate whorl, possessing a narrow lip that covers the umbilicus. The surface of the test is covered in scattered pseudomuricae. Jurassic (Oxfordian to Early Kimmeridgian; Plate 3.3, Figs. 4 and 8–10).

Family Sphaerogerinidae new family

This family consists of those genera of the Favuselloidea that do not possess a favose surface structure and have a compact test and a very narrow aperture. The aperture is intraumbilical, basal, low- to high-arcuate, and sometimes rimmed by a narrow lip. They only occur in the Late Triassic (Rhaetian).

- *Sphaerogerina* Korchagin *et al.*, 2003 (Type species: *Sphaerogerina tuberculata* Korchagin *et al.*, 2003). This form has a spherical, compactly coiled, irregularly microperforate test. It differs from *Compactogerina* (Plate 3.5, Figs. 13–20) by having a lower number of chambers in the last whorl and by the absence of a widely arcuate aperture that is characteristic of *Compactogerina*. Late Triassic (Rhaetian; Plate 3.2, Fig. 1).

3.3 Biostratigraphy and phylogenetic evolution

Identifying the first occurrence of truly planktonic foraminifera is complex and controversial, with many suggested planktonic forms subsequently being reinterpreted as benthic (see below).

Fuchs (1975) described over 20 species from the Triassic of Austria and the Jurassic of Poland. He referred to the Triassic forms as "Globigerina-like" and described them as planktonic. According to Fuchs's scheme, his Triassic genus *Oberhauserella* (Plate 3.1, Figs. 10–18) was the direct ancestor of the Jurassic genus *Conoglobigerina*, and that the Triassic *Schmidita* (Plate 3.1, Figs. 1–3 and 7–9) was the ancestor of the Callovian–Oxfordian *Mariannenina* (Plate 3.1, Figs. 4–6); he believed that this last genus evolved into the holoplanktonic genus *Hedbergella*. In their review of the evolution of early planktonic foraminifera, BouDagher-Fadel *et al.* (1997) dismissed the suggestion that the Triassic and Jurassic foraminifera described by Fuchs between 1964 and 1979 (see Plate 3.1, Figs. 1–18) were planktonic, but rather viewed them as benthic foraminifera, belonging to the family Oberhauserellidae. *Mariannenina* Fuchs is a genus of uncertain status (it may be congeneric with *Jurassorotalia* according to Loeblich and Tappan, 1988). Furthermore, it seems to have a double peripheral keel, a characteristic absent in Jurassic or even Early Cretaceous planktonic foraminifera. Many research workers have subsequently examined Fuchs's material (Drs. A. Görög of Hungary and F. Rögl of Austria, personal communication) and have concluded that it consists mainly of badly preserved and recrystallized benthic specimens. Examinations of his specimens strongly suggest that most of his taxa have a flat umbilical side and a closed umbilicus and are indeed benthic specimens.

Wernli (1995) described specimens from the Early Jurassic (Toarcian) of Switzerland that he named "protoglobigerina," and identified them as genera and species of Oberhauserellidae and to Fuchs's genus *Praegubkinella*, including a new species *Praegubkinella racemosa* (Plate 3.1, Figs. 19–20). Wernli believed *P. racemosa* to be morphologically transitional to *Conoglobigerina*. All of the species that were named by Wernli (1995) appear to have strongly concave or flattened umbilical sides, which were probably the attached sides of benthic forms. However, BouDagher-Fadel *et al.* (1997) suggested that of these various forms only *P. racemosa* may have been free living, probably with an umbilicate but otherwise convex ventral side, and it may well have been the ancestral, in Toarcian time, to the earliest (Bajocian) *Conoglobigerina*. Hart and co-authors (Hart, 2006; Hudson *et al.*, 2009) adopted the hypothesis outlined in BouDagher-Fadel *et al.* (1997) that the aragonitic Oberhauserellidae might have evolved a planktonic mode of life, but further speculated that the evolutionary impetus for the development of this planktonic mode of life was the Early Toarcian anoxic event. They adopted *Oberhauserella quadrilobata* as a possible ancestor. However, *O. quadrilobata* (reproduced in Plate 3.1, Figs. 16–18) has a strongly concave attached umbilical side and is clearly a benthic form. All but one of the oberhauserellids are, therefore, suggested as being benthic and belonging to the Robertinida, and only one species (*P. racemosa*) is seen as a transitional form existing between both families. It would, therefore, be attractive for this aragonitic member of the Oberhauserellidae to be the possible ancestor of the (aragonitic) Favuselloidea. However, more work needs to be done before a clearer picture can emerge.

Korchagin *et al.* (2003) recorded in the LateTriassic of the Crimea *Schmidita hedbergelloides* Fuchs, *Oberhauserella quadrilobata* Fuchs, *O. prarhaetica* Fuchs, "*Globuligerina*" *almensis* Korchagin and Kuznetsova, *Praegubkinellla turgescens* Fuchs, and *Wernliella explanata*. Korchagin *et al.* (2003) did not figure or describe their species *S. hedbergelloides* Fuchs, *O. quadrilobata* Fuchs, and *O. prarhaetica* Fuchs. However, *Oberhauserella* and *Schmidita* are both benthic forms belonging to the family Oberhauserellidae. Among the figured forms, Korchagin *et al.*(2003) identified "*Globuligerina*" *almensis* as a new species. Examination of their figures (their p. 485, fig. 2, 1–3) strongly suggests, however, that this form is a benthic foraminifera possibly belonging to the agglutinated genus *Trochammina*. The umbilical side is almost flat to concave as seen in side and umbilical views. The test is an internal mould, and the identity of this form remains questionable until better preserved specimens are found. Korchagin *et al.* (2003) also referred to their figs. 2.4–6 as *Wernliella explanata*. The genus *Wernliella* is described by Kuznetsova (2002) from the Toarcian of Turkey. The type species *W. toarcensis* is, according to Kuznetsova (2002), distinguished by a flap covering the umbilicus. However, the holotype is a damaged specimen and in fact resembles the benthic form *Oberhauserella parocula* described by Wernli and Görög (2007) from Bajocian–Bathonian of the Southern Jura Mountains of France. *W. explanata* has a flattened umbilical side, which again is probably the attached side of a benthic form. However, in association with these benthic forms, Korchagin *et al.* (2003) described a new genus, *Sphaerogerina*, including the species of *S. tuberculata* and *S. crimica* (Plate 3.2, Fig. 1). *Sphaerogerina* species, which first appeared in the Rhaetian, did not survive the End Triassic extinction event. They seem to be the oldest foraminifera that can be confidently identified as being truly planktonic probably representing the first cycle of zooplanktonic life evolving from a benthic oberhauserellid in the Late Triassic. Korchagin *et al.* (2003) placed *Sphaerogerina* in the Conoglobigerinidae, but since *Sphaerogerina* went extinct at the end of the Triassic, they cannot be co-sanguineous with the true Conoglobigerinidae, which did not appear until the Middle Jurassic. It is suggested here therefore that, on phylogenic grounds, *Sphaerogerina* should be placed in its own family, the Sphaerogerinidae.

Unpublished work by Apthorpe (2002) suggests that a previously undescribed species of *Conoglobigerina* is present in Triassic sediments off Northwest Australia, together with species of *Oberhauserella*. However, the specimens seem to be transitional forms between the benthic *Oberhauserella* and *Praegubkinella* (Hudson *et al.*, 2009). The oldest, clearly defined species of *Conoglobigerina* are the meroplanktonic forms from the

Jurassic Late Bajocian (namely, *C. avarica*, Plate 3.2, Figs. 7–9; *C. avariformis*, Plate 3.3, Figs. 19–21; *C. balakhmatovae*, Plate 3.2, Figs. 14–20; and *C. dagestanica*, Plate 3.2, Figs. 10–13; see also Figs. 3.3 and 3.4). It seems that the *Conoglobigerina* evolved quite separately from the Sphaerogerinidae of the Triassic. Subsequently, the Conoglobigerinidae became holoplanktonic and widespread with the first *Globuligerina*, which appeared in the Bathonian. There are no known species of *Conoglobigerina* in sediments older than Bajocian, but as so many species occur there, it is probable that their sudden appearance may relate to rising eustatic sea level in the Bajocian (Haq *et al.*, 1988), opening up new niches.

There appears to be a clear stratigraphical, morphological, and phylogenetic unity within the Jurassic Conoglobigerinidae. They all have early whorls of at least five to six chambers, while the last whorl has only four chambers. The early whorls are discorbid-like and were probably benthic, while the last whorls, with globular chambers, represent the meroplanktonic stage, suggesting a reproductive cycle, which probably paralleled that of the "Tretomphaloid" discorbids and cymbaloporids, shelf foraminifera, which have evolved a planktonic reproductive phase within the past million years (see Fig.2.6; Banner *et al.*, 1985; BouDagher-Fadel *et al.*, 1997). Benthic, rotaliid foraminifera are often resident on firm substrates (rock, algae, etc.) and the adherent sides of their tests become flattened. When such forms reproduce, the progeny are released into the seawater in the immediate neighbourhood of the parent test. When a planktonic mode of life is acquired, the chambers lose their adherence and become fully globular. Planktonic habitats acquired before reproduction enable the species to disperse its gametes more freely through the seawater and to spread progeny more widely. The surface tension, at the sea surface, would tend to bring both the breeding adults and their released gametes into close proximity. Many living rotaliid genera lose their benthic mode of life and become fully planktonic just before and during reproduction. In the Favuselloidea and subsequently the Globigerinoidea, the development of globular chambers enabled the pseudopodia to extrude equally from extrathalamous cytoplasm over every part of the perforated wall; in contrast, an adherent, umbilically flattened chamber would extrude pseudopodia dominantly from the opposite side only (Alexander and Banner, 1984).

The meroplanktonic mode of life (see Chapter1) seems to have been associated with sexual dimorphism, a physical character unknown in any holoplanktonic foraminifera. The early growth stages (the benthic ones) seem to have developed in two distinct ways in each species. One way involved the development a very high spire with many chambers in the earliest whorls (e.g., *Conoglobigerina conica*, Fig. 3.2); this would be the microspheric form. The other way was to grow a low dorsal spire with as many chambers in each whorl as in the adult; this would be the megalospheric generation (e.g., *Conoglobigerina terquemi*, Fig. 3.2). The chambers of the earliest whorls of *Conoglobigerina* spp. are not always easy to see because of the thickness of the chamber wall, but the presence of dimorphic high and low-spired forms is often recognizable. In some cases, the dimorphic pair of a single meroplanktonic species have been given different specific names, as in the case of the megalospheric, low-spired *C. terquemi* and the microspheric, high-spired *C. conica*. Both "species" are strictly contemporaries, both being found in Tithonian limestones near Stubel village, northeast Bulgaria, and are therefore likely in fact to be the same biological species. Similarly, the youngest *Conglobigerina* (Berriasian–Early Valanginian) has two forms as described by Gorbachik and Poroshina (1979): *C. gulekhensis* (Plate 3.3, Figs. 14–16) and its contemporary *C. caucasica* (Plate 3.3, Figs. 11–13), which is low spired and probably megalospheric. Both have the same distinctive chamber shape and wall sculpture.

The small, Jurassic, microperforate *Conoglobigerina* had a conspicuous surface structure of pseudomuricae and perforation cones. The pseudomuricae are often fused together to form short discontinuous ridges, and these are particularly well developed in the above mentioned stratigraphically youngest species of *Conoglobigerina* (the Berriasian–Early Valanginian species *C. caucasica*, and *C. gulekhensis*). Ultimately, the elongation of these ridges and their fusion to form reticulate favose patterns led to the evolution of the genus *Favusella* in the Cretaceous (see Chapter 4). It should be noted that in the rest of the fossil record, the only known microperforate and pseudomuricate genus of the *Globigerinida*, which resembles any member of the Favuselloidea, is *Globastica* Blow (Type species: *Globigerina daubjergensis* Brönnimann) of the Early Paleocene (Danian).

Conoglobigerina evolved (see Fig. 3.3) in the Late Bathonian into *Globuligerina* (*G. bathoniana*, Late Bajocian–Early Kimmeridgian, Plate 3.5, Figs. 1–8). This form had distinct globular chambers throughout its ontogeny and was the first truly holoplanktonic form. Its "bulimine" aperture is unique within the *Globigerinida*. It appears in the Bathonian, through the Callovian (*Globuligerina calloviensis*, Plate 3.3, Figs. 5–7) to the Oxfordian(*G. oxfordiana*, Plate 3.4, Figs. 1–6) with records in the Late Bajocian and Early Kimmeridgian (BouDagher-Fadel *et al.*, 1997), but left no descendants. In contrast, *Conoglobigerina*, with its low intraumbilical aperture, has a much greater stratigraphic range. The oldest known species, for example, *C. balakhmatovae* (Plate 3.2, Figs. 14–20) and *C. avarica* (Plate 3.2, Figs. 7–9), are known from the Late Bajocian to Early Callovian, while the youngest (e.g., *C. terquemi*, Fig. 3.2; *C. caucasica*, Plate 3.3, Figs. 11–13) are found in the Tithonian. The gap between the Callovian and the Tithonian might be due to lack of sampling or exposed material. *Conoglobigerina* of the Late Jurassic show a morphological and phylogenetic gradation into their favusellid Cretaceous descendants (see Chapter 4).

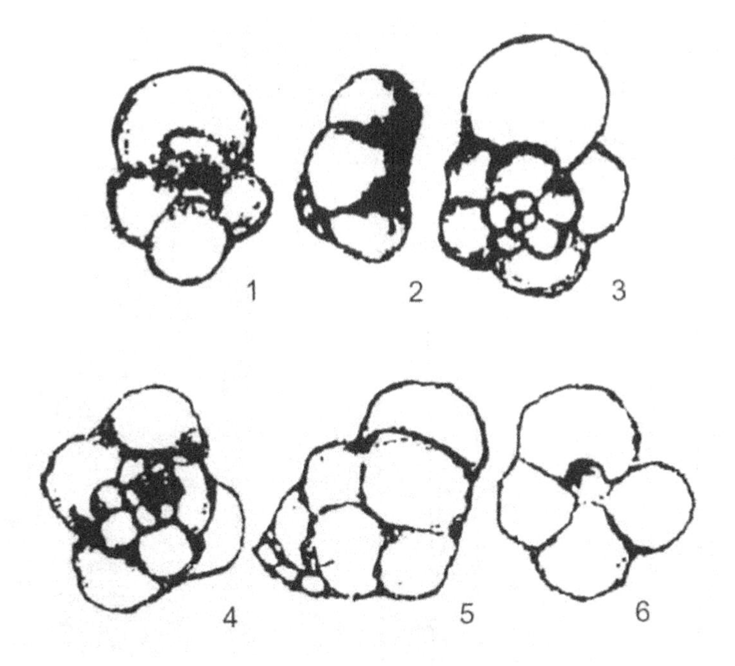

Figure 3.2. (1–3) *Conoglobigerina terquemi* (Ivočeva and Trifonova). Paratype figured by Ivočeva and Trifonova (1961), Tithonian, NW Bulgaria, x33. (4–6) *Conoglobigerina conica* (Ivočeva and Trifonova). Holotype, figured by Ivočeva and Trifonova (1961), Tithonian, NW Bulgaria, x33.

Age (Ma)
199.6
175.6
161.2
145.5
Period, Epoch, Stage
Triassic
Jurassic
Early
Middle
Late
Rha.
Hett.
Sine.
Plien.
Toar.
Aale.
Baj.
Bath.
Call.
Oxf.
Kim.
Tith.
Oberhauser-ellidae
Conoglobigerinidae
Sphaerogerina
Conoglobigerina
Globuligerina
Haueslerina
Compactogerina

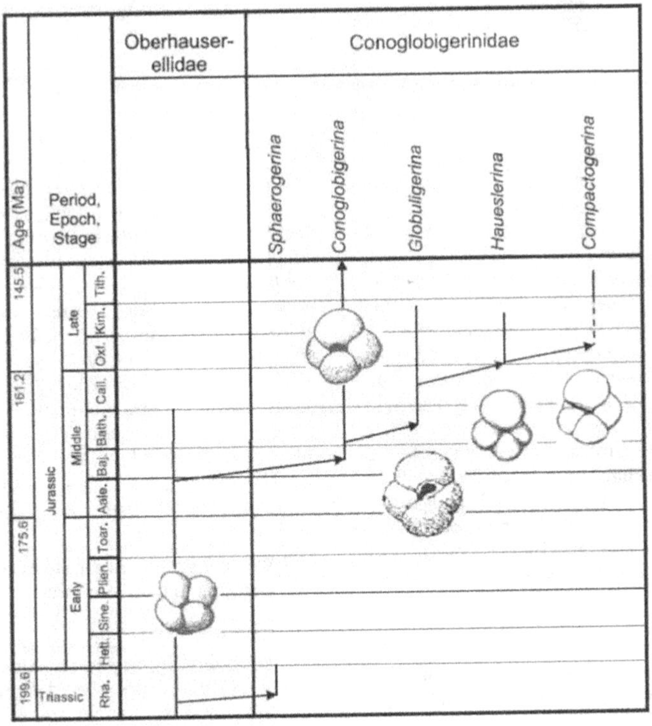

Figure 3.3. Phylogenetic evolution of the Jurassic planktonic foraminiferal genera

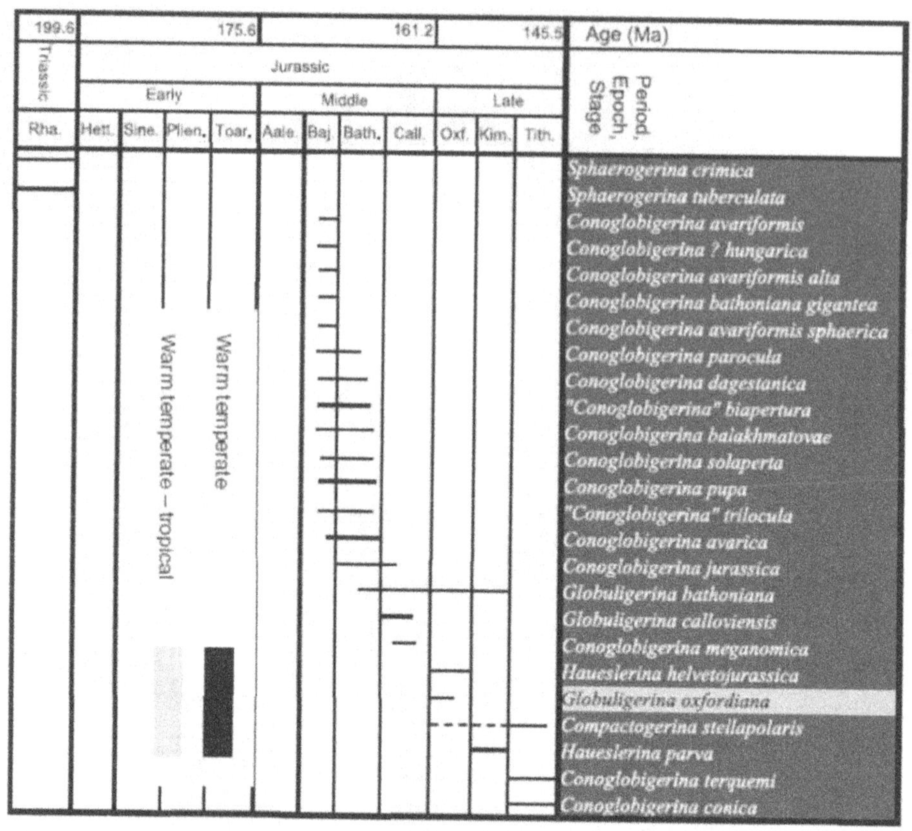

Figure 3.4. Range chart of the Jurassic planktonic foraminiferal species.

Whenever the aperture is intraumbilical, bullae are present on all Favuselloidea from the earliest to the last genera of the Conoglobigerinidae (see Plate 3.5, Fig. 12) and the Cretaceous Favusellidae (see Chapter 4). However, when the aperture is intra–extraumbilical (e.g., in *Haeuslerina*, *Compactogerina* in the Conoglobigerinidae, or *Ascoliella* in the Favusellidae), bullae are not present. This phenomenon is repeated again in the Cenozoic Globigerinoidea (e.g., *Catapsydrax*, with a bulla covering its umbilical aperture, compared to *Paragloborotalia* which has an intra–extraumbilical aperture and no bulla, see Chapters 5 and 6). It may be the bulla was a skeletal structure that covered the umbilicus to conceal a mass of extrathalamous but digestive cytoplasm. This is analogous to that seen in some living benthic rotaliines (Alexander and Banner, 1984). It is possible that with the extraumbilical extension of the aperture and narrowing of the umbilicus bullae became unnecessary. Bullae are not known in the Praehedbergellidae, the descendants of the Conoglobigerinidae (see Chapter 4), but umbilical coverings were developed in the macroperforate descendants of the Hedbergellidae (e.g., *Rotalipora*, *Globotruncana*; see Chapter 4).

Finally, the apparently phylogenetically unrelated *Compactogerina stellapolaris* (Plate 3.5, Figs. 13–20) was first described by Grigelis (1958) from the Timan–Pechora Basin in northern Russia. As such, it is the most northerly record of early planktonic foraminifera. *C. stellapolaris* is the only known species of the genus *Compactogerina* and dates from the Tithonian. *Compactogerina* sp. cf. *C. stellapolaris*, described from the Oxfordian clays of the Dorset Coast near Weymouth, UK (Oxford *et al.*, 2002), is a badly preserved, questionable form and not until better preserved material is found can the existence of this genus in the Jurassic of the UK be confirmed.

The stratigraphic ranges of the Late Triassic and Jurassic species are shown in Fig. 3.4.

3.4 Paleogeography and paleoecology of the Triassic–Jurassic planktonic foraminifera

By the Late Triassic, a permanent Tethyan seaway was already beginning to separate Africa from Europe and North America, with the shallow waters of Tethys covering much of Europe and Russia (Fig. 3.5). The global climate was dry and hot, except for the cooler and wetter environment in southern Gondwana, northern Europe, and Greenland. Warm temperate climates extended almost to the poles. Sea levels were low and the seas had little dissolved oxygen and were possibly alkaline (Woods, 2005). Sudden dissociation of the subsea gas hydrates may have caused anoxic events during the Late Triassic (Hesselbo *et al.*, 2002). Evidence of anoxia has also been documented in Triassic–Jurassic sections (Hallam and Wignall, 1997, 2000), and it is here suggested to be a significant factor in what might have prompted the benthic "protoglobigerines" to adopt a transitional mode of life from being attached benthic forms to becoming a floating planktonic form. As the Earth's oceans became significantly depleted in oxygen and the lower layers of the ocean waters became toxic, a sessile benthic life style became uncompetitive and a mutation that enabled a new form of reproduction involving a planktonic stage provided an evolutionary advantage. Marine regression during this time reduced the available shallow–marine habitats for the Triassic foraminifera, and consequent competition may also have been a forcing mechanism for introducing the meroplanktonic mode of life. By adopting this strategy, the Sphaerogerinidae, which possess slightly more inflated chambers than other forms, were able to reproduce in the still oxygenated surface layer of the oceans. This mode of life, with a benthic juvenile test, may explain the relatively restricted geographic limits of the Triassic Sphaerogerinidae as they are only found in the Crimea.

At the end of the Triassic, abrupt extinctions were seen in all groups in the marine realm (see BouDagher-Fadel, 2008 and references therein) and the newly evolved meroplanktonic foraminifera did not survive the Triassic–Jurassic crisis. This Triassic–Jurassic extinction is generally recognized as being one of the five largest in the Phanerozoic, and climate-induced changes occurring during the Rhaetian might have been the possible cause for mass the extinction. However, the Triassic–Jurassic boundary also coincides with large-scale volcanic eruptions that created a flood basalt province (the so-called Central Atlantic Magmatic Province) that covered at least 7 x10^6 km^2 of North America, South America, and Africa. This event triggered a large increase in atmospheric CO_2 and a marine biocalcification crisis (Whiteside *et al.*, 2010) that almost certainly played a major role in the extinction of most marine foraminifera including the Triassic sphaerogerinids.

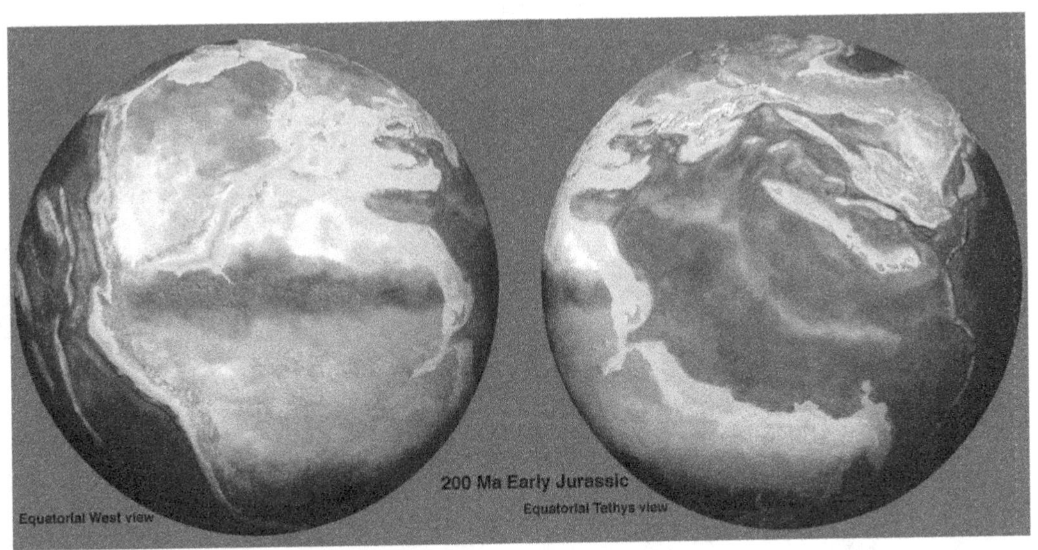

Figure 3.5. Paleogeographic and tectonic reconstruction of the Early Jurassic (by R. Blakey http://jan.ucc.nau.edu/~rcb7/paleo geographic.html).

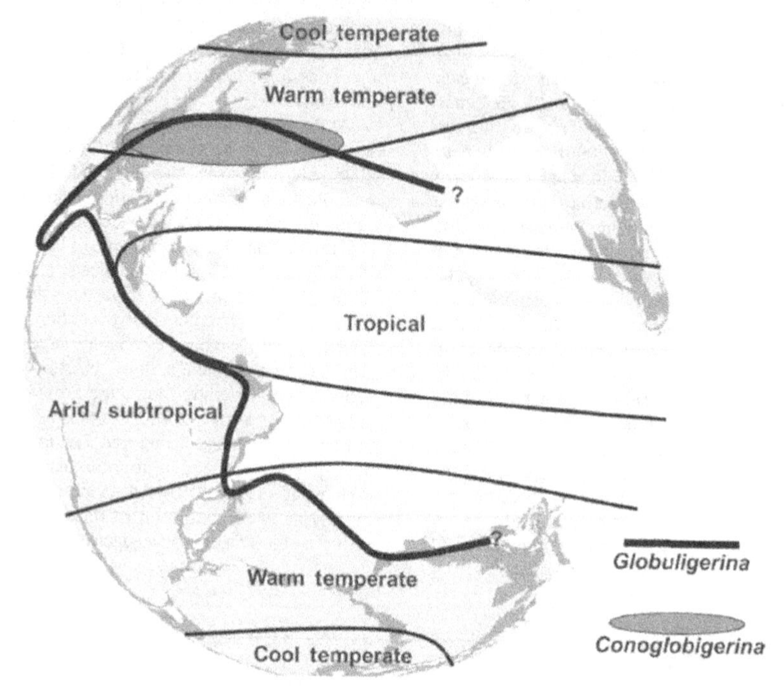

Figure 3.6. Climate zones in the Middle Jurassic (165 Ma), also showing the limited paleogeographic extent of the meroplanktonic conoglobigerinids and the more global coverage of the holoplanktonic *Globuligerina*.

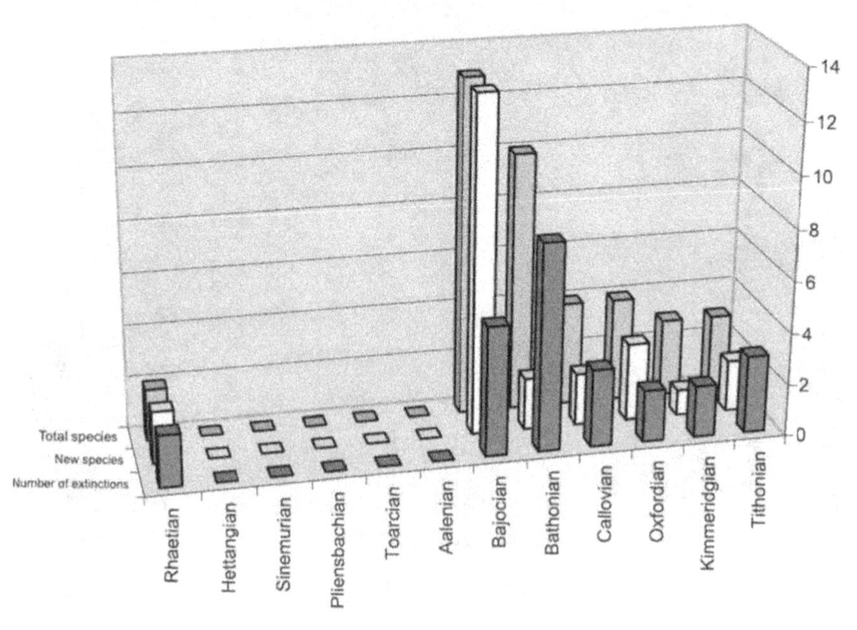

Figure 3.7. The total number of species, extinctions, and new appearances of planktonic foraminifera in each stage of the latest Triassic and Jurassic. The extinctions coincide with the end of each stage and the appearances with the beginning of the stage.

After the major End Triassic event, the Jurassic period saw warm tropical greenhouse conditions worldwide (Fig. 3.6). The sea level gradually rose (O'Dogherty *et al.*, 2000) and the shallow warm waters of Tethys and the Proto-Atlantic flooded large portions of the continents and spread across Europe. The Early Jurassic was a period of recovery. The diversity of Tethyan benthic foraminifera was poor, and assemblages were composed of small infaunal benthic foraminifera of textulariids (see BouDagher-Fadel, 2008) together with the Oberhauserellidae thriving in Eastern Europe. This period also saw the development of organic carbon rich shales in large parts of Western Europe and other parts of the world. The widespread occurrence of the Early Toarcian shales is explained by a widespread oceanic anoxic event (Jenkyns, 1988; Jenkyns and Clayton, 1997). After these highly unfavorable conditions for benthic foraminifera, the Bajocian saw the conoglobigerinids evolving by mimicking the meroplanktonic mode of life of the earlier Triassic species, which enabled them gradually to colonize the shallow waters of the continental shelf.

It is proposed therefore that there was a second transition from a benthic to a planktonic mode of life in the Jurassic, which occurred under conditions similar to those that triggered planktonic speciation in the Late Triassic. This suggestion is generally compatible with the hypothesis that the planktonic foraminifera are indeed polyphyletic, involving several originations of planktonic taxa from benthic ancestors (Darling *et al.*, 1996, 1999; Ujiié *et al.*, 2008). According to these authors, molecular evidence from modern triserial planktonic foraminifera suggests that they are not related to the Cretaceous–Paleogene triserial species, and that the sporadic occurrences in the fossil record are not the result of poor preservation, but reflect multiple transitions from benthic to planktonic mode of life.

The Late Bajocian witnessed a relatively diverse fauna of *Conoglobigerina* (see Figs. 3.4 and 3.7). However, as marine connections, other than eastward into Tethys, remained closed during the Bajocian (see Fig. 3.6), the conoglobigerinids still had a limited geographic distribution. They are still only known from eastern Europe (central and northern Tethys), but in the Late Bathonian, conoglobigerinids spread over a wider area of the Tethyan realm. The End Bajocian saw the extinction of a variety of closely related *Conoglobigerina*, which had limited geographic ranges that may have been triggered by local events.

In the Middle Jurassic, *Conoglobigerina* evolved into the wholly planktonic and cosmopolitan *Globuligerina* (Fig. 3.3; Plate 3.4, 3.5 and 3.6), which colonized deeper water environments. *Globuligerina* developed special features, such as globular chambers throughout ontogeny and was holoplanktonic, which helped them to colonize deeper waters than the contemporaneous *Haeuslerina* spp. and *Conoglobigerina*. *Globuligerina* diversified and expanded its range as new ocean basins developed. In places, such as the Carpathians of Southern Poland, some of these Jurassic limestones yield abundant assemblages that can only be described as Jurassic "*Globigerina* oozes" (Hudson *et al.*, 2009). This period saw the opening up of new habitats as the incipient North Atlantic began to widen as a result of the rifting induced by the CAMP volcanism at the end of the Triassic. *Globuligerina* became cosmopolitan and can be found in Eastern and Western Europe and Canada (Stam, 1986).

The End Bathonian is associated with high extinction rates (see Fig. 3.7). This Middle Jurassic epoch coincides with at least two major impact events that gave rise to the 80-km diameter Puchezh–Katunki crater in Russia and the 20-km diameter Obolon crater in the Ukraine. These events might also have contributed to enhanced environmental stress that could have been responsible for part of the enhanced extinction rate in these stages. The trend of high extinction rates continued in the Callovian with just a few species appearing locally, for example, *Conoglobigerina meganomica* (Plate 3.3, Figs. 1–3) and *Globuligerina calloviensis* (Plate 3.3, Figs. 5–7) both in the Crimea.

The transition from the Middle to the Late Jurassic was characterized by significant changes in oceanography and climate. These changes were accompanied by modifications in the global carbon cycle as shown in the carbon isotope record (Louis-Schmid *et al.*, 2007). They were triggered by the opening and/or widening of the Tethys–Atlantic–Pacific seaway, and a massive spread of shallow–marine carbonate production leading to higher P_{CO2}, and according to Louis-Schmid *et al.* (2007), this increase in P_{CO2} may have triggered changes in the biological carbon pump and in organic carbon burial in the Middle Oxfordian.

The Oxfordian and Kimmeridgian show another burst of new forms of conoglobigerinids; however, by end of Early Kimmeridgian, *Globuligerina* had died out and only sporadic conoglobigerinids were found across the Tithonian–Early Berriasian boundary (Figs. 3.4 and 3.7). The increased extinction rates at the end of the Early Kimmeridgian and Tithonian saw numbers of genera decline as the Jurassic came to an end. This general decline may be related to the final opening of the proto-North Atlantic, and a consequent change in global circulation patterns that seems to have affected many marine faunas (see BouDagher-Fadel, 2008).

All Jurassic conoglobigerinids are "globigerine" morphotypes that in more recent times inhabit the shallow waters of the continental shelves. *Compactogerina*, which appears to be a Jurassic homeomorph of the modern taxon *Neogloboquadrina*, is unlike any other Jurassic taxon and appears abruptly in the Tithonian. The thickened shell may be an adaptation to the relatively cold waters of that region (Fig3.6), similar to modern-day *Neogloboquadrina pachyderma* (Ehrenberg), see Chapter 6.

The conoglobigerinids seem to persist across the Jurassic–Cretaceous boundary, but their occurrence in the Upper Jurassic is rare, and across the Jurassic–Cretaceous boundary, the record is very incomplete. This gap in our knowledge is quite critical as the Jurassic taxa are regarded as having aragonitic tests, while those in the Lower Cretaceous are calcitic. Therefore, as will be discussed in Chapter 4, the phylogenetic link across the boundary must still be considered tenuous.

Plate 3.1. Figures 1–3 *Schmidita inflata* Fuchs. Holotype, figured by Fuchs (1967), Lower Carnian, Austria, (1) NHM P059810, (2) NHM P059958, (3) NHM P059766, x213. Figures 4–6 *"Mariannenina" nitida* Fuchs. ?Paratype, figured by Fuchs (1973), lowermost Oxfordian, Vienna, (1) NHM P059760, (2) NHM P059811, (3) NHM P059730, x213. Figures 7–9 *Schmidita ladinica* (Oberhauser). Paratype, figured by Oberhauser (1960), Ladinian, Vienna, (7) NHM P059761, (8) NHM P059812, (9) NHM P059731, x145. Figures 10–12 *Oberhauserella mesotriassica* (Oberhauser). Paratype, figured by Oberhauser (1960), Ladinian, Vienna, (10) spiral view, NHM P059762, (11) peripheral view, NHM P059814, (12) umbilical view, NHM P059732, x145. Figures 13–15 *Oberhauserella alta* Fuchs. Paratype, Rhaetian, Austria, (13) spiral view, NHM P059763, (14) peripheral view, NHM P059815, (15) umbilical view, NHM P059733, x200. Figures 16–18 *Oberhauserella quadrilobata* Fuchs. ?Paratype, figured by Fuchs (1967), Late Norian of Central Austria, (16) spiral view, NHM P059764, (17) umbilical view NHM P059740,(18) peripheral view NHM P059816, x260. Figures 19–20 *Praegubkinella racemosa* Wernli. Figured by Wernli (1995), Early Toarcian, Swiss Alps, (19) holotype, spiral view, (20) paratype, peripheral view, x260. Figure 21 *Praegubkinella fuschi* Wernli. Figured by Wernli (1995), Early Toarcian, Swiss Alps, side view, x260.

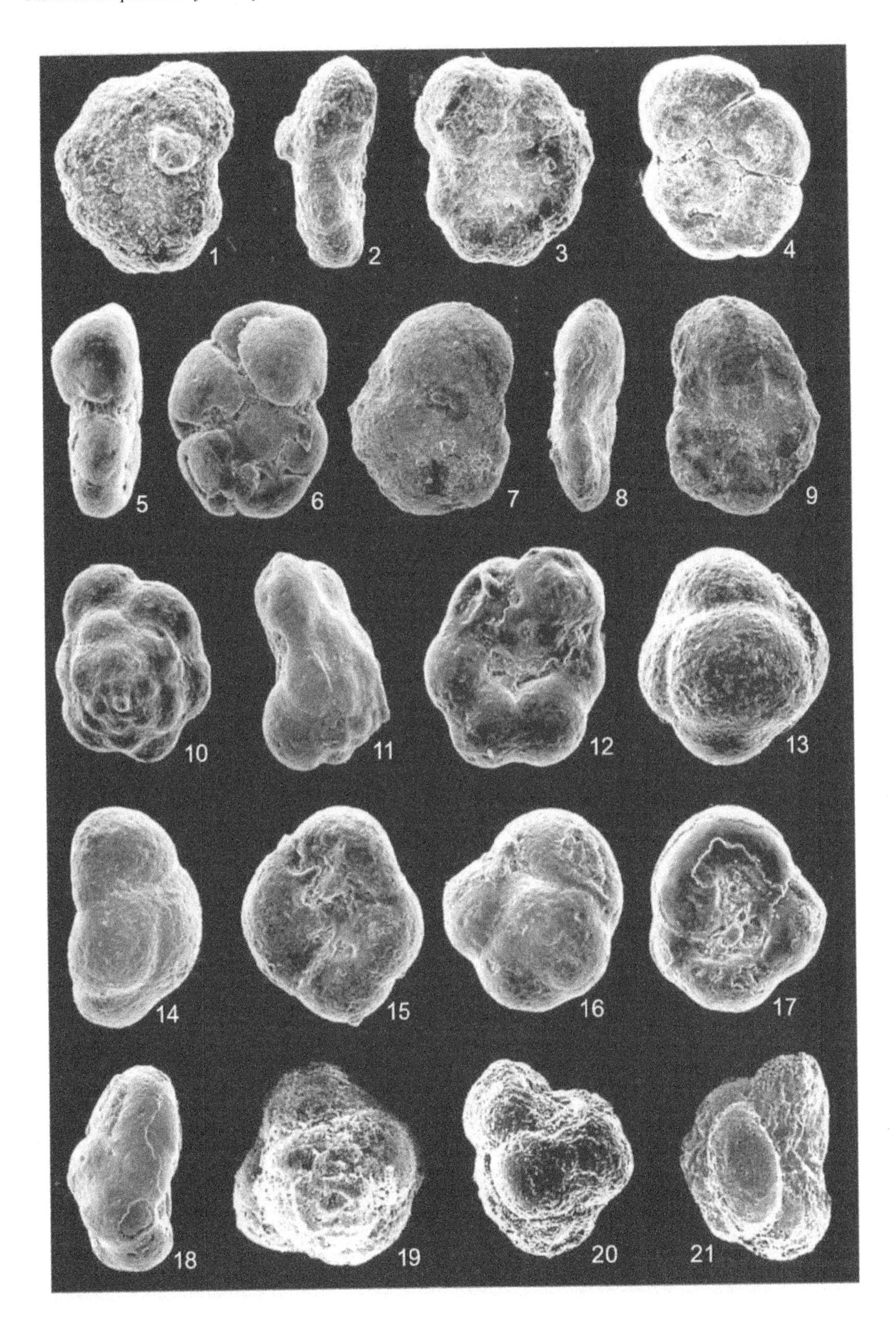
1
2
3
4
5
6
7
8
9
10
11
12
13
14
15
16
17
18
19
20
21

Plate 3.2. Figure1 *Sphaerogerina crimica* Korchagin and Kuznetsova. Figured by Korchagin *et al.* (2003), paratype, No. 4776/ 19, umbilical view, x160. Figures 2–6 *Conoglobigerina dagestanica* Morozova (= *Globuligerina araksi* Kasimova). Late Bajocian, Jurassic, Azerbaijan, (2) holotype, No. 3513/2, (3–5) paratypes, (6) enlargement of the wall surface to show pseudomuricae often fusing laterally into short bridges, x310. Figures 7–9 *Conoglobigerina avarica* Morozova. Early Bathonian, central Dagestan, (7) holotype, peripheral view showing a high spire, x235, (8) paratype, x330, (9) paratype, x220. Figures 10–13 *Conoglobigerina dagestanica* Morozova. Early Bathonian, central Dagestan, (10) x285, (12–13) x265. Figures 14–20 *Conoglobigerina balakhmatovae* (Morozova). Bathonian, central Dagestan, (11, 14–15, and 20) paratypes, x235, (16) enlargement of the surface showing microperforations and pseudomuricae, x1500, (17–19) holotype, x285. Figs 2-19 from BouDagher-Fadel *et al.* (1997).

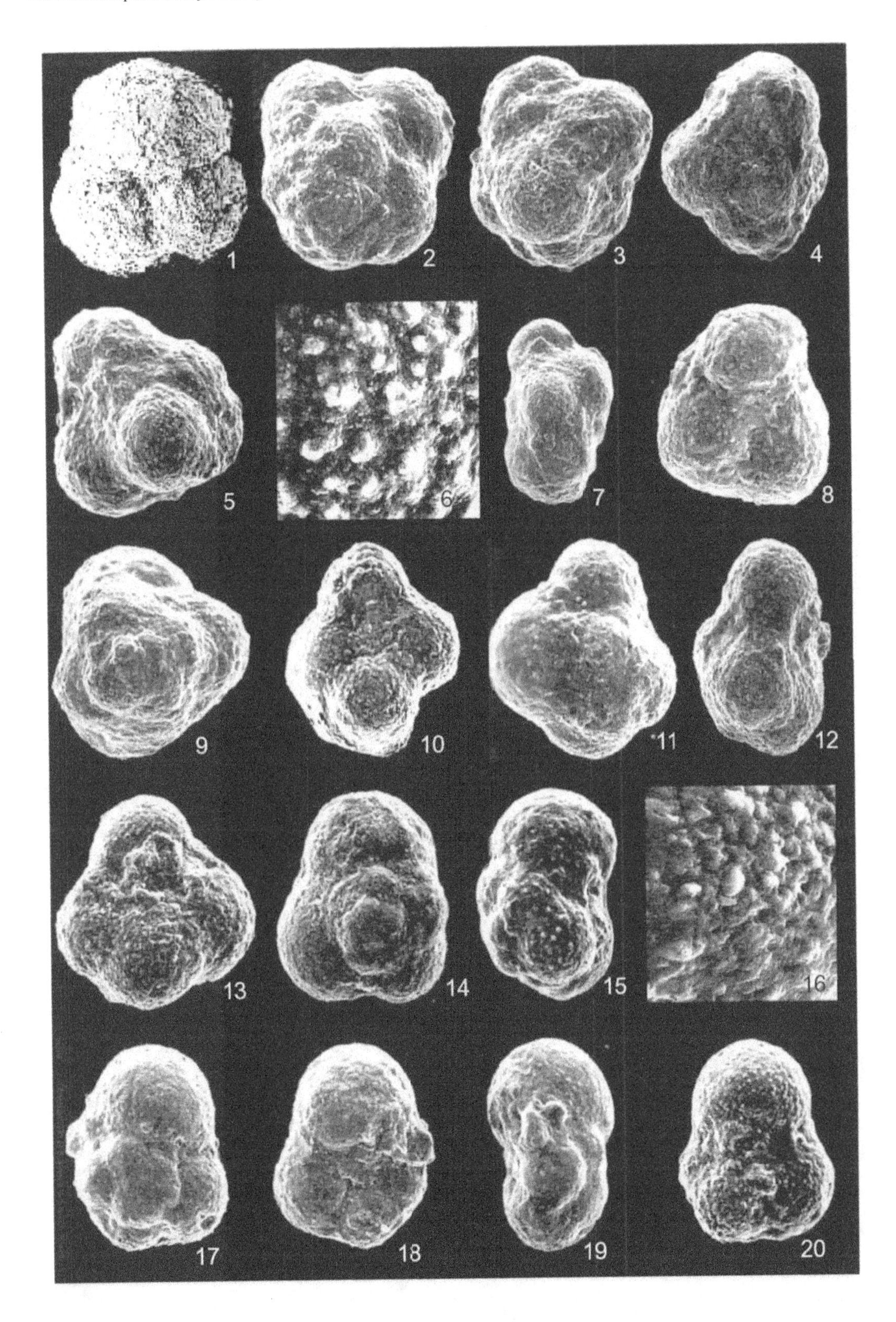
1
2
3
4
5
6
7
8
9
10
11
12
13
14
15
16
17
18
19
20

Plate 3.3. Figures 1–3 *Conoglobigerina meganomica* (Kuznetsova). Paratype, Late Callovian, North Crimea, x200. Figure 4 *Haeuslerina parva* (Kuznetsova). Metatype, Early Kimmeridgian, Crimea, x285. Figures 5–7 *Globuligerina calloviensis* Kuznetsova. Metatype, Early–Middle Callovian, Crimea, x220. Figures 8–10 *Haeuslerina hevetojurassica* (Hauesler). Topotype, figured by Stam (1986), Oxfordian, Switzerland, x365. Figures 11–13 *Conoglobigerina caucasica* (Gorbachik and Poroshina). Metatype, Early Berriasian, Azerbaijan, x250. Figures 14–16 *Conoglobigerina gulekhensis* (Gorbachik and Poroshina). Metatype, Early Berriasian, Azerbaijan, x235. Figures 17–18 *Conoglobigerina jurassica* (Hofman). Neotype, Early Callovian, Crimea, x150. Figures 19–21 *Conoglobigerina avariformis* Kasimova. Late Bajocian, Azerbaijan, x125. Figs 1-7 and 11-21 from BouDagher-Fadel *et al.* (1997).

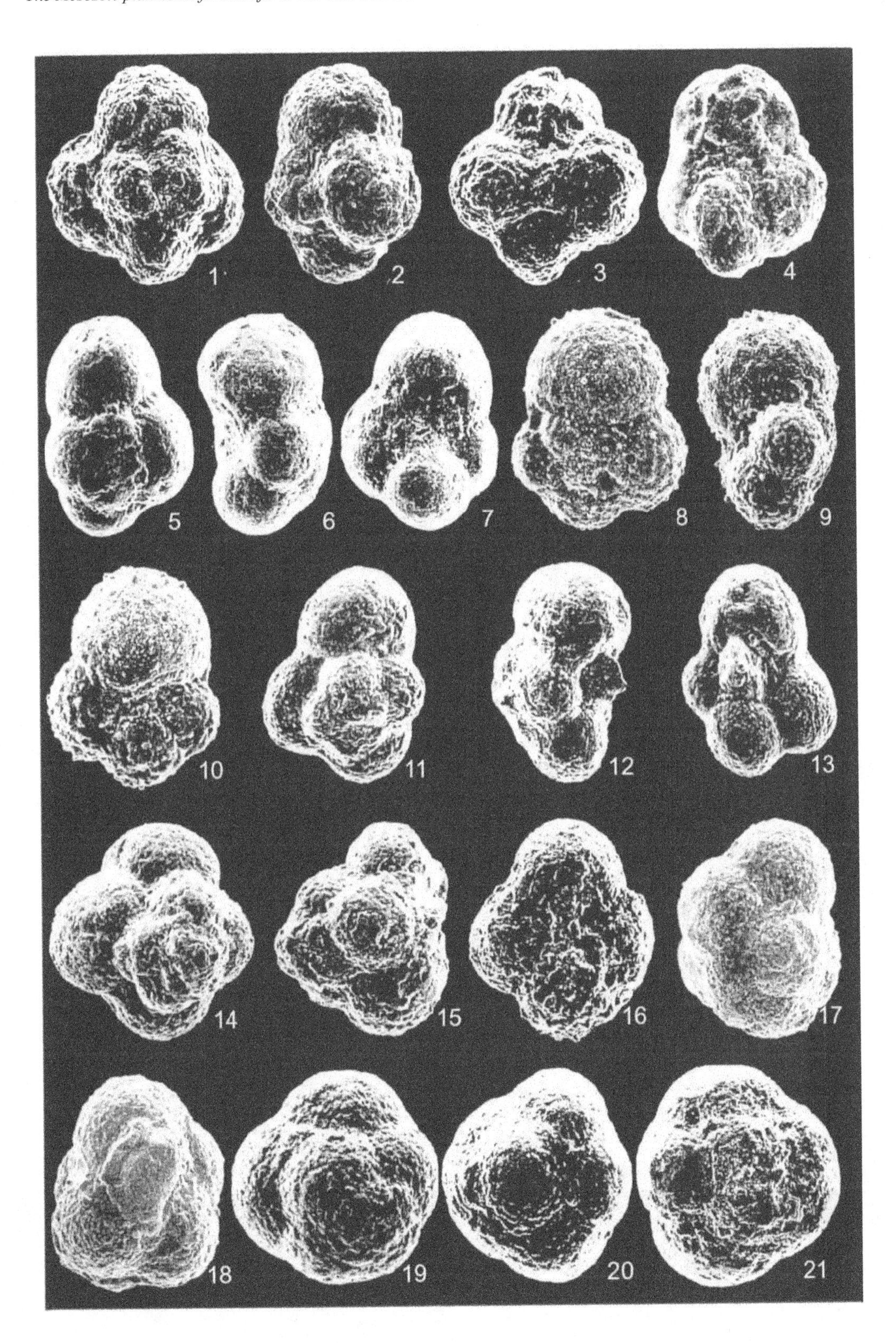
1
2
3
4
5
6
7
8
9
10
11
12
13
14
15
16
17
18
19
20
21

Plate 3.4. Figures 1–6 *Globuligerina oxfordiana* (Grigelis). Oxfordian, Upper Volga Basin, Russia, (1) x370, (2) enlargement of the tests showing scattered hollow muricae, x1202, (3) x600, (4) x500, (5–6) x900. Images from the UCL Collection.

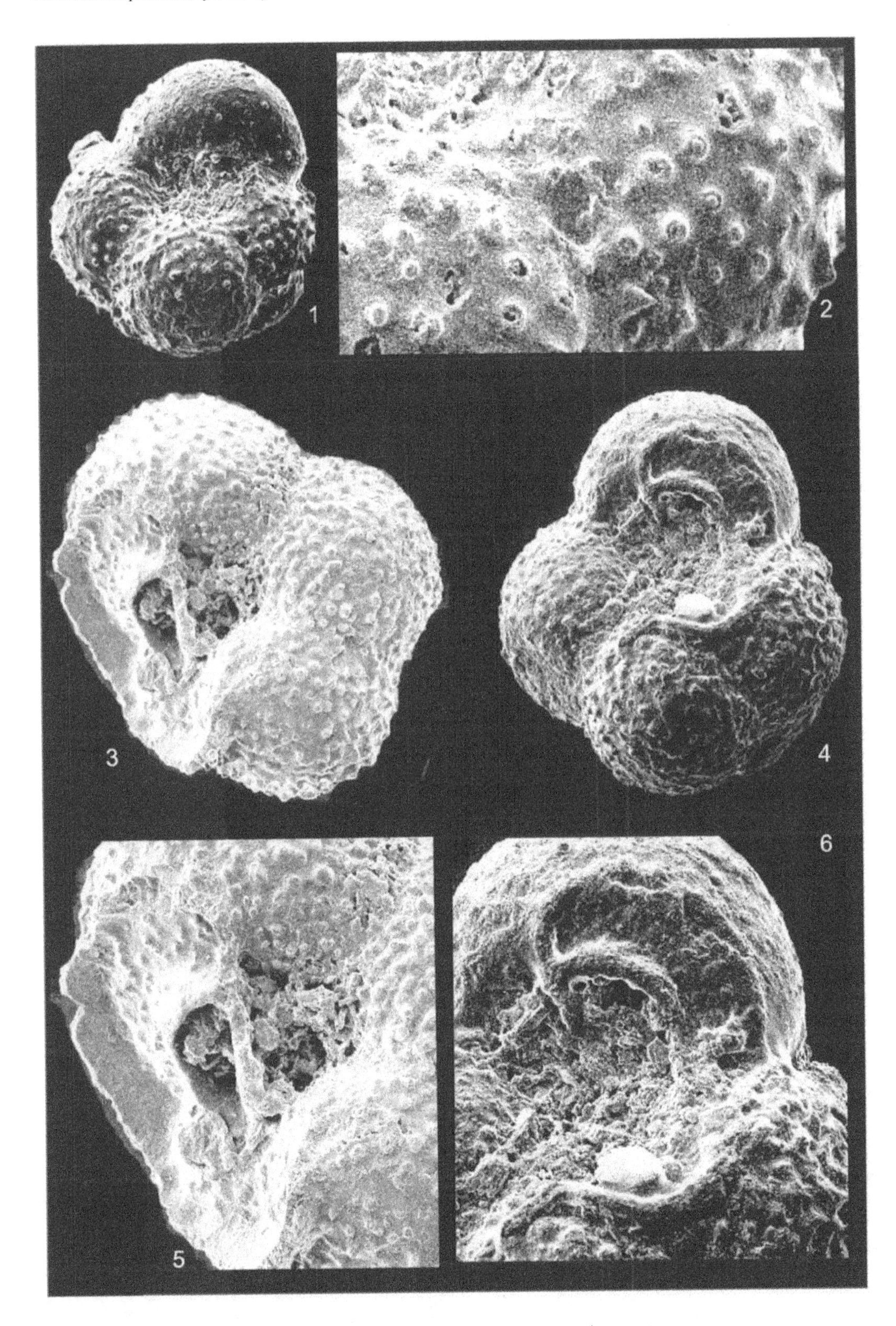
1
2
3
4
5
6

Plate 3.5. Figures 1–8 *Globuligerina bathoniana* (Pazdrowa). Metatypes, Middle Bathonian, Poland, (1–3) x250, (4) enlargement of the surface, x1250, (5–8) topotype, (5–7) x235, (8) x750. Figures 9–12 *Globuligerina oxfordiana* (Grigelis). Oxfordian, North of France, (9–11) x330, (12) x300. Figures 13–20 *Compactogerina stellapolaris* (Grigelis). Metatypes, Early Volgian, Northern Russia, (13) x650, (14–16) x170, (17–20) x155. All images from BouDagher-Fadel *et al.* (1997).

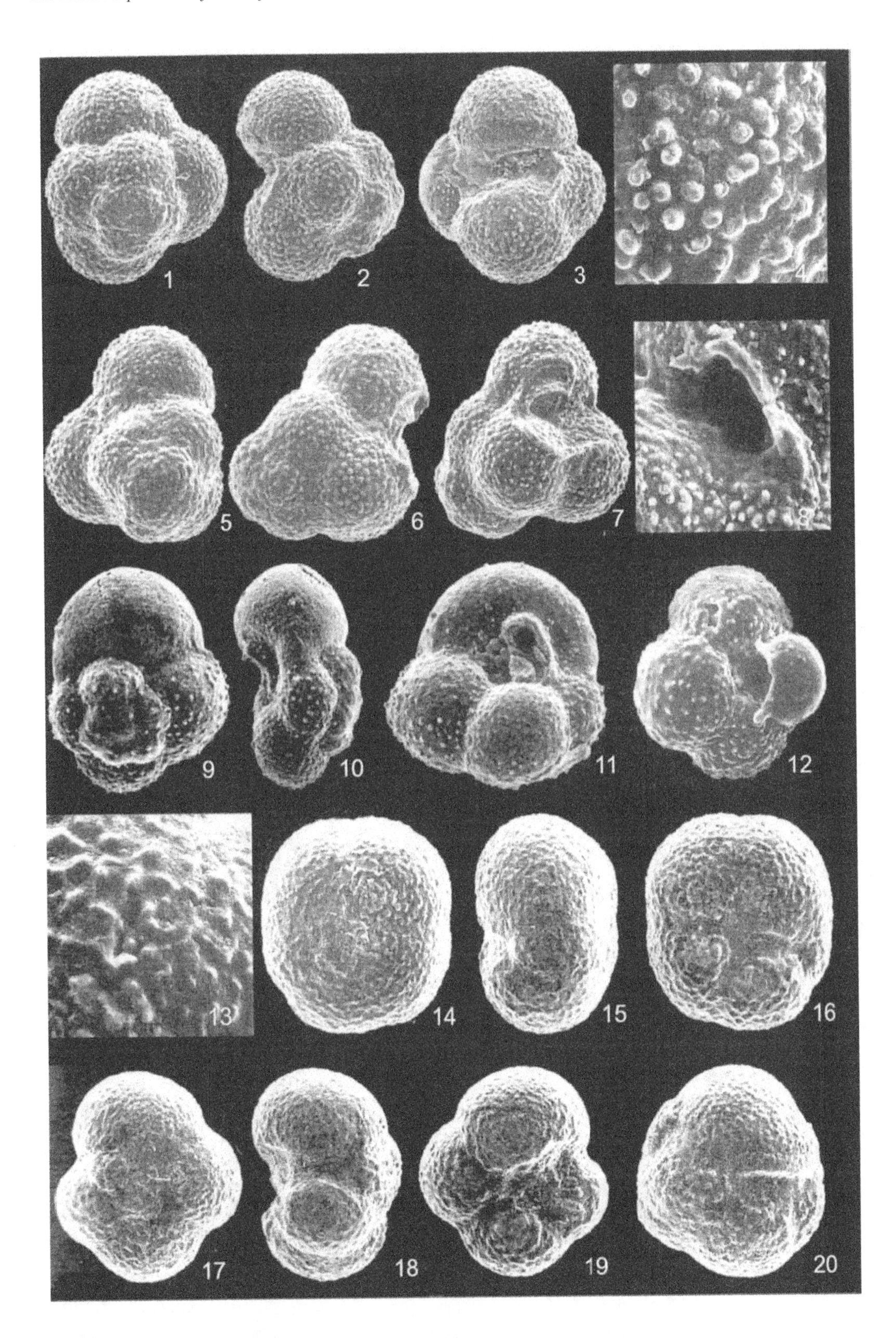
1
2
3
4
5
6
7
8
9
10
11
12
13
14
15
16
17
18
19
20

Plate 3.6. Figures 1–4 *Conoglobigerina avariformis* Kasimovain. Thin section, figured by Wernli and Görög (2000), Bajocian, Bakony Mountain, x80. Figure 5 *Globuligerina* sp. From the UCL Collection, Oxfordian, Dorset Coast, x75. Figure 6 *Globuligerina* sp. From the UCL Collection, Oxfordian, Dorset Coast, x240. Figure 7 *Globuligerina oxfordiana* (Grigelis). Figured by Wernli and Görög (2000), Bajocian, Bakony Mountain, x160. Figure 8 *Conoglobigerina avariformis* Kasimova forma sphaerica Wernli and Görög. Figured by Wernli and Görög (2000), Bajocian, Bakony Mountain, x160. Figure 9 *Conoglobigerina* avariformis Kasimova forma alta Wernli and Görög. Figured by Wernli and Görög (2000), Bajocian, Bakony Mountain, x160.

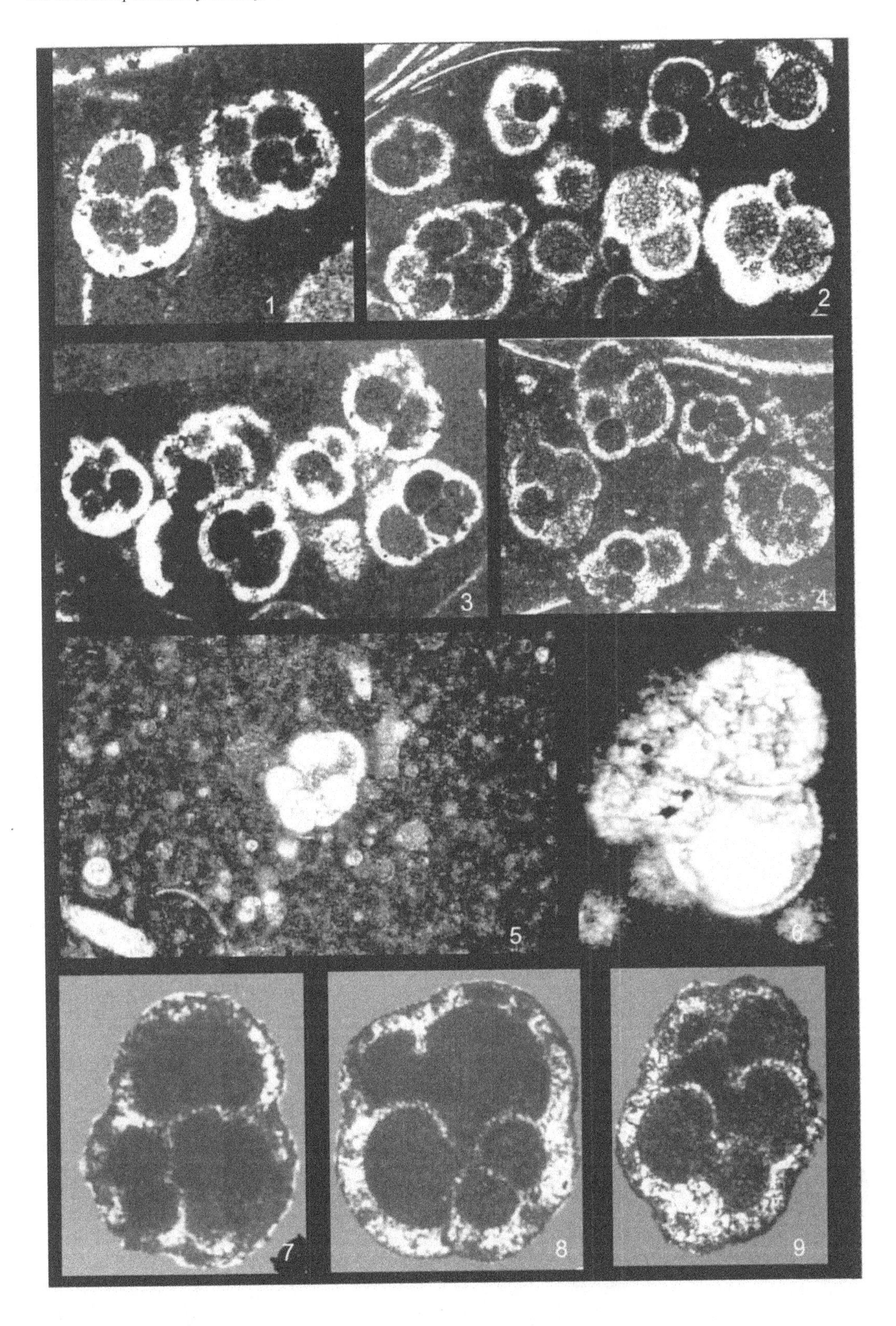
1
2
3
4
5
6
7
8
9

Chapter 4

The Mesozoic planktonic foraminifera: The Cretaceous

4.1 Introduction

Although most Jurassic planktonic foraminifera became extinct toward the end of that period, the meroplanktonic conoglobigerinids persisted into the Early Cretaceous and gave rise to the favusellids in the Berriasian. In turn, the favusellids, an aragonitic group, led to the calcitic praehedbergellids in the Valanginian (Fig. 4.1). The praehedbergellids typify the holoplanktonic foraminiferal fauna of the Early Cretaceous and were the first geographically widespread Globigerinoidea. In the Barremian, the praehedbergellids gave rise to the planispiral schackoinids, and in the Late Aptian to the hedbergellids, from which the other trochospiral lineages, characteristic of the Late Cretaceous, developed.

The Late Albian also witnessed the adaptation of the biserial and triserial elongate species of the benthic buliminds to the planktonic realm, thereby giving rise to the heterohelicids and guembelitriids. The heterohelicids evolved, therefore, independently from the Globigerinida and represent a separate lineage of planktonic foraminifera, as can be inferred from DNA studies on the living Guembelitriidae (Ujiié *et al.*, 2008). On these grounds therefore, it is proposed here that the heterohelicids should not be placed within the Globigerinida, but rather within the order Heterohelicida (Fig. 4.1), which was created by Fursenko (1958) to represent these elongated forms, which gave rise to many genera in the Late Cretaceous by multiplying their chambers in one or all planes.

In addition to the significant phylogentic developments of the planktonic foraminifera, the Cretaceous also sees a noticeable waxing and waning in the diversity of planktonic foraminifera assemblages. There are three major stages within this system (see Figs. 4.2–4.5 and 4.19):

- first, increasing diversification from the Early Valanginian culminating in a significant extinction event at the Aptian–Albian boundary (with approximately 82% of the species going extinct),
- an Albian recovery followed by a further marked extinction at the Late Cenomanian (with 47% of the species going extinct), and
- finally, a second recovery followed by the major global extinction characteristic of the End Cretaceous, Maastrichtian event (with 95% of the species going extinct).

Because of their widespread occurrence, Cretaceous planktonic foraminifera are one of the most useful biostratigraphic tools for correlating strata in this stratigraphic system. The Early Cretaceous forms have been revised by BouDagher-Fadel *et al.* (1997) and by Verga and Premoli Silva (2003), while those of the Late Cretaceous were carefully reviewed by the European Working Group on Planktonic Foraminifera, as part of Project Number 58 (the "Mid-Cretaceous Event") of the International Geological Correlation Programme. From this work, an atlas of species was published (Robaszynski and Caron, 1979) and a refined taxonomy and stratigraphy, based on the phylogeny of Late Cretaceous taxa, was developed (Robaszynski *et al.*, 1984). In this chapter, the taxonomy of the major Cretaceous planktonic foraminifera is summarized, and revised where necessary, and is then followed by a discussion of their biostratigraphic, phylogentic, paleoenvironmental, and paleogeographic significance.

4.2 Morphology and taxonomy of the Cretaceous planktonic foraminifera

The Cretaceous planktonic foraminifera are divided into four main superfamilies:

- the aragonitic trochospiral Favuselloidea,
- the calcitic trochospiral Globigerinoidea,
- the calcitic planispiral Planomalinoidea, and
- the biserial to multiserial Heterohelicoidea.

Many authors in the past have attempted to divide the Late Cretaceous calcitic trochospiral planktonic foraminifera into different superfamilies (e.g., Georgescu and Huber, 2006; Loeblich and Tappan, 1988). However, the characteristics of all of the proposed new superfamilies were, in our opinion, defined with a degree of ambiguity. To avoid confusion and to maximize utility, therefore, it is our recommendation that the Globigerinoidea should all be included in one superfamily, and differences should be considered at the family level. Therefore, we submit that the taxonomy of the Cretaceous genera can be best summarized as follows:

CLASS FORAMINIFERA Lee, 1990

ORDER GLOBIGERINIDA LANKASTER, 1885

Tests are planispiral or trochospiral, at least in the early stage, microperforate or macroperforate, smooth, muricate, or with spines. Apertures are terminal, umbilical, intra-extraumbilical, or peripheral, and walls are calcitic, but early forms may be aragonitic. Late Triassic (Rhaetian) to Holocene.

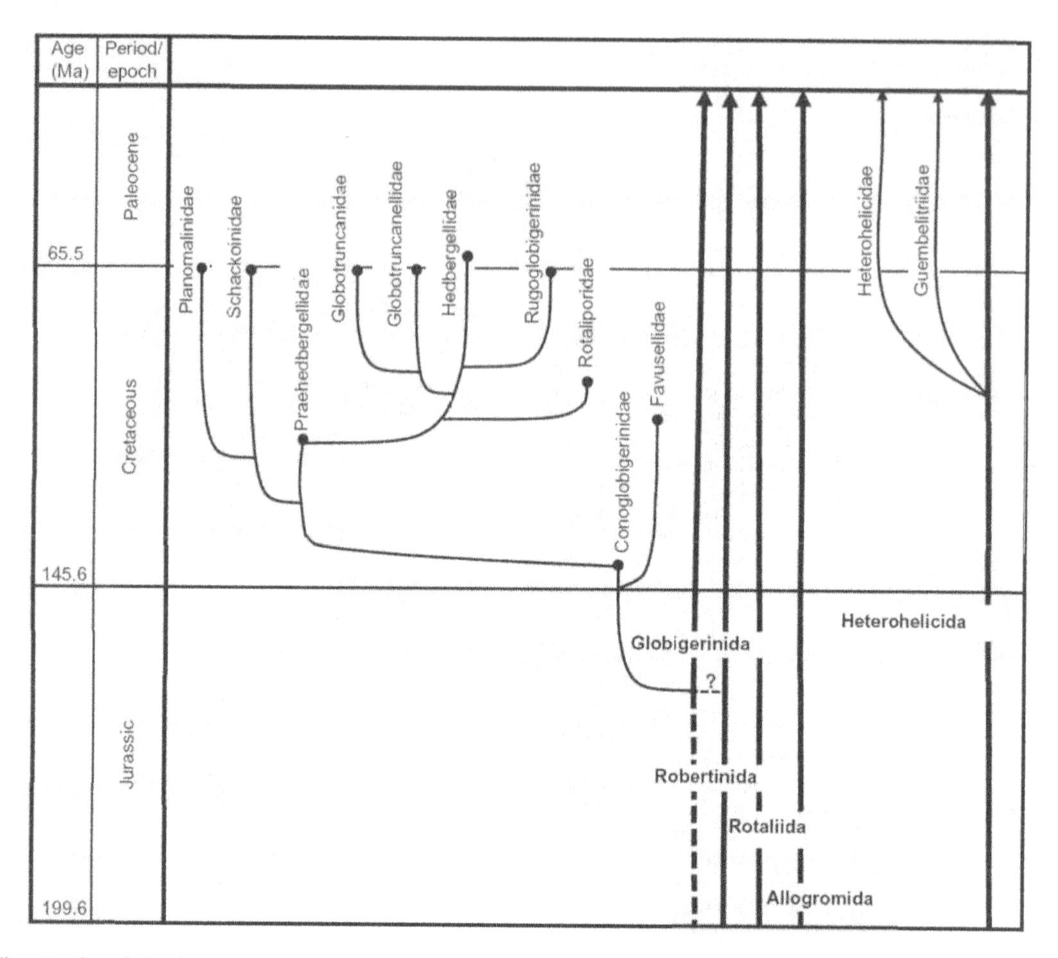

Figure 4.1. The evolution of the Cretaceous planktonic foraminiferal families (thin lines) from the Early Jurassic Conoglobigerinidae.

Superfamily FAVUSELLOIDEA Longoria, 1974 emend. Banner and Desai, 1988

For the members of this superfamily, the test is trochospiral with two or more whorls of subglobular chambers. The test surface is covered by microperforations and pseudomuricae. They may also have perforated cones or may possess a favose surface structure of fused pseudomuricae, forming an anastomizing reticulation. The spiral side is evolute, while the umbilical side is involute with a smallumbilicus. The aperture is intraumbilical, often covered by a bulla, or intra-extraumbilical and develops a high or low arch, which may be strongly asymmetric. The walls are believed to be composed primarily of aragonite. They range from the Late Triassic (Rhaetian) to Cretaceous (Early Cenomanian).

Family Favusellidae Longoria, 1974

This family consists of those genera of the Favuselloidea that possess rounded pseudomuricae, which fuse into ridges that in turn anastomose to form reticulations over the test surface. The aperture is intraumbilical or intra-extraumbilical, with a high or low arch, which may be strongly asymmetric. The intraumbilical aperture may be covered by a bulla and have a loop-shaped opening. Cretaceous (Berriasian to Early Cenomanian).

- *Favusella* Michael, 1973 (Type species: *Globigerina washitensis* Carsey, 1926). A favusellid with four to four and a half chambers in the final whorl, with an intraumbilical aperture. Cretaceous (Berriasian to Early Cenomanian) (Plate 4.1, Figs. 1–20; Plate 4.2, Figs. 1 and 19; Plate 4.20, Figs. 8, 10, 11, 16; Plate 4.23, Fig. 1).
- *Ascoliella* Banner and Desai, 1988 (Type species: *Ascoliella scotiensis* Banner and Desai, 1988). A favusellid with five to seven chambers in the final whorl and an intra-extraumbilical aperture. Cretaceous (Late Aptian to latest Albian) (Plate 4.2, Figs. 2–18, 20, 21).

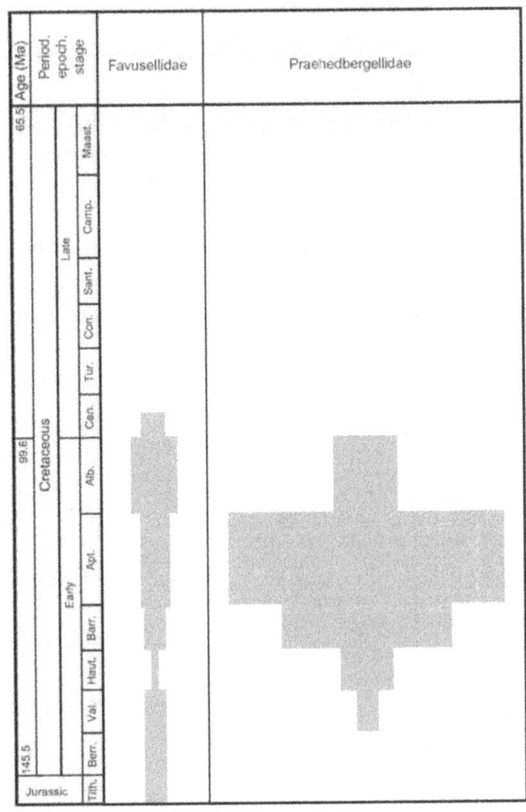

Figure 4.2. The biostrsatigraphic range and diversity of the main Early Cretaceous families.

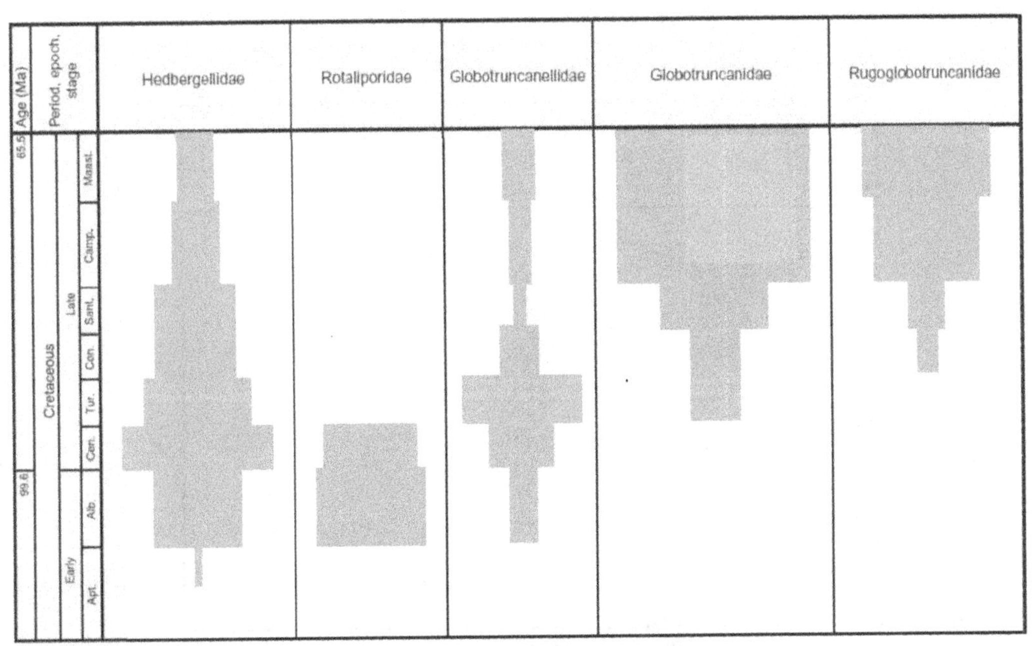

Figure 4.3. The biostrsatigraphic range and diversity of the main Late Cretaceous trochospiral families.

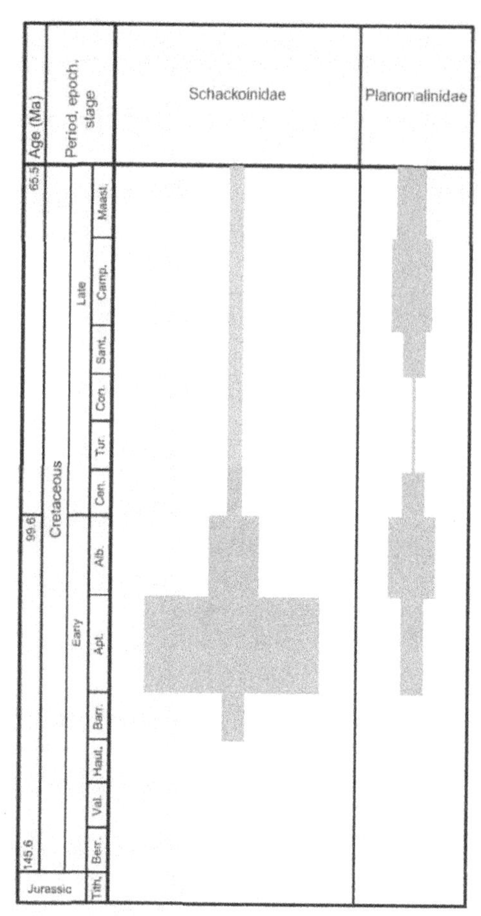

Figure 4.4. The biostrsatigraphic range and diversity of the main Cretaceous planispiral families.

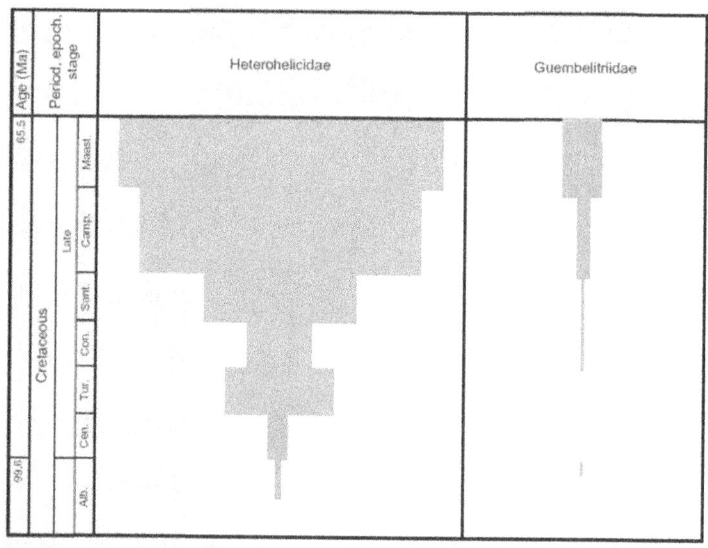

Figure 4.5. The biostrsatigraphic range and diversity of the main Cretaceous heterohelicids.

Superfamily GLOBIGERINOIDEA Carpenter, Parker and Jones, 1862

The members of this superfamily have a trochospiral with chambers that are rounded or angular with a peripheral keel or an imperforate band surrounded by a double keel. The primary aperture is umbilical with portici or is covered by tegilla with accessory apertures. When the portici are fused, accessory apertures or supplementary sutural apertures are formed. The wall is microperforate or macroperforate, and the surface may be smooth, with or without perforation cones, muricate, or spinose. The aperture is interiomarginal, umbilical, or intra-extraumbilical and bordered by a lip. Cretaceous to Holocene.

Family Praehedbergellidae Banner and Desai, 1988

The test is trochospiral throughout or initially trochospiral, later whorls may be streptospiral, microperforate lacking muricae, pseudomuricae, pustules or spine-bases, nonpustulate, nonspinose, but sometimes possessing perforation cones especially near the aperture. The primary aperture is intraumbilical or intra-extraumbilical. Cretaceous (Late Valanginian to Albian).

- *Blefuscuiana* Banner and Desai, 1988 (Type species: *Blefuscuiana kuznetsovae* Banner and Desai, 1988) = *Microhedbergella* Huber and Leckie, 2011. A praehedbergellid with five or more sub-globular chambers in the final whorl. Aperture: intra-extraumbilical. Cretaceous (Barremian to Coniacian) (Plate 4.5, Figs. 1–22; Plate 4.6, Figs. 1–20; Plate 4.7, Figs. 1–20; Plate 4.20, Fig. 4).
- *Gorbachikella* Banner and Desai, 1988 (Type species: *Globigerina kugleri* Bolli, 1959) = *Archaeokassabella* Maamouri and Salaj, 1995. A praehedbergellid with a smooth test, four to five chambers in the first whorl, and four globular chambers in the final whorl. The aperture is intraumbilical. Cretaceous (Late Valanginian to Early Aptian) (Plate 4.3, Figs. 1–20).
- *Praehedbergella* Gorbatchik and Moullade, 1973 (Type species: *Globigerina tuschepsensis* Antonova, 1964). A praehedbergellid with a smooth test but sometimes with perforation cones, four to four and a half chambers per whorl, with the chambers of the last whorl being globular or slightly depressed. The aperture is intra-extraumbilical. Cretaceous (Hauterivian to Late Aptian) (Plate 4.4, Figs. 1–21; Plates 4.5 and 4.6).
- *Lilliputianella* Banner and Desai, 1988 (Type species: *Lilliputianella longorii* Banner and Desai, 1988). Like *Blefuscuiana*, but with radially elongate chambers. Cretaceous (Latest Barremian to Late Aptian) (Plate 4.8, Figs. 1–17).
- *Lilliputianelloides* BouDagher-Fadel, Banner, and Whittaker, 1997 (Type species: *Clavihedbergella eocretacea* Neagu, 1975). A praehedbergellid with chambers of the last whorl radially elongated and being higher than they are wide. Cretaceous (Barremian) (Plate 4.7, Figs. 21 and 22).
- *Wondersella* Banner and Strank, 1987 (Type species: *Wondersella athersuchi* Banner and Strank, 1987). Initially trochospiral, with later whorls streptospiral, so that the umbilicus becomes closed because it is overlapped by later chambers. The chambers become high and elongate radially. Cretaceous (Late Aptian) (Plate 4.8, Figs. 18–20).

Family Hedbergellidae Loeblich and Tappan, 1961

The test is trochospiral, macroperforate (macroperforations may be very dense to create the appearance of "cancelations"), muricate with muricae fusing on the first whorls to form irregularly arranged rugosities. Perforation cones may be present on the early chambers of the last whorl, coalescing to form irregular imperforate ridges. The aperture is intra-extraumbilical, and the umbilicus has short unfused portici. Cretaceous (Late Aptian) to Paleocene (Danian).

- *Asterohedbergella* Hamoui, 1965 (Type species: *Hedbergella (A.) asterospinosa* Hamoui, 1965). The chambers become high with tubulospines, but not radially elongate or clavate. Cretaceous (Middle to Late Cenomanian) (Plate 4.9, Figs. 1 and 2).
- *Clavihedbergella* Banner and Blow, 1959 (Type species: *Hastigerinella subcretacea* Tappan, 1943) = *Pessagnoina* Georgescu, 2009 = *Pseudoclavihedbergella* Georgescu, 2009. The chambers are radially elongated and sometimes clavate in the last whorl, but with no tubulospines. The umbilicus is narrow with short narrow portici, and the surface is smooth but densely perforated. Cretaceous (Middle Albian to Early Santonian) (Plate 4.9, Fig. 3; Plate 4.16, Figs. 35–40).
- *Costellagerina* Petters, El-Nakhal, and Cifelli, 1983 (Type species: *Rugoglobigerina bulbosa* Belford, 1960) = *Paracostellagerina* Georgescu and Huber, 2006. The test has globular chambers, and the wall is covered by partially fused muricae forming short ridges. The aperture has a long and broad porticus. Cretaceous (Late Albian to Santonian) (Plate 4.9, Fig. 4; Plate 4.16, Fig. 28).
- *Hedbergella* Brönnimann and Brown, 1958 (Type species: *Anomalina lorneiana* d'Orbigny var. *trocoidea* Gandolfi, 1942) = *Muricohedbergella* Huber and Leckie, 2011 = *Planogyrina* Zakharova and Atabekyan, 1961 = *Brittonella* Kortchagin, 1989 = *Liuella* Georgescu, 2008 = *Hillsella* Georgescu, 2008. There are four to four and a half chambers per whorl, and short portici cover the umbilicus. Cretaceous (Late Aptian) to Paleocene (Danian) (Plate 4.9, Figs. 6–10; Plate 4.16, Figs. 11–27; Plate 4.20, Fig. 9A; Plate 4.21, Fig. 6; Plate 4.22,

Fig. 9A; Plate 4.24, Fig. 4).

- *Loeblichella* Pessagno, 1967 (Type species: *Praeglobotruncana hessi hessi* Pessagno, 1962). The test has a subacute periphery and a small sutural aperture on spiral side. Muricae are strongly developed on spiral side. Cretaceous (Cenomanian to Early Maastrichtian) (Fig. 4.8).
- *Planohedbergella* BouDagher-Fadel, Banner, Whittaker, and McCarthy, 1997 (Type species: *Planomalina ehrenbergi* Barr, 1961). The surface is wholly muricate. Cretaceous (Cenomanian to Maastrichtian) (Plate 4.9, Figs. 11–13; Plate 4.22, Fig. 14).
- *Whiteinella* Pessagno, 1967 (Type species: *Whiteinella archaeocretacea* Pessagno, 1967) = *Hebergellita* Maslakova, 1983 = *Fingeria* Georgescu, 2005. The test is almost equally biconvex and muricate. Muricae sometimes fuse into short ridges. The aperture has long and broad portici, but no tegillum, and the umbilicus is broad (~25% or more of test diameter) and many relict apertures are visible. A weak keel is seen on the periphery of later chambers only, but not extending into dorsal features. Cretaceous (Middle Cenomanian to Early Campanian) (Plate 4.9, Figs. 14–17).

Family Rotaliporidae Sigal, 1958

The test is trochospiral, macroperforate with a single keel. The primary aperture is intraextraumbilical, with a prominent apertural lip, and the umbilicus is covered with long portici, which sometimes fuse in the center to form supplementary apertures. The latter migrate toward the sutures to become sutural apertures in the advanced forms. Cretaceous (Late Aptian to Cenomanian).

- *Anaticinella* Eicher, 1973 (Type species: *Globorotalia multiloculata* Morrow, 1934). There are five to nine chambers in the final whorl. The periphery is broadly rounded but sometimes with a weakly developed keel on the periphery of last whorl. The umbilicus is wide with fused portici and supplementary apertures near the base of the sutures in an umbilical position. Cretaceous (Cenomanian) (Plate 4.9, Fig. 18).
- *Claviticinella* Banner, 1982 (Type species: *Claviticinella digitalis* Banner, 1982). Like *Ticinella* (see below) but with radially elongate to subclavate chambers, which are higher than they are long. The periphery is broadly rounded, but not keeled. The umbilicus is wide with fused portici and supplementary apertures. Cretaceous (Late Albian) (Fig. 4.9).
- *Paraticinella* Premoli Silva, Caron, Leckie, Petrizzo, Soldan, and Verga, 2009 (Type species: *Ticinella eubejaouaensis* Randrianasolo and Anglada, 1998, and *Ticinella bejaouaensis* as emended by Moullade, 1966). According to Premoli Silva *et al.* (2009), this genus is a homeomorph of *Ticinella* (see below); however, it possesses perforation cones that coalesce to form irregular imperforate ridges paralleling the coiling direction and is finely perforate. *Paraticinella* also lacks true umbilical supplementary apertures. The umbilical area may be covered by large flaps that fuse forming a weakly developed umbilical accessory aperture in the last chambers. Cretaceous (Late Aptian) (Plate 4.9, Figs. 20 and 21).
- *Pseudothalmanninella* Wonders, 1977, emend. Gonzalez Donoso *et al.*, 2007 (Type species: *Globotruncana ticinensis* forma *tipica* Gandolfi, 1942). Like *Ticinella* (see below) but with a biconvex or spiroconvex test and a well-developed keel on the early chambers of the spiral side. Sutures are depressed on the umbilical side. The umbilicus is narrow, bordered by periumbilical ridges, and with visible supplementary apertures. *Parathalmanninella* differs from *Pseudothalmanninella* in having triangular chambers on the umbilical side and a sutural position for the last supplementary apertures. Cretaceous (Late Albian to Early Cenomanian) (Fig. 4.9).
- *Thalmanninella* Sigal, 1948 (Type species: *Thalmanninella brotzeni* Sigal, 1948) = *Pseudoticinella* Longoria, 1973 = *Parathalmanninella* Lipson-Benitah, 2008. The test has an angular, keeled peripheral margin, which bifurcates on the umbilical side along the curved and raised umbilical sutures. The spiral sutures are raised and curved. Well-developed ornamentations are present on the umbilical side. The umbilicus is rimmed with supplementary apertures, becoming sutural in the last chambers, sometimes bordered by periumbilical ridges. Cretaceous (Late Albian to Cenomanian) (Plate 4.23, Fig. 25).
- *Ticinella* Reichel, 1950 (Type species: *Anomalina roberti* Gandolfi, 1942). The test has a rounded, not carinate periphery. The umbilical area is large with long portici fusing in the center to leave open supplementary apertures. Cretaceous (Late Aptian to Albian) (Plate 4.10, Figs. 11 and 12).
- *Rotalipora* Brotzen, 1942 (Type species: *Rotalipora turonica* Brotzen, 1942) = *Pseudorotalipora* Ion, 1983. The periphery is acutely angled with a single keel. Chambers are angular with elevated sutures on the spiral and umbilical sides. All the supplementary apertures are sutural, formed by flap-like extensions from the chambers. Cretaceous (Middle Cenomanian to Late Cenomanian) (Plate 4.23, Figs. 23 and 24).

Family Globotruncanellidae Maslakovae, 1964

The test is trochospiral, plano-convex, or concavo-convex with compressed chambers. Walls are macroperforate, with muricae strongly developed on both spiral and umbilical sides, concentrating near the periphery where they might fuse to form a keel or two keels surrounding an imperforate band. Apertures are intra-extraumbilical, and the umbilicus is covered by portici or tegilla. Cretaceous (Late Albian to Maastrichtian).

- *Dicarinella* Porthault, 1970 (Type species: *Globotruncana indica* Jacob and Sastry, 1950). The test is very compressed, with elevated spiral sutures, and depressed umbilical sutures. The periphery has closely spaced, appressed keels. The umbilicus is open wide with short portici. It might be a senior synonym of the very closely related *Concavatotruncana* (see below). Cretaceous (Turonian to Santonian) (Plate 4.10, Figs. 13 and 14).
- *Concavatotruncana* Korchagin, 1982 (Type species: *Rotalia concavata* Brotzen, 1934). The test is planoconvex, with the spiral side being very flat and the umbilical side very convex. Sutures are elevated on the spiral side but depressed on the umbilical side. The periphery has closely spaced, appressed keels. The umbilicus has short unfused portici. Cretaceous (Late Cenomanian to Santonian) (Plate 4.23, Fig. 15).
- *Falsotruncana* Caron, 1981 (Type species: *Falsotruncana maslakovae* Caron, 1981). A concavo-convex test with widely separated keels. Cretaceous (Turonian to Early Coniacian) (Fig. 4.10).
- *Globotruncanella* Reiss, 1957 (Type species: *Globotruncana citae* Bolli, 1951). The test is compressed with four to five petaloid to trapezoidal chambers in the final whorl, and muricate with muricae strongest at periphery. The axial periphery is acute; the sutures are depressed. The aperture is bordered by a large porticus, covering most of the narrow umbilicus. Cretaceous (Late Campanian to Maastrichtian) (Plate 4.10, Figs. 15–17; Plate 4.23, Fig. 22).
- *Helvetoglobotruncana* Reiss, 1957 (Type species: *Globotruncana helvetica* Bolli, 1945). The test has a strongly convex umbilical side and a flat spiral side so that peripheral angles in axial view are obtuse. The periphery has one keel extending from the angled periphery into sutures of the spiral side. The aperture has a long and broad porticus. The umbilicus is broad, with several relict apertures. Cretaceous (latest Late Cenomanian to Middle Turonian) (Plate 4.10, Figs. 18–21; Plate 4.16, Figs. 1–6; Plate 4.20, Fig. 9B).
- *Praeglobotruncana* Bermudez, 1952 (Type species: *Globorotalia delrioensis* Plummer, 1931) = *Rotundina* Subbotina, 1953. The periphery has muricocarina, in which muricae may be fused to form a single keel. Spiral sutural limbations are weak. Umbilical sutures are depressed. The aperture has a porticus. The umbilicus is narrow with short portici and relict apertures. Cretaceous (Late Albian to Middle Turonian) (Plate 4.11, Figs. 1–6; Plate 4.16, Figs. 7–10; Plate 4.23, Fig. 21).

Family Globotruncanidae Brotzen, 1942

The test is trochospiral, macroperforate, muricate with muricae fusing on the first whorls to form irregularly arranged rugosities. The peripheral margin is angular, truncated with a peripheral keel or two keels. The umbilicus is widely open with short unfused portici or covered by a tegillum. The primary aperture is umbilical or intra-extraumbilical with a well-developed lip. Cretaceous (Turonian to Maastrichtian).

- *Abathomphalus* Bolli, Loeblich, and Tappan, 1957 (Type species: *Globotruncana mayaroensis* Bolli, 1951). The test is flattened, and the peripheral margin has two keels. The primary aperture is covered by a tegilla. Muricae may align and fuse into short, discontinuous ridges (costellae) on chamber sides. Cretaceous (Maastrichtian) (Plate 4.11, Figs. 7–10; Plate 4.21, Figs. 2 and 4; Plate 4.22, Fig. 18; Plate 4.23, Fig. 6; Plate 4.24, Fig. 3).
- *Contusotruncana* Korchagin, 1982 (Type species: *Pulvinulina arca* var. *contusa* Cushman, 1926) = *Rosita* Caron, Donoso, Robaszynski, and Wonders, 1984. The test is biconvex with a lobate to rounded outline. Chambers are initially globular becoming more crescentic. They often possess inner marginal depressions or undulationary spiral surfaces. Intercameral sutures can be raised, sigmoidal, elevated, or depressed on both spiral and umbilical sides. Double keeled with an usually narrow imperforate peripheral band, which may reduce to a single keel in the final chamber. The primary aperture is umbilical and with short portici. Cretaceous (Coniacian to Maastrichtian) (Plate 4.11, Figs. 11–15; Plate 4.23, Figs. 17 and 18).
- *Globotruncana* Cushman, 1927 (Type species: *Pulvinulina arca* Cushman, 1926). The test is biconvex, lobate to circular in outline, and may be truncate in axial view. The chamber form is variable from globular to crescentic. Intercameral sutures are raised on the spiral side and raised or depressed on the umbilical side. This form is double keeled with an intermediate imperforate band. The keels may become appressed and may reduce to a single keel in the final chamber. The primary aperture is intraumbilical in position and covered by a tegillum. Cretaceous (Late Coniacian to Maastrichtian) (Plate 4.11, Figs. 16–20; Plate 4.12, Figs. 1–6; Plate 4.21, Figs. 6, 8–12; Plate 4.22, Figs. 1, 4A, 11, 12, 18, 19; Plate 4.23, Figs. 11, 12, 27, 30–32; Plate 4.24, Figs. 2, 6, 11, 16).
- *Globotruncanita* Reiss, 1957 (Type species: *Rosalina stuarti* De Lapparent, 1918). The test is biconvex and lobate to circular in outline. The peripheral margin is angular and truncated. Chambers are variable in shape, from trapezoidal to petaloid to crescentic on the spiral side.

Intercameral sutures on both sides are elevated but sometimes may be depressed on the umbilical side. A single keel is formed from fused spiral and umbilical keels. The primary aperture is umbilical with a porticus. Cretaceous (Late Santonian to Maastrichtian) (Plate 4.12, Figs. 7–12; Plate 4.20, Figs. 17 and 18; Plate 4.21, Figs. 1, 6, 7; Plate 4.22, Figs. 2 and 16; Plate 4.23, Figs. 9, 14, 19, 20; Plate 4.24, Fig. 1B, 8, 14, 15B).

- *Kassabiana* Salaj and Solakius, 1984 (Type species: *Globotruncana falsocalcarata* Kerdany and Abdelsalam, 1960). The test is planoconvex with small globular early chambers followed by triangular chambers with peripheral spines. The periphery has a single keel. Sutures are raised on both sides and around the wide and deep umbilicus on the umbilical side. The primary aperture is intraumbilical with portici fusing irregularly to form accessory apertures. Cretaceous (Late Maastrichtian) (Fig. 4.11).
- *Marginotruncana* Hofker, 1956 (Type species: *Rosalina marginata* Reuss, 1845). The test is planoconvex to biconvex, with umbilical sutures depressed and sigmoidal. Both umbilical and spiral sutures are keeled. The periphery has an imperforate band between two keels. The primary aperture is intra-extraumbilical with portici. It might be a senior synonym of the very closely related *Sigalitruncana*. Cretaceous (Middle Turonian to Santonian) (Plate 4.12, Figs. 13 and 14; Plate 4.22, Figs. 6, 7, 15; Plate 4.23, Fig. 7).
- *Radotruncana* El-Naggar, 1971 (Type species: *Globotruncana calcarata* Cushman, 1927). The test is planoconvex with a flattened spiral side and a strongly convex umbilical side. Chambers are rhomboidal in section, and the five to seven chambers of the final whorl each have a tubulospine at the proximal end of the chamber and in the plane of the spiral side. Sutures are depressed to slightly elevated and may be nodose. The periphery has a single keel. The surface is rugose along the sutures. The umbilicus is wide. The primary aperture is umbilical, bordered by a large porticus. Cretaceous (Campanian) (Plate 4.12, Figs. 15 and 16; Plate 4.23, Fig. 13).
- *Sigalitruncana* Korchagin, 1982 (Type species: *Globotruncana sigali* Reichel, 1950). The test is planoconvex, with sutures thickened and elevated on the spiral side, but sinuate and depressed around the small umbilicus on the umbilical side. The peripheral keel is formed by two rows of closely spaced pustules in the early part and may grade into a simple imperforate band on the final chamber. The primary aperture is intra-extraumbilical and has a porticus. The portici of successive chambers overlap and project into the umbilicus, bordering or covering it. Cretaceous (Turonian to Coniacian) (Plate 4.12, Figs. 17 and 18; Plate 4.22, Fig. 8; Plate 4.23, Fig. 16).

Family Rugoglobigerinidae Subbotina, 1959

The test is trochospiral with a broadly rounded or anglular periphery, which may have a keel or an imperforate band bordered by two keels. The test is macroperforate and muricate. Muricae may fuse to form short ridges or costellae. Sutures are always depressed. The aperture is intraumbilical with fused portici or tegilla. Cretaceous (Coniacian to Maastrichtian).

- *Archaeoglobigerina* Pessagno, 1967 (Type species: *ArchaeoGlobigerina blowi* Pessagno, 1967). Four to six globular chambers in final whorl. The peripheral margin has an imperforate band between two slightly developed keels. The umbilicus is large and covered by a delicate tegillum. Cretaceous (Coniacian to Maastrichtian) (Plate 4.12, Figs. 19 and 20).
- *Bucherina* Brönnimann and Brown, 1956 (Type species: *Bucherina sandidgei* Brönnimann and Brown, 1956). This form has a flat spiral side and a convex umbilical side with a single keel and a broadly angled periphery. Muricae are laterally fused to form meridionally aligned costellae. The umbilicus is deep, open with a short porticus. Cretaceous (Maastrichtian) (Fig. 4.12).
- *Gansserina* Caron, Gonza΄lez Donoso, Robaszynski, and Wonders, 1984 (Type species: *Globotruncana gansseri* Bolli, 1951). A planoconvex test has a convex umbilical side and a flattened spiral side. The peripheral angle in axial view is broadly acute or a right angle. A single peripheral keel, which fills all the intercameral sutures of the spiral side, has a ventral base on the terminal face, but umbilical sutures are depressed. Umbilical sutures are depressed, but spiral sutures are elevated. The umbilicus is broad, with several relict apertures, and their portici are visible. The primary aperture is covered by a tegillum. Cretaceous (Late Campanian to Maastrichtian) (Plate 4.13, Figs. 1 and 2; Plate 4.23, Fig. 8; Plate 4.24, Figs. 10 and 17).
- *Plummerita* Brönnimann, 1952 (Type species: *Rugoglobigerina (Plummerella) hantkeninoides hantkeninoides* Brönnimann, 1952). The test has inflated triangular chambers, and those of the final whorl are radially elongate and ending in a tubulospine. Sutures are depressed. The umbilicus is small provided with tegilla and having both proximal and distal accessory apertures. Cretaceous (Maastrichtian) (Plate 4.13, Figs. 3 and 4).
- *Rugoglobigerina* Brönnimann, 1952 (Type species: *Globigerina rugosa* Plummer, 1927) = *Kuglerina* Brönnimann and Brown, 1956. There are four to five inflated chambers in the final whorl that are globular on both sides of the test, which is covered in coarse to fine meridionally

arranged costellae. Sutures are depressed on both sides. The umbilicus is covered with a tegillum. Cretaceous (Late Santonian to Maastrichtian) (Plate 4.13, Figs. 8–16; Plate 4.22, Figs. 3, 9B; Plate 4.23, Figs. 26 and 29).

- *Rugotruncana* Brönnimann and Brown, 1956 (Type species: *Rugotruncana tilevi* Brönnimann and Brown, 1956, junior synonym of *Globigerina circumnodifer* Finlay, 1940). It should be noted that a topotype figured by Loeblich and Tappan (1989) is not representative of the genus. There are four to five inflated chambers in the final whorl. Sutures are thickened, curved, and depressed on the spiral side, but radial and depressed on the umbilical side. The periphery is double keeled with a variable imperforate band. Chamber surfaces are covered in coarse costellae in a meridional pattern which may be less intense on the spiral side. The umbilicus is variable in width, usually deep set with a short large portico which may fuse to form a tegillum with accessory apertures. Cretaceous (Late Campanian to Maastrichtian) (Plate 4.13, Fig. 17; Plate 4.21, Fig. 3).
- *Trinitella* Brönnimann, 1952 (Type species: *Trinitella scotti* Brönnimann, 1952). Similar to *Rugoglobigerina* but with compressed and flattened last chambers. Cretaceous (Maastrichtian) (Fig. 4.12).

Superfamily PLANOMALINOIDEA Bolli, Loeblich and Tappan, 1957

In members of this superfamily, the test is calcitic, planispiral, but may tend to become trochospiral. The walls are microperforate to macroperforate, and surfaces may be smooth with or without perforation cones or muricate. Apertures are equatorial with bordered lips. Cretaceous (Barremian to Maastrichtian).

Family Schackoinidae Pokorný, 1958

The test is planispiral, microperforate, and biumbilicate. Apertures are interiomarginal, equatorial, and symmetrical, extending laterally into each umbilicus. The final aperture and relict apertures are usually furnished with portici. The final aperture is umbilical–extraumbilical and equatorial. Cretaceous (Late Barremian to Maastrichtian).

- *Blowiella* Kretzchmar and Gorbachik, 1971, emend. Banner and Desai (1988) (Type species: *Planomalina blowi* Bolli, 1959). A schackoinid with four to five chambers per whorl. Chambers are longer than or as long as they are high, not radially and narrowly elongate. Portici are usually broadest at their posterior or at their extreme anterior. Cretaceous (Barremian to Aptian) (Plate 4.14, Figs. 1–14; Plate 4.24, Fig. 9).
- *Claviblowiella* BouDagher-Fadel *et al.*, 1996 (Type species: *Claviblowiella saundersi* Bolli, 1975). A schackoinid with four to four and a half chambers per whorl. Chambers are radially elongate, higher than they are long, but broadly clavate. Cretaceous (Late Aptian) (Plate 4.14, Figs. 15–18).
- *Globigerinelloides* Cushman and Ten Dam, 1948 (Type species: *Globigerinelloides algerianus* Cushman and Ten Dam, 1948). A schackoinid with 8–10 chambers in the last whorl, with broad umbilici and longer portici. Chambers are compressed laterally but not radially elongate. Cretaceous (Late Barremian to Early Albian) (Plate 4.14, Figs. 19–21; Plate 4.20, Fig. 15; Plate 4.21, Figs. 2 and 4; Plate 4.24, Fig. 17).
- *Leupoldina* Bolli, 1957, emend. Banner and Desai, 1988 (Type species: *Leupoldina protuberans* Bolli, 1957). A schackoinid with four to six chambers per whorl, with later chambers becoming radially and broadly elongate, sometimes with two or more branches, terminally clavate or bulbous, but without tubulospines. Cretaceous (Late Barremian to Early Albian) (Plate 4.15, Figs. 4 and 5).
- *Schackoina* Thalmann, 1932 (Type species: *Siderolina cenomana* Schacko, 1897). A schackoinid with four to five radially elongated chambers in the last whorl, each ending with a narrow thin-walled tubulospine, occasionally two, at the end of the last chamber. The final aperture has a porticus which broadens equatorially and/or laterally. Cretaceous (Late Aptian to Maastrichtian) (Plate 4.15, Figs. 7–10).
- *Pseudoplanomalina* Moullade, 2002 (Type species: *Planulina cheniourensis* Sigal, 1952). The test is planispiral, bi-evolute, and microperforate The periphery of the chambers is not rounded as in *Globigerinelloides* but subangular, with a distinct break in the peripheral round, better marked on the first chambers of the last whorl by a concentration of perforation cones giving it a peripheral pseudo-keel. Cretaceous (Late Aptian) (Plate 4.15, Figs. 11 and 12).

Family Planomalinidae Bolli, Loeblich and Tappan, 1957

The test is nearly involute to planispiral and biumbilicate, macroperforate, and may be muricate. The muricae may fuse into a peripheral keel. The aperture is interiomarginal, equatorial, symmetrical, extending laterally into each umbilicus. The final aperture and relict apertures are usually furnished with portici. Cretaceous (Late Aptian to Maastrichtian).

- *Alanlordella* BouDagher-Fadel, 1995 (Type species: *Alanlordella banneri* BouDagher-Fadel, 1995). Macroperforations may be sunk into perforation depressions. Early whorls may be muricate. The periphery is rounded, sometimes compressed in the early stage, but never developing a keel. Cretaceous (Late Aptian to Maastrichtian) (Plate 4.15, Figs. 13–16; Plate 4.23, Fig. 28; Plate 4.24, Fig. 16B).

- *Biglobigerinella* Lalicker, 1948 (Type species: *Biglobigerinella multispina* Lalicker, 1948). Like *Globigerinelloides* (see above) but with the final chamber becoming very thick with the interiomarginal aperture divided by its closure over the periphery of the penultimate whorl. The portici are shorter radially than they are long antero-posteriorly. Cretaceous (Cenomanian to Maastrichtian) (Fig. 4.13).
- *Biticinella* Sigal, 1956 (Type species: *Anomalina breggiensis* Gandolfi, 1942). The test is pseudo-planispiral; portici are fused distally, leaving intralaminal accessory apertures proximally. Cretaceous (Albian) (Plate 4.15, Fig. 19; Plate 4.23, Fig. 2).
- *Eohastigerinella* Morozova, 1957 (Type species: *Hastigerinella watersi* Cushman, 1931). The test is nearly involute; chambers are globular in the earlier stage and clavate to strongly radially elongate with bulbous projections at the chamber distal edge. Cretaceous (Late Turonian to Santonian) (Fig. 4.3).
- *Hastigerinoides* Brönnimann, 1952 (Type species: *Hastigerinella alexanderi* Cushman, 1931). The test is planispiral and biumbilicate. Chambers taper distally in the final whorl and sutures are depressed. Scattered muricae are often present on the earlier chambers of the test. The aperture is bordered by a well-developed porticus. Cretaceous (Coniacian to Santonian) (Plate 4.15, Figs. 20–22).
- *Planomalina* Loeblich and Tappan, 1946, emend. BouDagher-Fadel *et al.*, 1997 (Type species: *Planulina buxtorfi* Gandolfi, 1942). The test is muricate, with the muricae fusing at the compressed and acutely angled periphery and on the sutures to form a murico-carina. Nine or more chambers form the whorl with a relatively narrow umbilicus (~20% of the test diameter). The portici are shorter radially than they are long antero-posteriorly. Cretaceous (latest Aptian to Late Cenomanian) (Plate 4.15, Figs. 23 and 24; Plate 4.23, Figs. 4 and 5; Plate 4.16).

ORDER HETEROHELICIDA FURSENKO, 1958

Tests are biserial or triserial, at least in the early stage. They may be reduced to uniserial in later stages, microperforate or macroperforate, and smooth or muricate. Apertures are terminal, with a low to high arch. Walls are calcitic. Cretaceous (Aptian) to Holocene.

Superfamily HETEROHELICOIDEA Cushman, 1927

The test is mainly planispiral in the early stage, then biserial to triserial and possibly multiserial, rarely becoming uniserial in the adult stage. Apertures have a high to low arch at the base of the final stage or are terminal in the uniserial stage. Walls are calcitic, smooth, or muricate. Cretaceous (Aptian) to Holocene.

Family Guembelitriidae Montanaro Gallitelli, 1957

The test is triserial throughout with a straight axis of triseriality, or becoming multiserial in the adult, with globular inflated chambers. The aperture is a simple arch bordered by a rim, symmetrical about the equatorial plane, or there may be more than one aperture per chamber in the multiserial stage. Walls may be muricate. Cretaceous (Late Albian) to Holocene.

- *Chiloguembelitria* Hofker, 1978 (Type species: *Chiloguembelitria danica* Hofker, 1978) = *Jenkinsina* Haynes, 1981. The test is microperforate and triserial throughout. The wall is muricate. Cretaceous (Late Campanian) to Late Oligocene (Chattian) (Plate 4.19, Figs. 37 and 38).
- *Guembelitria* Cushman, 1940 (Type species: *Guembelitria cretacea* Cushman, 1933). The test is macroperforate and triserial throughout. The aperture is a simple arch at the base of the final chamber, bordered with a lip. The wall has "pore mounds" or perforation cones. Cretaceous (Late Albian) to Paleocene (Early Danian) (Plate 4.19, Figs. 1–3).
- *Guembelitriella* Tappan, 1940 (Type species: *Guembelitriella graysonensis* Tappan, 1940). The test is microperforate, and triserial in the early stage, but becomes multiserial in the adult. The wall is microperforate and muricate. Multiple apertures occur in the final stage. Cretaceous (Early Cenomanian) (Plate 4.17, Figs. 30 and 31).

Family Heterohelicidae Cushman, 1927

The test is planispiral in the early stages, then biserial, or the biserial stage may be followed by a multiserial stage in the adult. Walls are microperforate to macroperforate. The primary aperture is asymmetrical about the equatorial plane. Cretaceous (Late Albian) to Eocene (Priabonian).

- *Bifarina* Parker and Jones, 1872 (Type species: *Dimorphina saxipara* Ehrenberg, 1854). The test is macroperforate, initially biserial, becoming uniserial, with the terminal aperture becoming areal. Chambers have no lateral extensions but remain subglobular, reniform, or ovoid. Cretaceous (Late Cenomanian) to Middle Eocene (P12) (Plate 4.18, Figs. 29–33).
- *Gublerina* Kikoine, 1948 (Type species: *Gublerina cuvillieri* Kikoine 1948). The test is compressed and flabelliform, and the initial part of the test is a planispiral coil, followed by biserially arranged diverging chambers; enlarged apertural flanges connect subsequent chambers and form a narrow central cavity with chambers of each pair of biseries becoming separated by a broad, nonseptate, median area with no supplementary, lateral or median, sutural apertures. Later chambers are proliferate in the plane of the early biseries. Internal sutures are depressed. Enlarge apertural flaps connect the diverging chambers. The final aperture is an arch

at the base of the final chamber. However, bi-apertural terminal chambers may be present in some species, such as *Heterohelix* globulosa (Plate 4.17, Figs. 19 and 20) and *Heterohelix* planata (Plate 4.17, Figs. 6 and 7). Cretaceous (Late Campanian to Maastrichtian) (Plate 4.17, Figs. 28 and 29).

- *Heterohelix* Ehrenberg, 1843 (Type species: *Textilaria americana* Ehrenberg, 1843) = *Spiroplecta* Ehrenberg, 1844 = *Laeviheterohelix* Nederbragt, 1990 = *Protoheterohelix* Georgescu and Huber, 2009 = *Planoheterohelix* Georgescu and Huber, 2009 = *Globoheterohelix* Georgescu and Huber, 2009. The test is biserial microperforate, with subglobular chambers, with or without an initial planispiral coil. The wall is striate. Incipient multiserial growth may be present as a terminal stage. The aperture is simple and symmetrical, bordered by a lip. Cretaceous (Late Albian to Maastrichtian) (Plate 4.17, Figs. 1–27; Plate 4.20, Fig. 1; Plate 4.21, Figs. 1 and 6; Plate 4.22, Figs. 5, 9C, 10, 13, 19; Plate 4.24, Figs. 8, 12B, 15A).
- *Lunatriella* Eicher and Worstell, 1970 (Type species: *Lunatriella spinifera* Eicher and Worstell, 1970).The test is macroperforate, biserial, becoming very loosely biserial terminally, with a trend to uniseriality. Final chambers have long and broad lateral extensions. The aperture is a high, narrow arch and interiomarginal. Cretaceous (Turonian to Campanian) (Plate 4.18, Figs. 34 and 35).
- *Planoglobulina* Cushman, 1927 (Type species: *Guembelina acervulinoides* Egger, 1899). The test has a biserial early part, but later subglobular, or ovoid or reniform chambers proliferate in the equatorial plane. Sutures are depressed. The test becomes fan shaped. Surfaces are smooth or costellate or striate or costate, always parallel to long axis. Apertures occur on both margins of the chambers in the multiserial stage. *Planoglobulina* is similar to *Ventilabrella* (see below), but in *Planoglobulina*, the early stage is much thicker than the early stage of *Ventilabrella*. Cretaceous (Late Campanian to Maastrichtian) (Plate 4.19, Figs. 4–8, 13–15, 24, 25; Plate 4.22, Fig. 4A).
- *Platystaphyla* Masters, 1976 (Type species: *Planoglobulina brazoensis* Martin, 1972). The test is macroperforate, initially biserial, with chamber pairs as thick as they are broad, so that the test is initially subconical in shape. Later chambers proliferate only in the plane of the early biseries so that the test becomes laterally compressed terminally. Cretaceous (Maastrichtian) (Plate 4.18, Figs. 7 and 8).
- *Pseudoguembelina* Brönnimann and Brown, 1953 (Type species: *Guembelina excolata* Cushman, 1926) = *Huberella* Georgescu, 2007. The test is biserial, macroperforate, with longitudinal costae. Opposed chambers of the biseries are separated by portici or a plate. Chambers are broader in the equatorial than in the axial plane. Later chambers have secondary apertures. Cretaceous (Turonian to Maastrichtian) (Plate 4.18, Figs. 9, 11–21, 36).
- *Pseudotextularia* Rzehak, 1891 (Type species: *Cuneolina elegans* Rzehak, 1891). The test is biserial throughout, or with an initial planispiral stage, and macroperforate. Chambers are kidney shaped, strongly increasing in thickness. Apertures are a broad wide arch, bordered by a lip. Cretaceous (Coniacian to Maastrichtian) (Plate 4.18, Figs. 22–28; Plate 4.24, Fig. 13).
- *Racemiguembelina* Montanaro Gallitelli, 1957 (Type species: *Guembelina fructicosa* Egger, 1899). The test is subconical in the adult stage, macroperforate, with a biserial initial part, becoming multiserial, with chambers arranged in three dimensions in a plane perpendicular to the biserial one. Apertures have a cover plate with accessory apertures. Cretaceous (Maastrichtian) (Plate 4.17, Fig. 32; Plate 4.18, Figs. 1–6, 10; Plate 4.21, Fig. 5; Plate 4.24, Fig. 1A).
- *Sigalia* Reiss, 1957 (Type species: *Guembelina deflaensis* Sigal, 1952). The wall is macroperforate. The test is biserial throughout or with ephebic chambers proliferating only in the equatorial plane. Chambers are subglobular to reniform, with each pair wider than it is thick. Sutures are limbate and often raised. Chamber surfaces are striate or costellate initially but may become smooth terminally. Cretaceous (Late Santonian) (Plate 4.19, Figs. 16–22).
- *Ventilabrella* Cushman, 1928 (Type species: *Ventilabrella eggeri* Cushman, 1928). The wall is macroperforate. The initial part of the test is biserial, followed by a number of multiserial sets. Indi¬vidual chambers are globular. As in *Planoglobulina*, the number of relapsed chambers per set may occasionally vary. Sutures between multiserial chambers are usually depressed. *Ventilabrella* is similar to *Planoglobulina*, but the early stage in *Ventilabrella* is similar to that of *Heterohelix*, while that of *Planoglobulina* is similar to that of *Pseudotextularia*. Cretaceous (Santonian to Maastrichtian) (Plate 4.19, Figs. 9–11, 26–36; Plate 4.24, Fig. 7).
- *Zeauvigerina* Finlay, 1939, emend. Huber and Boersma, 1994 (Type species: *Zeauvigerina zelandica* Finlay, 1939). The test is microperforate. The surface is covered by irregularly scattered pustules. Chambers are biserially arranged in the early part, with a tendency to become uniserial in the later part, increasing gradually in size. The aperture is terminally positioned, circular, or oval shaped and may be produced on a short neck. Late Cretaceous (Maastrichtian, 3b) to Eocene (Priabonian, P17a) (Plate 5.10, Fig. 22).

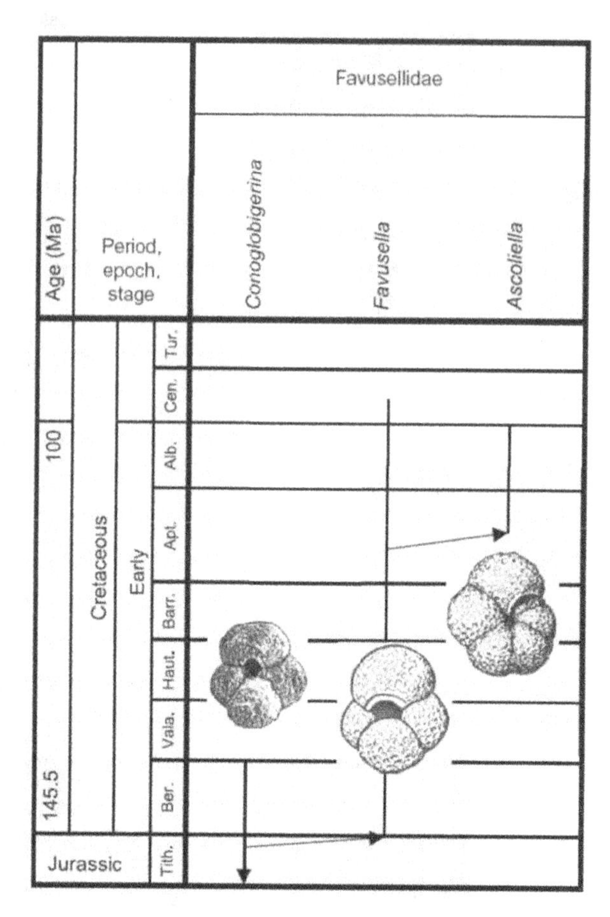

Figure 4.6. Phylogenetic evolution of the Cretaceous Favusellidae genera.

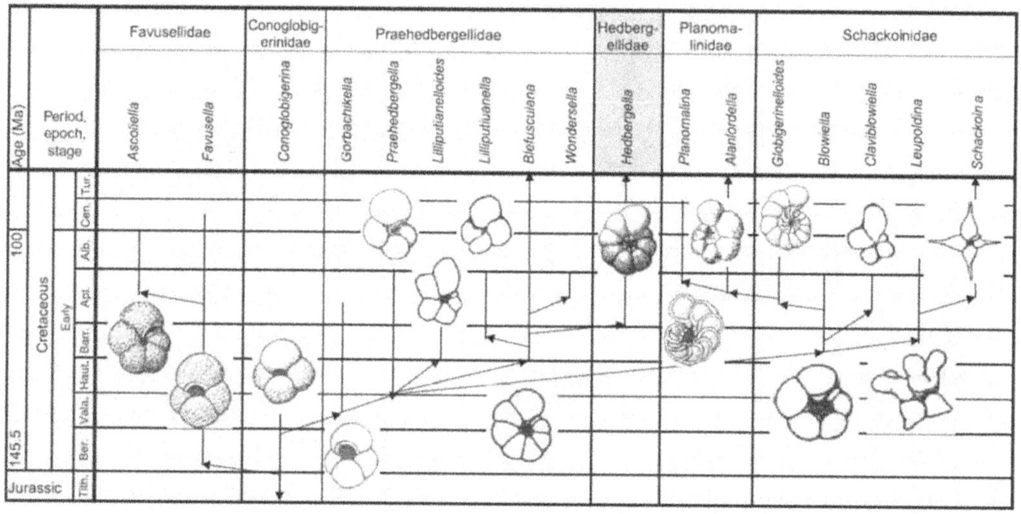

Figure 4.7. Phylogenetic evolution of the main Early Cretaceous families.

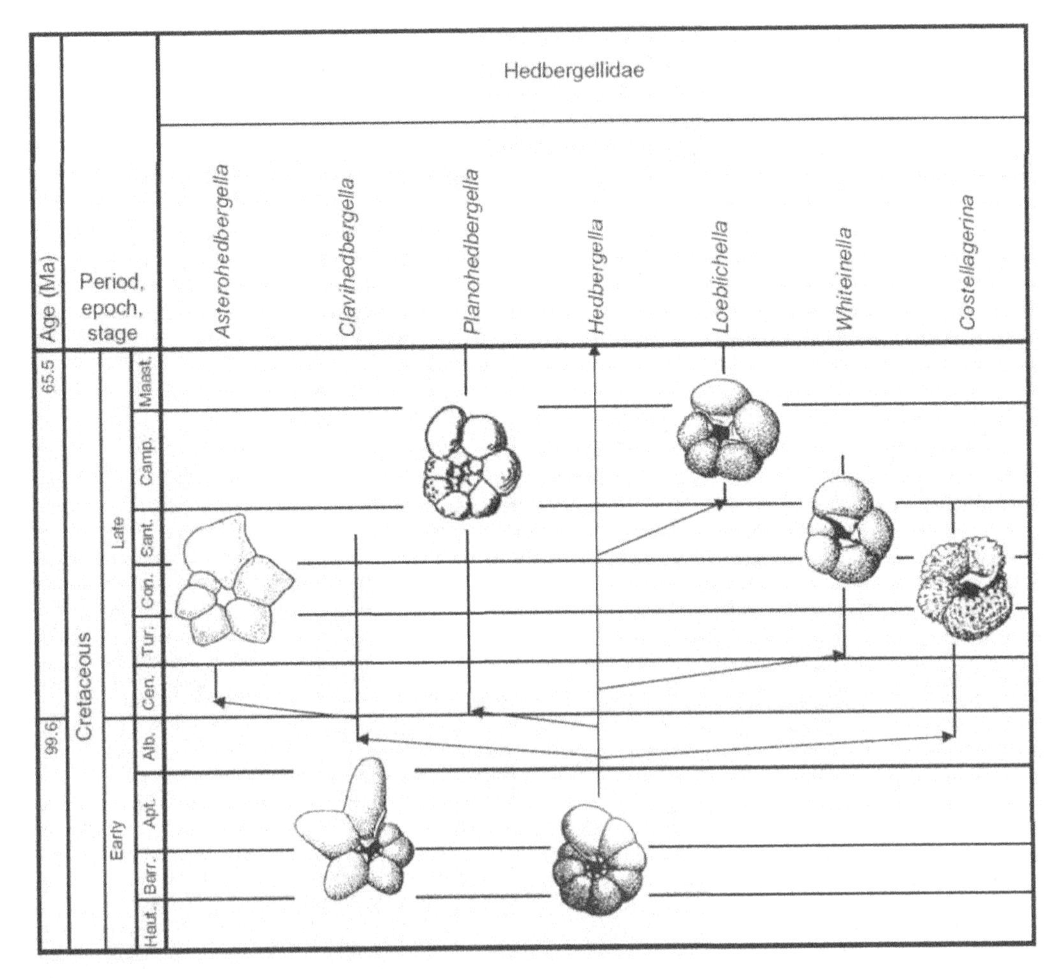

Figure 4.8. Phylogenetic evolution of the Cretaceous Hedbergellidae genera.

4.3 Biostratigraphy and phylogenetic evolution

During the Cretaceous, planktonic foraminiferal assemblages exhibit short biostratigraphic ranges and so enable relatively high-resolution stratigraphic correlations to be made (e.g., see Figs. 4.6–4.15). The systematics and biostratigraphy of these forms have been extensively studied, and a number of zonations based on species-level occurrences have been erected (see Charts 4.1–4.4, http://dx.doi.org/10.14324/111.9781910634257).

The four main superfamilies of Cretaceous planktonic foraminifera flourished at different stages but overlapped to some extent with each other:

- The **favusellids**, which evolved from forms originating in the Jurassic (Fig. 4.6).
- The trochospiral **globigerinids**, which are represented by:
  - the microperforate **praehedbergellids** in the Early Cretaceous (Fig. 4.7),
  - the macroperforate **hedbergellids** (Fig. 4.8), which replaced many of the praehedbergellids at the Aptian–Albian boundary,
  - the **rotaliporids**, which evolved from the hedbergellids in the Albian and Cenomanian (Fig. 4.9),
  - the keeled and compressed **globotruncanellids**, which first evolved from the hedbergellids in the Late Albian (Fig. 4.10),
  - the keeled **globotruncanids**, which evolved from the hedbergellids in the Turonian (Fig. 4.11), and
  - the costellate **rugoglobigerinids**, which evolved from the hedbergellids in the Coniacian (Fig. 4.12).
- The planispiral **planomalids** (Fig. 4.13), which are represented by:
  - the microperforate **schackoinids** and
  - the macroperforate **planomalinids**.

- The morphologically distinct **heterohelicids** (Fig. 4.14) and **guembelitriids** (Fig. 4.15), which appear in the Late Albian and evolved into many lineages in the Late Cretaceous.

The phylogentic evolution and biostratigraphic significance of each of these groups will be discussed in detail below.

4.3.1 The favusellids of the Cretaceous

It was suggested by Gorbachik and Kusnetsova (1986), and later confirmed by BouDagher-Fadel *et al.* (1997), that the favusellids originally possessed walls which were primarily aragonitic (see Chapter 3). This mineralogical characteristic is unique within the Globigerinida and may explain the unique pseudomuricae and the favose reticulation of the Favusellidae, which has no parallel elsewhere within the Globigerinida. However, Carter and Hart (1977) and Hart *et al.* (2002) have stated that *Favusella* spp. are to be found well preserved in the UK chalk facies of the Albian and Cenomanian and that these later forms do not appear to be aragonitic. If these later occurrences of calcitic forms are not due to recrystallization, then the favusellids must have made the transition from being aragonitic to calcitic sometime before the Albian.

The earliest Cretaceous favusellids appeared in the Early Berriasian of southeastern Europe. They were essentially descendants from the Jurassic favusellids and persisted to the Early Valanginian, with the last dimorphic pair with a meroplanktonic mode of life being *Conoglobigerina*, the higher spired *C. gulekhensis* (see Plate 3.3, Figs. 14–16) and the lower spired *C. caucasica* (see Plate 3.3, Figs. 11–13). These two species were covered by pseudomuricae which often fused together to form short, discontinuous ridges. In the Berriasian, these discontinuous ridges fused to form a reticulate, pseudomuricate coating over the surface of the oldest test of *Favusella* (Plate 4.1, Figs. 1–20; Plate 4.2, Figs. 1 and 19; Plate 4.20, Figs. 8, 10–14; Plate 4.23, Fig. 1).

Favusella, showing no dimorphism, became globally cosmopolitan. The holoplanktonic mode of life had at last been achieved in this evolutionary lineage (although the separate and stratigraphically earlier *Globuligerina* had also become holoplanktonic, see Chapter 3).

The cells within the reticulations of *Favusella* were initially small, and their bases were curved and concave. The earliest known *Favusella* is *F. hoterivica* (Plate 4.1, Figs. 1–13). It possesses a continuous reticulum but with narrow inner ridge spaces, containing only two or three microperforations. As the genus evolved, the reticulations became broader containing clusters of about 5–10 microperforations, as in *Favusella stiftia* (Plate 4.1, Fig. 14). From the Aptian, the inner-ridge spaces of the reticulum had broadened and the majority are flat floored, containing at least 10 microperforations. *Favusella washitensis* (Plate 4.1, Figs. 16–20), in the Late Aptian–Early Cenomanian, has very broad inner-ridge spaces of the reticulum, and these are flat floored, with 15–20 or more microperforations in each space. Therefore, the gradation from short discontinuous ridges in *Conoglobigerina* (see Fig. 4.6), to complete fine reticulation in the Berriasian to Hauterivian forms of *Favusella* (Plate 4.1, Figs. 1–13), to broadly reticulate forms in the Albian to Cenomanian, appears to be continuous (Plate 4.1, Figs. 16–20).

All species of *Favusella* possessed three and a half to four chambers that are visible on the umbilical side of the last whorl and an intraumbilical aperture. Occasionally, the aperture is covered by a bulla (Plate 4.1, Fig. 18), or sometimes only a trace of it can be seen. Similar bullae occurred in *Globuligerina* and *Conoglobigerina* (see Chapter 3), but they differ from the bullae of the Cenozoic species in often possessing short discontinuous ridges formed from pseudomuricae. All these bullae would have allowed contact with the exterior only through a single accessory, infralaminal aperture at their margins. It should be noted that the bullae of *Catapsydrax* (see Chapter 6) in the Cenozoic were perforate, not muricate, and had from one to four accessory, infralaminal apertures.

In the Late Aptian, the genus *Ascoliella* evolved from *Favusella* resulting in the migration of the anterior end of the aperture beyond the umbilicus (Fig. 4.6) and is the only known example in the Favusellidae of an evolutionary trend that parallels the evolution of *Praehedbergella* (Plate 4.4, Figs. 1–21) from *Gorbachikella* (Plate 4.3, Figs. 1–20) in the Praehedbergellidae (see Fig.4.7 and below). This intra-extraumbilical aperture allowed the species of *Ascoliella* (Plate 4.2, Figs. 2–18, 20, 21) to develop five or six chambers in the last whorl, but no bullae are known to have occurred. In the Conoglobigerinidae (see Chapter 3), intra-extraumbilical apertures were formed in the genera *Haueslerina* and *Compactogerina*, but these taxa are not known to have developed the large number of chambers in the final whorl displayed by species of *Ascoliella*. *Ascoliella* exhibted the broadest inner-space ridges between the reticulum in the Late Albian; *A. voloshinae* (Plate 4.2, Figs. 20 and 21) from the Late Albian of NE Mexico has a coarse reticulum.

All the Favuselloidea, from the Bajocian to the Cenomanian, are essentially pseudomuricate, and the pseudomuricae (or the ridges derived there from) are distributed almost uniformly over each chamber surface. In the Globigerinoidea, muricae only appeared on Hedbergellidae and in the Planomalinoidea on Planomalinidae from the Middle Aptian; then and thereafter, the muricae are concentrated in the approaches to the apertures and on the test periphery. There is such a phylogenetic gap, however, between the development of pseudomuricae on fasvusellids and on the muricae on globigerinids that the pseudomuricae and muricae of each superfamily cannot be homologous and may not even have been analoguous in function (see below).

4.3.2 The globigerinids of the Cretaceous

The morphology of the praehedbergellid test is very simple, but it shows variations that reflect biological differences which distinguish the species. These specific variations have a short stratigraphic range and have been useful in enabling broad stratigraphic correlations to be made (see Chart 4.1 online and Fig. 4.7). They have been systematically studied on a regional scale by a number of authors and used extensively in industrial biostratigraphy. However, while taxonomists prefer to "split" the morphogroups into different genera and subsequently species, industrial biostratigraphers rely more on "lumping" the different species into well-established genera in order to simplify biostratigraphic data. This enables them to use the biostratigraphically short-ranged species to date the assemblages without the systematic confusion created by many authors who split the genera on morphological characteristics not easily identified by optical microscopy.

In the revision of the Early Cretaceous Globigerinida, BouDagher-Fadel *et al.* (1997) retained the principle of the pioneering works of the Russian school but relied more heavily on the external aspect of the test, as observed under the optical microscope. They assessed the praehedbergellids by recognizing the morphological characters common to different groups of species in order to trace phylogenetic lineages that enable the ready identification of the species themselves.

This morphological definition of the Early Cretaceous forms, however, was rejected later by workers such as Moullade *et al.* (2002) and Verga and Premoli Silva (2003), who argued that splitting forms at the generic level following the number and/or form of their chambers is not practical and can lead to intense taxonomic proliferation. However, given that the shape and wall structure of the Early Cretaceous species are in fact practically very useful in biostratigraphical analysis, we consider the division is on balance justified and is the scheme we adopt here.

The most primitive of the praehedbergellids, *Gorbachikella*, appeared in the Late Valanginian of North Africa and were, like the favusellids, microperforate but superficially smooth, lacking all muricae and surface ridges. All forms had four subglobular chambers in the last whorl. The aperture was an intraumbilical arch, furnished with a narrow lip, and the umbilicus itself was open and deep. Forms similar to *Gorbachikella* were called Caucasella by Longoria (1974) and many later authors, but this is a homonym as its type species was *Globigerina hoterivica*, which we now know to be a *Favusella* (Plate 4.1, Figs. 1–13).

Gorbachikella grandiapertura (Plate 4.3, Figs. 12–16), which has clearly five chambers in the early whorls, was the immediate descendant of *Conoglobigerina* in the Late Valanginian. On the other hand, the modification of the anteriorly facing aperture in *Gorbachikella anteroapertura* (Plate 4.3, Figs. 5–8), which occurs in beds as old as Late Hauterivian, might have led to the formation of the intra-extraumbilical aperture typical of *Praehedbergella*.

The original definition of *Praehedbergella* by Gorbachik and Moullade (1973) was interesting but very difficult to use. Those authors considered *Praehedbergella* to be a subgenus of *Clavihedbergella* Banner and Blow (1959), and to differ from it by its "globigeriniform" (not digitiform) chambers and by its thinner wall (of "monolamellar structure," as compared to the "bilamellar structure" of *Hedbergella*). *Hedbergella* (Plate 4.9, Figs. 6–10; Plate 4.16, Figs. 11–27; Plate 4.20, Fig. 9A; Plate 4.21, Fig. 6; Plate 4.22, Fig. 9A; Plate 4.24, Fig. 4) and *Clavihedbergella* (Plate 4.9, Fig. 3; Plate 4.16, Figs. 35–40) are macroperforate, and both are of Late Aptian and younger Cretaceous age, while *Praehedbergella* is microperforate and smooth, and of Early Cretaceous (Aptian and older) age. *Praehedbergella* was redefined by Banner and Desai (1988), and by Banner *et al.* (1993), to include all those microperforate forms which have chambers on the umbilical side abutting the umbilicus.

In the Early Barremian, the trochospiral taxon *Praehedbergella*, with four chambers abutting the umbilicus, gave rise to *Blefuscuiana* as the number of chambers in the whorl increased (see Fig. 4.6). By Aptian times, there was a diversity of many-chambered species of *Blefuscuiana*, and the genus lasted into the Coniacian. The globular or slightly depressed chambers of *Blefuscuiana* became elongate in the Barremian. These elongated species, still trochospiral in coiling mode, were united by BouDagher-Fadel *et al.* (1997) into the genera *Lilliputianella* (Plate 4.8, Figs. 1–17) and *Lilliputianelloides* (Plate 4.7, Figs. 21 and 22). The genus *Lilliputianella* clearly evolved from *Blefuscuiana* in the latest Barremian, while *Lilliputianelloides* derived directly from Praedhedbergella in the later part of the Early Barremian. These two genera have different morphologies, different stratigraphic ranges, and different geographic distributions. Chamber elongation also occurred in the latest Aptian of the Middle East in *Wondersella* (Plate 4.8, Figs. 18–20), a genus which is also characterized by streptospiral coiling in the final stages of growth.

Around the latest Early Aptian times, small microperforate *Blefuscuiana* evolved into larger macroperforate *Hedbergella* (BouDagher-Fadel, 1996) with macroperforations becoming sometimes very dense, so as to create the appearance of cancelation (Fig. 4.8; Chart 4.2 online). The muricae soon became partially fused forming short ridges as early as the Late Albian genus, *Costellagerina* (Plate 4.9, Fig. 4; Plate 4.16, Fig. 28).

Parallel to this line of evolution, *Hedbergella* developed radially elongated chambers in the last whorl in the Late Albian to the Early Santonian (in *Clavihedbergella*) and high chambers with tubulospines in the Late Cenomanian, *Asterohedbergella* (Plate 4.9, Figs. 1 and 2). Small sutural apertures appeared dorsally in the Campanian–Maastrichtian, *Loeblichella* (Fig. 4.8).

Planohedbergella (Plate 4.9, Figs. 11–13; Plate 4.22, Fig. 14; Fig. 4.8) arose in the Cenomanian from *Hedbergella* (Fig. 4.8) by becoming planispiral. The earliest whorls of *Planohedbergella* may still possess low trochospirality, but the adult final two or more whorls became clearly planispiral. Like the ancestral *Hedbergella*, the chamber surfaces of *Planohedbergella* were not only macroperforate but also uniformly and completely muricate; however, *Planohedbergella* never acquired a keel either in the intercameral sutures or on its periphery. Unlike the *Alanlordella–Planomalina* phylogeny, the evolving species of *Planohedbergella* did not become more compressed and more multichambered; instead the only descendant of *Planohedbergella* was the morphologically conspicuous but rare planomalinid, *Biglobigerinella* (see Fig. 4.13).

In the latest Aptian, the first Rotaliporidae appeared, their macroparforate and muricate tests reflecting their ancestry in *Hedbergella* (e.g., *H. planispira*, Plate 4.9, Figs. 5, 7, 8), but being distinguished by having accessory apertures that were never seen in the Praehedbergellidae.

The evolution of the rotaliporids (Fig. 4.9) is marked by a gradual increase of the size of the test, an increase in the number of chambers in the whorl, a widening of the umbilicus which reached its maximum in *Ticinella* (e.g., *T. roberti*, Plate 4.10, Figs. 11 and 12), where the portici are distally fused to form accessory intralaminal apertures or completely fused to develop infralaminal accessory apertures, as in *Paraticinella* (Plate 4.9, Figs. 20 and 21).

In the Late Albian, *Ticinella* gradually acquired a peripheral keel by fusing the muricae over the periphery of the test, periumbilical ridges, and accessory and supplementary apertures by fusing the portici in the umbilical area (Fig. 4.9). This was accompanied by the development of raised sutures on the spiral side (*Pseudothalmanninella*) and the gradual acquisition of raised sutures on both umbilical and spiral sides (*Thalmanninella*, Plate 4.23, Fig. 25), which displays a dextral-coiling preference (Desmares *et al.*, 2008). *Pseudothalmanninella* died out in the Early Cenomanian, while *Thalmanninella* did not disappear before the Late Cenomanian. This morphological trend of acquiring a peripheral keel and raised sutures was reversed in the Turonian species of *Anaticinella* (Plate 4.9, Fig. 18) that are very similar to *Ticinella*. However, these two genera have different stratigraphical ranges, and their phylogenetic separation is therefore justified. On the other hand, *Anaticinella multiloculata* is always associated with *Thalmanninella greenhornensis*, and there are intermediate forms between them (Eicher, 1973). This has led some authors, such as Gonzàlez-Donoso *et al.* (2007), to consider *Anaticinella* specimens as simply, extreme variants of the *Thalmanninella* in the Western Interior Seaway of North America. In the Late Cenomanian, and possibly quite independently from *Ticinella*, species of *Praeglobotruncana* (Plate 4.11, Figs. 1–6; Plate 4.16, Figs. 7–10; Plate 4.23, Fig. 21) progressively developed a single peripheral keel and raised spiral sutures. In the rotaliporid lineage, the portici of *Praeglobotruncana* fused completely in the umbilical area, and the supplementary openings migrated toward the sutures. The suggested development of *Rotalipora* (Plate 4.23, Figs. 23 and 24) from of *Praeglobotruncana* is supported by the same proportionate-coiling direction in both *Praeglobotruncana* and *Rotalipora* (Desmares *et al.*, 2008).

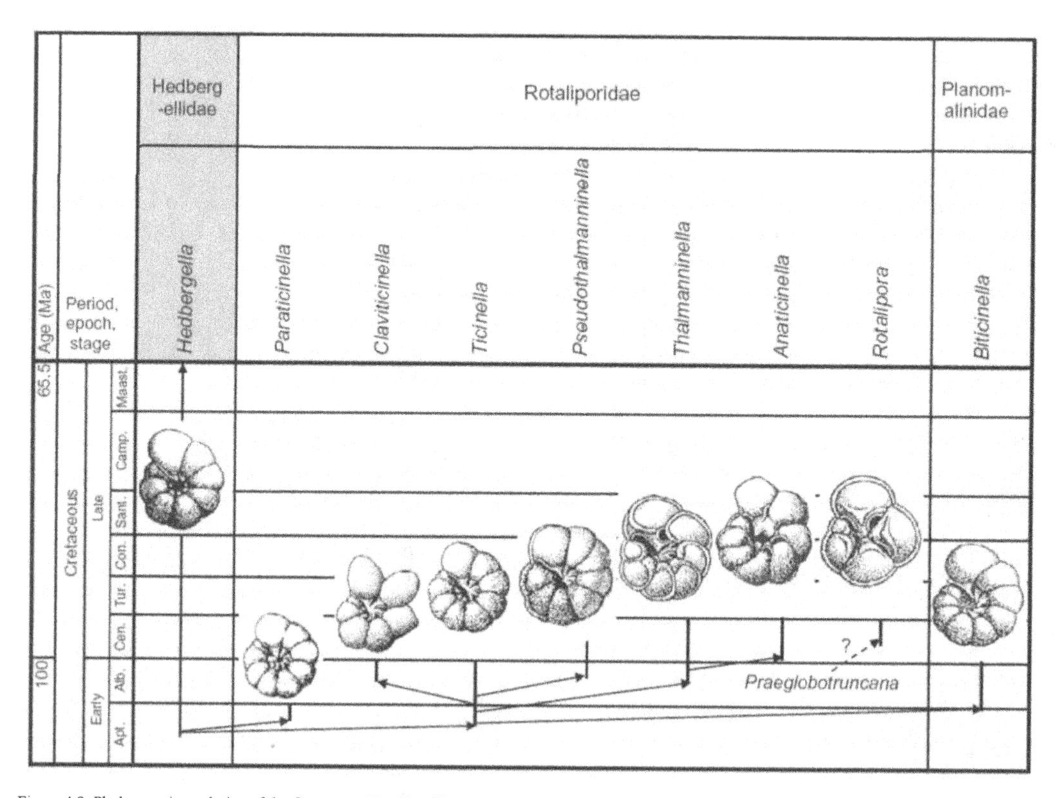

Figure 4.9. Phylogenetic evolution of the Cretaceous Rotaliporidae genera.

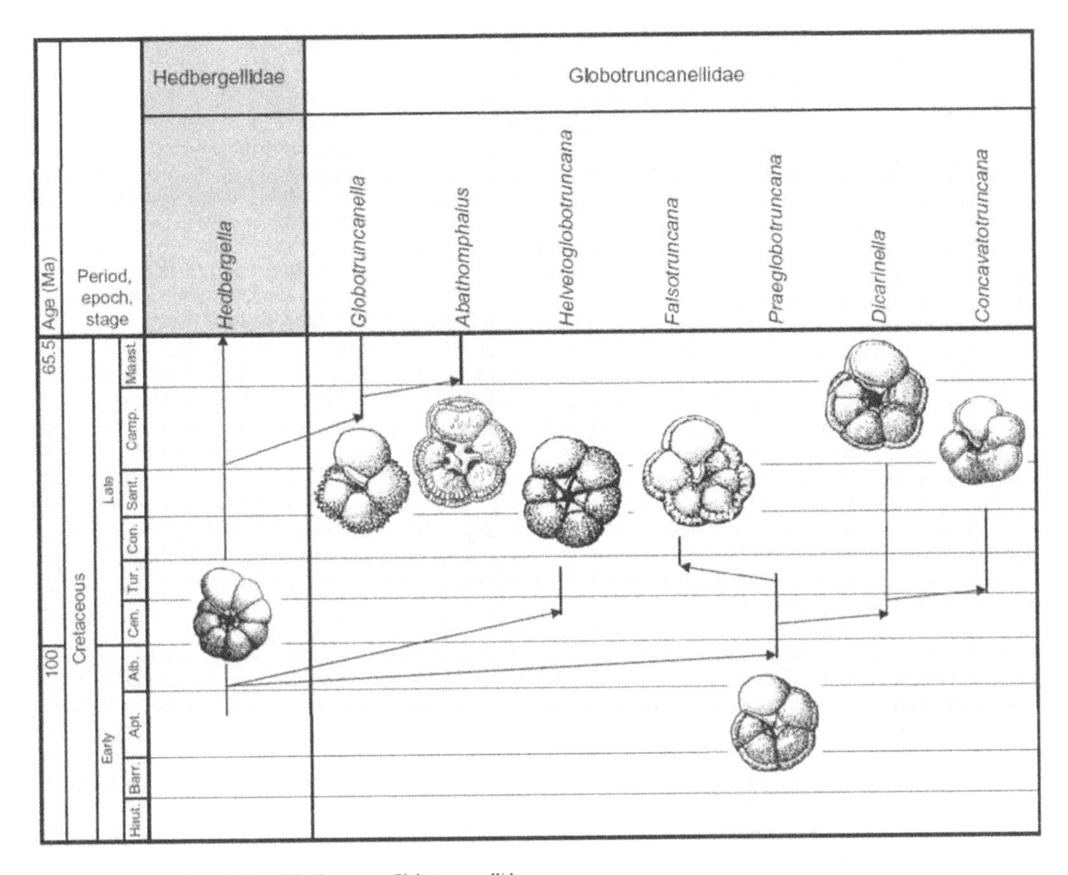

Figure 4.10. Phylogenetic evolution of the Cretaceous Globotruncanellidae genera.

In the Late Aptian, *Hedbergella* developed concentrations of muricae round the periphery of the test and along the spiral sutures, so gradually forming a peripheral and spiral keel (muriococarina) in *Praeglobotruncana* (Fig. 4.10). These early keeled members of the globotruncanellids evolved into forms with two closely spaced, peripheral appressed keels in the Late Cenomanian (*Dicarinella*, Plate 4.10, Figs. 13–15), or with widely separated keels in the Middle Turonian to the Middle Coniacian (*Falsotruncana*, Fig. 4.10). In the Turonian, *Dicarinella* evolved into planocovex forms (*Concavatotruncana*, Plate 4.23, Fig. 15). In the Campanian, *Hedbergella* gave rise to a form with a compressed and strongly muricate test, with muricae strongest near the imperforate peripheral band. The aperture became bordered by a large porticus, covering most of the narrow umbilicus, to give rise to *Globotruncanella* (Plate 4.10, Figs. 16–18; Plate 4.23, Fig. 22). In the Maastrichtian, species of *Globotruncanella* evolved into *Abathomphalus* (Plate 4.11, Figs. 7–10) by becoming more compressed with less inflated chambers and developing a double peripheral keel by the fusion of the strong muricae surrounding the imperforate peripheral band.

In the Turonian, *Dicarinella* gave rise to *Sigalitruncana* (Fig. 4.11) with a planoconvex test, sinuate umbilical sutures, and portici of successive chambers overlapping and projecting into the umbilicus. From *Sigalitruncana* (Plate 4.12, Figs. 17 and 18; Plate 4.22, Fig. 8; Plate 4.23, Fig. 16), three lineages evolved. The *Marginotruncana* (Plate 4.12, Figs. 13 and 14; Plate 4.22, Figs. 6 and 7, 15; Plate 4.23, Fig. 7) evolved in the Middle Turonian to Santonian, where the portici fuse laterally. In the Campanian to Maastrichtian, the intercameral sutures on both sides became elevated, the aperture evolved to be completely umbilical, and the portici completely fused to form a tegillum, as in *Globotruncana* (Plate 4.11, Figs. 16–20). Parallel to this lineage, *Globotruncanita* (Plate 4.12, Figs. 7–12) evolved in the Late Santonian as the two pressed keels fused into a thick peripheral keel, which extends on both spiral and umbilical sides. The umbilicus remained open with short portici. In the Campanian, *Radotruncana* (Plate 4.12, Figs. 15 and 16; Plate 4.23, Fig. 13) developed a planoconvex test with tubulospines at the proximal end of the chambers of the last whorl, while in the Maastrichtian, *Kassabiana* had a neanic *Plummerita* stage with tubulospines and a globotruncanid ephebic stage. In the Late Coniacian, *Sigalitruncana* evolved into *Contusotruncana* (Plate 4.11, Figs. 11–15) with the primary aperture opening entirely into the umbilicus and being surrounded by short portici. In the Late Cenomanian, the primary aperture of *Hedbergella* opened completely into the broad umbilicus to give the short-lived *Helvetoglobotruncana* (Plate 4.16, Figs. 1–6). The latter has a keel which extends into the sutures of the flat spiral side.

From the Middle Cenomanian to Early Campanian, quite independently from *Helvetoglobotruncana*, muricate Hedbegella with an intra-extraumbilical aperture gave rise to forms with an intraumbilical aperture through *Whiteinella* (Plate 4.9, Figs. 14–17) and *Archaeoglobigerina* in the Coniacian to the Maastrictian (Fig. 4.12). The latter in turn evolved into *Rugoglobigerina* (Plate 4.13, Figs. 8–16; Plate 4.22, Figs. 3, 9B; Plate 4.23, Figs. 26 and 29) where the umbilicus is covered with a tegillum and the muricae fuse meridionally across the test surface at right angles to the periphery to form costellae. In the Late Campanian, *Rugoglobigerina* developed, at the periphery, a double keel with a variable imperforate band, giving rise to *Rugotruncana* (Plate 4.13, Fig. 17; Plate 4.21, Fig. 3). In the latest Campanian–Maastrichtian, *Rugoglobigerina* gave rise to short-lived varieties, *Gansserina* (Late Campanian to Maastrichtian, Plate 4.13, Figs. 1 and 2; Plate 4.23, Fig. 8; Plate 4.24, Figs. 10 and 17), with a planoconvex test similar to Hevetoglobotruncana; *Bucherina*, with a single keel at the broadly angled periphery; *Trinitella* Brönnimann, with compressed and flattened last chambers; and *Plummerita* (Plate 4.13, Figs. 3 and 4), having chambers in the final whorl that were radially elongate and ending in a tubulospine (Fig. 4.12).

4.3.3 The planomalinids of the Cretaceous

The trochospiral praehebergellids were also immediately ancestral to the planispiral forms, the Schackoinidae, which appeared in the Early Barremian (Figs. 4.7). The first schackoinids were simple species with four to five chambers in the last whorl, *Blowiella* (Fig. 4.7) as emended by Banner and Desai (1988). This genus was later rejected by some authors, such as Verga and Premoli Silva (2003), as they tried to simplify the nomenclature of the Early Cretaceous. But it is retained here because of its stratigraphical importance and utility. The five-chambered *Blowiella* (Plate 4.14, Figs. 1–14; Plate 4.24, Fig. 9) first appeared in the Barremian and gradually evolved into the many-chambered *Globigerinelloides* (Plate 4.14, Figs. 19–21; Plate 4.20, Fig. 15; Plate 4.21, Figs. 2 and 4; Plate 4.24, Fig. 17; Fig. 4.7) in the Late Aptian. The high chambered forms are grouped into the short-lived genus *Claviblowiella* (Plate 4.14, Figs. 1–14; Plate 4.24, Fig. 9; Fig. 4.7) by BouDagher-Fadel (1996), which became extinct at the close of the Aptian or soon after. In the Barremian, chambers became radially elongate, ending with bulbous chamber extensions, as in *Leupoldina* (Plate 4.15, Figs. 4 and 5; Fig. 4.7), which ranged into the Early Albian, or elongate slender tubulospines as in the long-ranging *Schackoina* (Plate 4.15, Figs. 7–10; Figs. 4.7 and 4.13), which appeared in the Late Aptian and ranged up into the Maastrichtian. The characteristic of the genus *Schackoina* appears to be one of great morphological variability and with individual variations. Some species possess only one tubulospine in the final chamber, but others may have two or three, or even more.

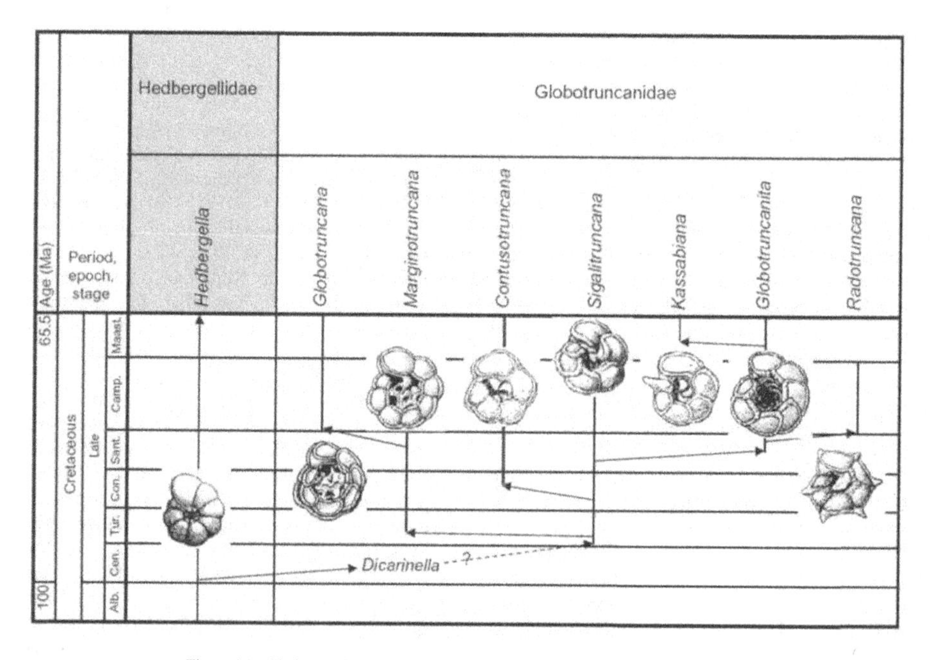

Figure 4.11. Phylogenetic evolution of the Cretaceous Globotruncanidae genera.

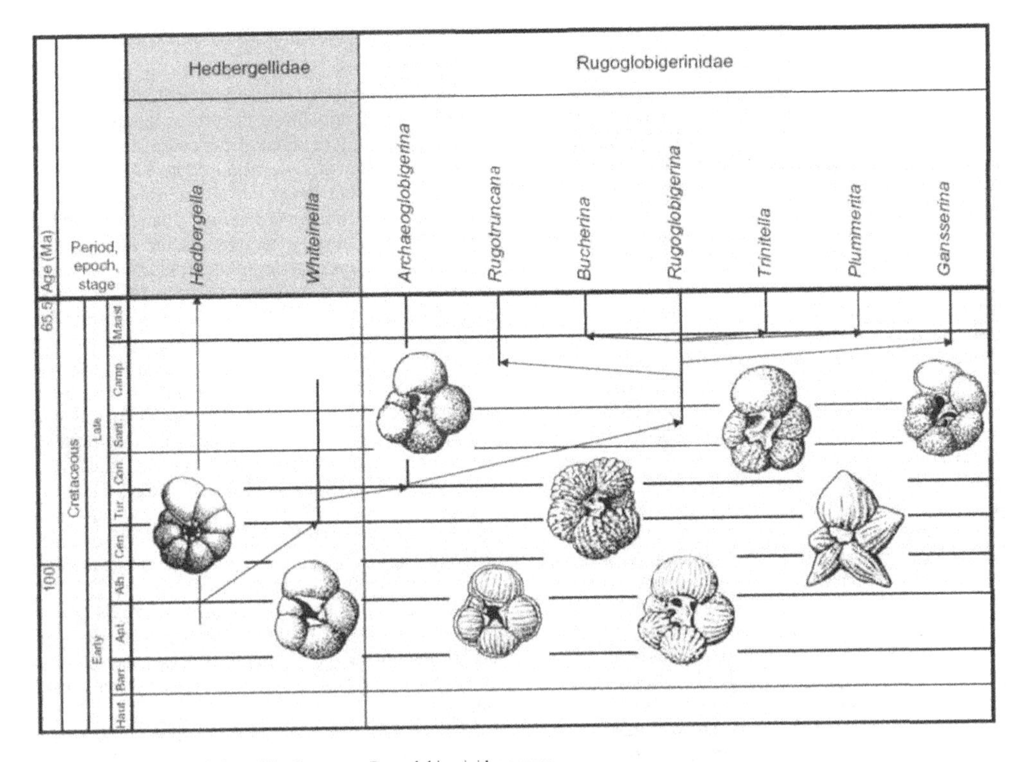

Figure 4.12. Phylogenetic evolution of the Cretaceous Rugoglobigerinidae genera.

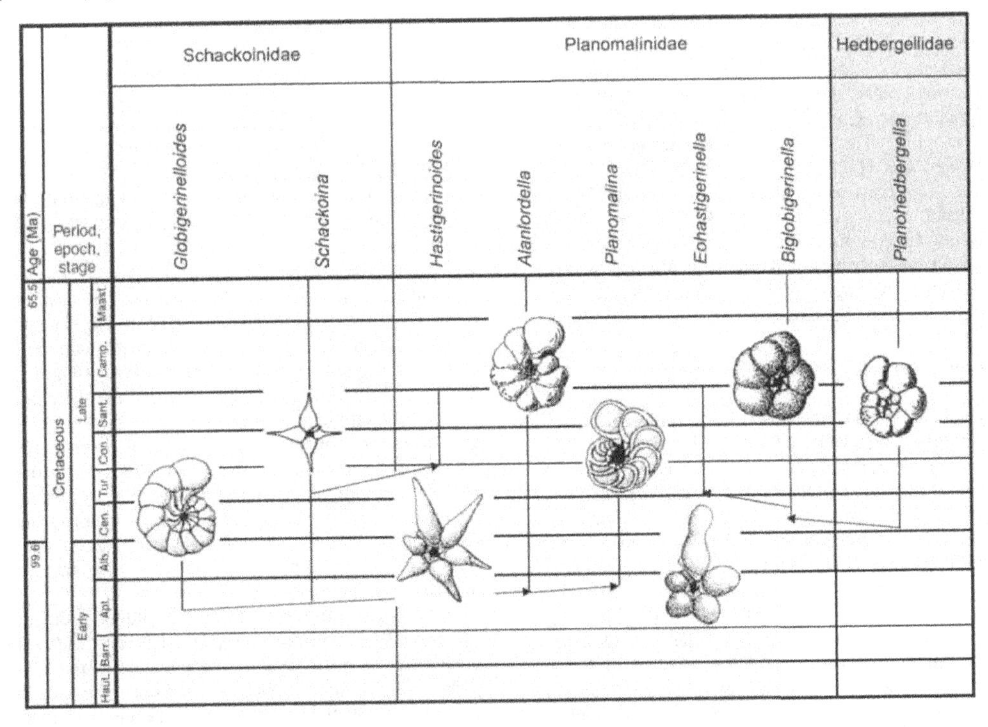

Figure 4.13. Phylogenetic evolution of the Cretaceous planispiral families.

In the Planomalinidae, *Alanlordella* (Plate 4.15, Figs. 13–16; Plate 4.23, Fig. 28; Plate 4.24, Fig. 16B; Fig. 4.7) arose separately, in the latest Aptian, from *Globigerinelloides* (Figs. 4.7 and 4.13) by gaining macroperforations (*A. bentonensis,* Plate 4.15, Fig. 15), and then in the Albian by also developing muricae, first on the surface of the early chambers of the last whorl (*A. banneri*, Plate 4.15, Figs. 13 and 14) and then in the intercameral sutures (*A. praebuxtorfi*, Plate 4.15, Fig. 16). The muricae fused to produce a keel on the periphery of the adult test, which extends into the sutures in *Planomalina* (*P. buxtorfi*, Plate 4.15, Figs. 23 and 24; Plate 4.23, Figs. 4 and 5). Throughout this evolution, the tests become more compressed and acquire more and more chambers in the last whorl. The spread of the muricae during the Aptian–Albian–Cenomanian, from a partial keel, as in *Alanlordella*, to the fully keeled *Planomalina* (Fig. 4.13), is essentially gradual.

Quite separately from *Planomalina*, the hedbergellid *Planohedbergella* (Fig. 4.8) gave rise to *Biglobigerinella* (Fig. 4.13), in which the test became less compressed with growth, the final chambers became very thick, and the interiomarginal aperture became divided by its closure near the periphery of the penultimate whorl, so that separate apertures opened in each umbilicus, and then even the last chambers themselves divided, to produce a brief, planispirally coiled, biserial stage. These bilobate forms evolved repeatedly over time and may have been selected for as a result of local ecological conditions. They appear to be confined to the Americas and westernmost Tethys. Berggren (1962), Pessagno (1967), and Moullade *et al.* (2002) among many others regarded *Biglobigerinella* as a junior synonym of *Globigerinelloides*. But it is retained here because of its stratigraphic and geographic value. *Biglobigerinella* gave rise in the Turonian to the nearly involute *Eohastigerinella* (Fig. 4.13) with clavate chambers and bulbous projections at the chambers distal edge. On the other hand, *Hastigerinoides* (Fig. 4.13) evolved in the Coniacian directly from planispiral Globigerinoides by elongating and tapering distally their chambers in the final whorl. Both *Eohastigerinella* and *Hastigerinoides* died out at the end of the Santonian.

4.3.4 The heterohelicids of the Cretaceous

In the Late Albian, the biserial compressed benthic foraminifera, *Brizalina*, with a microperforate test, imperforate longitudinal costae and a basal loop aperture, evolved so that their mode of life changed from being benthic dwellers in seafloor sediments of continental shelves to being planktonic inhabitants of the surface waters of the open ocean. This mechanism of planktonic forms evolving from benthic ancestors may have occurred multiple times in the history of the heterohelicids (see Chapter 1). The early primitive heterohelicids were represented by *Heterohelix moremanni* (Plate 4.20, Fig. 1) in the Late Albian–Cenomanian (Fig. 4.14). It was a small opportunist species with a subglobular and elongate parallel-sided test. It evolved in the Late Cenomanian into *Bifarina* (Plate 4.18, Figs. 27–31), a macroperforate heterohelicid initially biserial but with ephebic chambers becoming uniserial. Unlike the rest of the heterohelicids, *Bifarina* did not go extinct at the end of the Cretaceous. During the Turonian, *Lunatriella* (Plate 4.18, Figs. 32 and 33), with long and broad lateral extensions on the final chambers, and *Pseudoguembelina* (Plate 4.18, Figs. 9, 11–17), with secondary apertures in the final chambers, made their first appearance. In the Coniacian, *Heterohelix* evolved into *Pseudotextularia* (Plate 4.18, Figs. 20–26; Plate 4.24, Fig. 13) with chambers strongly increasing in thickness. However, it was not until the Late Santonian to the Maastrichtian that *Heterohelix* started to give rise to forms that develop chamber proliferation and an increase in size. This became a common feature in the Campanian to Maastrichtian. This multiserial chamber proliferation is common in larger benthic foraminifera, such as the orbitoids in the Cretaceous and the miogypsinids in the Cenozoic (see BouDagher-Fadel, 2008), but is less common in planktonic foraminifera. It may occur as a gerontic stage in some genera or species, for example, the triserial guembelitriids (Kroon and Nederbragt, 1990) or in the Late Cretaceous planomalinids, *Biglobigerinella* (Fig. 4.13); however, the Heterohelicidae are unique among the planktonic foraminifera, in that multiserial chamber proliferation developed repeatedly and in various lineages.

In the Santonian, different species of *Heterohelix* increased in size by multiplying the number of their chambers in the biserial plane to evolve into *Sigalia* (Plate 4.19, Figs. 16–21) and *Ventilabrella* (Fig. 4.14). *Gublerina* (Plate 4.17, Figs. 28 and 29) evolved in the Campanian from *Pseudoguembelina* (Fig. 4.14) by forming a fan-shaped test through enlargement of the apertural face and median plate separating the biseries.

In the Late Campanian and through to the Maastrichtian, *Pseudotextularia* (Fig. 4.14) developed a subconical test in the adult stage, with multiserial chambers arranged in the biserial plane, *Planoglobulina* (Plate 4.19, Figs. 4–8, 13–15, 23, 24; Plate 4.22, Fig. 4A), and in the Maastrichtian in three dimensions in a plane perpendicular to the biserial one, *Racemiguembelina* (Plate 4.17, Fig. 32; Plate 4.18, Figs. 1–6, 10; Plate 4.21, Fig. 5; Plate 4.24, Fig. 1A), with the test becoming laterally compressed terminally, *Platystaphyla* (Plate 4.18, Figs. 7 and 8). The multiserial growth in the flabelliform test of the heterohelicids was achieved by the evolution of a bi-apertural chamber after the biserial juvenile stage, called the progressive chamber (Van Hinte, 1965). In the first proliferation step, both apertures of the progressive chamber are each covered by one subsequent chamber (Nederbragt, 1990). This pattern is repeated until the individual is completed. In *Gublerina*, the pattern of multiserial chamber arrangement is essentially the same as in *Planoglobulina*, with the test extending in the plane of biseriality (Brown, 1969). However, *Racemiguembelina* shows extensions perpendicular to the biserial plane. Most of the heterohelicids died out at the end of the Cretaceous, together with most of the other planktonic foraminifera.

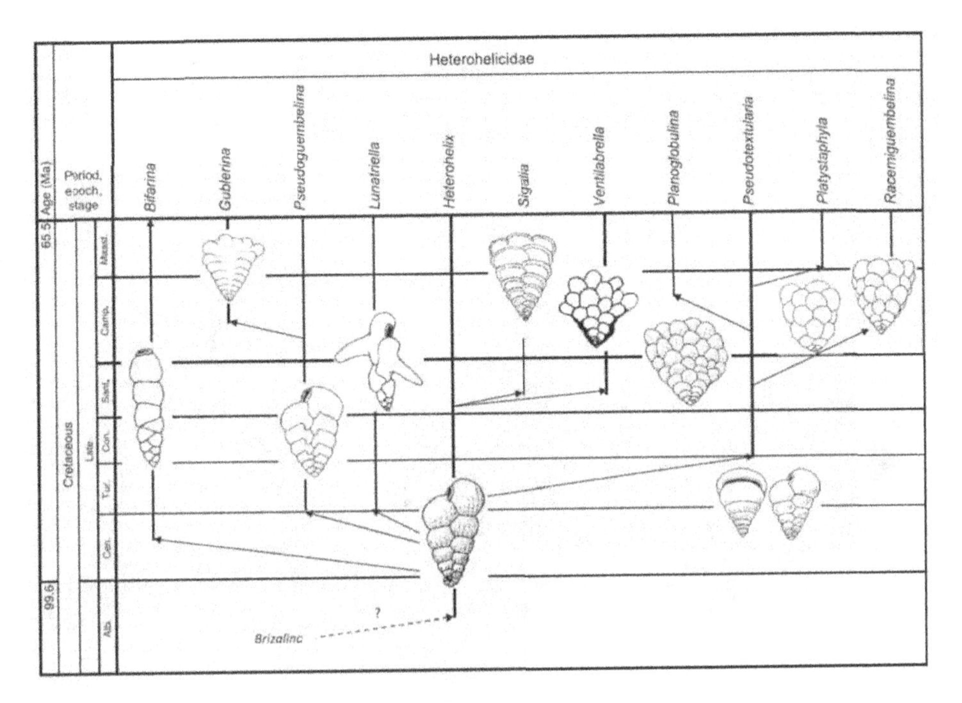

Figure 4.14. Phylogenetic evolution of the Cretaceous Heterohelicidae genera.

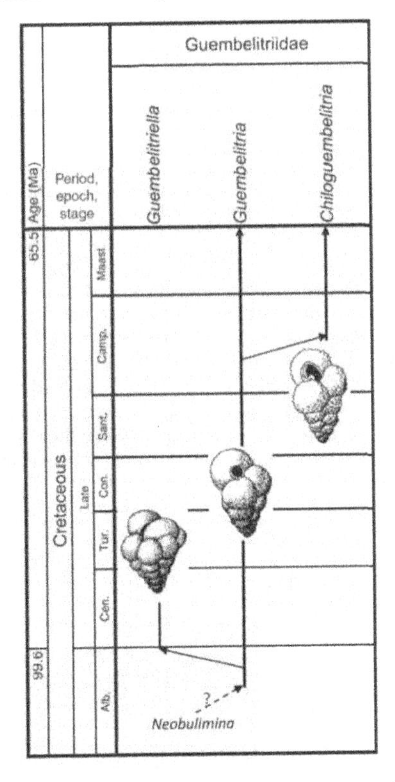

Figure 4.15. Phylogenetic evolution of the Cretaceous Guembelitriidae genera.

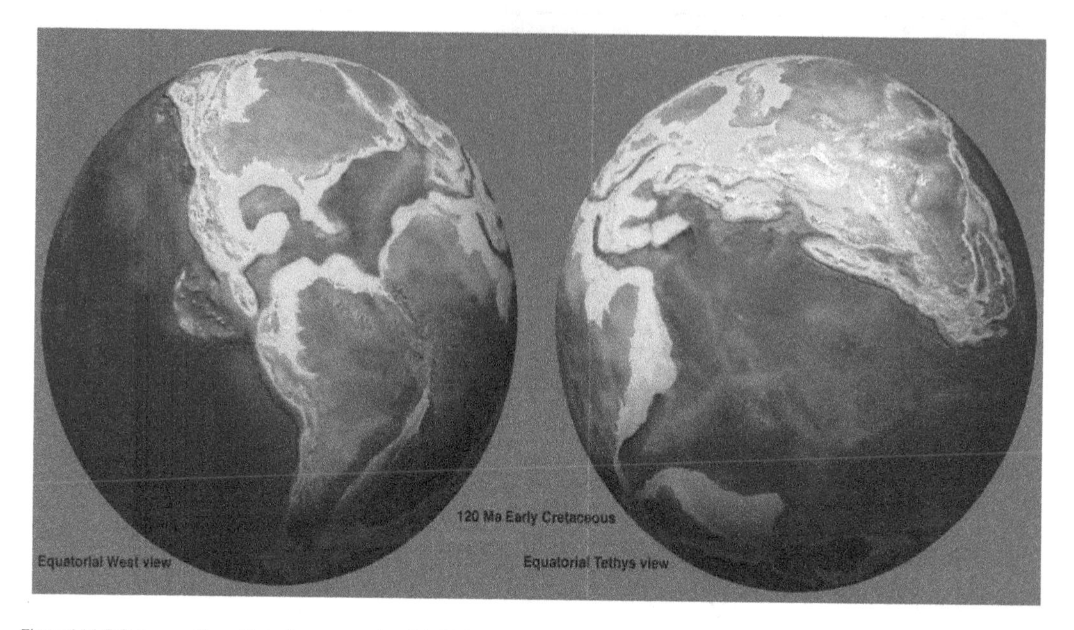

Figure 4.16. Paleogeographic and tectonic reconstruction of the Early Cretaceous (by R. Blakey http://jan.ucc.nau.edu/~rcb7/paleogeographic.html).

The triserial guembelitriids may have evolved from the Cretaceous benthic form *Neobulimina*, which has subglobular triserially arranged chambers (Fig. 4.15). The adaptation from benthic to planktonic domains occurred in the Late Albian at the same time as for the heterohelicids. As a planktonic mode of life evolved, *Guembelitria* (Plate 4.19, Figs. 1–3) developed perforation cones that turned into muricae in the Campanian to Maastrichtian, *Chiloguembelitria* (Plate 4.19, Figs. 37 and 38). They only developed multiseriality in the Early Cenomanian with the short-ranging *Guembelitriella* (Plate 4.17, Figs. 30 and 31). The small-sized disaster/opportunists triserial guembelitriids had a distribution comparable with that of the heterohelicids, but interestingly unlike the latter, most of them survived the end Cretaceous extinction event.

4.4 Paleogeography and paleoecology of the Cretaceous planktonic foraminifera

The Early Cretaceous saw the breakup of the super-continent Pangaea, which had begun during the Jurassic. Many of the continental masses were covered by shallow or inland seas. Much of Europe, Asia, North Africa, and North America were a series of islands. The Cretaceous saw the lengthening and widening of the proto-North Atlantic Ocean, which began to spread further southward, while the Alps began to form in Europe. India broke free of Gondwana and became an island continent. Africa and South America began to split apart, Africa moving north and closing the gap that was the Tethys. The continents began to take on their modern forms (see Fig. 4.16).

These tectonic developments contributed to the first geographical dispersion of the Globigerinoidea in the form of the Praehedbergellidae and Schackoinidae, which typify the holoplanktonic foraminiferal fauna of the Early Cretaceous. A major turnover of the Globigerinoidea during the Cretaceous contributed to the deposition of extensive chalk on the deeper shelves and epeiric sea and the open ocean in Europe, in the Middle East, Australia, and parts of North America during this period.

Continental rifting and regional volcanic activity continued throughout the early Berriasian and into the Hauterivian. This may have resulted in higher CO_2 concentrations in the atmosphere and a warming of the climate from the middle of the Berriasian, which triggered an increase in the level of oxygen that is inferred (see Ward, 2006) to steadily increase throughout this period. As rifting continued, three biogeographical provinces were established in the Cretaceous world, the northern and southern boreal (temperate) and Tethyan (tropical) provinces (Fig. 4.17).

During the Berriasian, the Praehedbergellidae had yet to evolve, but the favusellids were present (see Chart 4.1 online). Nearly all of them were from the Jurrasic *Conoglobigerina* stock (75%) with the recently evolved Cretaceous *Favusella* making its first appearance, but still as a minority form (~25%). *Conoglobigerina* species, restricted to the temperate climate zone of Eastern Europe (see Fig. 4.17), disappeared in the Early Valanginian, whereas *Favusella* persisted into the Cenomanian, thriving in the relatively shallow temperate, subtropical, and tropical waters. Koutsoukos *et al.* (1989) suggested that all the various *Favusella* taxa were ecophenotypic expressions of a single stock suited to shallow, warm, hypersaline, and carbonate environments. However, this is unlikely to be the case as the favusellid taxa have different stratigraphic ranges and different regional distributions. *Ascoliella*, for example, is not known in Tethys even though the stratigraphically latest species of *Favusella* is geographically widespread.

At the end of the Berriasian, there were a few specific level extinctions, but although the number of cosmopolitan species was not great, they all survived the Berriasian to Valanginian transition (see Fig. 4.18). The Early Valanginian saw a final shift from the relatively cool climate of the end of the Jurassic and earlier Berriasian to the "greenhouse" world that continued for the rest of the Cretaceous. This change coincided with the disappearance of the last of the conoglobigerinids. In the Late Valanginian, small globular *Gorbachikella* species, the ancestor stock of the Praehedbergellidae, made their first appearance in North Africa (making up 74% of the whole Valanginian component, see Fig. 4.2) and migrated to Eastern Europe and Central America (including the Caribbean and Mexico) from the Late Valanginian to the Early Aptian. These globular forms are considered to be indicators of warm sea-surface temperatures (Coccioni *et al.*, 1998).

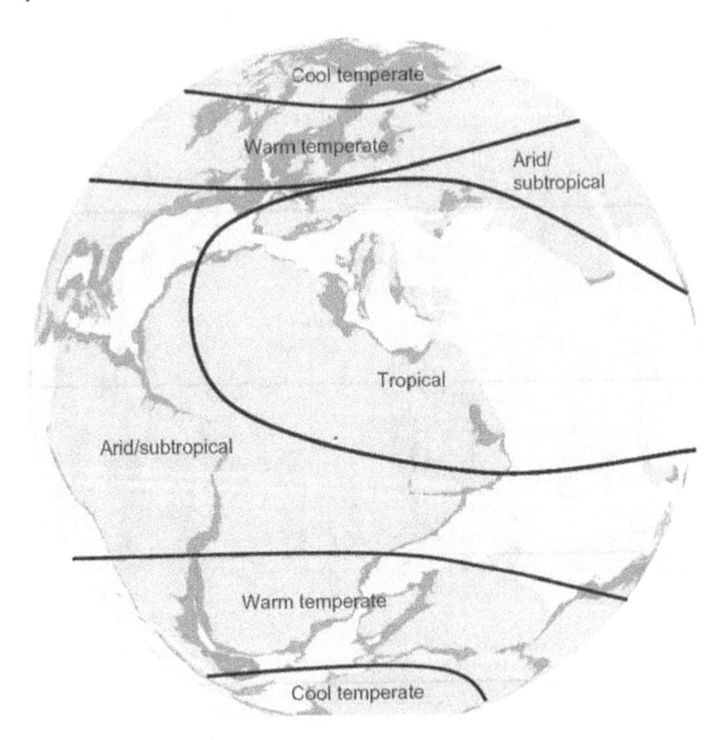

Figure 4.17. Climate zones in the Early Cretaceous (140 Ma).

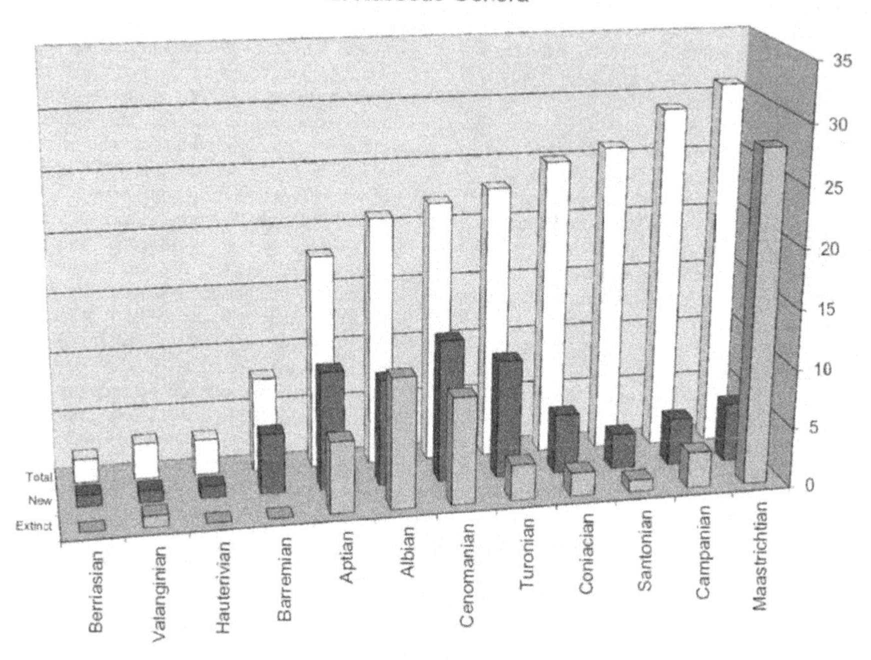

Figure 4.18. The total number of genera, extinctions, and new appearances of planktonic foraminifera in each stage of the Cretaceous. The extinctions coincide with the end of each stage and the appearances with the beginning of the stage.

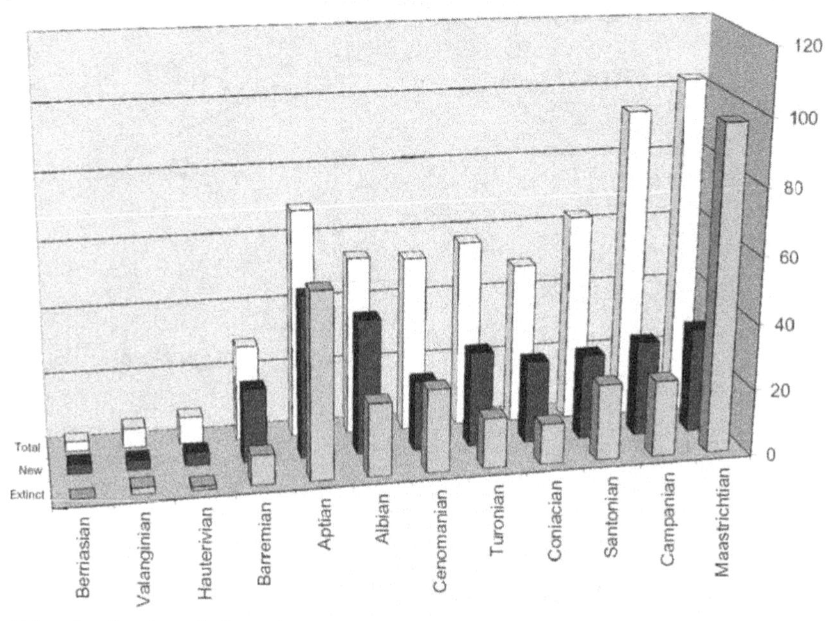

Figure 4.19. The total number of species, extinctions, and new appearances of planktonic foraminifera in each stage of the Cretaceous. The extinctions coincide with the end of each stage and the appearances with the beginning of the stage.

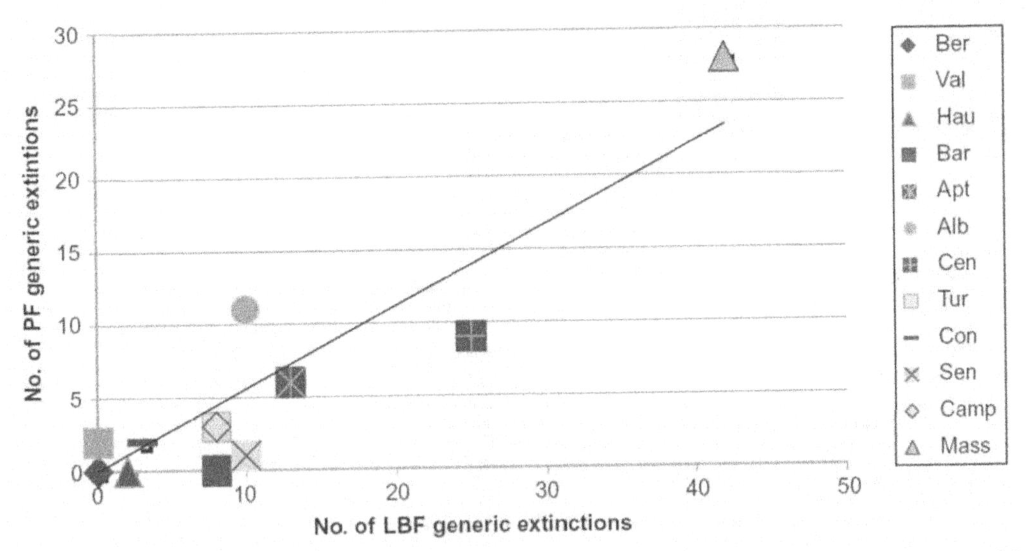

Figure 4.20. The total number of extinctions of planktonic foraminifera compared to those of the larger benthic foraminifera in each stage of the Cretaceous.

Although the end of the Valanginian and Early Hauterivian saw considerable volcanic activity, such as the development of the Paraná traps in South America, together with their smaller rifted counterpart of the Etendeka traps in Namibia and Angola (Courtillot and Renne, 2003), the end Valanginian did not witness any major extinctions of planktonic forms. Similarly, the shallow water and reef building larger benthic foraminifera were also largely unaffected at this stage (see BouDagher-Fadel, 2008). This is in contrast to the statement made by Courtillot and Renne (2003) that the end of the Valanginian is a prominent extinction horizon (although there may have been a major crisis for bryozoan faunas). The Valanginian may have also coincided with a cooling event, as noted by Walter (1989) that was followed by a warming event around (133–132 Ma) in the Early Hauterivian. The latter, it would appear was an important phase in the development and distribution of the praehedbergellids, the ancestor stocks of the Middle and Late Cretaceous globigerinids. *Praehedbergella* made its first appearance in the Early Hauterivian but unlike *Gorbachikella* had a wide geographic distribution. It ranged in age from the Early Hauterivian to the Late Aptian, evolving gradually into many forms that adapted to the changing environment of that time. Its species are recorded from localities as far as northern Mexico, the North Sea area, and North Yorkshire of England, to as far south as the Walvis Ridge in the South Atlantic (see BouDagher-Fadel *et al.*, 1997).

A short-lived anoxic event in Western Tethys during the Late Hauterivian interval (~127.5 Ma) called the "Faraoni Event" (Baudin, 2005) may have caused a small number of extinctions of planktonic foraminifera (12% of species but no genera, see Figs. 4.18 and 4.19). This correlates well with the relative number of extinctions of the larger benthic foraminifera also seen at this time (see Fig. 4.20 and BouDagher-Fadel, 2008). This event coincides with the spreading and diversification of small globular opportunistic *Gorbachikella*. A sea-level rise in the Early Hauterivian is rapidly followed by a regressive trend and a sequence boundary in the final Hauterivian (Ha7 of Hardenbol *et al.*, 1998).

At the beginning of the Barremian, foraminifera flourished and many new species of planktonic foraminifera appeared (76% of the planktonic foraminifera in the Barremian were new, see Fig. 4.19). This diversification in the planktonic domain was accompanied by a similar one in the reefal realm (BouDagher-Fadel, 2008). The small Praehedbergellidae, with subglobular chambers such as *Gorbachikella*, *Praehedbergella*, *Blefuscuiana*, are likely to have inhabited both the most superficial waters and those which (although below the depths of sea/air mixing due to wave action) diurnally gained dissolved oxygen from algal photosynthesis (the greatest depth to which fossil calcareous algae were believed to exist is 85 m; BouDagher-Fadel *et al.*, 1997). Some of these Early Cretaceous species of the praehedbergellids developed perforation cones on the surface of their tests that enabled them to disaggregate their food particles prior to ingestion. This was achieved by the extrathalamous cyto-plasm, as it deposited a new surface of lamella of calcite over the earlier test at each new chamber formation. The penultimate chamber adds one extra lamella of calcite so that earlier chambers have more and more lamellae, thickening the test and strengthening the cones. Of necessity, therefore, this leads to the formation of the most prominent cones near the aperture. The only other analogues to perforation cones are found in the heterohelicids (see below), but there are none in any Holocene globigerinids.

The Late Barremian was a time for major oceanic anoxic events (OAEs) and also of global sea-level rise (Courtillot and Renne, 2003). The anoxic events could have been the results of, or triggered by, the Ontong-Java submarine flood basalt eruption, which gave rise to the largest of all oceanic basaltic plateaus that in places reaches a thickness of 40 km (Courtillot and Renne, 2003).

However, many planktonic genera were unaffected by these events, and similarly, there were only a relatively modest number of extinction of reef forming larger benthic foraminifera at that time (see Fig. 4.20 and BouDagher-Fadel, 2008). Of the 30% of the planktonic species that did become extinct, the majority were species of the subspherical *Blefuscuiana*. The survivors were species of *Blowiella*, *Lilliputianella*, and *Leupoldina* that had a competitive shape, suitable for adaptation in increasingly dysoxic seawater. The planispiral genus *Blowiella* occupied the same niches as *Blefuscuiana* but was more tolerant toward oligotrophication (Premoli Silva and Sliter, 1999). The broad radial and flattened elongated chambers of *Lilliputianella* and *Leupoldina*, with a reduced chamber volume and relatively larger proportion of extrathalamous cytoplasm, are thought to have also enhanced their ability to survive in waters lower in oxygen. The praehedbergellid *Lilliputianella* is known from Central America, northern and southern Europe, and North Africa, but the occurrences of the schackoinid *Leupoldina* are sporadic and rare in the fossil record. They were found in Trinidad, Crimea, and Tunisia.

The Aptian was an eventful, long stage, with a rapid rate of oceanic spreading in the Atlantic. The Atlantic Ocean opened wide enough to allow significant mixing of waters across the equator that seemed to have been associated with a series of OAEs, which continued periodically for 35 Ma, into the Santonian. The Kerguelen plateau (see BouDagher-Fadel, 2008), the second largest oceanic plateau after the Ontong-Java plateau, erupted in the southern Indian Ocean in the Late Aptian (around 118 Ma), which is also the time of eruption of the Rajmahal traps in eastern India. The Kerguelen plateau, Rajmahal traps, and eruptions in western Australia all occurred at a similar time and were roughly coeval with the breakup of eastern Gondwana (Courtillot and Renne, 2003). During these eruptions, the sea floors rose, forcing the sea levels to rise worldwide, flooding up to 40% of the continents. Sea levels began to rise to be over 200 m higher than present-day levels. Such a high rate of volcanic activity released massive amounts of carbon dioxide into the ocean and ultimately the atmosphere. This quantity of carbon dioxide would have made the oceans relatively low in oxygen. The evidence for this is displayed by the abundance of black shale and petroleum-rich formations from this stage. Black shales, signaling deep ocean anoxia, have been found in both the Central Tethys (Europe) and the Western Pacific provinces. The dramatic rises in temperature recorded in the Early Aptian to Middle Aptian (Jenkyns, 2003; Jenkyns and Wilson, 1990) coincided with a significant turnover of planktonic foraminifera worldwide, the highest in the Cretaceous period. This is also mirrored in the increase of diversity of the shallow-water larger benthic foraminifera throughout the Aptian in Tethys (BouDagher-Fadel, 2008).

In the Late Aptian, the Schackoinidae evolved near-trapezoidal chambers with multiple spines, *Schackoina*. These again increased the surface area of the chambers and therefore the potential proportion of extrathalamous cytoplansm, enhancing their ability to survive in dysoxic seawater. The elongate chambers in the schackoinids (*Schackoina*, *Leupoldina*) and praehedbergellids (*Lilliputianella*, *Claviblowiella*, *Leupoldina*, etc.) would have also allowed them to occupy both the shallow, fully oxygenated waters and deeper oxygen-depleted waters, below the level of wave induced air mixing. On the other hand, some of the small microperforate, smooth species of *Blefuscuiana* that lived in the surface waters began developing large perforations and muricae that gradually concentrated near the aperture and then along the sutures. The developments of large, macroperforate and muricate forms allowed the hedbergellids to colonize deeper, subsurface waters as these morphological characters were advantageous in gathering and disaggregating the larger, heavier food particulates which had sunk from the upper waters and represented an evolutionary step that was important in the development of many phylogenetic lineages of hedbergellids from the Middle to Late Cretaceous.

The Late Aptian saw a significant number of planktonic foraminifera extinctions (63% of total species, see Fig. 4.19), but these were on the whole compensated for by the establishment of large number of new genera at the Aptian–Albian boundary (38% of the total species and 47% of the total genera of the Albian, see Fig. 4.19). These extinctions coincided with a short-lived but rapid sea-level fall at the Aptian–Albian boundary, which also affected the larger benthic foraminifera causing the widespread collapse of reef ecosystems (BouDagher-Fadel, 2008; Walliser, 1996).

After developing muricae in the Late Aptian, the hedbergellids and planomalinids evolved further in the Albian as the muricae fused and formed muricocarinae/keels that were even more advantageous for life in deeper waters. It is known that the spines of living globigerinids are covered by extrathalamous extensions of the cytoplasm, capable of not only carrying vacuoles and transporting symbionts (Hemleben and Splinder, 1983), but also performing the feeding and exchange function of expodia when the latter are lost (Adshead, 1980). The carinae of some predatory, biconvex benthic Rotaliina (e.g., *Amphistegina*) have been observed to form the foundation of the "takeoff" point of food gathering pseudopodia (BouDagher-Fadel, 2008; BouDagher-Fadel *et al.*, 1997). For this reason, it is possible that the peripheral muricae, developed between the Albian and the Late Cretaceous, of the descendants of the Hedbergellidae functioned similarly, namely they formed a peripheral rim from which nutrient-gathering pseudopodia could extend in a disc-like fan. Therefore, species that could extrude pseudopodia in such a way would have had a considerable advantage in their collection of available suspended particulate nutrients and would be able to thrive in deeper waters. The repeated competitive success of such taxa with evolved keels led to them supplanting the praehedbergellids and schackoinids from their dominant position.

During the Late Albian, an anoxic event (the "Breistroffer event") developed across Eastern and

Western Tethys (at 100 Ma). This event was associated with a rise in sea level and CO_2-induced global warming (Leckie *et al.*, 2002) and may have been correlated with the Hess Rise volcanic event in the North Pacific Ocean (Eldholm and Coffin, 2000). During this time, hedbergellids developed in addition to the muricae and peripheral keel, umbilical cover plates that partially close the umbilical digestive area (e.g., *Ticinella*, *Rotalipora*), which seems to be advantageous for extrathalamous digestion of disaggregated particles.

In addition, the development of anoxic ocean bottom conditions seems to have triggered a change on some infaunal benthic, biserial and triserial foraminifera, as they evolved their bottom dwelling life into a planktonic one. The resulting heterohelicids seem therefore to have evolved their planktonic life strategy independently from the preexisting globigerinids. The Late Albian to Cenomanian heterohelicids are rare in sediments representing open marine conditions but are found abundantly in sediments from marginal environments (Nederbragt, 1990).

During the Late Albian, the ocean temperatures may have reached up to 32°C (Ando *et al.*, 2010), but this was followed by a cooling period in the Early Cenomanian (Petrizzo *et al.*, 2008; Wilson and Norris, 2001). Across the Albian–Cenomanian boundary, the cooling event induced a reduction in the upper-ocean stratification (Ando *et al.*, 2010). Forty-three percentage of the Albian planktonic foraminifera species became extinct at this boundary (representing 16% of the Albian genera, see Fig. 4.19). The Cenomanian is marked by a short-term turnover in planktonic foraminifera, with 21 new species belonging to 7 genera appearing in middle and low latitudes. The foraminifera which survived the Albian–Cenomanian boundary and thrived in the relatively deeper dysoxic waters of the Cenomanian had the selective advantage of strong morphological features, such as peripheral keels and larger umbilical plates covering the umbilical area, which helped in digesting larger particles. The development of the deeper water anoxia favoured the growth of populations with a reduced keel, such as *Anaticinella*, which, having no keel, could colonize near surface waters. Small opportunistic triserial guembelitriids made their first appearance in the Late Albian, only apparently to disappear in the Early Cenomanian but reappear in the Coniacian. It is not clear, however, whether these Coniacian occurrences are "Lazarus forms" or whether they represent an independent example of the evolution of the planktonic life strategy from another Coniacian benthic form.

The Cenomanian–Turonian boundary coincided with the onset of the major anoxic event of the Cretaceous, referred to in Europe as the "Bonarelli Event." It is marked by the collapse of paleotropical reef ecosystems (Walliser, 1996) and strongly affected the reefal building larger benthic foraminifera (see Fig. 4.19 and BouDagher-Fadel, 2008). Johnson *et al.* (1996) proposed that the collapse of Middle Cretaceous reef ecosystems may be attributed to the collapse of the Tethyan ocean–climate system. All of the rotaliporids became extinct, and indeed, all of the Early Cretaceous favuselloids failed to survive the end of the Cenomanian. These extinctions globally left many empty planktonic and benthic niches. Seventy-five percentage of the surviving taxa of planktonic foraminifera were globular, surface water-dwelling hedbergellids that had tests which were already adapted to low-oxygen conditions, having as they did enlarged perforations (e.g., *Hedbergella*, *Whiteinella*) and/or elongate chambers (e.g., *Asterohedbergella*). For the rare flattened *Praeglobotruncana* and *Dicarinella*, which migrated over considerable vertical distances within the upper water column, survival is likely to have been due to their selective adaptation of strong keels that enabled the capture and disaggregation of the large particles found at these depths. Among the other survivors of the Cenomanian–Turonian boundary anoxic events were the biserial *Heterohelix*, an oxygen-minimum zone dweller, and the triserial *Guembelitria* (Plate 4.19, Figs. 1–3), a eutrophic surface dweller. Bambach (2006) considered this peak of extinction to be due to the major OAE, but these events might also have been affected by another major submarine volcanic eruption (the Wallaby eruption) in the Indian Ocean (Eldholm and Coffin, 2000), which again would have been associated with the high emission of CO_2 contributing to the global warming peak and greenhouse climates during that period.

In the Turonian, few new planktonic foraminifera forms appeared (50% of the species representing 43% of the Turonian genera, see Figs. 4.18 and 4.19). However, toward the end of the Turonian, 26% of the species became extinct and many of them that had survived the Albian–Cenomanian crisis disappeared. The restriction of reef ecosystems might have contributed to larger benthic foraminifera being slightly more affected (eight larger benthic foraminifera genera went extinct compared to three planktonic foraminifera genera; see Fig. 4.20 and BouDagher-Fadel, 2008). Kerr (2006) has proposed that volcanism-related CO_2 increases at this time led to an enhanced greenhouse climate and eventually to the extinction of the most vulnerable reefal communities of larger foraminifera and rudists. Courtillot and Renne (2003) stated that the volcanism, which occurred between 91 and 88 Ma in a number of short, discrete events, may be the cause of the Late Turonian extinction events. The Caribbean–Colombian Cretaceous Igneous province, the Madagascar event (see BouDagher-Fadel, 2008), and Phase 2 of the Ontong-Java event all occurred within this short interval and would have certainly contributed to environmental stress that could have triggered the extinctions at the end of the Turonian. Recent studies have revealed an oxygen isotope anomaly at 91.3 Ma (Bornemann *et al.*, 2008), which has been interpreted as being indicative of a glaciation event in the middle of the Turonian, despite it being one of the warmest periods of the Mesozoic. If this event occurred, it may have been triggered by short-term atmospheric changes associated by flood basalt events in the Caribbean and Madagascar regions.

The paleoecology of Middle Cretaceous planktonic foraminifera (~95–86 Ma) remains controversial, as much of the tropical marine record is preserved as chalk and limestone, which may have experience geochemical overprinting. In contrast to models that associate ontogenetic changes with depth (e.g., Caron and

Homewood, 1983; Leary and Hart, 1988), mid-Cretaceous isotopic data show no evidence for depth migration of keeled foraminifera. Rather, they suggest (Bornemann and Norris, 2007; Norris and Wilson, 1998) that keeled species, such as *Marginotruncana*, grew primarily in surface waters along with globular, non-keeled species, such as *Whiteinella*, and did not change their depth habitat substantially during their life cycle. It is not until the Campanian or Maastrichtian that $\delta^{18}O$ and $\delta^{13}C$ size-related trends are observed in keeled foraminifera. However, isotopic data might not be very reliable in the Cretaceous, and whole assemblages should be studied in order to assess paleoenvironment of material from the Middle Cretaceous.

In the Late Cretaceous, the Atlantic Ocean was expanding as the Americas were gradually moving westward (Fig. 4.21). This contributed to a major shift in the distribution of tropical and subtropical planktonic foraminifera. New keeled groups progressed to fill the empty niches left by the two successive extinction events at the end of the Cenomanian and end of the Turonian. At the Coniacian boundary, new planktonic foraminifera appeared (see Figs. 4.18 and 4.19), and about 51% of the Coniacian species of planktonic foraminifera (five genera) were new, and only 24% of the species (two genera) became extinct toward the end of this stage. The foraminifera turnover in the Coniacian takes place according to Walliser (1996) during a global flooding interval and the termination of the second regional oxygen-depletion event, which is recognized in many places by organic-rich dark shales. Although the percentage of the new genera was only 12%, the Santonian saw the appearance of a significant number of new species (41%); however by the end of the Santonian, a high percentage of foraminifera species again became extinct (38%), but they represented only three genera, mainly in the tropical–subtropical realm. This may again have been triggered by renewal of the Kerguelen, Broken Ridge volcanic event that generated over 9 million km^3 of eruptive material (Eldholm and Coffin, 2000).

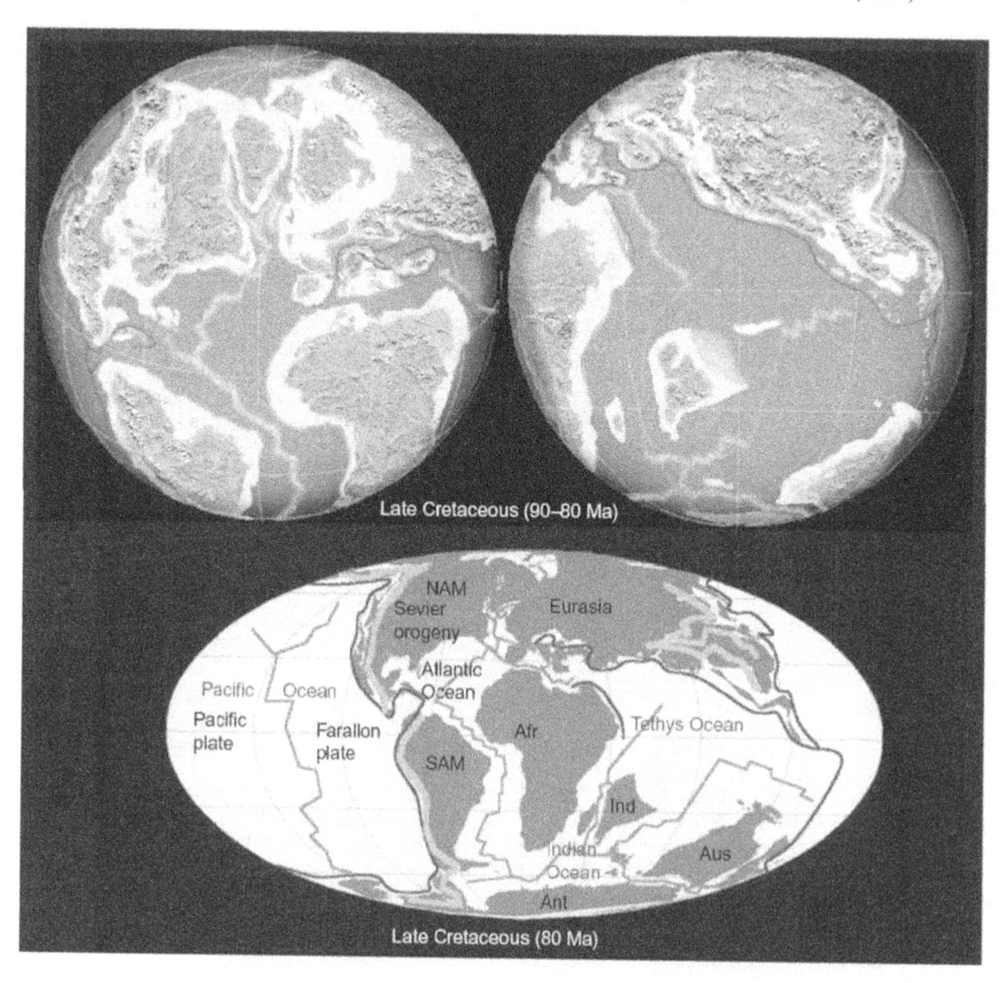

Figure 4.21. Paleogeographic and tectonic reconstruction of the Late Cretaceous (by R. Blakey http://jan.ucc.nau.edu/~rcb7/ paleogeographic.html).

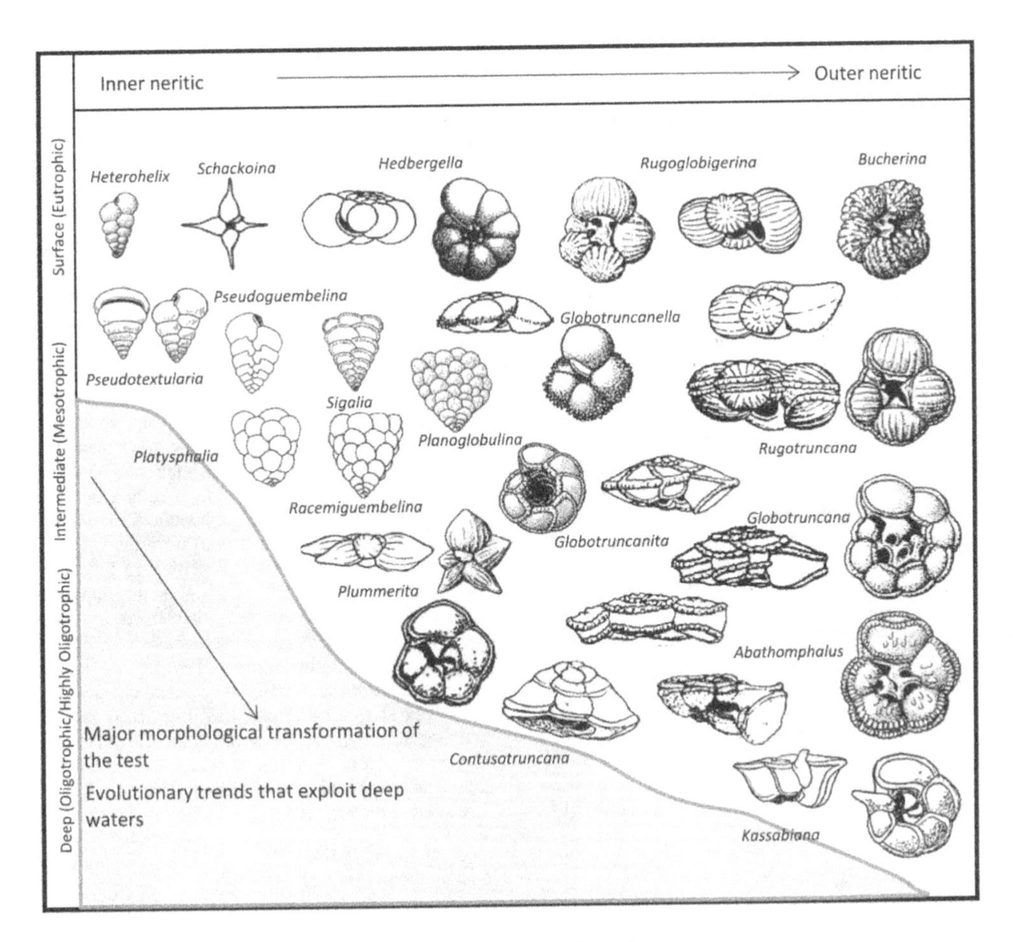

Figure 4.22. The distribution of the Late Cretaceous planktonic foraminiferal in the neritic environment.

In the Santonian, assemblages were represented mainly by earlier Coniacian forms as only 31% of the planktonic foraminifera species were new and represented by just four genera (Figs. 4.18 and 4.19). Among these genera were *Rugoglobigerina*, *Globotruncanita*, and *Ventilabrella*, which became prominent in the latest Cretaceous assemblages, while *Sigalia* straddled the Santonian–Campanian boundary. The multiserial growth which was adopted by the heterohelicids *Ventilabrella* and *Sigalia* allowed them to colonize deeper waters (see Fig. 4.22) and niches which were previously occupied by keeled globotruncanelliids (e.g., *Dicarinella*). Many of the new species (88%) were species of biserial heterohelicids, which were mainly distributed in temperate and subtropical waters.

The Late Cretaceous was still a warm period, with tropical sea-surface temperatures at least 28–32°C (Pearson *et al.*, 2001), and as a result of high sea levels, isolated land masses were surrounded by inland seas (see BouDagher-Fadel, 2008). The Campanian saw a rapid turnover of new species (30%, see Fig. 4.18), many of which colonized deeper waters than their ancestors, the praehedbergellids and the hedbergellids. Of these new appearances, 60% belonged to the heterohelicids, which by now had spread to tropical waters. The new heterohelicids exhibited multiserial growth, which resulted in their tests sinking into deeper waters than the simple biserial forms (see Fig. 4.22). Appearances in the tropical waters might have been the result of the filling of empty niches previously occupied by the keeled globotruncanellids (e.g., *Concavatotruncana*) and some globotruncaniids (e.g., *Marginotruncana*)which had died out in the Middle Cretaceous. Globular hedbergellids lived in the near-surface waters, while mixed assemblages of globular hedbergellids and rugoglobigerinids and keeled globotruncanids are indicative of near surface or inner neritic environments. However, assemblages dominated by keeled foraminifera are suggestive of deeper outer neritic environments (Fig. 4.22).

Toward the end of the Campanian, 25% of the planktonic foraminifera became extinct (Fig. 4.19). This might have been related by two volcanic events in the Atlantic Ocean, the Sierra Leone Rise, and the Maud Rise (Eldholm and Coffin, 2000). These were again accompanied by a regional oxygen-reduction event (Walliser, 1996) occurring during the Campanian, and the beginning of global sea level falls.

The beginning of the Maastrichtian was a period of high turnover, and 31% of the Maastrichtian planktonic foraminifera had their first appearance at the Campanian–Maastrichtian boundary (see Figs. 4.18 and 4.19). The planktonic foraminifera in the Maastrichtian were highly developed with morphological features such as double peripheral keels (e.g., *Globotruncana*), multiserial growth (e.g., *Racemiguembelina*), and spines (e.g., *Trinitella*) that allowed them to sink into deeper waters than earlier forms and so exploit different niches (Fig. 4.22). Diversity and size of the foraminifera increased during the Maastrichtian pointing to warm temperatures and stable environments.

These cosmopolitan assemblages of the Maastrichtian were brought to an end by one of the greatest mass extinctions of all time, the "K-T" (or now more correctly the K-P) event, or terminal Mesozoic extinction. The K-P extinction has been the subject of many studies. Some have argued that it really began before the end of the Cretaceous, between 67.5 and 68 Ma, with the abrupt extinction of rudist bivalve-dominated reef ecosystems (Johnson and Kauffman, 1990), with radiometric dates putting this extinction around 68–68.5 (Walliser, 2003). Others associate the K-P extinction with a single or multiple impacts at ~65.5 Ma (Alegret *et al.*, 2005; Keller *et al.*, 2003; Kuroda *et al.*, 2007; Schulte *et al.*, 2010; Stüben *et al.*, 2005). There is certainly evidence supporting multiple impact events at the end Maastrichtian, as in addition to the well-documented Chixulub Crater, an impact crater of 24 km diameter also occurred in Ukraine (the Boltysh Crater) at around 65.17 Ma (Kelley and Gurov, 2002). Also at about the same time as these impact events, there was a major volcanic event that formed the Deccan Traps. Courtillot and Renne (2003) pointed to the Ir-bearing layer related to the Chixculub impact occurring within the traps, which indicate that Deccan volcanism began prior to this impact and straddled it in time.

These differing hypotheses have recently been critically reviewed by Schulte *et al.* (2010). They found that a globally distributed single ejecta-rich deposit, which is compositionally linked to the Chicxulub impact, can be found at the K-P boundary around the world. They conclude that the temporal match between the ejecta layer and the onset of the extinctions suggests that the Chicxulub impact was indeed the trigger for the global mass extinction. As a result of this event, about 90% of the planktonic foraminifera became extinct (compared to about 83% of the Maastrichtian larger benthic foraminifera that died out; see Fig. 4.20 and BouDagher-Fadel, 2008). As with the larger benthic foraminifera, the surviving species (5% of the total) were the smallest and most robust of the planktonic foraminifera forms. About half of the survivors were small, globular, opportunistic shallow-dwelling hedbergellids that lived in eutrophic conditions (e.g., *Hedbergella monmouthensis*, Plate 4.16, Figs. 22–24), and the other half were the small-sized disaster/opportunists triserial guembelitriids (e.g., *Guembelitria cretacea*, Plate 4.19, Figs. 1–3) and biserial to uniserial forms (e.g., *Bifarina*, Plate 4.18, Figs. 27–31). These forms went on to provide the stock from which the Cenozoic planktonic foraminifera derived, as will be discussed in the next chapter.

Plates 4.1–4.24

Plate 4.1. Figures 1–13: *Favusella hoterivica* (Subbotina). (1, 2) Paratype, Hauterivian, northern Caucasus, Crimea, x200. (3, 4) Holotype, x211. (6, 7) (= *Globigerina tardita* Antonova), topotype, Late Barremian–Early Aptian, along the Tusheps River, northwestern Caucasus, (6) x200, (7) x263. (8) (= *Globigerina tardita* Antonova), holotype, figured by Antonova (1964), x174. (9, 10) (= *Globigerina quadricamerata* Antonova), topotype, Early Aptian, along the Tusheps River, northwestern Caucasus, x177. (11–13) Figured by Wernli *et al.* (1995), from offshore wells, Eastern Canada, (11) Early Valanginian, x196, (12, 13) Berriasian, x160. Figure 14: *Favusella stiftia* Rösler, Lutze and Pflaumann. Holotype, Late Hauterivian, DSDP Site 397, eastern North Atlantic, x280. Figure 15: *Favusella orbiculata* Michael (*nomen dubium*). Holotype, a deformed specimen of *Favusella*, Late Albian, Weno Formation, Washita Group, Texas, USA, x143. Figures 16–20: *Favusella washitensis* (Carsey). (16) (= *Favusella confusa* Longoria and Gamper, 1977). Holotype, Late Albian, northern Mexico (magnification not given). (17–19) Neotype, Early Cenomanian, Street Bridge, Austin, Texas, USA, x68. (20) Early Cenomanian, Paw Paw Formation, Washita Group, Grayson County, Texas, USA, NHM P52139, x77. All images from BouDagher-Fadel *et al.* (1997).

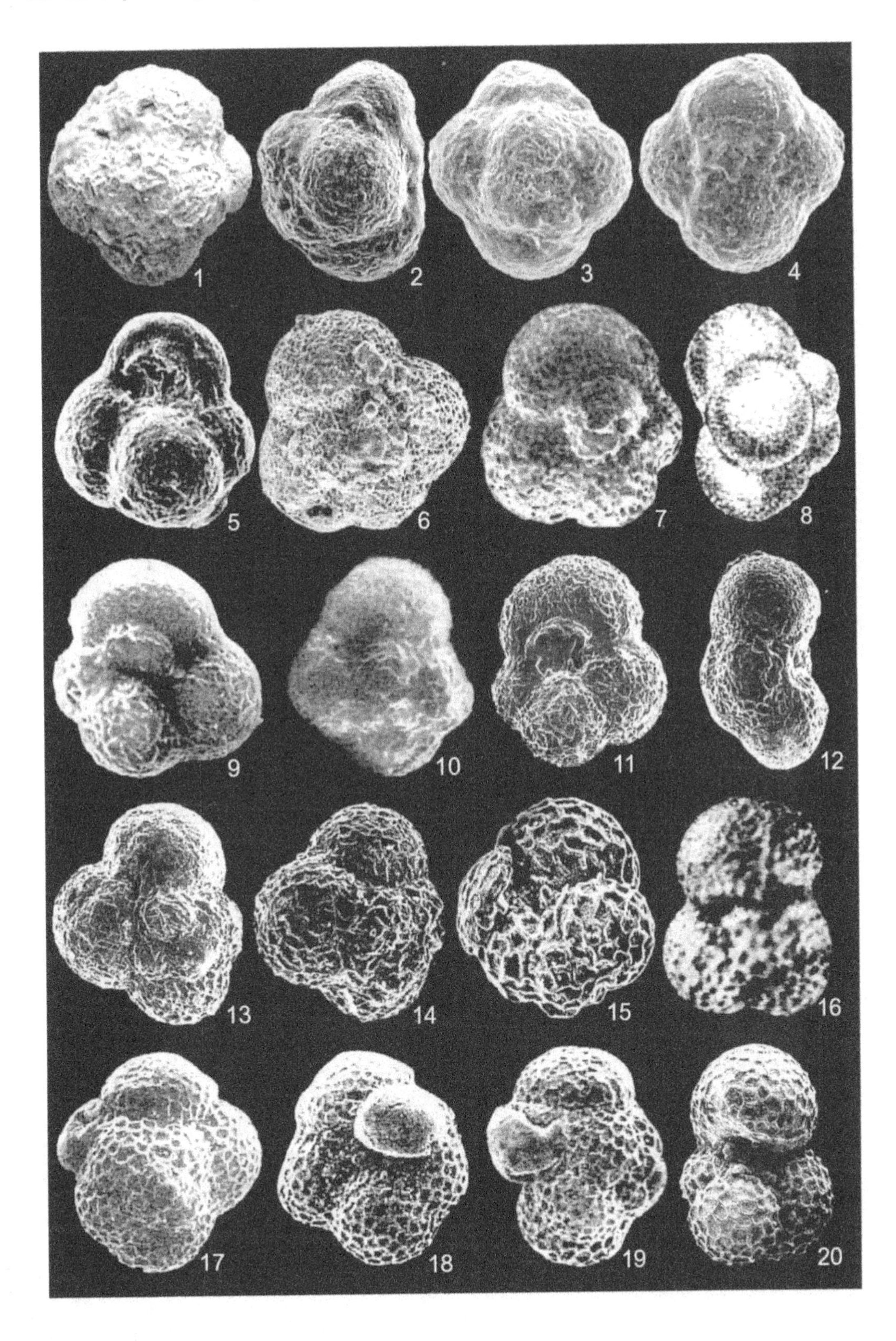
1
2
3
4
5
6
7
8
9
10
11
12
13
14
15
16
17
18
19
20

Plate 4.2. Figure 1: *Favusella* sp. Cenomanian, Libya, UCL A145 5420–5430, x80. Figures 2–10, 18: *Ascoliella nitida* (Michael). (2–4) Holotype, Late Albian, Duck Creek Formation, Washita Group, Grayson County, Texas, USA, USNM172793, x90. (5–7) (= *A. scotiensis* Banner and Desai, 1988), Aptian, offshore Canada, deposited in the GSC 53770, x100. (8–10) Holotype (= *Favusella pessagnoi* Michael), Late Albian, Duck Creek Formation, Washita Group, Grayson County, Texas, USA, USNM172797, x100. (18) (= *Favusella hedbergellaeformis* Longoria and Gamper), holotype, Late Albian, Lower Cuesta del Cura Formation, Mexico, USNM207239, magnification not given. Figures 11–13: *Ascoliella quadrata* (Michael). Holotype, Late Albian, Duck Creek Formation, Washita Group, Grayson County, Texas, USA, USNM172799, x100. Figures 14–17: *Ascoliella scitula* (Michael). (14–16) Holotype, Late Albian, Duck Creek Formation, Washita Group, Grayson County, Texas, USA, USNM172801, x115. (17) (= *Favusella papagayosensis* Longoria and Gamper), paratype, Albian, Tamaulipas Formation, Mexico, magnification not given. Figure 19: *Favusella hiltermanni* (Loeblich and Tappan). Holotype, Early Cenomanian, Hamburg, Germany, x100. Figures 20 and 21: *Ascoliella voloshinae* (Longoria and Gamper). Holotype, Late Albian, northeastern Mexico, USNM207237, magnification not given. All images from BouDagher-Fadel (1997).

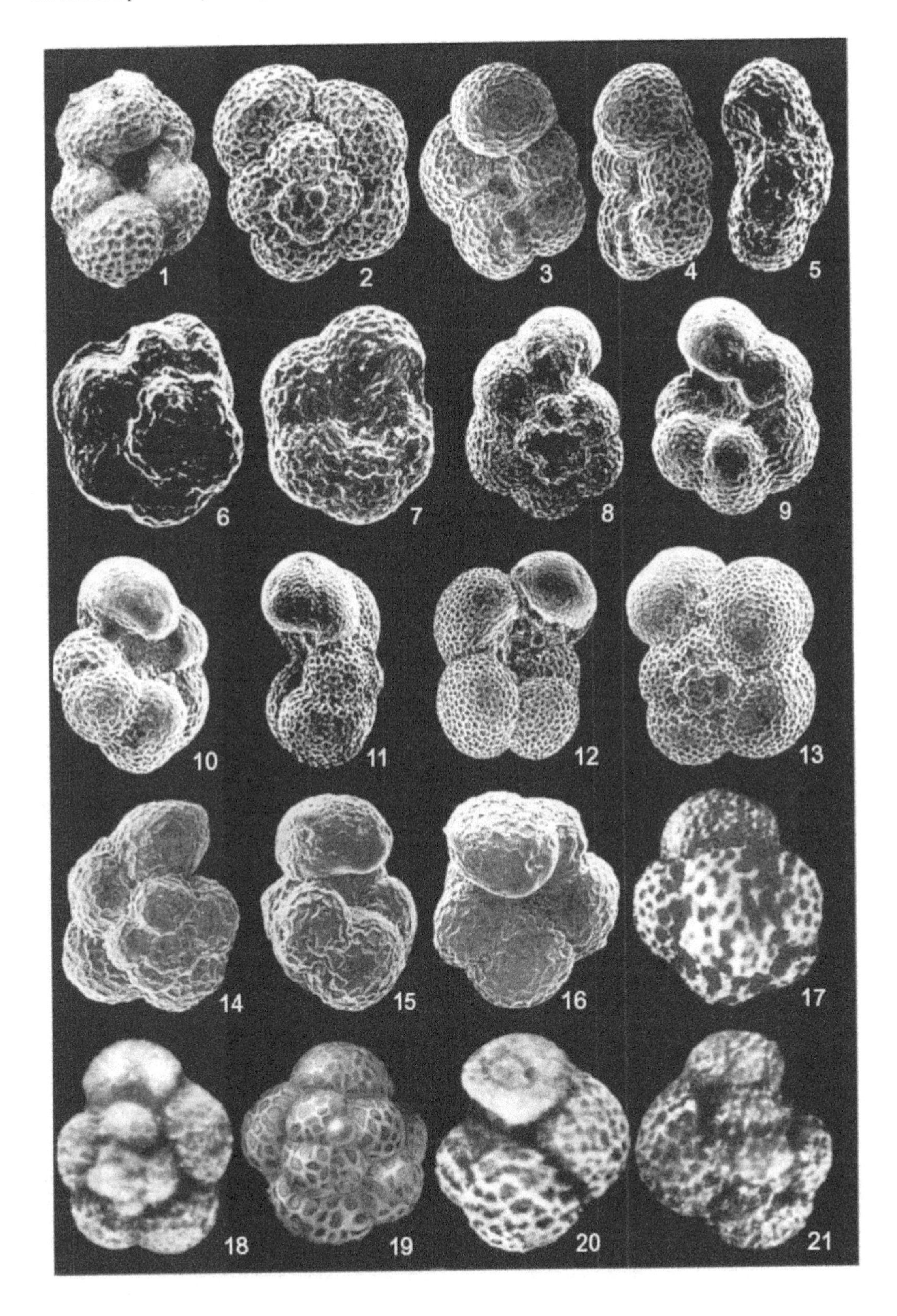
1
2
3
4
5
6
7
8
9
10
11
12
13
14
15
16
17
18
19
20
21

Plate 4.3. Figures 1–4: *Gorbachikella kugleri* (Bolli). (1, 2) Early Barremian, Beauvoir-I Well, Tunisia, NHM P52989, x165. (3, 4) Late Hauterivian, Beauvoir-I Well, Tunisia, NHM P52986, x100. Figures 5–8: *Gorbachikella anteroapertura* BouDagher-Fadel, Banner, Bown, Simmons and Gorbachik. Early Barremian, Beauvoir-I Well, Tunisia. (5) Holotype, x178. (6–8) Paratype, x176. Figures 9–11: *Gorbachikella depressa* BouDagher-Fadel, Banner, Bown, Simmons and Gorbachik. Early Barremian, Jebel Oust-III Well. (9–10) Holotype, NHM P52990, x156. (11) Paratype, NHM P52991, x162. Figures 12–16: *Gorbachikella grandiapertura* BouDagher-Fadel, Banner, Bown, Simmons and Gorbachik. (12–14) Holotype, Early Barremian, Beauvoir-I Well, Tunisia, NHM P52994, x150. (15, 16) (= *Globuligerina compressa* Maamouri and Salaj, 1995), holotype, Late Valanginian, Jebel Oust, x200. Figures 17, 18: *Gorbachikella depressa* BouDagher-Fadel, Banner, Bown, Simmons and Gorbachik (= *Globuligerina mejezensis* Maamouri and Salaj, 1995). Holotype, Late Valanginian, Jebel Oust, x248. Figures 19 and 20: *Gorbachikella neili* (Maamouri and Salaj). (19) Late Valanginian, Jebel Oust, x212. (20) Early Barremian, Beauvoir-I Well, Tunisia, x200. All images from BouDagher-Fadel (1997).

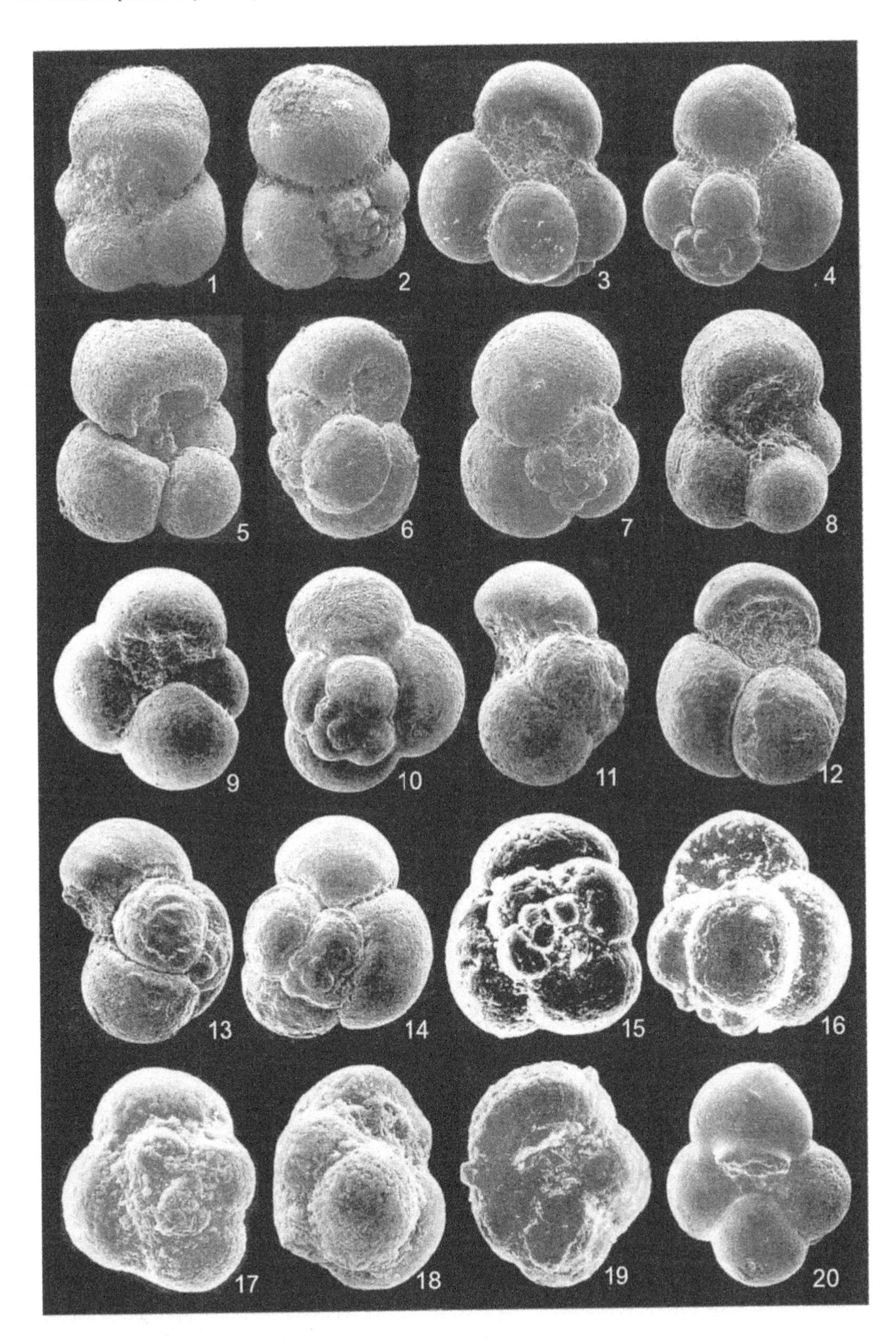
1
2
3
4
5
6
7
8
9
10
11
12
13
14
15
16
17
18
19
20

Plate 4.4. Figures 1 and 2: *Praehedbergella tuschepsensis* (Antonova). Holotype, Tuscheps River, northwestern Caucasus, x213. Figures 3 and 4: *Praehedbergella perforare* Banner, Copestake, and White. Holotype, Early Aptian, Well 20/2-2, Central North Sea, NHM P52758, x192. Figure 5: *Praehedbergella grigelisi* (Banner and Desai). Holotype, Late Aptian, Speeton Cliff, NE England, NHM P52129, x195. Figures 6 and 7: *Praehedbergella handousi* (Salaj). Holotype, Early Hauterivian, Jebel Oust, x160. Figures 8 and 9: *Praehedbergella pseudosigali* Banner, Copestake, and White. Holotype, Late Barremian, Well 20/ 2-2, Central North Sea, NHM P52735, x261. Figures 10 and 11: *Praehedbergella ruka* Banner, Copestake, and White. Holotype, Early Aptian, Well 16/28-6RE, Central North Sea, NHM P52739, x280. Figures 12 and 13: *Praehedbergella contritus* Banner, Copestake, and White. Holotype, Late Aptian, Well 16/28-6RE, Central North Sea, NHM P52737, x315. Figure 14: *Praehedbergella papillata* Banner, Copestake, and White. Holotype, Early Aptian, Well 16/28-6RE, Central North Sea, NHM P52736, x257. Figures 15, 16, 19: *Praehedbergella sigali* Moullade. (15, 16) Holotype, Early Barremian, Well 16/28-6RE, Central North Sea, NHM P52742, x288. (19) Late Barremian, Kacha River, southwestern Crimea, Russia, x355. Figures 17 and 18: *Praehedbergella compacta* Banner, Copestake, and White. Holotype, Early Barremian, Well 20/2-2, Central North Sea, NHM P52746, x324. Figure 20: *Praehedbergella tatianae* Banner and Desai, Holotype, Late Aptian, Speeton Cliff, NE England, x240. Figure 21: *Praehedbergella yakovlevae* BouDagher-Fadel, Banner, and Whittaker. Holotype, Early Aptian, Well 16/28-6RE, Central North Sea, NHM P52733, x213. All images from BouDagher *et al.*(1997).

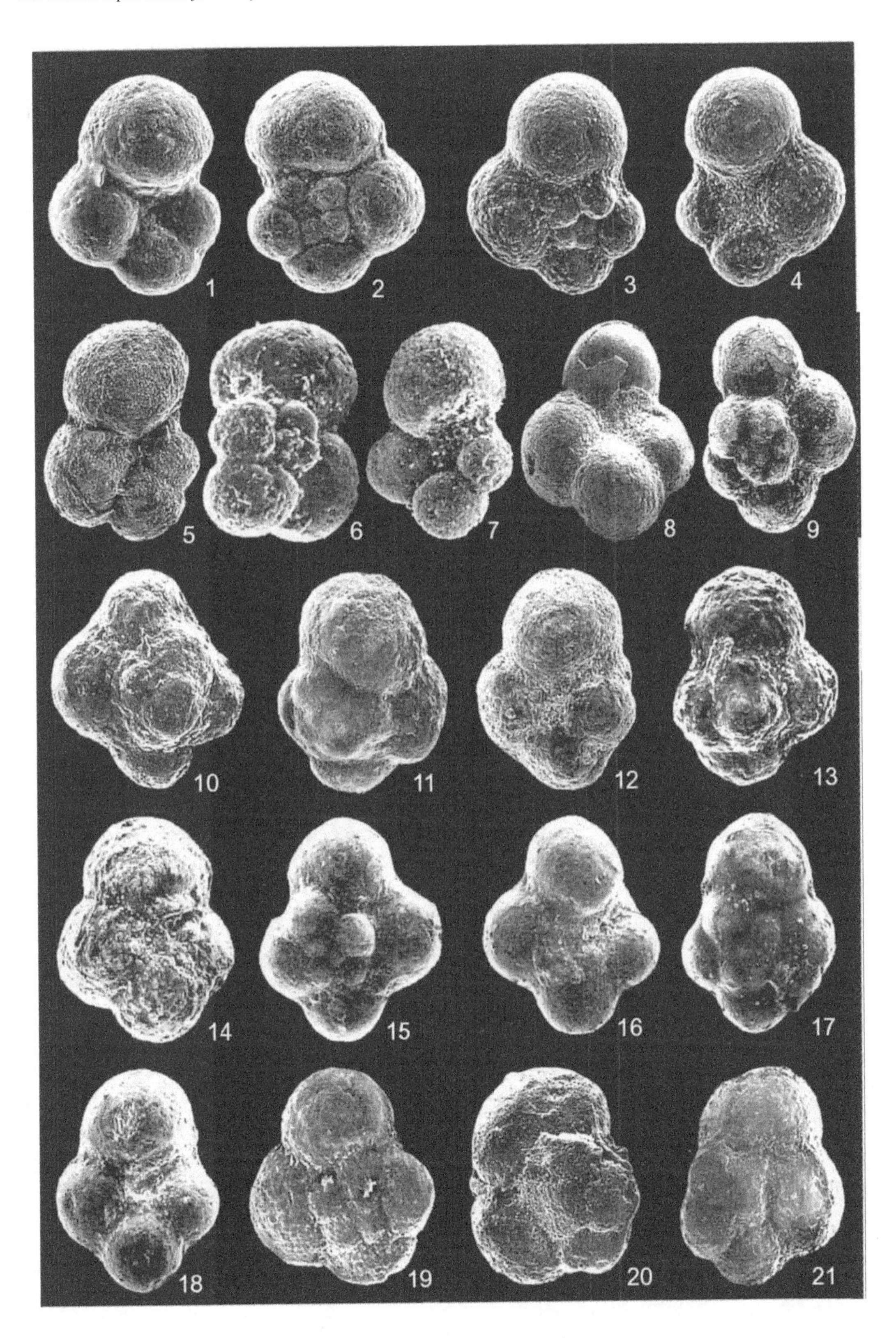
1
2
3
4
5
6
7
8
9
10
11
12
13
14
15
16
17
18
19
20
21

Plate 4.5. Figures 1–3: *Blefuscuiana kuznetsovae* Banner and Desai. Late Aptian, Speeton Cliff, NE England. (1) Holotype, NHM P52093, x160. (2–3) Paratype, NHM P52093, x146. Figures 4–7: *Blefuscuiana albiana* BouDagher-Fadel, Banner, Gorbachik, Simmons and Whittaker. (4, 5) Holotype, Late Albian, SW Crimea, x155. (6, 7) Late Albian, offshore Cote d'Ivoire, x133. Figs. 8–10. *Blefuscuiana aptiana* (Bartenstein). Late Aptian, Speeton Cliff, NHM P52082, x213. Figures 11 and 12: *Blefuscuiana orientalis* BouDagher-Fadel, Banner, Gorbachik, Simmons and Whittaker. Holotype, Early Aptian, NE Azerbaijan, NHM P53013, x145. Figures 13 and 14: *Blefuscuiana convexa* (Longoria). Holotype, Early Late Aptian, La Drôme region, SE France, USNM185019, x340. Figures 15–17: *Blefuscuiana daminiae* Banner, Copestake, and White. Early Aptian, Well 15/30-3, Central North Sea (15) x200. (16, 17) Holotype, NHM P52696, x190. Figures 18–20: *Blefuscuiana excelsa* (Longoria). Holotype, Early Aptian, La Drôme region, SE France, USNM184998, x260. Figures 21 and 22: *Blefuscuiana cumulus* Banner, Copestake, and White. Holotype, Early Aptian, Well 15/30-3, Central North Sea, NHM P52708, x220.

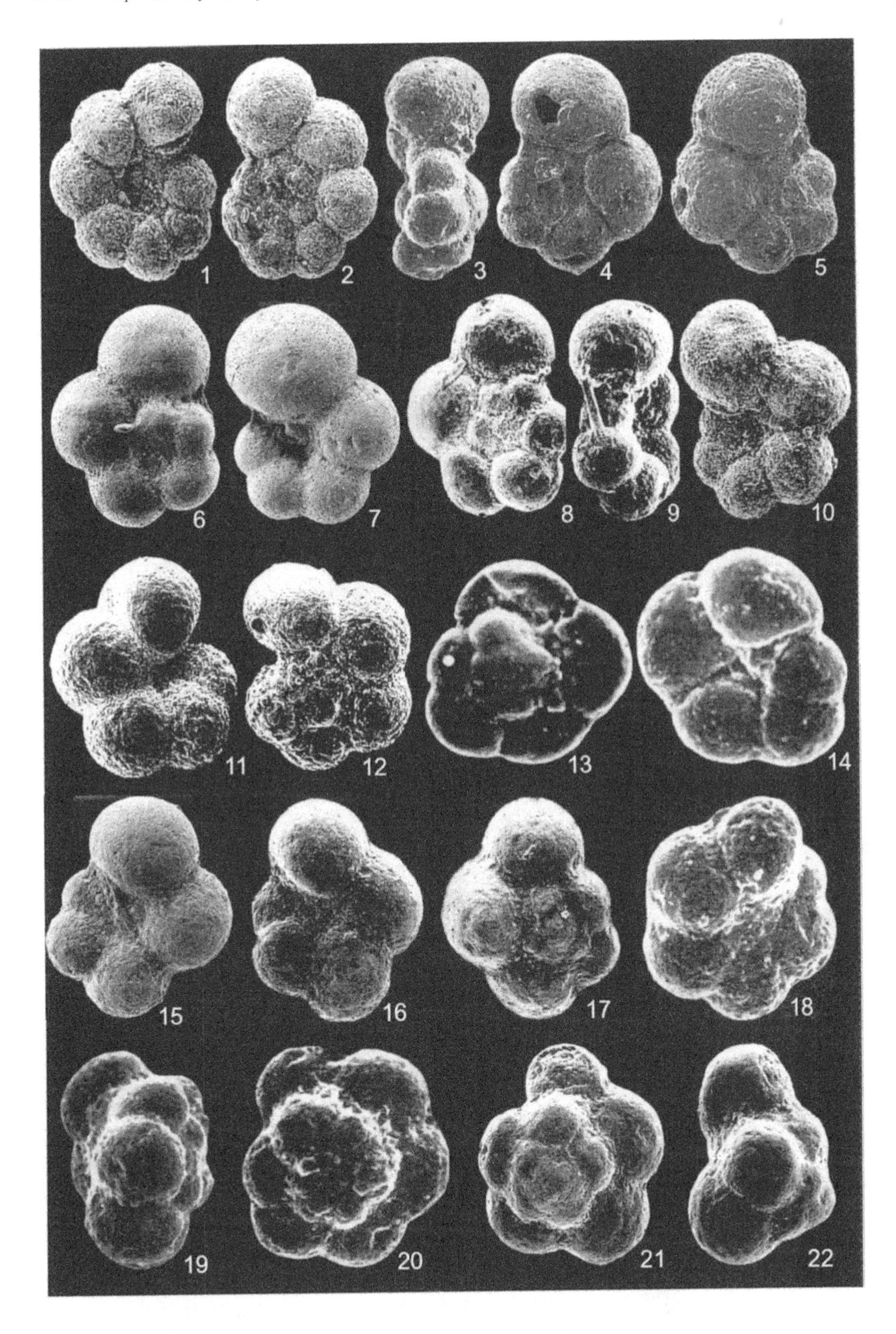
1
2
3
4
5
6
7
8
9
10
11
12
13
14
15
16
17
18
19
20
21
22

Plate 4.6. Figures 1 and 2: *Blefuscuiana cumulus* Banner, Copestake, and White. (1) Holotype, Early Aptian, Well 15/30-3, Central North Sea, NHM P52708, x220. (2) Early Aptian, Kacha River, Crimea, NHM P53074, x204. Figures 3 and 4: *Blefuscuiana gorbachikae* (Longoria). Early Aptian, Jebel Chenanrafa, Tunisia, x204. Figures 5 and 6: *Blefuscuiana hexacamerata* BouDagher-Fadel, Banner, and Whittaker. Holotype, Late Aptiahn, Kacha River, SW Crimea, NHM P53079, x178. Figures 7 and 8: *Blefuscuiana hispaniae* (Longoria). Holotype, Late Aptian, Sierra de la Silla, Mexico, USNM185002, x187. Figures 9 and 10: *Blefuscuiana infracretacea* (Glaessner). Late Aptian, Germany, NHM P52102, x225. Figures 11 and 12: *Blefuscuiana occidentalis* BouDagher-Fadel, Banner, and Whittaker. (11) Holotype, Early Aptian, North Sea Well 20/2-2, NHM P52102, x173. (12) Late Aptian, Speeton Cliff, NE England, NHM P52098, x220. Figures 13 and 14: *Blefuscuiana laculata* Banner, Copestake, and White. Holotype, Early Barremian, Well 20/2-2, Central North Sea, NHM P52724, x255. Figures 15 and 16: *Blefuscuiana alobata* Banner, Copestake, and White. Holotype, Early Barremian, Well 20/ 2-2, Central North Sea, NHM P52726, 240. Figure 17: *Blefuscuiana mitra* Banner and Desai. Holotype, Late Aptian, Speeton Cliff, NE England, NHM P52089, x200. Figures 18–20: *Blefuscuiana multicamerata* Banner and Desai. Holotype, Late Aptian, Speeton Cliff, NE England, NHM P52090, x400. All images from BouDagher-Fadel *et al.* (1997).

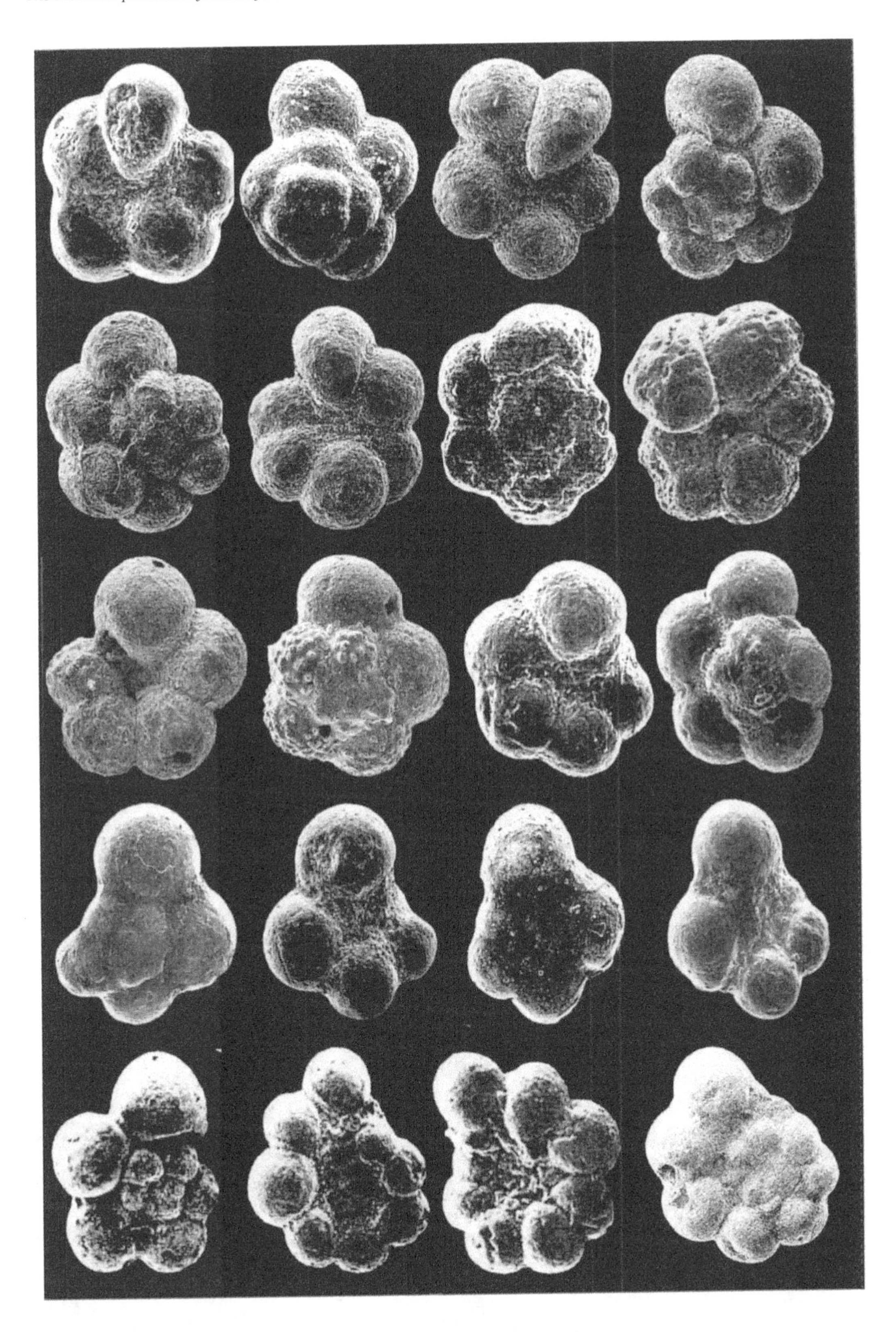

Plate 4.7. Figures 1 and 2: *Blefuscuiana occulta* (Longoria). Holotype, Late Aptian, La Peña Formation, La Boca Canyon, Mexico, USNM185011, x160. Figure 3: *Blefuscuiana perforocculta* Banner, Copestake and White. Holotype, Early Barremian, Well 20/2-2, Central North Sea, NHM P52678, x180. Figures 4 and 5: *Blefuscuiana laculata* Banner, Copestake, and White. Holotype, Early Barremian, Well 20/2-2, Central North Sea, NHM P52691, 204. Figures 6–10: *Blefuscuiana praetrocoidea* (Ketchmar and Gorbachik). Paratypes, Early Aptian, Kacha River, SW Crimea, (6–8) x204, (9) x188, (10) x230. Figure 11: *Blefuscuiana primare* (Kretchmar and Gorbachik). Topotype, Late Aptian, Kacha River, SW Crmea, x408. Figures 12 and 13: *Blefuscuiana rudis* Banner, Copestake, and White. Holotype, Early Barremian, Well 16/28-6RE, Central North Sea, NHM P52713, x160. Figure 14: *Blefuscuiana pararudis* BouDagher-Fadel, Banner, and Whittaker. Holotype, Early Aptian, Central North Sea, Well 20/2-2, NHM P52715, 333. Figures 15 and 16: *Blefuscuiana speetonensis* Banner and Desai. Holotype, Early Late Aptian, Speeton Cliff, NE England, NHM P52075, x178. Figures 17 and 18: *Blefuscuiana tunisiensis* BouDagher-Fadel. Holotype, Early Albian, Beauvoir-VI Well, Tunisia, x194. Figures 19 and 20: *Blefuscuiana daminiae* Banner, Copestake and White. Early Aptian, Well 15/30-3, Central North Sea, NHM P52717, x232. Figures 21 and 22: *Lilliputianelloides eocretaceus* (Neagu). Late Barremian, Beauvoir-I Well, Tunisia, NHM P62923, x190. All images from BouDagher-Fadel *et al.* (1997).

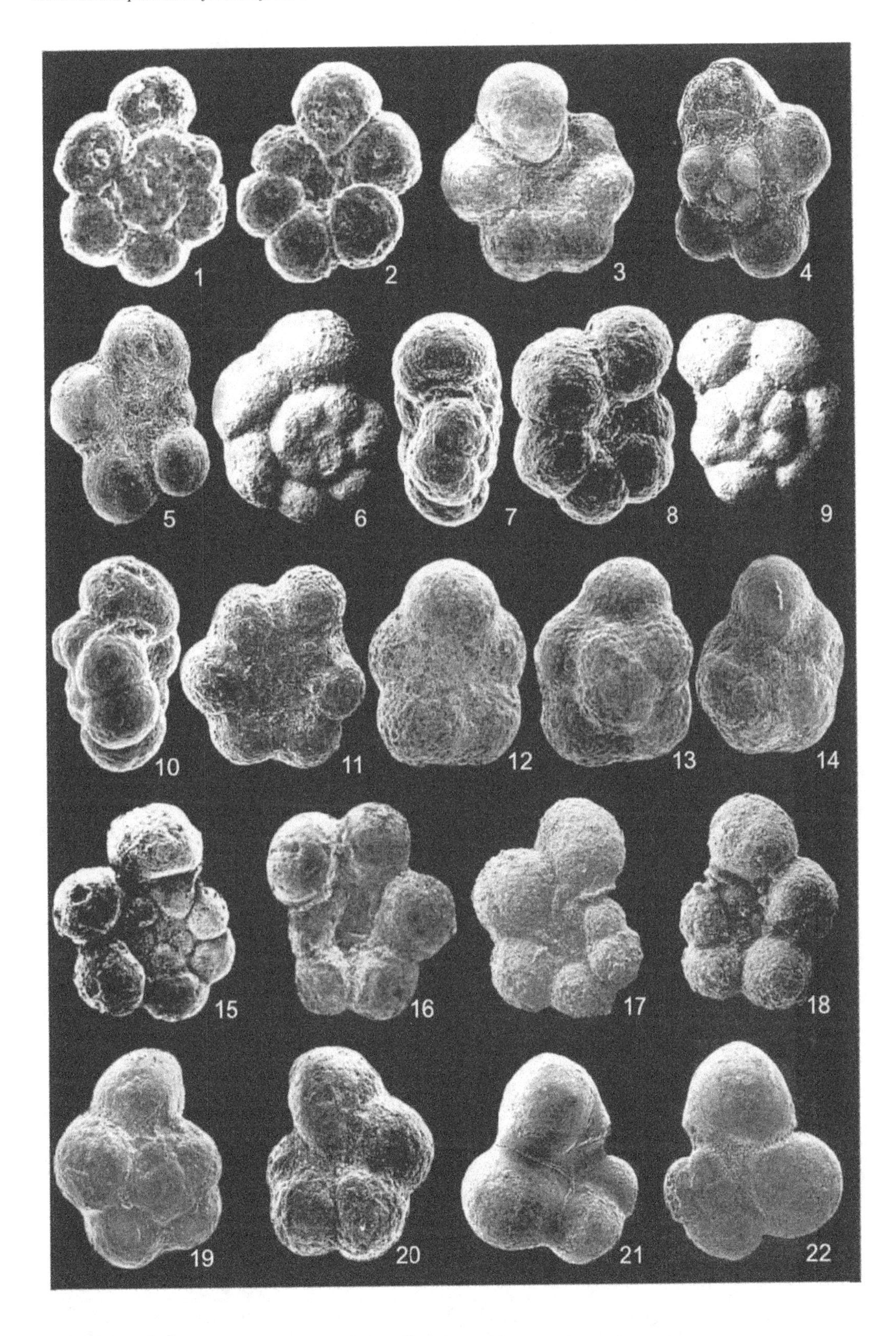
1
2
3
4
5
6
7
8
9
10
11
12
13
14
15
16
17
18
19
20
21
22

Plate 4.8. Figures 1–4: *Lilliputianella globulilifera* (Kretchmar and Gorbachik). (1) Topotype, Late Aptian, Krasnaya, SW Crimea, NHM P62913, x185. (2, 3) (= *Hedbergella maslakovae* Longoria), holotype, Late Aptian, La Boca Canyon, Sierra de la Silla, Mexico, USNM1855009, 135. (4) (= *Hedbergella similis* Longoria). Holotype, La Peña Formation, La Boca Canyon, Sierra de la Silla, Mexico, x140. Figures 5 and 6: *Lilliputianella labocaensis* (Longoria). Paratype, Late Aptian, La Boca Canyon, Sierra de la Silla, Mexico, x145. Figures 7 and 8: *Lilliputianella longorii* Banner and Desai. Holotype, Late Aptian, Speeton Cliff, NE England, NHM P52117, x160. Figures 9 and 10: *Lilliputianella kuhryi* (Longoria). Paratype, Late Aptian, Argos Formation, SE Spain, NHM P52124, x250. Figures 11 and 12: *Lilliputianella roblesae* (Obergón de la Parra). Late Aptian, Krasnaya, SW Crimea, NHM P62921, x240. Figures 13–17: *Lilliputianella bizonae* (Chevalier). (13, 14) Late Aptian, Kacha River, SW Crimea, NHM P53085, x120. (15–17) (= *Hedbergella bollii* Longoria). Holotype, Late Aptian, La Drôme section, France, USNM184994, x145. Figures 18–20: *Wondersella athersuchi* Banner and Strank. Paratypes, Late Aptian, highest Shuaiba Formation, Abu-Dhabi, NHM P52041-3, (18) x400, (19, 20) x160.

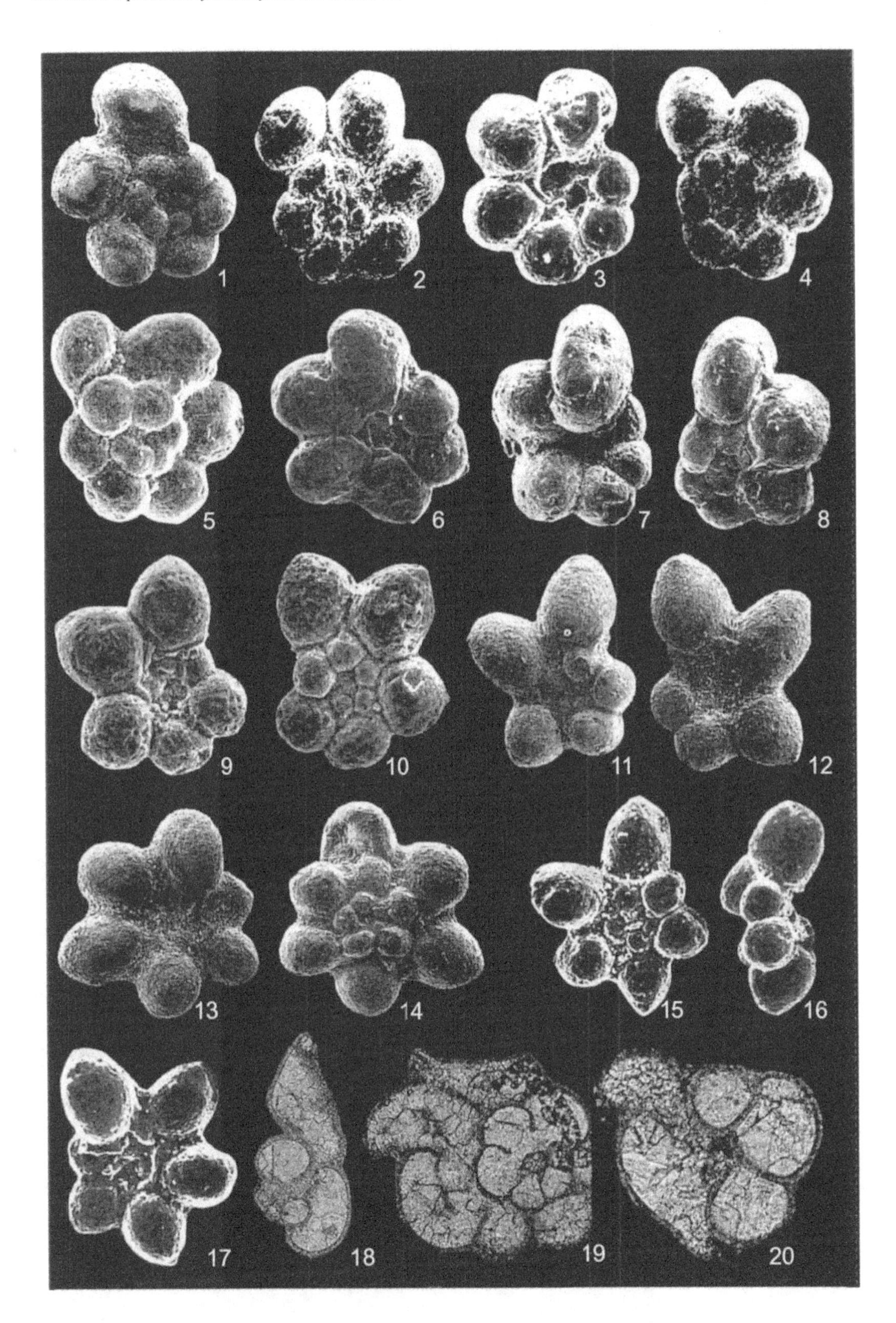
1
2
3
4
5
6
7
8
9
10
11
12
13
14
15
16
17
18
19
20

Plate 4.9. Figures 1 and 2: *Asterohedbergella asterospinosa* (Hamaoui). Late Cenmanian, Israel, Banner's Collection UCL, x266. Figure 3: *Clavihedbergella subcretacea* (Tappan). Late Cenomanian, Eagle Ford Group, Lake Waco Formation, Texas, USA, Banner's Collection UCL, x107. Figure 4: *Costellagerina lybica* (Barr). La Drôme section, France, UCL Collection, x260. Figures 5, 7, 8: *Hedbergella planispira* (Tappan). Topotype, Cenomanian, Grayson Formation, Danton County, Texas, USA, figured by Robaszynski and Caron (1979), x168. Figure 6: *Hedbergella simplex* (Morrow). Albian, Gault Clay, Arlesey, England, UCL Collection, x100. Figures 9 and 10: *Hedbergella trocoidea* (Gandolfi). Late Aptian, Simferopol, Crimea, Russia, UCL Collection, x124. Figures 11–13: *Planohedbergella ehrenbergi* (Barr). Holotype, Coniacian–Santonian, Culver Cliff, Isle of Wight, S. England, NHM P44608, x148. Figures 14 and 15: *Whiteinella archaeocretacea* Pessagno. Cenomanian, Dover, England, UCL Collection, x100. Figures 16 and 17: *Whiteinella baltica* Douglas and Rankin. Turonian, Dover, England, UCL Collection, x100. Figure 18: *Anaticinella multiloculata* (Morrow). Late Cenomanian, San Juan Island, USA, figured by Georgescu (2010), x57. Figure 19: *Thalmanninella greenhornensis* (Morrow). Late Cenomanian, San Juan Island, USA, figured by Georgescu (2010), x57. Figures 20 and 21: *Paraticinella eubejaouensis* (Randrianasolo and Anglada). Late Aptian, ODP Hole 1049C-12-6, figured by Premoli Silva *et al.* (2009), x200.

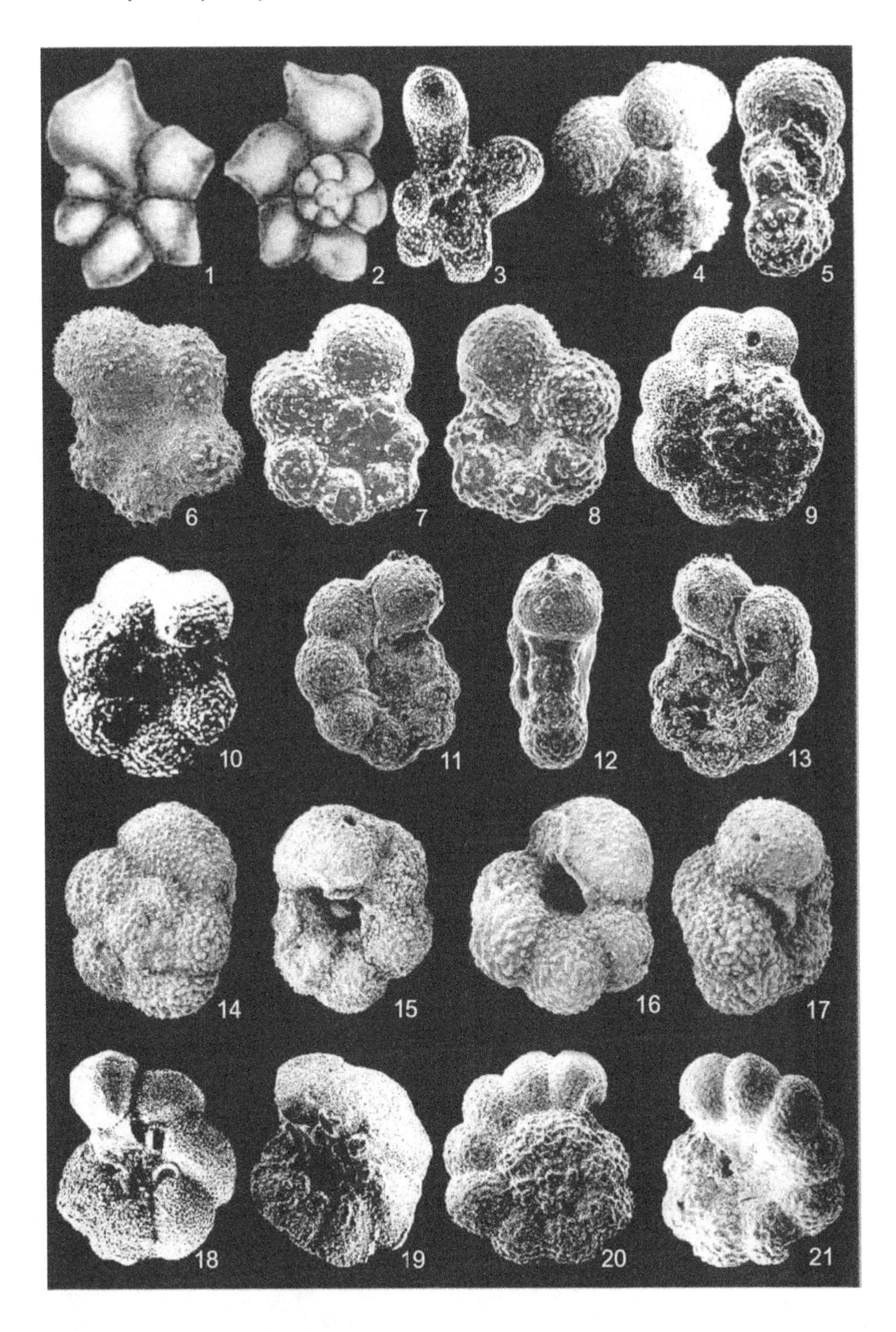
1
2
3
4
5
6
7
8
9
10
11
12
13
14
15
16
17
18
19
20
21

Plate 4.10. Figures 1–3: *Rotalipora appenninica* (Renz). Cenomanian, figured by Robaszynski *et al.* (1984), x67. Figures 4–6: *Rotalipora reicheli* Monrod. Middle Cenomanian, France, figured by Robaszynski and Caron (1979), x73. Figures 7 and 8: *Rotalipora greenhornensis* (Morrow). Cenomanian, ODP 7664, UCL Collection, x67. Figures 9 and 10: *Rotalipora cushmani* (Morrow). Late Cenomanian, Tunisia, figured by Robaszynski and Caron (1979), x72. Figures 11 and 12: *Ticinella roberti* (Gandolfi). Topotype, Middle Albian, Gorge of Breggia River, Switzerland, UCL Collection, x100. Figures 13 and 14: *Dicarinella hagni* (Scheibnerova). Early Turonian, Plenus Marls, Dover, UCL Collection, x100. Figure 15: *Globotruncanella minuta* (Caron and Gonzalez). Maastrichtian, Navarro County, Texas, USA, UCL Collection, x133. Figure 16: *Globotruncanella pshadae* (Keller). Late Maastrichtian, Navarro County, Texas, USA, UCL Collection, x133. Figure 17: *Globotruncanella petaloidea* (Gandolfi). Late Maastrichtian, ODP Hole 1210B-294, UCL Collection, x133. Figures 18–20: *Helvetoglobotruncana praehelvetica* Trujillo. Early Turonian, Dover, England, UCL Collection, (18) Transitional form from *Whiteinella*, x100, (19, 20) x130. Figure 21: *Helvetoglobotruncana helvetica* (Bolli). Banner Collition UCL, x67.

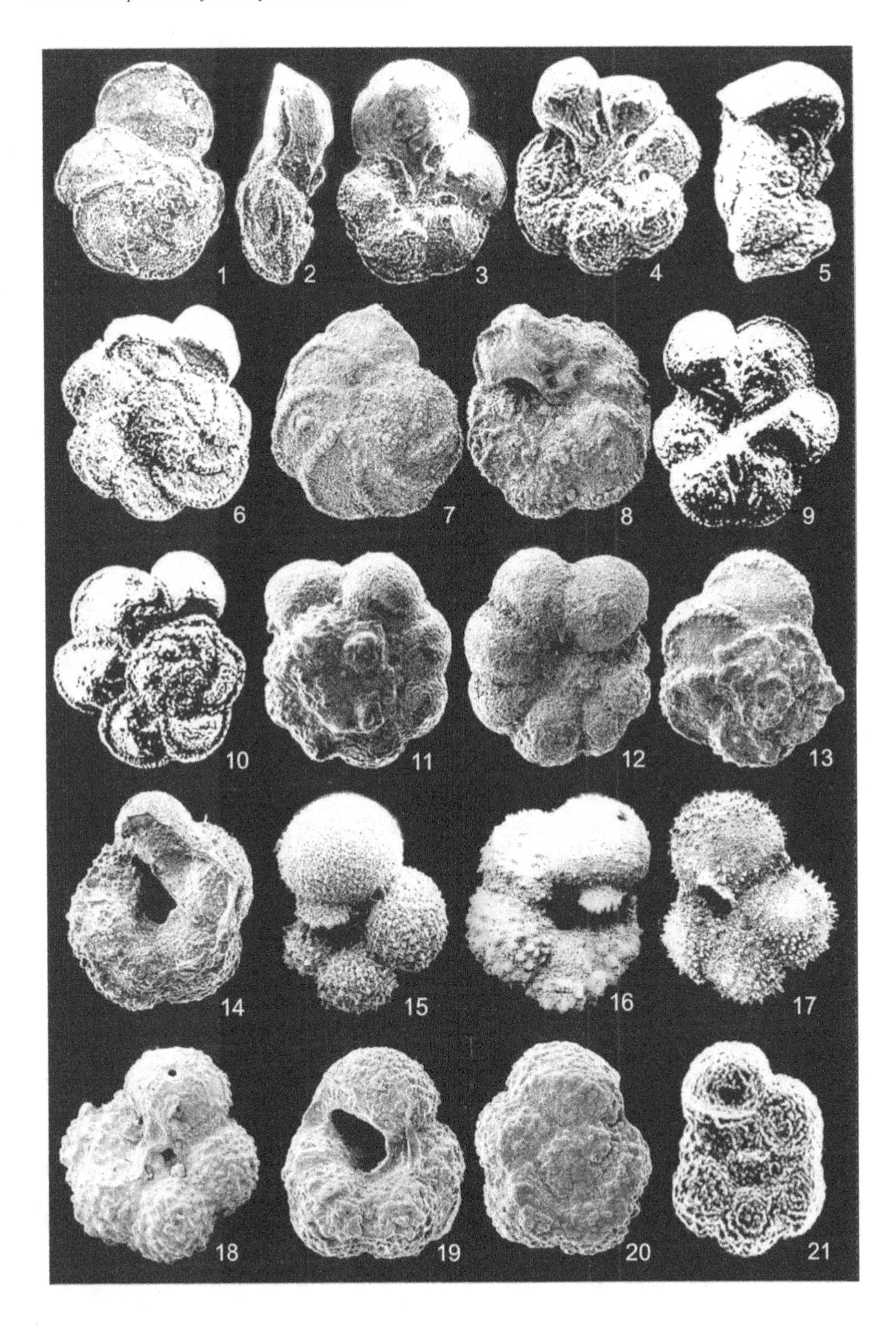
1
2
3
4
5
6
7
8
9
10
11
12
13
14
15
16
17
18
19
20
21

Plate 4.11. Figures 1–3: *Praeglobotruncana delrioensis* (Plummer). Topotype, Cenomanian, Grayson Formation, Danton County, figured by Robaszynski and Caron (1979) and BouDagher-Fadel *et al.* (1997), x128. Figures 4–6: *Praeglobotruncana stephani* (Gandolfi). Topotype, Cenomanian, de la Breggia, Switzerland, figured by Robaszynski and Caron (1979) and BouDagher-Fadel *et al.* (1997), x85. Figures 7–10: *Abathomphalus mayaroensis* Bolli. Topotype, Late Maastrichtian, Guayaguayare, Trinidad, UCL Collection, x67. Figures 11 and 12: *Contusotruncana patelliformis* (Gandolfi). Early Maastrichtian, Tunisia, UCL Collection, x80. Figures 13–15: *Contusotruncana* sp. Late Campanian, Upper Taylor Marl, Texas, USA, UCL Collection, x80. Figures 16 and 17: *Globotruncana arca* (Cushman). Early Campanian, Havana, Cuba, Brönnimann Collection UCL, x67. Figures 18 and 19: *Globotruncana falsostuarti* Sigal. Early Maastrichtian, Algeria, UCL Collection, x80. Figure 20: *Globotruncana rosetta* (Carsey). Campanian–Masstrictian, Aquitaine, UCL Collection, x67.

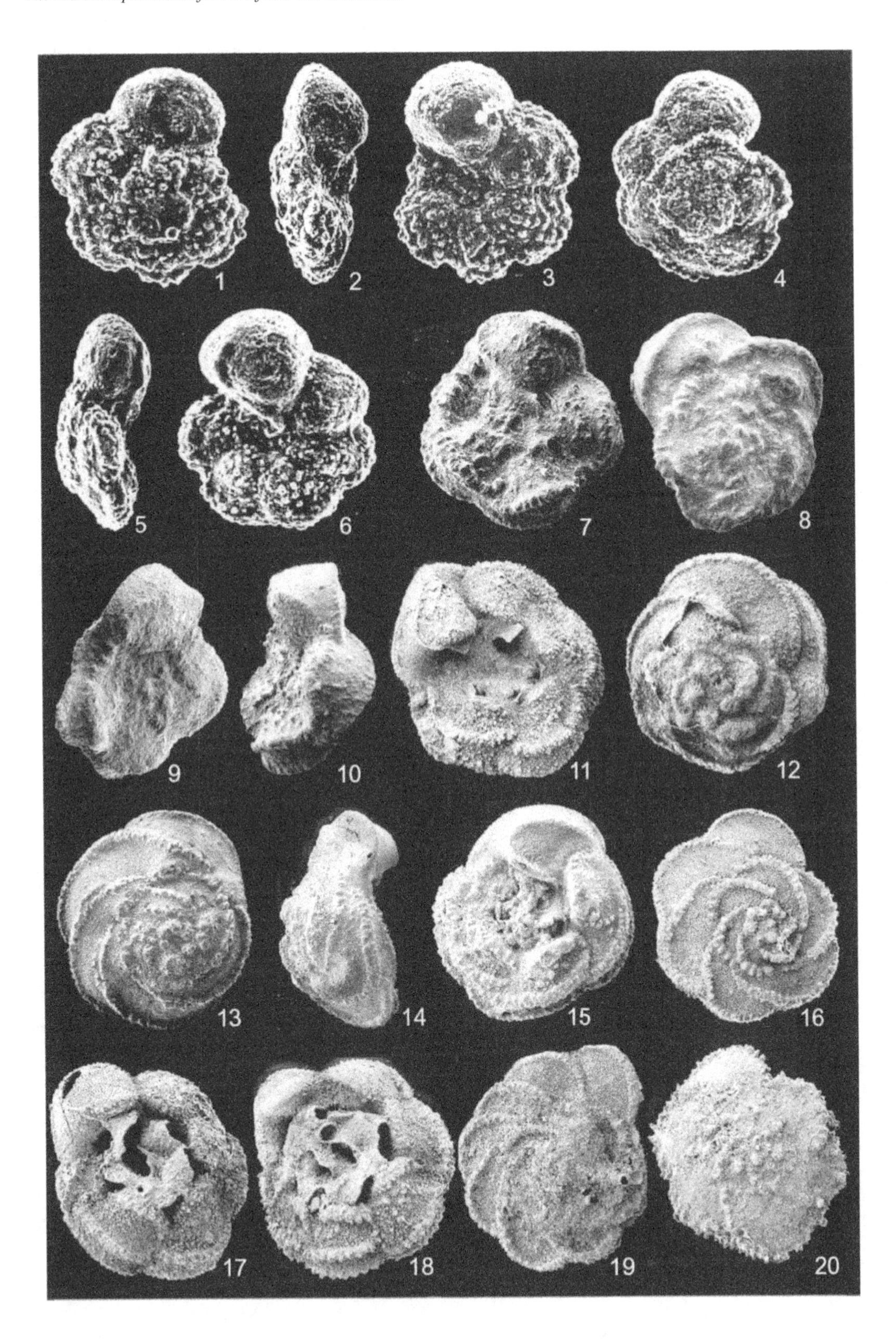
1
2
3
4
5
6
7
8
9
10
11
12
13
14
15
16
17
18
19
20

Plate 4.12. (Figs 1-16, 19-20 from UCL Collection; Figs 17-18 from Banner's Collection UCL). Figures 1 and 2: *Globotruncana rosetta* (Carsey). Campanian–Masstrictian, Aquitaine, x67. Figures 3 and 4: *Globotruncana linneiana* (d'Orbigny). Campanian, Tanzania, x100. Figures 5 and 6: *Globotruncana ventricosa* White. Late Campanian, Upper Taylor Marls, Texas, USA, x67. Figures 7 and 8: *Globotruncanita stuartiformis* (Dalbiez). Campanian, Tanzania, x67. Figures 9 and 10: *Globotruncanita conica* (White). Late Maastrichtian, Central Range Olistolith, Gautier River, Trinidad, x100. Figures 11 and 12: *Globotruncanita angulata* (Tilev). Late Maastrichtian, ODP Hole 1210B-294, x67. Figures 13 and 14: *Marginotruncana renzi* (Gandolfi). Turonian, Dover, England, x80. Figures 15 and 16: *Radotruncana calcarata* (Cushman). Campanian, Tanzania, x80, (16) enlargement of the wall showing macroperforations and muricae, x346. Figures 17 and 18: *Sigalitruncana sigali* (Reichel). Early Turonian, Algeria, x67. Figures 19 and 20: *Archaeoglobigerina cretacea* (d'Orbigny). Late Campanian, Paris Basin, x100.

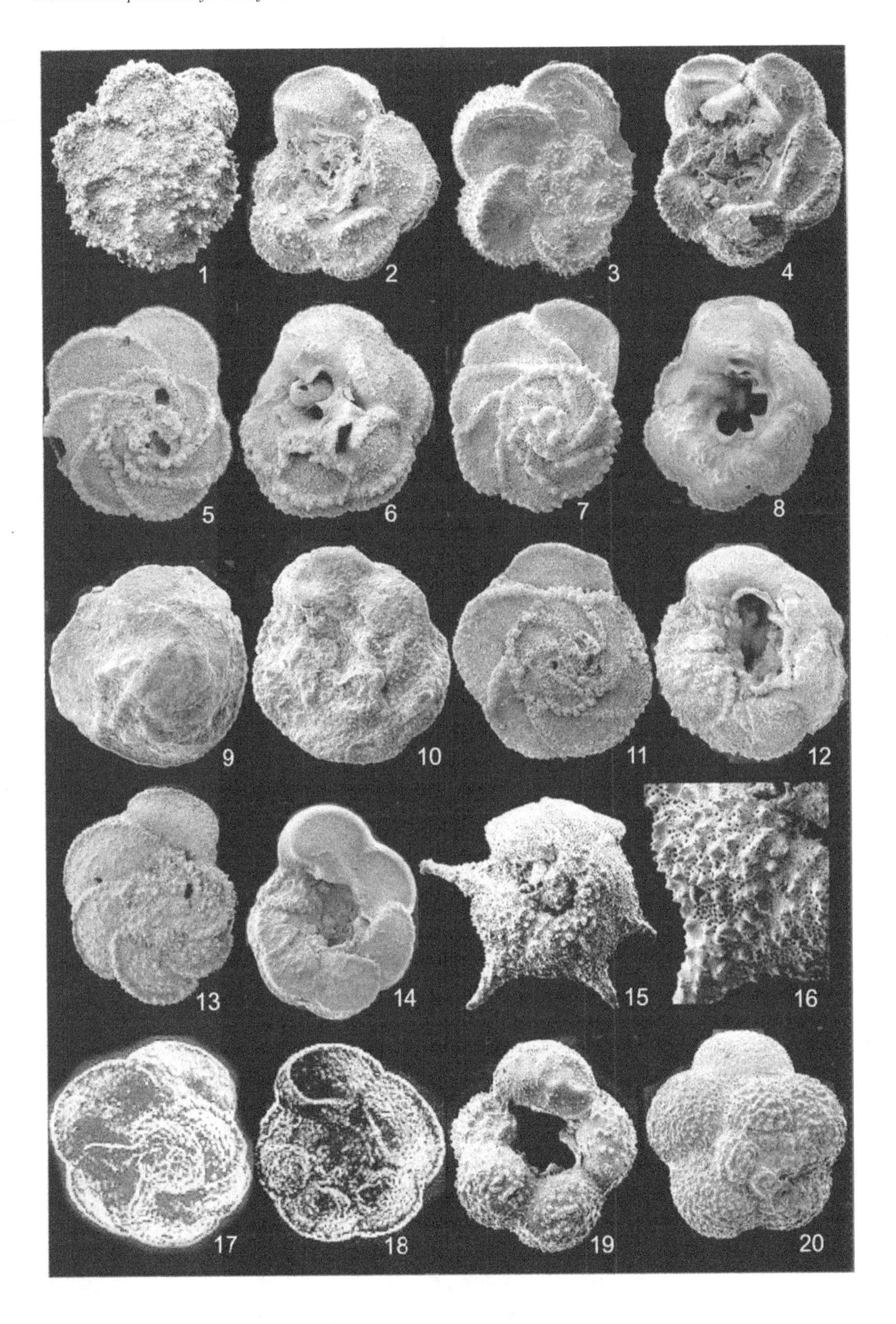
1
2
3
4
5
6
7
8
9
10
11
12
13
14
15
16
17
18
19
20

Plate 4.13. (All figures from UCL Collection). Figures 1 and 2: *Gansserina gansseri* (Bolli). Early Maastrichtian, Guayaguayare, Trinidad, x100. Figures 3 and 4: *Plummerita reicheli* (Brönnimann). Topotype, Late Maastrichtian, Guayaguayare, Trinidad, x100. Figures 5–7: *Kuglerina rotundata* (Brönnimann). Topotype, Late Maastrichtian, Guayaguayare, Trinidad, x100. Figures 8 and 9: *Rugoglobigerina* sp. Late Maastrichtian, Navarro County, Texas, USA, x70. Figures 10 and 11: *Rugoglobigerina rugosa* (Plummer). Topotype, Late Maastrichtian, Guayaguayare, Trinidad, x100. Figures 12 and 13: *Rugoglobigerina hexacamerata* Brönnimann. Early Maastrichtian, Guayaguayare, Trinidad, x100. Figures 14 and 15: *Rugoglobigerina pennyi* Brönnimann. Topotype, Late Maastrichtian, Guayaguayare, Trinidad, x80. Figure 16: *Rugoglobigerina tradinghousensis* Pessagno. Topotype, Late Maastrichtian, Guayaguayare, Trinidad, x80. Figure 17: *Rugotruncana subpenny* (Gandolfi). Early Maastrichtian, Navarro County, Texas, USA, x100. Figures 18 and 19: *Alanlordella subcarinatus* (Brönnimann). Early Maastrichtian, Navarro County, Texas, USA, (18) x80, (19) x200. Figure 20: *Alanlordella voluta* (White). Late Maastrichtian, Guayaguayare, Trinidad, x133.

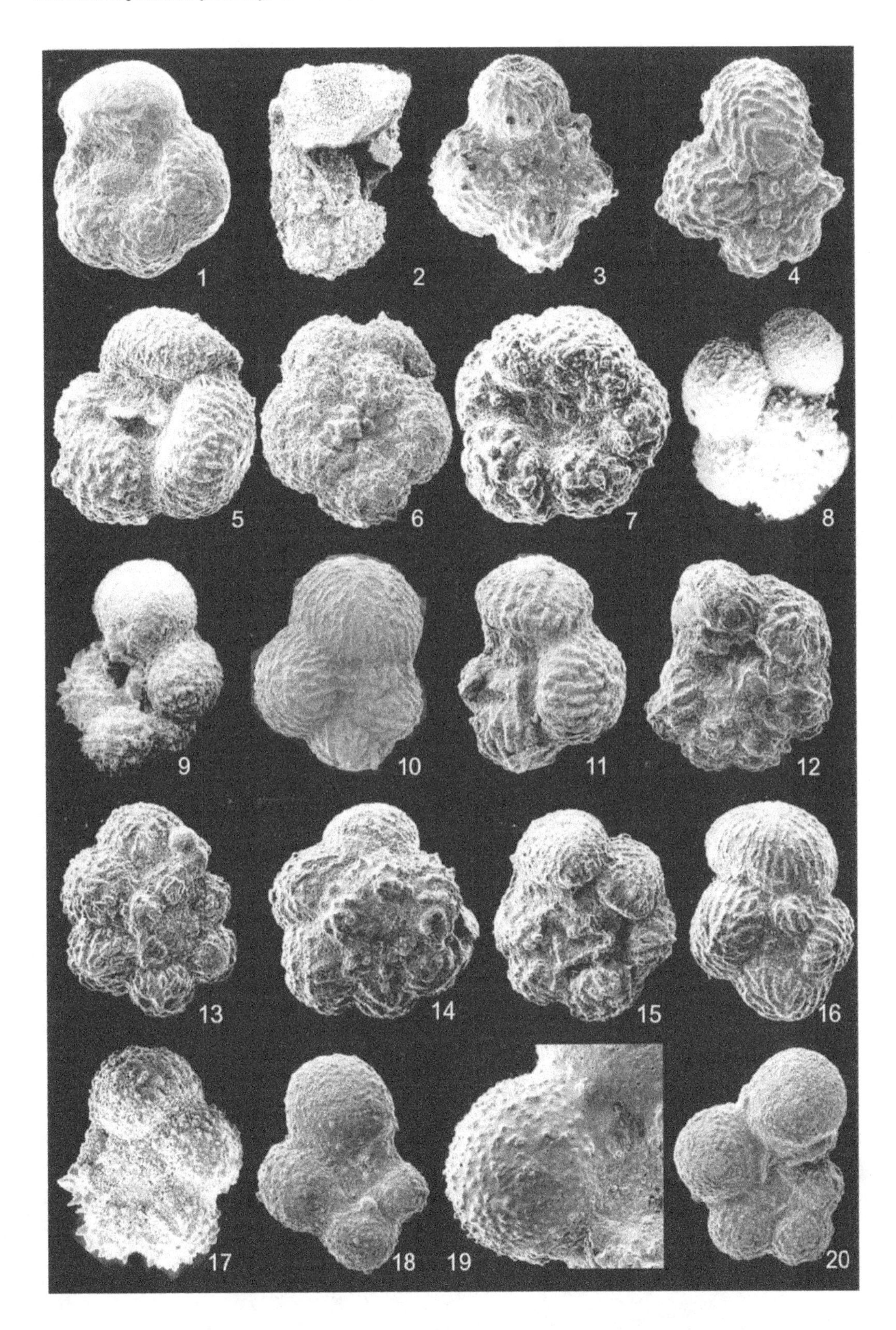
1
2
3
4
5
6
7
8
9
10
11
12
13
14
15
16
17
18
19
20

Plate 4.14. Figures 1 and 2: *Blowiella blowi* (Bolli). Early Aptian, Kacha River, SW Crimea, NHM P53082, x270. Figures 3 and 4: *Blowiella duboisi* (Chevalier). Late Aptian, Speeton Cliff, NE England, x215. Figures 5 and 6: *Blowiella gottisi* (Chevalier) Late Aptian, Speeton Cliff, NE England, x185. Figures 7–10: *Blowiella maridalensis* (Bolli). (7, 8) Late Aptian, Speeton Cliff, NE England, x165. (9, 10) Late Aptian, Kacha River, SW Crimea, (9) x133, (10) x200. Figures 11 and 12: *Blowiella moulladei* BouDagher-Fadel. Holotype, Early Aptian, Jebel Chenanrafa, Tunisia, NHM P53141, x173. Figures 13 and 14: *Blowiella solida* Kretchmar and Gorbachik. Topotype, Late Aptian, Krasnaya, SW Crimea, NHM P62920, x148. Figures 15 and 16: *Claviblowiella saudersi* (Bolli). Late Aptian, SW Crimea, NHM P62922, x120. Figures 17 and 18: *Claviblowiella sigali* (Longoria). Holotype, Late Aptian, La Peña Formation, La Boca Canyon, Mexico, USNM185024, (dimension not given). Figures 19–21: *Globigerinelloides algerianus* Cushman and Dam. (19, 20) Late Aptian, Krasnaya, SW Crimea, NHM P62910, x80. (21) Paratype, Late Aptian, Jebel Menaouer, Algeria, x62. All images from BouDagher-Fadel *et al.* (1997).

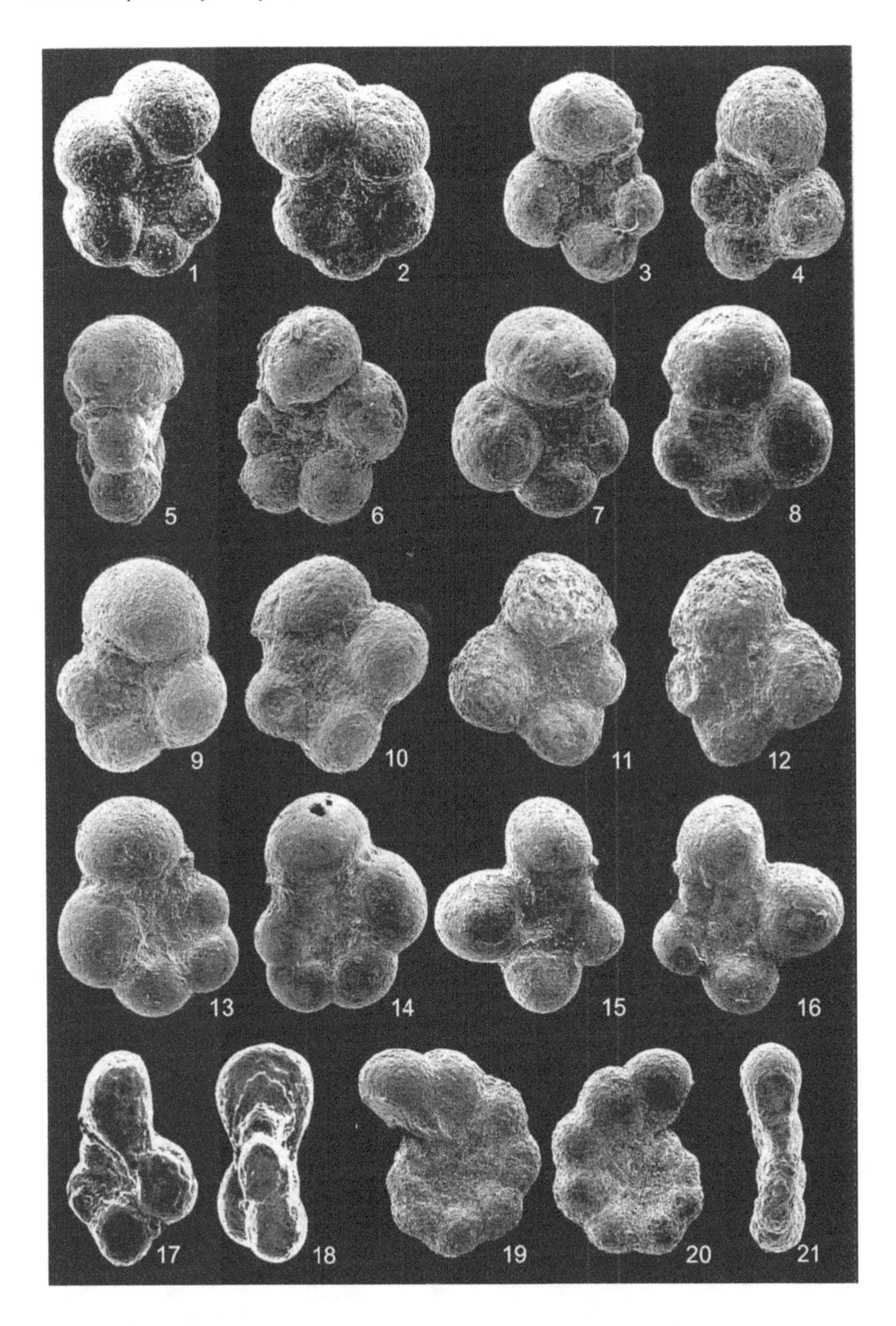
1
2
3
4
5
6
7
8
9
10
11
12
13
14
15
16
17
18
19
20
21

Plate 4.15. Figures 1–3: *Globigerinelloides ferreolensis* (Moullade). (1) Topotype, Late Aptian, Saint-Ferréol-Trente-Pas, La Drôme, SE France, x124. (2, 3) (= *Globigerinelloides macrocameratus* Longoria). Holotype, USNM185022, x89. Figures 4 and 5: *Leupoldina protuberans* Bolli. Late Aptian, Krasnaya, SW Crimea, NHM P62911, x130. Figure 6: *Schackoina cabri* Sigal, 1952. Early Late Aptian, sample 8109, NHM 52132, previously figured by BouDagher-Fadel *et al.* (1997). Figure 7: *Schackoina cenomana* (Schacko). Albian, Beauvoir-IV Well, Tunisia, NHM P62927, x117. Figures 8 and 9: *Schackoina pentagonalis* Reichel. Late Aptian, Jebel Oust-III Well, NHM P53148, x102. Figure 10: *Schackoina cepedai* (Obregón de la Parra). Aptian, Argos Formation, Spain, figured in BouDagher-Fadel *et al.* (1997), x142. Figures 11 and 12: *Pseudoplanomalina cheniourensis* (Sigal). Late Aptian, Jebel Oust, Tunisia, figured in BouDagher-Fadel *et al.* (1997), x60. Figures 13 and 14: *Alanlordella banneri* BouDagher-Fadel. Holotype, Early Albian, Beauvoir-III Well, Tunisia, figured in BouDagher-Fadel *et al.* (1997), x68. Figure 15: *Alanlordella bentonensis* (Morrow). Albian, Gault Clay, Arlesey, England, figured in BouDagher-Fadel *et al.* (1997), x30. Figure 16: *Alanlordella praebuxtorfi* (Wonders). Late Albian, SE Indian Ocean, NHM P62962, x106. Figures 17 and 18: *Globigerinelloides barri* (Bolli, Loeblich, and Tappan). Late Aptian, La Peña Formation, Mexico, figured by Longoria (1974), x110. Figure 19: *Biticinella breggiensis* (Gandolfi, 1942). Late Albian, Blake Plateau, western North Atlantic, figured by Georgescu (2010), x80. Figure 20: *Hastigerinoides alexanderi* (Cushman). Holotype, Santonian, Austin Chalk, Howe, Grayson County, Texas, USA, figured by Georgescu and Huber (2008), USNM309309, x100. Figures 21 and 22: *Hastigerinoides atlanticus* Georgescu and Huber. Santonian, DSDP Leg 39, Site 356, Sao Paolo Plateau, figured by Georgescu and Huber (2008), x57. Figures 23 and 24: *Planomalina buxtorfi* (Gandolfi). Late Albian, Lindi area, Tanzania, figured in BouDagher-Fadel *et al.* (1997), x93.

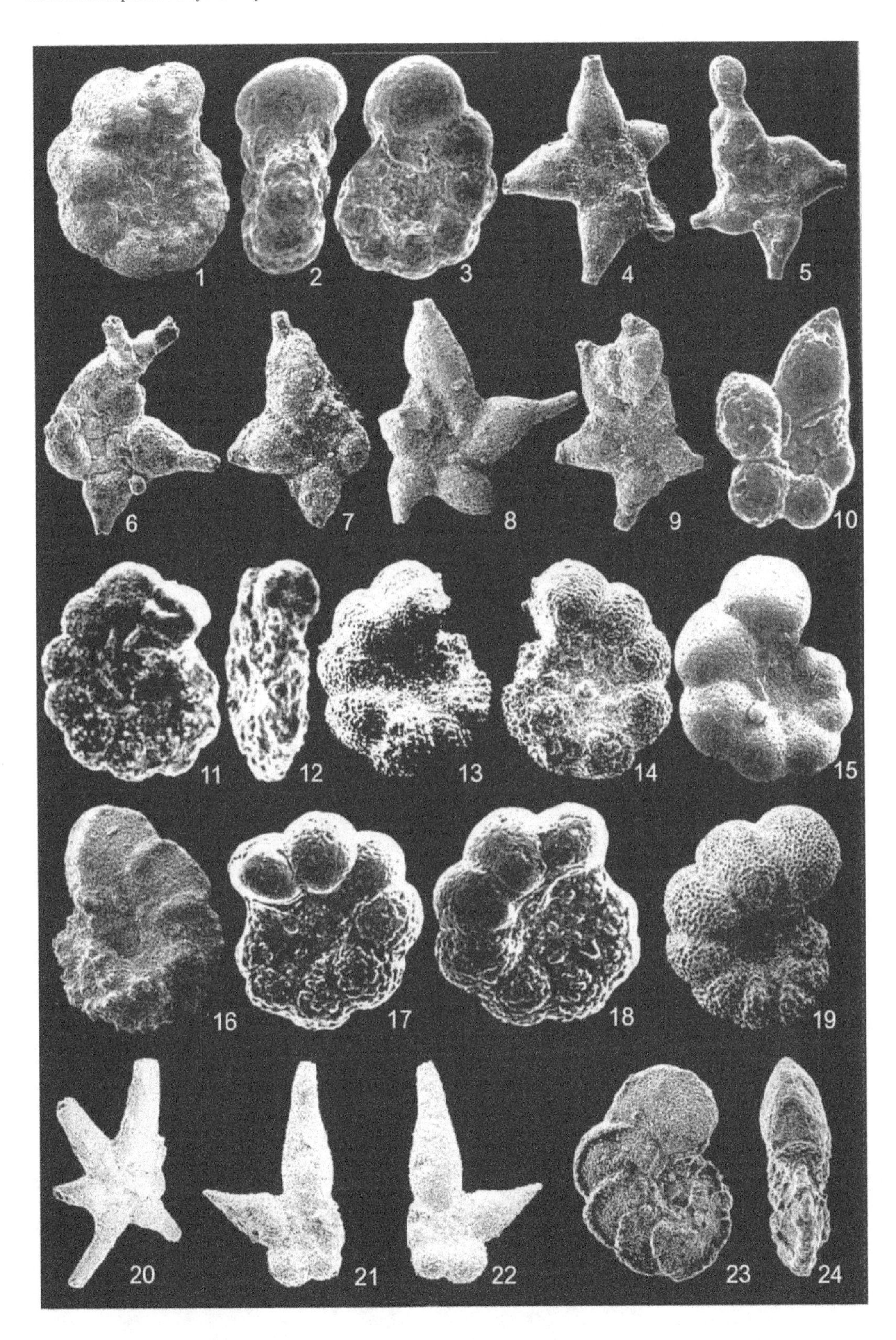
1
2
3
4
5
6
7
8
9
10
11
12
13
14
15
16
17
18
19
20
21
22
23
24

Plate 4.16. (All images are taken from original publications). Figures 1–3: *Helvetoglobotruncana praehelvetica* (Trujillo). Late Cenomanian–Middle Turonian, Salt Creek, California, x50. Figures 4–6: *Helvetoglobotruncana hevetica* (Bolli). Middle Turonian, Switzerland, x43. Figures 7 and 8: *Praeglobotruncana stephani* (Gandolfi). Late Albian–Middle Turonian, SE Switzerland, x45. Figures 9 and 10: *Praeglobotruncana turbinata* (Reichel). Late Cenomanian–Middle Turonian, SE Switzerland, x45. Figures 11–13: *Hedbergella delrioensis* (Carsey). Late Aptian–Early Santonian, Texas, USA, x100. Figures 14–16: *Hedbergella holmdelensis* Olsson. Maastrichtian, Navesink Formation, New Jersey, x100. Figure 17: Hedebergella *intermedia* Michael. Late Albian, Texas, USA, x75. Figure 18: *Hedbergella lata* Petters. Early Santonian, New Jersey, x100. Figures 19–21: *Hedbergella hoelzli* (Hagn and Zeil). Turonian, Germany, x60. Figures 22–24: *Hedbergella monmouthensis* (Olson). Maastrichtian, New Jersey, x100. Figures 25–27: *Hedbergella planispira* (Tappan). Late Cenomanian, Grayson Formation, Texas, USA, x100. Figure 28: *Costellagerina bulbosa* (Belford). Late Santonian, Western Australian, x100. Figures 29–31: *Globotruncanella pshadae* (Keller). Middle to Late Maastrichtian, Western Caucasus, x75. Figures 32–34: *Globotruncanella kefennsoura* Solakius. Maastrichtian, Kefwn Nsoura, Tunisia, x75. Figure 35: *Clavihedbergella subcretacea* (Tappan). Late Turonian, Duck Creek Formation, Texas, USA, x100. Figures 36 and 37: *Clavihedbergella simplex* (Morrow). Late Coniacian, Greenhorn Formation, Texas, USA, x75. Figures 38–40: *Clavihedbergella moremani* (Cushman). Late Turonian, Eagleford, Texas, USA, x100.

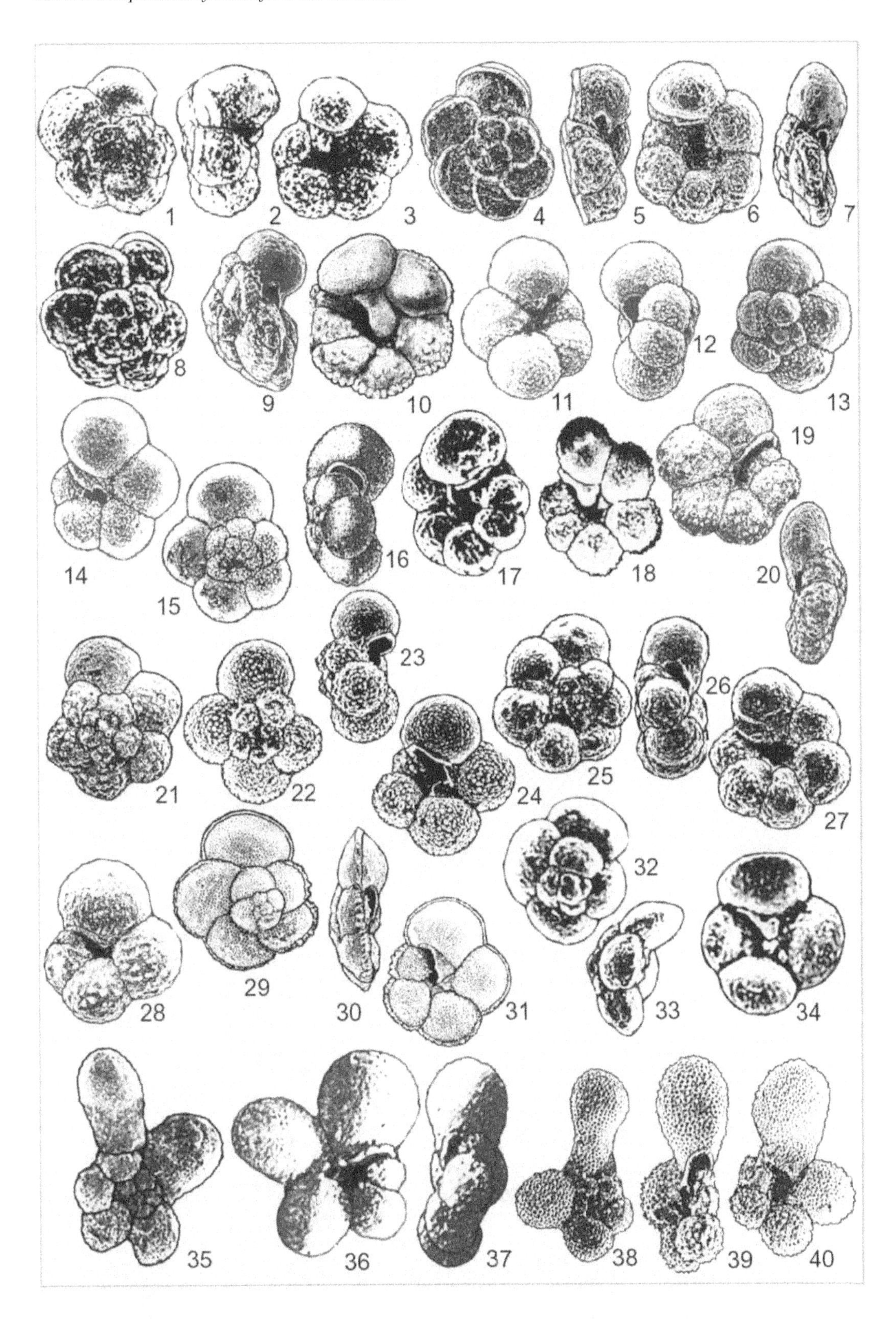
1
2
3
4
5
6
7
8
9
10
11
12
13
14
15
16
17
18
19
20
21
22
23
24
25
26
27
28
29
30
31
32
33
34
35
36
37
38
39
40

Plate 4.17. (All images are taken from original publications except where indicated). Figures 1 and 2: *Heterohelix sphenoides* Masters. Late Santonian–Campanian, Alabama, x80. Figures 3–5: *Heterohelix striata* (Ehrenberg). (3) Maastrichtian, Navaro, Texas, USA, UCL Collection, x100. (4, 5) Early Santonian–Late Maastrichtian, Poland, x133. Figures 6 and 7: *Heterohelix planata* (Cushman). Early Campanioan–Late Maastrichtian, Texas, USA, x60. Figures 8–12: *Heterohelix americana* (Ehrenberg). Early Santonian–Late Maastrictian, North America, x100. Figures 13 and 14: *Heterohelix moremani* (Cushman). Late Albian–Early Santonian, Texas, USA, x100. Figures 15 and 16: *Heterohelix semicostata* (Cushman). Late Campanian–Early Maastrichtian, Taylor Marl, Texas, USA, x133. Figures 17 and 18: *Heterohelix postsemicostata* (Vasilenko). Late Campanian–Late Maastrichtian, Russia, x100. Figures 19 and 20: *Heterohelix globulosa* (Ehrenberg). (19) Cenomanian–Late Maastrichtian, Germany, x100. (20) Late Cenomanian, Côte d'Ivoire, x150. Figures 21 and 22: *Heterohelix turgida* (Nederbragt). Santonian, Tunisia, x100. Figures 23 and 24: *Heterohelix dentata* (Stenestad). Late Campanian–Late Maastrichtian, Denmark, x200. Figures 25 and 26: *Heterohelix flabelliformis* (Nederbragt). Turonian, Tunisia, x133. Figure 27: *Heterohelix pulchra* (Brotzen). Middle Turoinian–Late Maastrichtian, Sweden, x133. Figure 28: *Gublerina cuvillieri* Kikoine. Middle–Late Maastrichtian, France, x58. Figure 29: *Gublerina robusta* de Klasz. Late Maastrichtian, Mexico, x60. Figures 30 and 31: *Guembelitriella graysonensis* Tappan. Early Cenomanian, Texas, USA, x133. Figure 32: *Racemiguembelina fructicosa* (Egger). Middle Maastrichtian, Havana, Cuba, UCL Collection, x100.

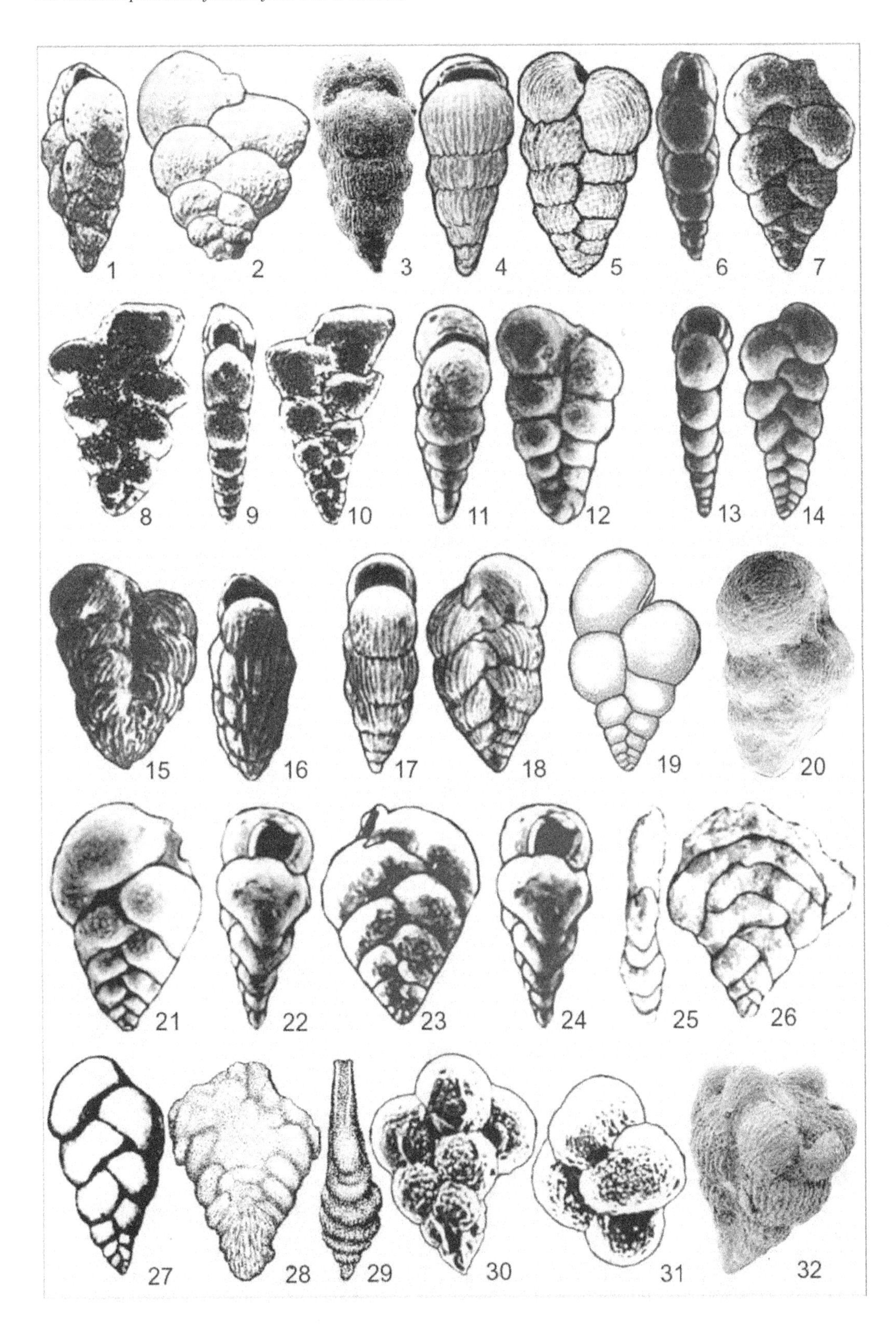
1
2
3
4
5
6
7
8
9
10
11
12
13
14
15
16
17
18
19
20
21
22
23
24
25
26
27
28
29
30
31
32

Plate 4.18. (All images are taken from original publications except where indicated). Figures 1 and 2, 10: *Racemiguembelina fructicosa* (Egger). (1, 2) Middle to Late Maastrichtian, Germany, x70. (10) Middle Maastrichtian, Havana, Cuba, UCL Collection, x75. Figures 3 and 4: *Racemiguembelina powelli* Smith and Pessagno. Middle to Late Maastrichtian, Texas, USA, x100. Figures 5 and 6: *Racemiguembelina intermedia* (de Klasz). Maastrichtian, Germany, x65. Figures 7 and 8: *Platystaphyla brazoensis* (Martin). Middle to Late Maastrichtian, Walter Creek, Texas, USA, x80. Figures 9, 15 and 16: *Pseudoguembelina excolata* (Cushman). Topotype, Maastrichtian, Mendez Shale Mexico, UCL Collection, (9) x50, (15, 16) x67. Figures 11 and 12: *Pseudoguembelina costulata* (Cushman). Middle Campanian–Late Maastrichtian, Taylor Marl, Texas, USA, x80. Figures 13 and 14: *Pseudoguembelina kempensis* Esker. Middle–Late Masstrichtian, Texas, USA, x133. Figure 17: *Pseudoguembelina palpebra* Brönnimann and Brown. Middle–Late Maastichtian, Cuba, x80. Figures 18 and 19: *Pseudoguembelina costellifera* Masters. Late Campanian–Maastrichtian, Alabama, x75. Figures 20 and 21: *Pseudoguembelina tessera* (Ehrenberg). Campanian, Tanzania, UCL Collection, x80. Figures 22 and 23: *Pseudotextularia punctulata* (Cushman). Early Campanian–Maastrichtian, Taylor Marl, Texas, USA, x80. Figures 24, 25, 28: *Pseudotextularia nuttalli* (Voorwijk). Late Coniacian–Maastrichtian, Havana, Cuba, (24-25) x67, (28) UCL Collection, x60. Figures 26 and 27: *Pseudotextularia elegans* (Rzehak). Middle Campanian–Late Maastrichtian, Austria, x90. Figure 29: *Bifarina texana* (Cushman). Late Cenomanian–Early Santonian, Sharon, Massachusetts, x120. Figure 30: *Bifarina hispidula* (Cushman). Early Santonian, Sharon, Massachusetts, x120. Figure 31: *Bifarina cretacea* (Cushman). Maastrictian, Sharon, Massachusetts, x180. Figure 32: *Bifarina bohemica* (Sulc). Coniacian–Campanian, Czechoslovakia, x180. Figure 33: *Bifarina ballerina* Eicher and Worstell. Early–Middle Turonian, Kansas, USA, x120. Figures 34 and 35: *Lunatriella spinifera* Eicher and Worstell. Early Turonian, Kansas, USA, (32) x133, (33) x180. Figure 36: *Pseudoguembelina cornuta* Seiglie. Early Maastrichtian, Trinidad, UCL Collection, x88.

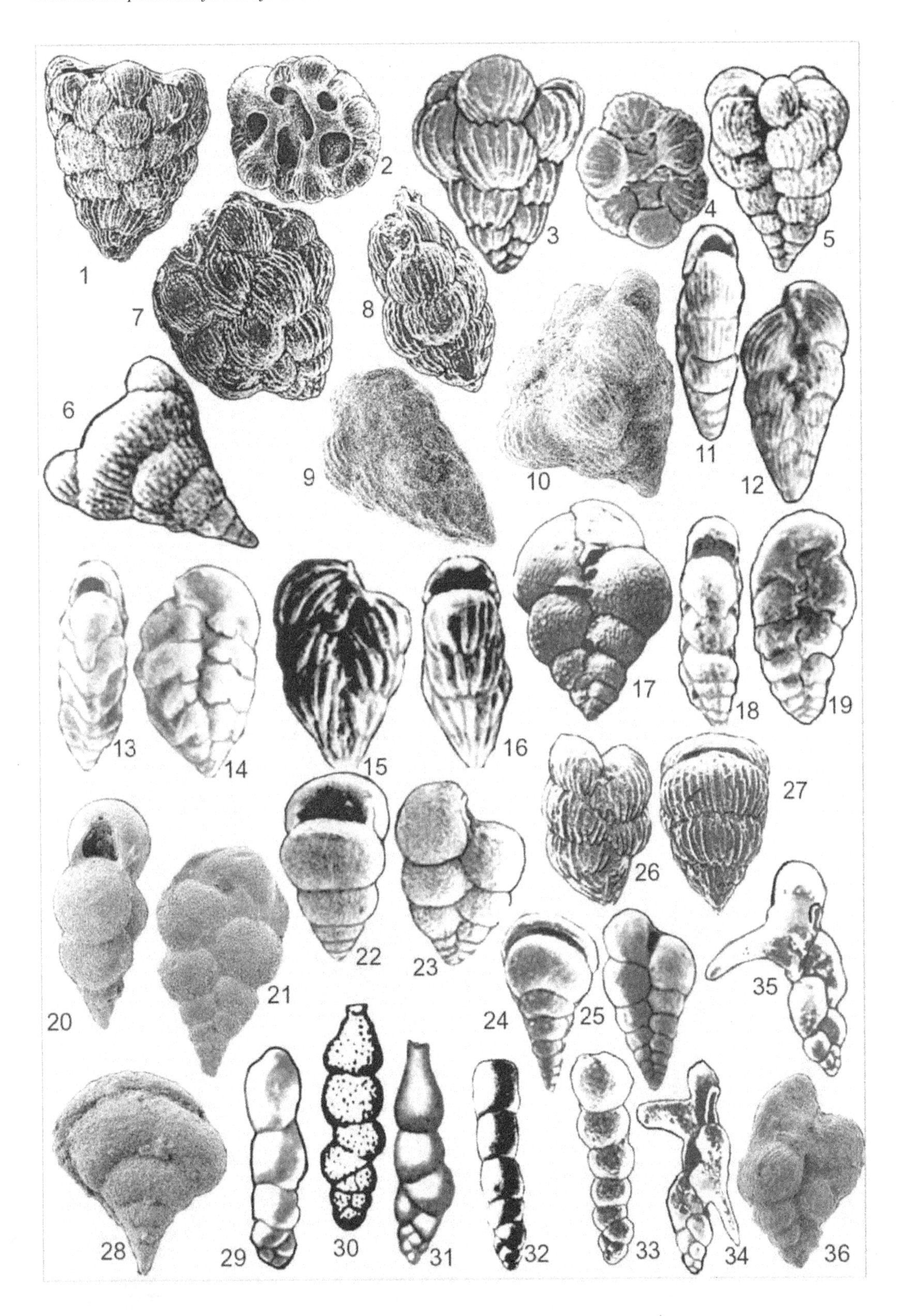
1
2
3
4
5
6
7
8
9
10
11
12
13
14
15
16
17
18
19
20
21
22
23
24
25
26
27
28
29
30
31
32
33
34
35
36

Plate 4.19. (All images are taken from original publications except where indicated). Figure 1: *Guembelitria harrisi* Tappan. Late Albian–Early Cenomanian, Grayson Formation, Texas, USA, x150. Figure 2: *Guembelitria cenomana* (Keller). Cenomanian–Early Turonian, Russia, x100. Figure 3: *Guembelitria cretacea* Cushman. Coniacian, Sharon, Massachusetts, x150. Figure 4: *Planoglobulina meyerhoffi* Seigli. Late Campanian–Late Maastrichtian, Cuba, x75. Figures 5–8: *Planoglobulina ornatissima* (Cushman and Church). Santonian–Campanian, California, x75. Figures 9–11: *Ventilabrella eggeri* (Cushman). Early Campanian, Taylor Marl, Texas, USA, x75. Figure 12: *Planoglobulina alpina* (de Klasz). Late Santonian–Middle Campanian, Germany, x85. Figures 13–15: *Planoglobulina hariaensis* (Nederbragt). Late Maastrichtian, Tunisia, x60. Figures 16 and 17: *Sigalia carpatica* (Salaj and Samuel). Santonian, Czechoslovakia, x75. Figures 19 and 20: *Sigalia decoratissima* (de Klasz). Middle Late Santonian, Germany, x75. Figures 18, 21, and 22: *Sigalia deflaensis* (Sigal). Late Santonian, Algeria, x60. Figures 24 and 25: *Planoglobulina carseyae* (Plummer). Middle–Late Maastrichtian, Texas, USA, x60. Figures 26–28: *Ventilabrella glabrata* (Cushman). Campanian, Texas, USA, x75. Figure 29: *Ventilabrella manuelensis* (Martin). Late Campanian–Late Maastrichtian, Texas, USA, x75. Figures 30 and 31: *Ventilabrella browni* Martin. Late Santonian–Middle Campanian, Texas, USA, x100. Figures 32 and 33: *Ventilabrella multicamerata* (de Kalsz). Late Maastrichtian, Austria, x50. Figures 34–36: *Ventilabrella austiniana* (Cushman). Late Santonian–Late Campanian, Texas, USA, x150. Figures 37 and 38: *Chiloguembelitria turrita* (Kroon and Nederbragt). Late Campanian–Early Maastrichtian, Grand Banks, x200.

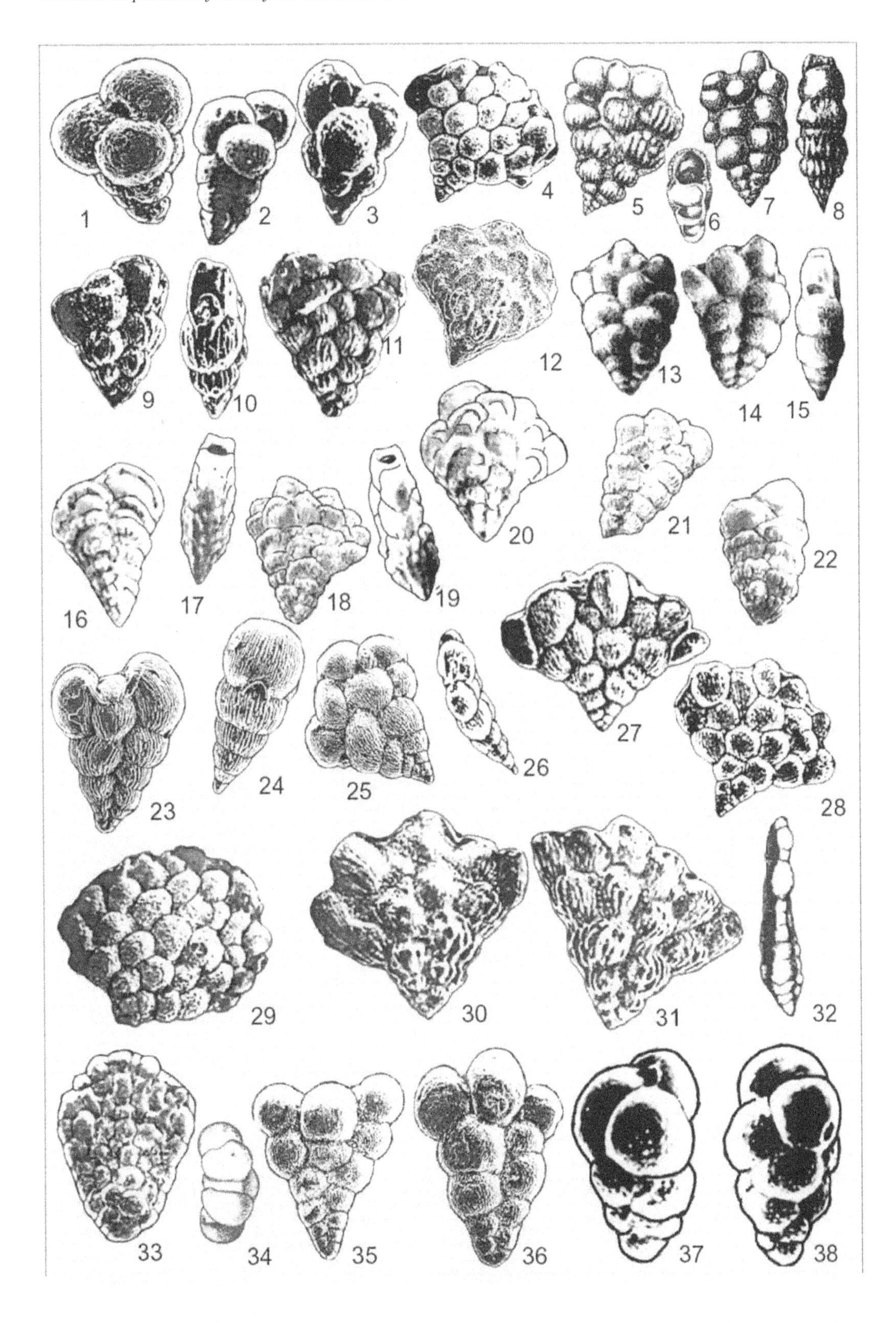
1
2
3
4
5
6
7
8
9
10
11
12
13
14
15
16
17
18
19
20
21
22
23
24
25
26
27
28
29
30
31
32
33
34
35
36
37
38

Plate 4.20. (All figures from the UCL Collection). Figure 1: *Heterohelix* moremani (Cushman). Late Albian, Dorset, England, x175. Figures 2, 3, 14. *Ticinella primula* Luterbacher. Late Albian, Dorset, England, (2) x65, (3) x75, (14) x55. Figure 4: *Blefuscuiana infracretacea* (Glaessner). Early Albian, Dorset, England, x60. Figure 5: *Ticinella madecassiana* Sigal. Late Albian, Melton Quarry, Devon, England, x65. Figure 6: *Ticinella roberti* (Gandolfi). Late Albian, Melton Quarry, Devon, England, x65. Figure 7: *Globigerinelloides barri* (Bolli, Loeblich and Tappan). Late Aptian, Dorset, England, x145. Figures 8, 10, 11, 16: *Favusella washitensis* (Carsey). Early Cenomanian, Melton Quarry, Devon, England, (8) x53, (10, 11, 16) x70. Figure 9: (A) *Hedbergella delrioensis* (Carsey), (B) *Helvetoglobotruncana praehelvetica* (Trujillo). Turonian, Dorset, England, x55. Figures 12 and 13: *Favusella* sp. 1. Late Albian, Dorset, England, x85. Figure 15: *Globigerinelloides ferreolensis* Moullade. Aptian, Tunisia, x180. Figure 17: *Globotruncanita elevata* (Brotzen). Campanian, Syria, x67. Figure 18: *Globotruncanita stuarti* (De Lapparent). Campanian, Syria, x67.

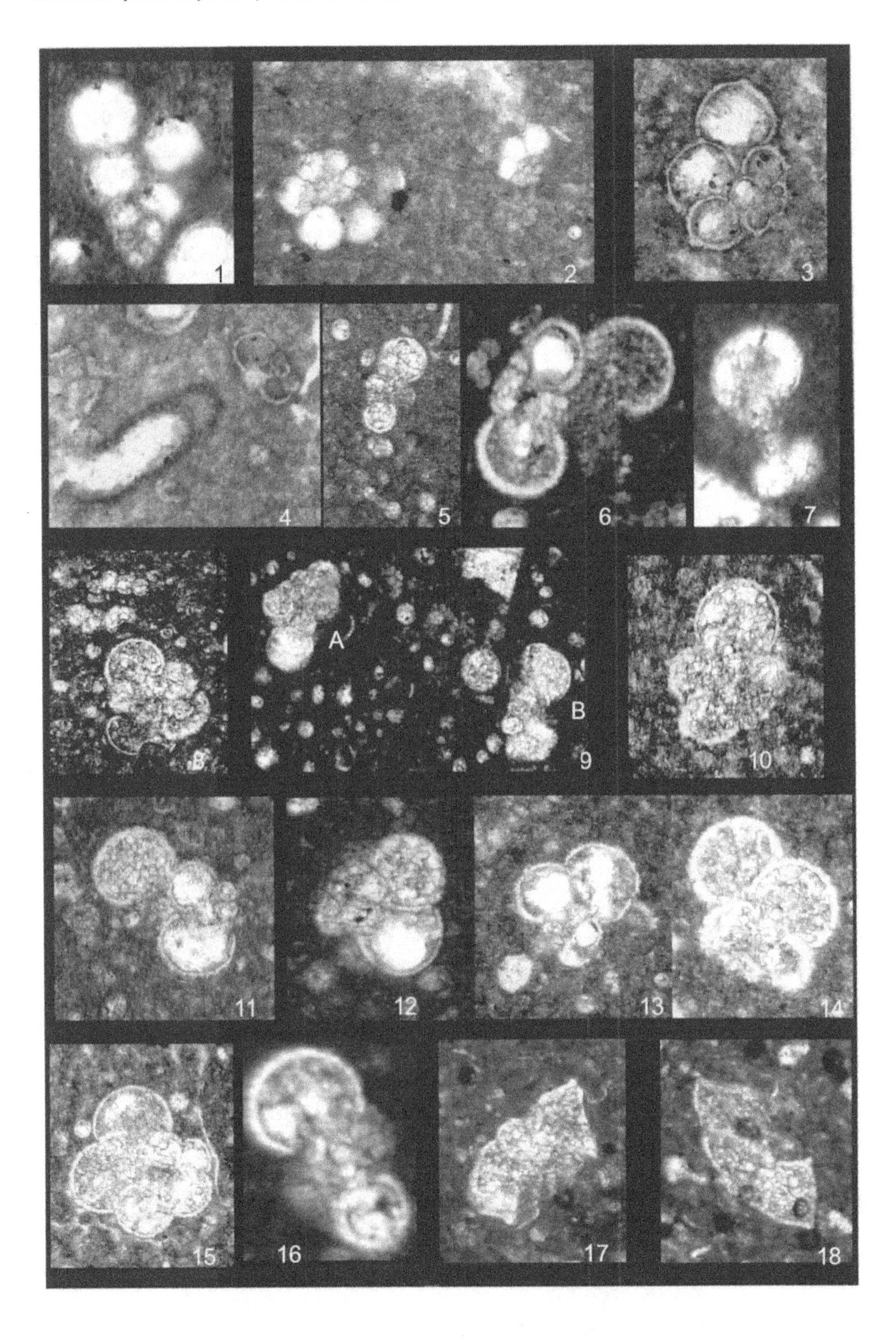
1
2
3
4
5
6
7
8
A
B
9
10
11
12
13
14
15
16
17
18

Plate 4.21. Figure 1: Thin section photomicrographs of *Globotruncanita stuarti* (De Lapparent) and *Heterohelix* spp. Campanian, Syria, UCL Collection, x25. Figures 2 and 4: Thin section photomicrographs of *Abathomphalus mayaroensis* Bolli, *Globigerinelloides* sp. Late Maastrichtian, Kasgan River, Iran, UCL Collection, x35. Figure 3: *Rugotruncana subcircumnodifer* (Gandolfi). Campanian, Syria, UCL Collection, x78. Figure 5: *Racemiguembelina fructicosa* (Egger). Late Maastrichtian, Kasgan River, Iran, UCL Collection, x120. Figure 6: Thin section photomicrographs of Late Cretaceous planktonic foraminifera packstone with *Globotruncana* sp., *Globotruncanita stuartiformis* (Dalbiez), *Heterohelix* sp., and *Hedbergella* spp., UCL Collection, x35. Figure 7: *Globotruncanita stuartiformis* (Dalbiez). Late Maastrichtian, Kasgan River, Iran, UCL Collection, x100. Figure 8: *Globotruncana mariei* Banner and Blow. Campanian, Syria, UCL Collection, x112. Figure 9: *Globotruncana rosetta* (Carsey). Campanian, Syria, UCL Collection, x124. Figure 10: *Globotruncana bulloides* Vogler. Campanian, Syria, UCL Collection, x75. Figure 11: *Globotruncana linneiana* (d'Orbigny). Campanian, Syria, UCL Collection, x72. Figure 12: *Globotruncana arca* (Cushman). Campanian, Syria, UCL Collection, x85.

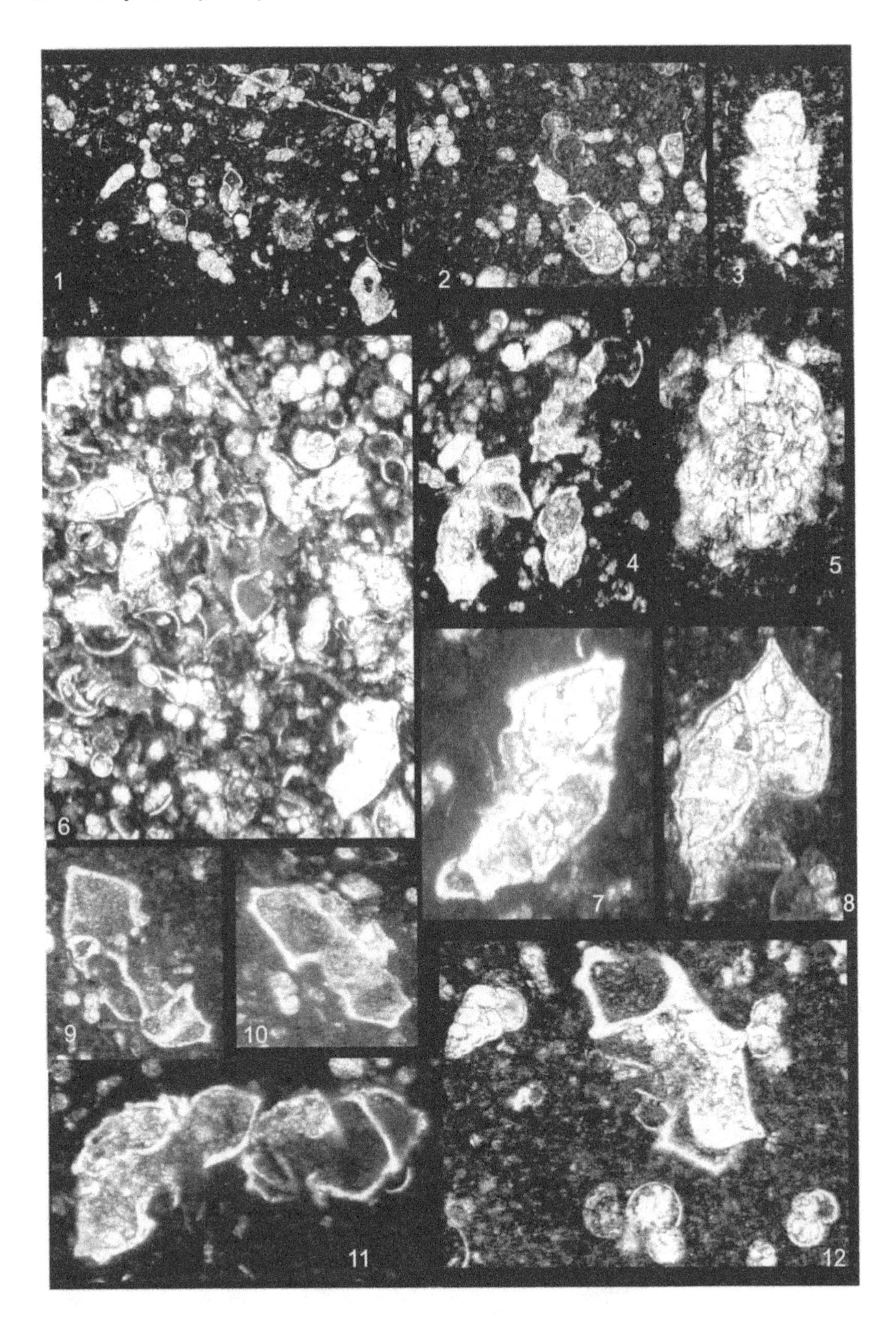
1
2
3
4
5
6
7
8
9
10
11
12

Plate 4.22. Figure 1: *Globotruncana aegyptiaca* Nakkady. Maastrichtian, Kasgan River, Iran, UCL Collection, x50. Figure 2: *Globotruncanita stuarti* (De Lapparent). Late Campanian, Syria, UCL Collection, x83. Figure 3: *Rugoglobigerina pennyi* Olsson. Late Campanian, Syria, UCL Collection, x83. Figure 4: (A) *Globotruncana arca* (Cushman), (B) *Planoglobulina* sp. Campanian, Syria, UCL Collection, x64. Figures 5 and 10: *Heterohelix globulosa* (Ehrenberg). Maastrichtian, Kasgan River, Iran, UCL Collection, x130. Figures 6, 7, 15: *Marginotruncana marginata* Reuss. Coniacian, Kasgan River, Iran, UCL Collection, x83. Figure 8: *Sigalitruncana sigali* (Reichel). Early Santonian, Kasgan River, Iran, UCL Collection, x100. Figure 9: (A) *Hedbergella monmouthensis* (Olson), (B) *Rugoglobigerina rugosa* (Plummer), (C) *Heterohelix* sp. Late Campanian, Syria, UCL Collection, x33. Figure 11: *Globotruncana ventricosa* White. Late Campanian, Syria, UCL Collection, x45. Figure 12: *Globotruncana rosetta* (Carsey). Campanian, Syria, UCL Collection, x78. Figure 13: *Heterohelix reussi* (Cushman). Campanian, Syria, UCL Collection, x120. Figure 14: *Planohedbergella* sp. Campanian, Syria, UCL Collection, x83. Figure 16: *Globotruncanita elevata* (Brotzen). Campanian, Syria, x58. Figure 17: *Contusotruncana contusa* (Cushman). Maastrichtian, Tunisia, x62. Figure 18: *Abathomphalus mayaroensis* Bolli. Late Maastrichtian, Tunisia, x33. Figure 19: *Globotruncana aegyptiaca* Nakkady. Maastrichtian, Kasgan River, Iran, UCL Collection, x62. Figure 20: Thin section photomicrographs of *Globotruncana aegyptiaca* Nakkady, *Globotruncana arca* (Cushman), *Heterohelix* sp. Maastrichtian, Kasgan River, Iran, UCL Collection, x76.

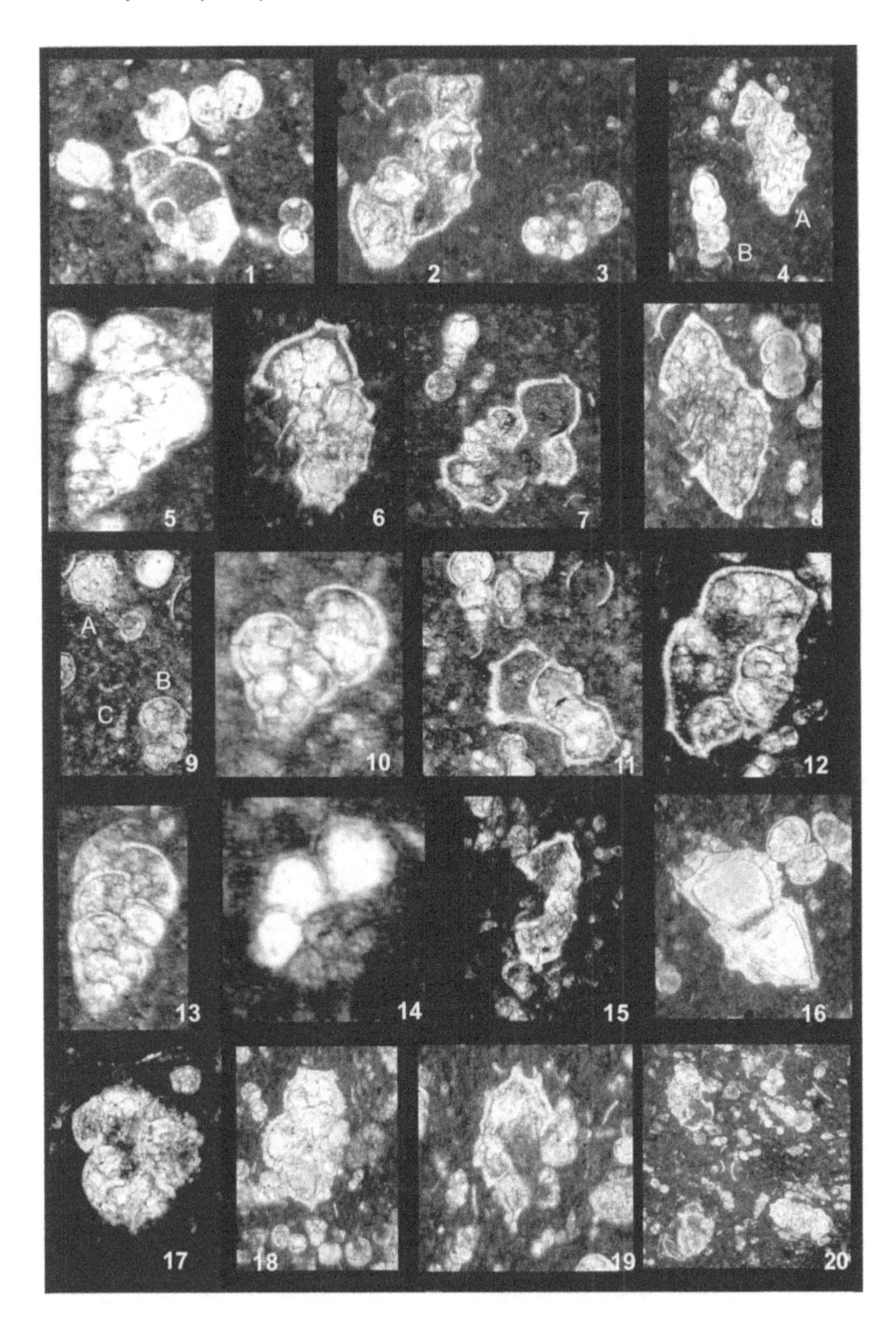
1
2
3
A
B
4
5
6
7
8
A
B
C
9
10
11
12
13
14
15
16
17
18
19
20

Plate 4.23. Figure 1: *Favusella washitensis* (Carsey). Cenomanian, Texas, USA, figured by Postuma (1971), magnification not given. Figure 2: *Biticinella breggiensis* (Gandolfi, 1942). Late Albian, Tunisia, figured by Postuma (1971), magnification not given. Figure 3: *Ticinella roberti* (Gandolfi). Middle Albian, Tunisia, figured by Postuma (1971), magnification not given. Figures 4 and 5: *Planomalina buxtorfi* (Gandolfi). Late Albian, Tunisia, figured by Postuma (1971), magnification not given. Figure 6: *Abathomphalus mayaroensis* Bolli. Late Maastrichtian, Tunisia, figured by Postuma (1971), magnification not given. Figure 7: *Marginotruncana renzi* (Gandolfi). Turonian, Tunisia, figured by Postuma (1971), magnification not given. Figure 8: *Gansserina gansseri* (Bolli). Early Maastrichtian, Turkey, figured by Postuma (1971), magnification not given. Figure 9: *Globotruncanita stuarti* (De Lapparent). Campanian, Tunisia, figured by Postuma (1971), magnification not given. Figure 10: *Helvetoglobotruncana helvetica* (Bolli). Turonian, Tunisia, figured by Postuma (1971), magnification not given. Figures 11, 30–32. *Globotruncana arca* (Cushman). (11) Early Campanian, Tunisia, figured by Postuma (1971), magnification not given. (30–32) Campanian, Kasgan River, Iran, UCL Collection, x80. Figure 12: *Globotruncana bulloides* Vogler. Campanian, Tunisia, figured by Postuma (1971), magnification not given. Figure 13: *Radotruncana calcarata* (Cushman). Campanian, Tunisia, figured by Postuma (1971), magnification not given. Figure 14: *Globotruncanita conica* (White). Late Maastrichtian, Tunisia, figured by Postuma (1971), magnification not given. Figure 15: *Concavatotruncana concavata* (Brotzen). Coniacian, Tunisia, figured by Postuma (1971), magnification not given. Figure 16: *Sigalitruncana sigali* (Reichel). Early Turonian, Tunisia, figured by Postuma (1971), magnification not given. Figure 17: *Contusotruncana fornicata* (Plummer). Campanian, Tunisia, figured by Postuma (1971), magnification not given. Figure 18: *Contusotruncana contusa* (Cushman). Maastrichtian, Tunisia, figured by Postuma (1971), magnification not given. Figure 19: *Globotruncanita conica* (White). Late Maastrichtian, Tunisia, figured by Postuma (1971), magnification not given. Figure 20: *Globotruncanita stuarti* (De Lapparent). Campanian, Tunisia, figured by Postuma (1971), magnification not given. Figure 21: *Praeglobotruncana stephani* (Gandolfi). Topotype, Cenomanian, Tunisia, figured by Postuma (1971), magnification not given. Figure 22. *Globotruncanella citae* (Bolli). Campanian, Tunisia, figured by Postuma (1971), magnification not given. Figure 23: *Rotalipora appenninica* (Renz). Cenomanian, Tunisia, figured by Postuma (1971), magnification not given. Figure 24: *Rotalipora reicheli* Monrod. Late Cenomanian, Tunisia, figured by Postuma (1971), magnification not given. Figure 25: *Thalmanninella greenhornensis* (Morrow). Late Cenomanian, Tunisia, figured by Postuma (1971), magnification not given. Figure 26: *Rugoglobigerina rugosa* (Plummer). Late Maastrichtian, Trinidad, figured by Postuma (1971), magnification not given. Figure 27: *Globotruncana lapparenti* Bolli. Santonian, Tunisia, figured by Postuma (1971), magnification not given. Figure 28: *Alanlordella messinae* (Brönnimann). Campanian, Trinidad, UCL Collection, x100. Figure 29: *Rugoglobigerina pennyi* Brönnimann. Campanian, Kasgan River, Iran, UCL Collection, x86.

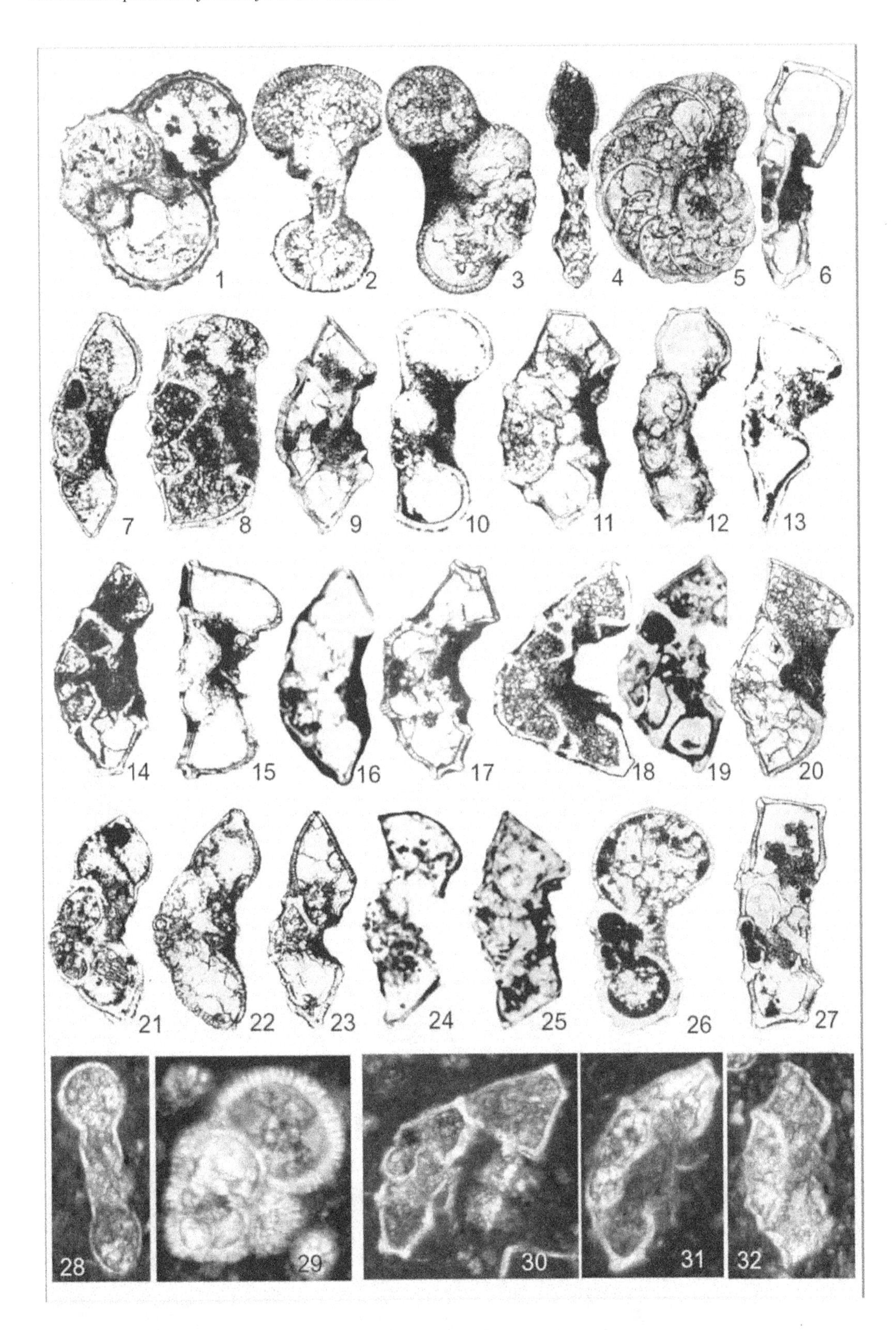
1
2
3
4
5
6
7
8
9
10
11
12
13
14
15
16
17
18
19
20
21
22
23
24
25
26
27
28
29
30
31
32

Plate 4.24. Figure 1: (A) *Racemiguembelina fructicosa* (Egger), (B) *Globotruncanita stuarti* (De Lapparent) (Bolli). Late Maastrichtian, Iran, UCL Collection, x35. Figure 2: *Globotruncana linneiana* (d'Orbigny). Early Maastrichtian, Syria, UCL Coolection, x75. Figure 3: *Abathomphalus mayaroensis* Bolli. Topotypes, Late Maastrichtian, Iran, UCL Collection, x78. Figure 4: Section photomicrographs of planktonic foraminifera with *Globotruncana arca* (Cushman) and *Hedbergella* sp. Maastrichtian, Iran, UCL Collection, x50. Figure 5: *Globotruncana ventricosa* White. Late Campanian, Iran, UCL Collection, x75. Figure 6: Section photomicrographs of planktonic foraminifera with *Globotruncana* sp. and *Contusotruncana contusa* (Cushman). Maastrichtian, Iran, UCL Collection, x23. Figure 7: *Ventilabrella eggeri* Cushman. Early Campanian, Iran, UCL Collection, x30. Figure 8: *Globotruncanita stuarti* (De Lapparent), *Heterohelix* sp. Maastrichtian, Iran, UCL Collection, x58. Figure 9: *Blowiella blowi* (Bolli). Early Aptian, Dorset England, UCL Collection, x100. Figure 10: *Gansserina gansseri* (Bolli). Maastrichtian, Syria, UCL Collection, x70. Figure ·11: *Globotruncana arca* (Cushman). Maastrichtian, Syria, UCL Collection, x87. Figure 12: (A) *Abathomphalus intermidius* (Bolli), (B) *Heterohelix* sp. Maastrichtian, Syria, UCL Collection, x33. Figure 13: *Pseudotextularia elegans* (Rzehak). Middle Campanian–Maastrichtian, Syria, UCL Collection, x86. Figures 14: *Globotruncanita stuarti* (De Lapparent). Maastrichtian, Iran, UCL Collection, x85. Figure 15: (A) *Heterohelix globulosa* (Ehrenberg), (B) *Globotruncanita stuarti* (De Lapparent). Maastrichtian, Iran, UCL Collection, x70. Figure 16: Section photomicrograph of planktonic foraminifera with (A) *Globotruncana arca* (Cushman), (B) *Alanlordella* sp. Maastrichtian, Syria, UCL Collection, x58. Figure 17: Section photomicrograph of planktonic foraminifera with *Globigerinelloides* sp. and *Gansserina gansseri* (Bolli). Maastrichtian, Syria, UCL Collection, x40.

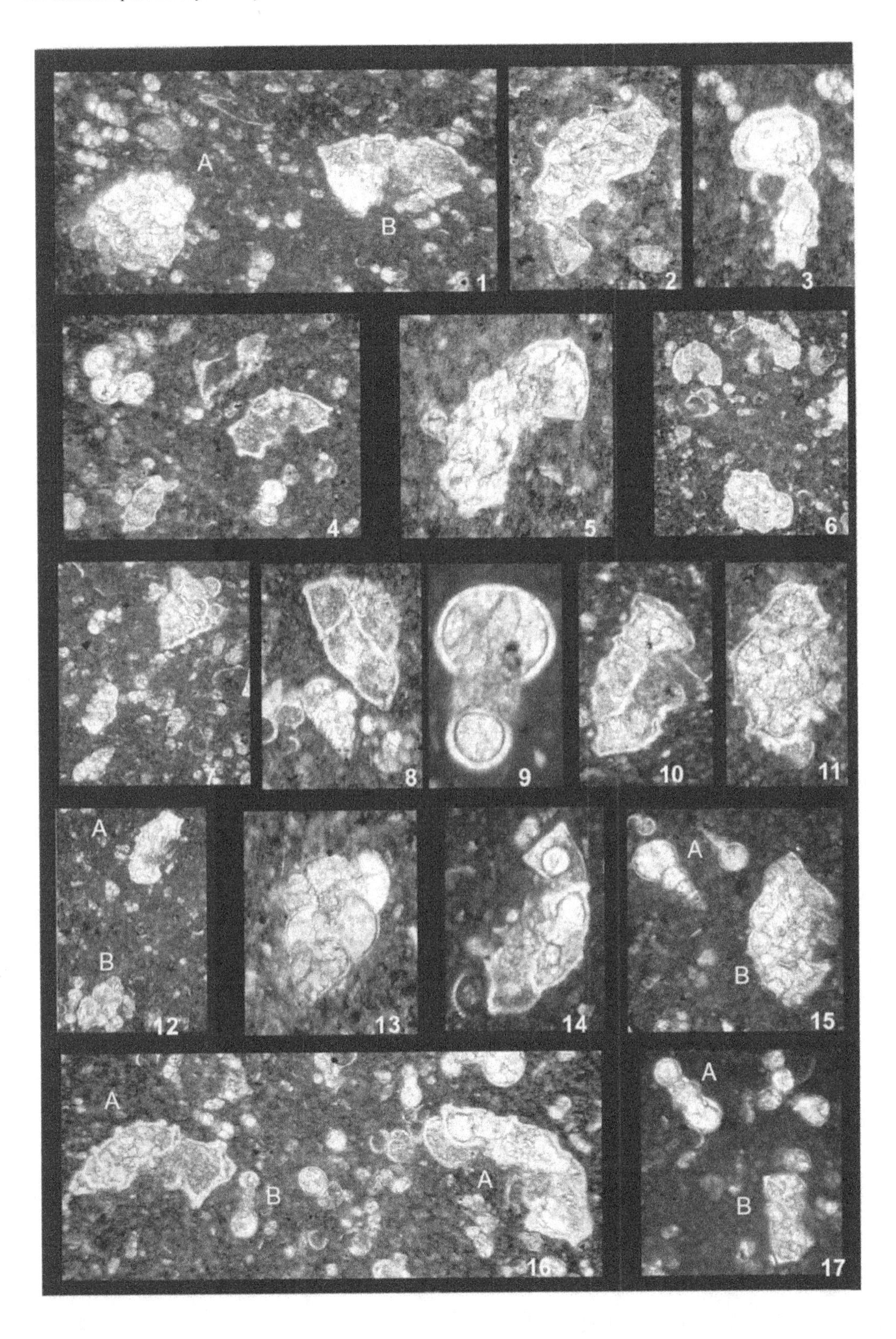
A
B
1
2
3
4
5
6
7
8
9
10
11
A
B
12
13
14
A
B
15
A
B
A
16
A
B
17

Chapter 5

The Cenozoic planktonic foraminifera: The Paleogene

5.1 Introduction

As seen in the previous chapter, the Cretaceous–Paleogene crisis wiped out over 95% of the Maastrichtian planktonic foraminifera. The Early Paleocene was a recovery period for planktonic foraminifera. As is seen during the recovery stage following other extinctions, the planktonic foraminifera that survived the K–P event were morphologically small, exhibiting the post-crisis, ecologically stressed "Lilliput effect" (characterized by a temporary, within-lineage, size decrease after an extinction event; Twitchett, 2006).

Only three of the few survivor species (*Hedbergella monmouthensis, H. holmdelensis*, and *Guembelitria cretacea*) went on to provide the stock from which the subsequent Cenozoic planktonic foraminifera devolped. These ancestors of the Paleogene lineages made their first appearance in the Late Danian (see Fig. 5.1) and included the microperforate globanomalinids (which became extinct in the Early Oligocene); the spinose, cancellate eoglobigerinids (which proliferated through the Paleocene and Eocene, but exhibited reduced numbers in the Oligocene; see Fig. 5.2); and the praemuricate ancestors of the muricate, photosymbiotic truncorotalids, *Acarinina, Morozovelloides*, and *Morozovella* (however, it was not before the Late Paleocene that the latter expanded and diversified worldwide; see Fig. 5.3). The spinose globigerinids made their first appearance in the Early Eocene.

In the tropical and subtropical climates, the globigerinids gave rise to the porticulasphaerids (which did not survive the cooling trend of the Late Eocene; see Fig. 5.3), while the globanomalinids led to the evolution of the keeled planorotalids, and the digitate hantkeninids (which also became extinct as the temperature dropped sharply at the end of the Eocene; see Fig. 5.2). The globigerinids continued into the Miocene where they became diverse, as did the globorotaliids.

The appearance of the tenuitellids from the globanomalinids in the Late Eocene is an example of convergent evolution, as they evolved morphological features which were similar to those of the extinct Early Paleocene *Parvularugoglobigerina* (Plate 5.1, Figs. 4 and 5). The tenuitellids never exhibited the abundance shown by the globigerinids, but they were prominent in the Late Eocene and Oligocene as their counterparts, the globigerinids, had adapted to cooler temperate climates (Fig. 5.4). The triserial guembelitriids and biserial heterohelicids were rare but consistently present throughout the Paleogene, giving rise to the short-lived globoconusids in the Early Paleocene and the tightly coiled, involute, trochospiral cassigerinellids in the Middle Eocene (Fig. 5.5).

Paleogene planktonic foraminifera were mainly cosmopolitan; however, during the warm-climate interlude of the Early to Middle Eocene, specialized tropical and subtropical forms developed. In the Late Eocene, planktonic foraminifera were again more cosmopolitan, being widely distributed in tropical to temperate climates, but it was not until the global temperature began to rise again in the Late Oligocene did specialized forms again become more common.

Paleogene planktonic foraminifera from different localities around the world have been studied in detail by many workers. A detailed zonal biostratigraphy of the Danian and Montian Stages, as recognized in the Crimea, North Caucasus, and "Boreal" areas (Russian Platform and Precaspian Basin), was developed by Morozova (1959, 1960, 1961). Paleocene planktonic foraminiferal biostratigraphy in the West was essentially developed in the form of a detailed zonation defined by Bolli (1957, modified in 1966) for the Paleocene and lower Eocene of Trinidad, which was followed by a zonal scheme for subtropical regions by Berggren (1969, 1971, modified by Berggren and Miller, 1988, and Blow, 1979). Premoli Silva and Bolli (1973) made minor changes to the earlier zonation of Bolli (1957), and Jenkins (1971) formulated a relatively broad zonal biostratigraphic scheme for the Paleocene (as part of a larger Cenozoic study) of New Zealand.

With the recognition that Paleocene low-latitude, subtropical zonations are not fully applicable at high latitudes, Stott and Kennett (1990) developed a zonal biostratigraphy for high (austral) latitudes of the Antarctic. Berggren and Norris (1993) and Berggren *et al.* (1995a, b) redefined the zonal boundary definitions of the five Paleocene zones (and their subdivisions) and also included a magnetochronologic calibration and an estimated timescale. Modifications to this scheme have been proposed by Lu and Keller (1995) based on their study of DSDP Site 577. Additional information on the history of Paleogene planktonic foraminiferal biostratigraphy may be found in Berggren and Miller (1988). More recently, a major contribution to phylogenetic reconstructions came from the Paleogene Planktonic Foraminifera Working Group, notably the publications by Olsson *et al.* (1992, 1999) and Berggren and Norris (1997) for the Paleocene, and by Pearson *et al.* (2006a,b) for the Eocene.

In this chapter, however, in order to resolve some of the outstanding problems with the existing classification of the

Paleogene planktonic foraminifera, a revision of the taxonomy of these forms is presented in which three new superfamilies, three new families, and three new genera are introduced. The biostratigraphic, phylogenetic, paleoenvironmental, and paleogeographic significance of all super-families are then discussed. The plates in this chapter contain new images as well as those taken from original sources. In some cases, images from the late Prof. Banner's Collection (now in UCL) have had morphological features highlighted for pedagogical purposes.

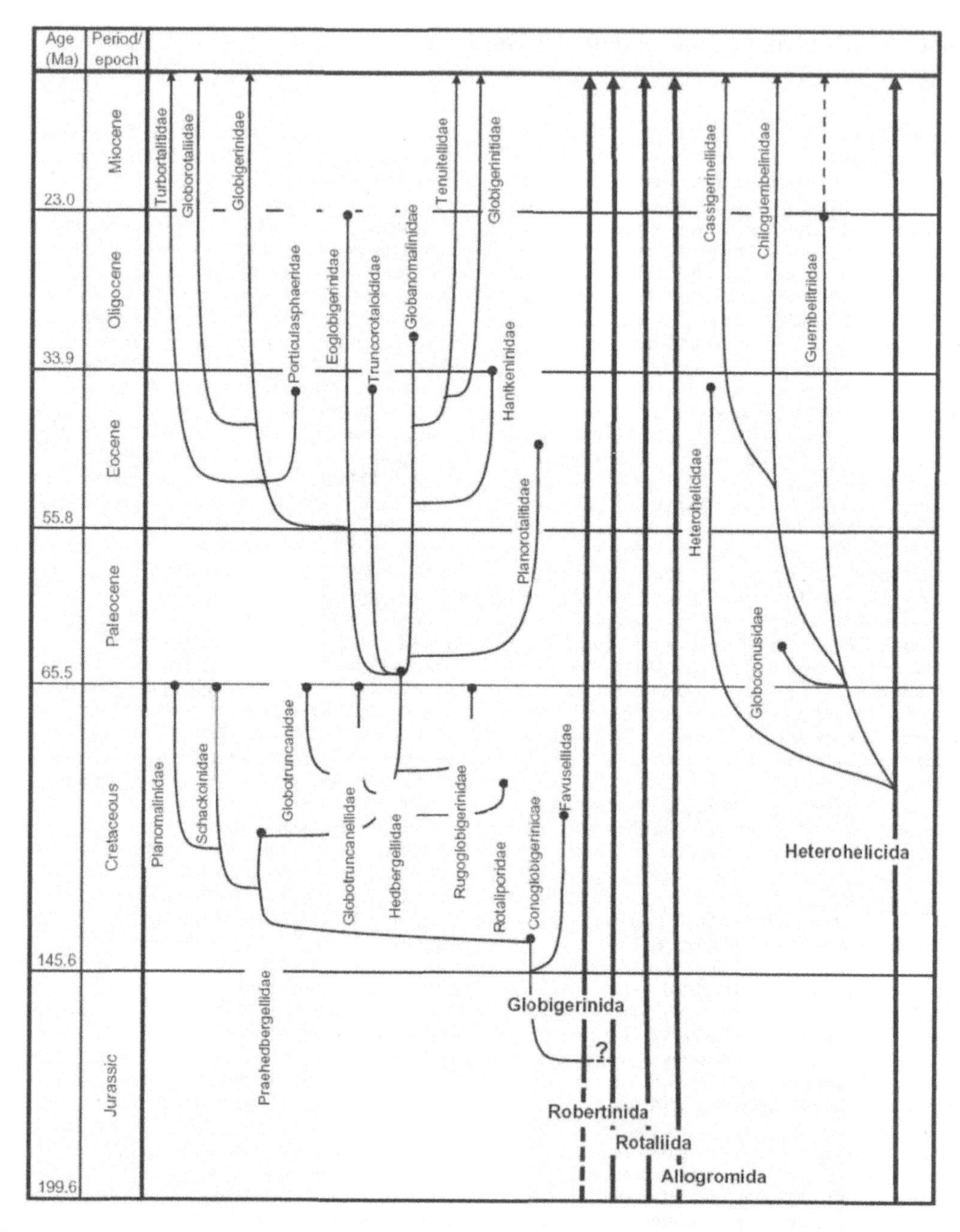

Figure 5.1. The evolution of the Paleogene planktonic foraminiferal families (thin lines) from their Cretaceous ancestors.

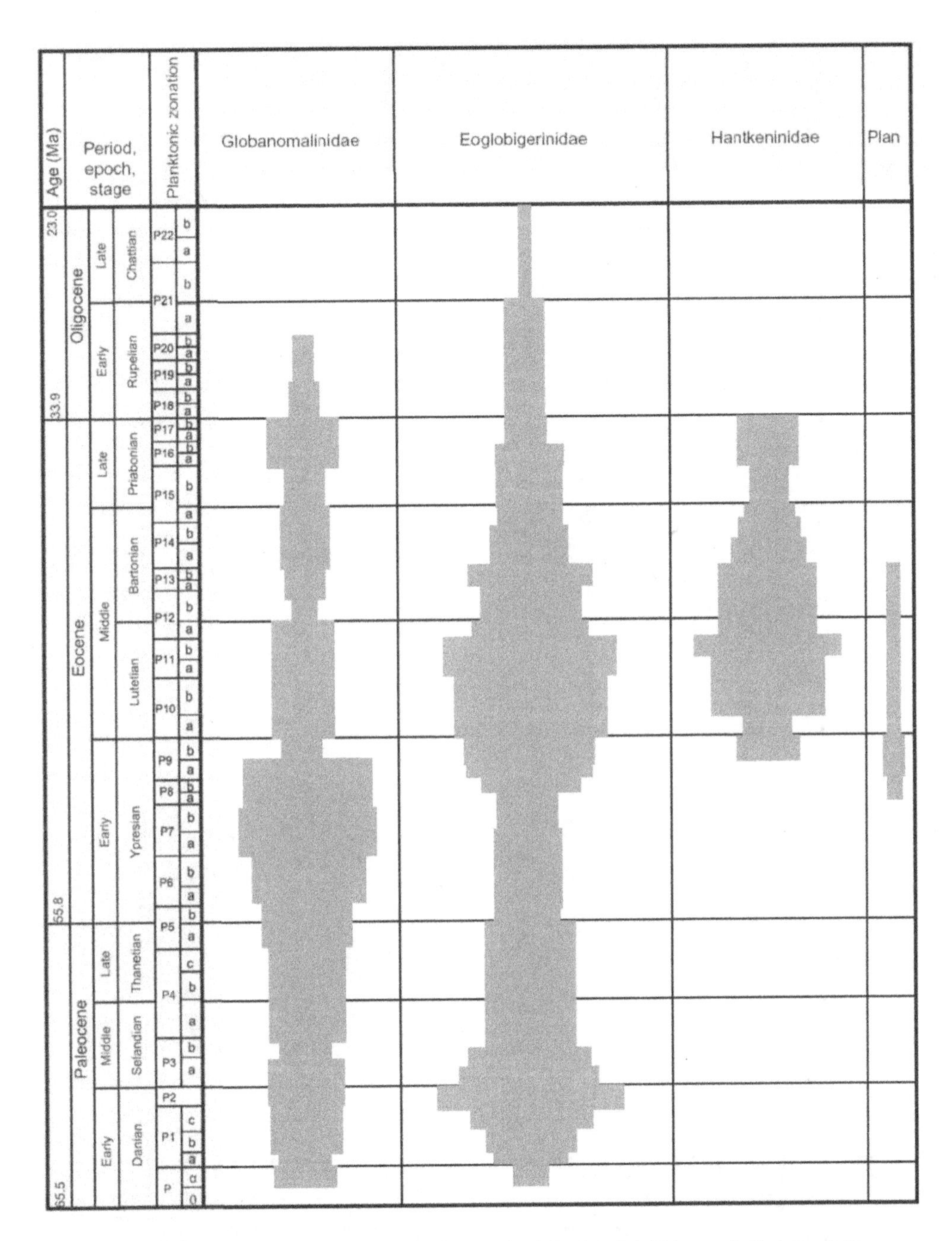

Figure 5.2. The biostratigraphic range and diversity of the main nonspinose, trochospiral, and planispiral Paleogene planktonic foraminifera.

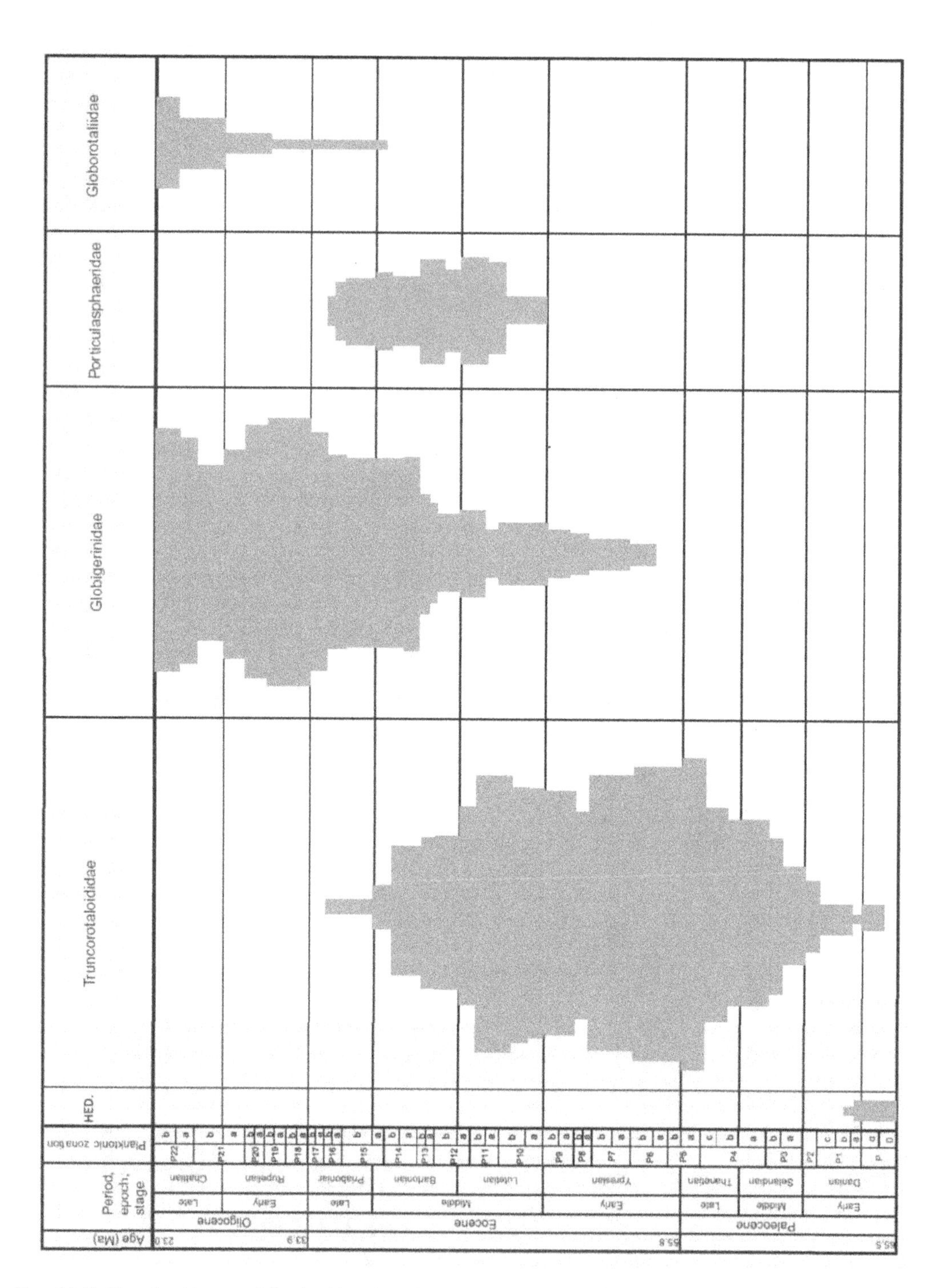

Figure 5.3. The biostratigraphic range and diversity of the main muricate or spinose Paleogene planktonic foraminifera.

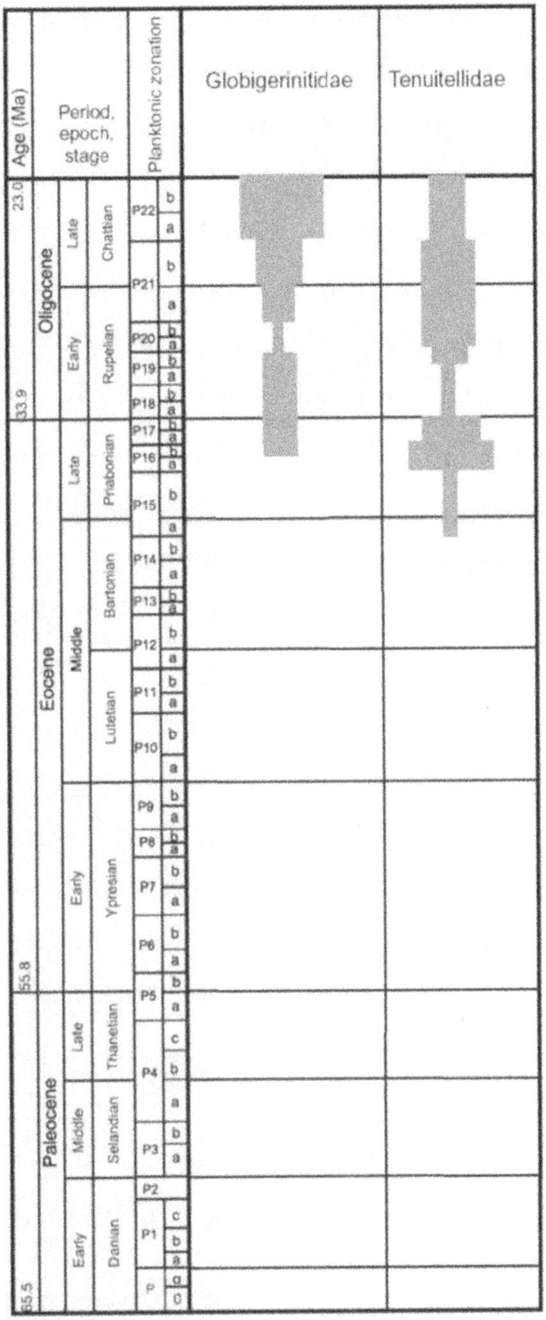

Figure 5.4. The biostratigraphic range and diversity of the main microperforate, pustulose, low trochospiral tenuitellids, and globigerinitids.

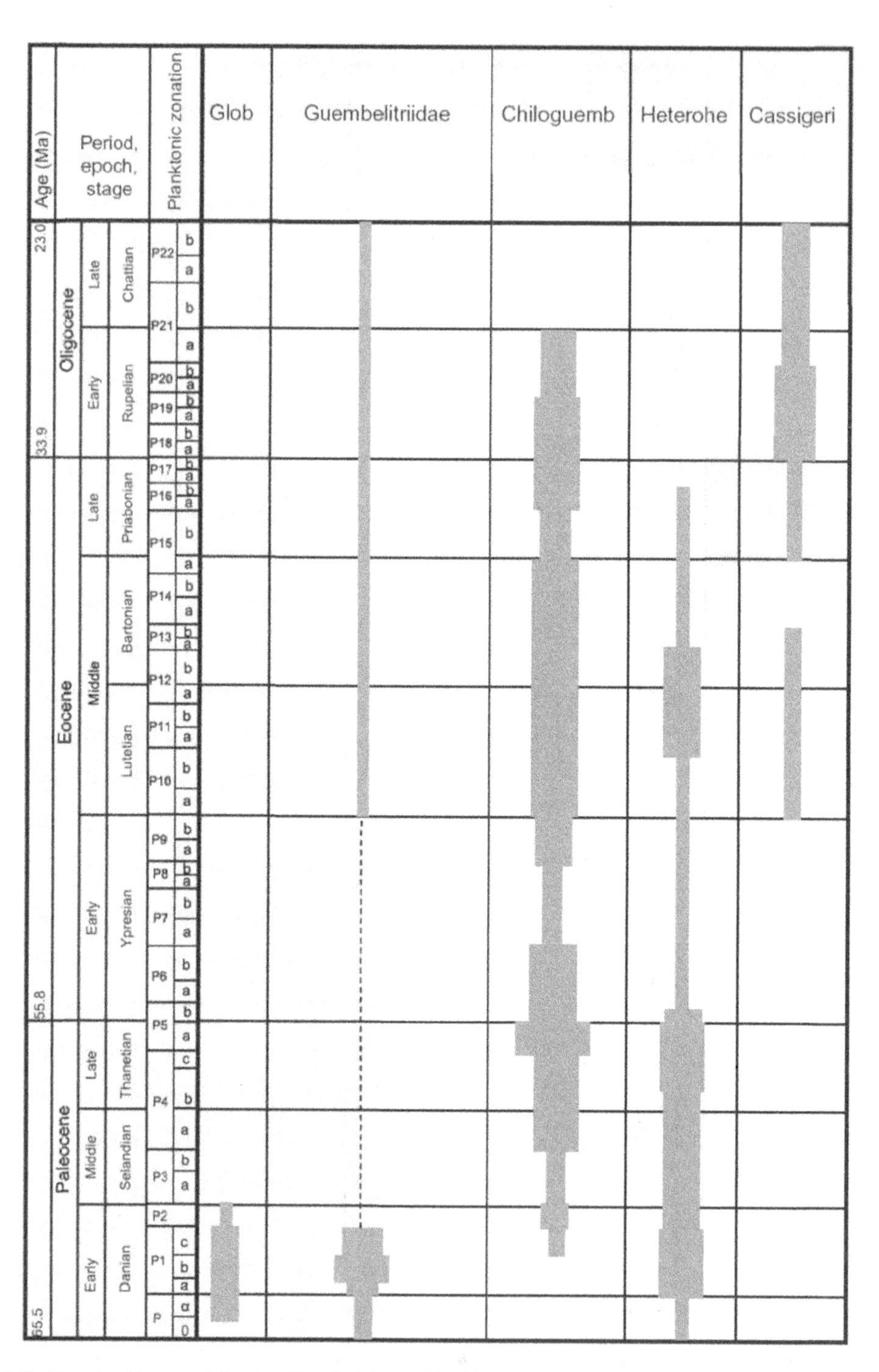

Figure 5.5. The biostratigraphic range and diversity of the main globoconusids and heterohelicids families.

5.2 Morphology and taxonomy of the Paleogene planktonic foraminifera

Here, we suggest that the Paleogene planktonic foraminifera should be divided into seven superfamilies:

- the Globigerinitoidea
- the Hantkeninoidea
- the newly defined Eoglobigerinoidea
- the Globigerinoidea
- the newly defined Truncorotaloidinoidea
- the newly defined Globoconusoidea
- the Heterohelicoidea

The key morphological features which enable the taxonomic classification of the Paleogene planktonic forams are the structural and textural differences in the test wall (Hemleben *et al.*, 1991; Olsson *et al.*, 1992). We have applied these features to define these seven superfamilies as follows.

CLASS FORAMINIFERA Lee, 1990

ORDER GLOBIGERINIDA LANKASTER, 1885

The test is planispiral or trochospiral, at least in the early stage, microperforate or macroperforate, smooth, muricate, or with spines. Apertures are terminal, umbilical, intra-extraumbilical, or peripheral. Walls are calcitic, but early forms may be aragonitic. Late Triassic (Rhaetian) to Holocene.

Superfamily GLOBIGERINITOIDEA BouDagher-Fadel, 2012

The test is calcitic, microperforate, low trochospiral to planispiral, nonspinose, without continuous encrustation, but may have blunt pustules. The umbilicus is open except when it is covered by bulla. Sutural pores may be present. The aperture is intraumbilical or intra-extraumbilical. Paleocene to Holocene.

Family Globanomalinidae Loeblich and Tappan, 1984

The test is trochospiral to nearly planispiral, without carinae or muricae, or spines, or pustules. The wall is smooth or with shallow perforation pits. The aperture is a low to high arch, umbilical–extraumbilical, but not distally elongate. Paleocene to Oligocene.

- *Carinoturborotalia* new genus (Type species: *Globorotalia cerroazulensis cunialensis* Toumarkine and Bolli, 1970). The test is calcitic, trochospiral, finely microperforate, and mostly smooth but may have pustules in the apertural region. About five chambers form the final whorl. The aperture is ventral, and umbilical–extraumbilical. No supplementary apertures are present. The periphery is acutely angular in axial view, with imperforate carina, at least in part of the test. Spiral sutures are limbate, with prominent carina (e.g., *C. cunealensis*) or narrowly depressed forms may exist with carina only developed in the last whorl (e.g., *C. cocoaensis*). The latter evolved from *Turborotalia cerroazulensis* (see Figs. 5.6 and 5.7). Eocene (Bartonian, P14 to Priabonian, P17) (Plate 5.12, Figs. 12–13).
- *Globanomalina* Haque, 1956 (Type species: *Globanomalina ovalis* Haque, 1956). The test is smooth or with shallow perforation pits. The aperture is intra-extraumbilical, but not elongate distally. Paleocene (Early Danian, Pa) to Eocene (Ypresian, P10) (Plate 5.1, Figs. 1–3).
- *Parvularugoglobigerina* Hofker, 1978 (Type species: *Globigerina eugubina* Luterbacher and Premoli Silva, 1964). The test has five to six subglobular chambers per whorl. The wall is smooth. The apertureis intra-extraumbilical and thin, elongated distally, highin the terminal face witha narrow, rim-like lip. The umbilicus is small and shallow. Paleocene (Early Danian, P1, to Early Selandian, P3a) (Plate 5.1, Figs. 4–5).
- *Pseudohastigerina* Banner and Blow, 1959 (Type species: *Nonion micra* Cole, 1927). The aperture is low, with a porticus. Umbilici have very few or no relict apertures. Surfaces may be muricate on early chambers. Chambers are low, not high or radially elongate, with no tubulospines. Accessory apertures are not present. Eocene (Early Ypresian, P6a) to Oligocene (Middle Rupelian, P19b) (Plate 5.2, Figs. 5–11; Plate 5.11, Fig. 5).
- *Pseudomenardella* new genus (Type species: *Globorotalia ehrenbergi* Bolli, 1957). Holotype (Plate 5.1, Figs. 6 and 7). The test has peripheral carina developed on the last chambers, but earlier ones are also acute peripherally. On the spiral side, chambers of the last whorl are two to three times higher than those of the penultimate whorl. The surface is smooth, without muricae or spines, but may have scattered pustules. The aperture is a low umbilical–extraumbilical slit, and it is not elongated digitally in the terminal face. This form evolves into *Planorotalites pseudomenardii* s.s. Paleocene (Selandian, P3a) to Eocene (Late Ypresian, P8) (Plate 5.1, Figs. 6–11; Plate 5.11, Fig. 3).
- *Turborotalia* Cushman and Bermudez, 1949 (Type species: *Globorotalia centralis* Cushman and Bermudez, 1949). The test has four to five or more chambers in the last whorl, and is planoconvex with a flattened spiral side and convex umbilical side, so becoming oviform in axial view. The wall is finely perforate, mostly smooth or with pustules in the apertural region only. The aperture is umbilical to extraumbilical with no supplementary apertures. The periphery becomes subangular or angular in axial view but does not have an imperforate carina. Eocene (Ypresian, P7) to Oligocene (Rupelian, P19a) (Plate 5.2, Figs. 8 and 13–19; Plate 5.4, Fig. 8; Plate 5.11, Figs. 1 and 2; Plate 5.12, Fig. 3B and 4).

Family Planorotalitidae BouDagher-Fadel, 2012, emended

The test is trochospiral, macroperforate, compressed, approximately equally biconvex, with a sharply angled periphery in axial view, with muricocarina which may have tubulospines. Apertures are umbilical, intra-extraumbilical. Paleocene to Eocene.

- *Planorotalites* Morozova, 1957 (Type species: *Globorotalia pseudoscitula* Morozova, 1957). A planorotalitid with no peripheral tubulospines. Paleocene (Selandian, P3b) to Eocene (Bartonian, P13)(Plate 5.2, Figs. 2–4; Plate 5.12, Fig. 12).
- *Astrorotalia* Turnovsky, 1958 (Type species: *Globorotalia (Astrorotalia) stellaria* Turnovsky, 1958). The test has peripheral, muricocarinate tubulospines oriented radially, one per chamber. Eocene (Late Ypresian, P8 to P9) (Fig. 5.6).

Family Tenuitellidae new family

The test is low trochospiral to planispiral. The wall is microperforate in the adult, with perforations irregularly scattered. The surface is unencrusted, but with scattered pustules over the wall surface or only in intercameral sutures, but has no spines. The aperture is interiomarginal, intra-extraumbilical, with a narrow lip (or lips) of uniform breadth. Few or no sutural pores are present. The umbilicus is open except when it is covered by a bulla. Eocene to Holocene.

- *Praetenuitella* Li Qianyu, 1987 (Type species: *Praetenuitella praegemma* Li Qianyu, 1987). The test is low trochospiral, microperforate in the adult, and macroperforate only in the juvenile. The aperture is intra-extraumbilical with no bulla. The surface is smooth or weakly pustulate. Eocene (Late Bartonian, P15a, to Priabonian, P17) (Plate 5.2, Figs. 18 and 19).
- *Tenuitella* Fleisher, 1974 (Type species: *Globorotalia gemma* Jenkins, 1966). The test is low trochospiral, with a pustulate wall. The aperture is intra-extraumbilical, lacking a bulla. Oligocene (Rupelian, P18) to Holocene (N23) (Plate 5.2, Figs. 20 and 21).

Family Globigerinitidae Loeblich and Tappan, 1984

The test is microperforate, trochospiral. The surface is smooth or pustulate. Primary aperture is single. Aperture is intra or intra-extraumbilical and may have a bulla with or without accessory apertures. The umbilicusis open except when coveredbybulla, and few or no sutural pores arepresent. The surface sometimes has pustules. Eocene to Holocene.

- *Antarcticella* Loeblich and Tappan, 1988 (Type species: *Candeina antarctica* Leckie and Webb, 1985). The test is smooth, microperforate, with pustules only in the intercameral sutures. The aperture is covered by a large bulla with small accessory apertures occurring in its suture between the many, closed, tunnel-like marginal extensions of the bulla. Oligocene (Chattian, P22) to Miocene (Serravallian, P12) (Plate 5.1, Figs. 19 and 20).
- *Globigerinita* Brönnimann, 1951 (Type species: *Globigerinita naparimaensis* Brönnimann, 1951). The test is trochospiral, microperforate. The primary aperture is intraumbilical, but bullate. There are accessory apertures in the untunnelled, smooth bulla marginal suture. Pustules are scattered over wall surfaces, but there are no spines or muricae. Oligocene (Chattian, P21b) to Holocene (N23b) (Plate 5.3, Fig. 1).
- *Tenuitellinata* Li, 1987 (Type species: *Globigerinita angustiumbilicata* Bolli, 1957). Holotype (Plate 5.3, Figs. 3 and 4). The wall is smooth or pustulate. The aperture is simple, intraumbilical, with no bulla. Eocene (Priabonian, P16b) to Holocene (Plate 5.3, Figs. 1–5).

Superfamily HANTKENINOIDEA Cushman, 1927

Members of this superfamily have calcitic, planispiral, biumbilicate, macroperforate tests. Walls may be spinose only in the early ontogenetic stage. Chambers are rounded to radially elongate, those of the final whorl may be clavate or with a distinct tubulospine arising from the peripheral margin. The aperture is a high interiomarginal and equatorial opening or may become cribrate. Eocene.

Family Hantkeninidae Cushman, 1927

The test is planispiral, biumbilicate, macroperforate, and smooth, without muricae, or perforation pits, or sutural pores. Chambers are rounded to radially elongate, those of the final whorl may be elongate, clavate, or with a distinct tubulospine arising from the peripheral margin. Adult surfaces are smooth but neanic chambers may be muricate, with spines or spine-bases. The aperture is intra-extraumbilical–peripheral, sometimes also cribrate-areal and always with a porticus. Eocene.

- *Aragonella* Thalmann, 1942 (Type species: *Hantkenina mexicana* Cushman var. Aragonensis Nuttall, 1930) = *Applinella* Bolli *et al.*, 1957. The aperture is high, with a porticus which broadens laterally. Umbilici are small, with few or no relict apertures. Chambers are high, radially elongate, not clavate, with tubulospines which are smooth proximally and bluntly digitate distally, not pointed. Eocene (Lutetian, P10a–P12a)(Plate 5.2, Figs. 5 and 6).
- *Cribrohantkenina* Thalmann, 1942 (Type species: *Hantkenina (Cribrohantkenina) bermudezi* Thalmann, 1942). The aperture is low, with a broad porticus which is penetrated by areal accessory apertures. Chambers are low, not radially elongate or clavate, but with a thick-walled tubulospine per chamber. Eocene (Priabonian, P16 and P17) (Plate 5.2, Figs. 5 and 6; Plate 5.11, Fig. 29).
- *Hantkenina* Cushman, 1925 (Type species: *Hantkenina alabamensis* Cushman) = *Hantkinella* Bolli *et al.*, 1957. The aperture is high, with a porticus which broadens laterally. The umbilici are small, with few or no relict

apertures. Chambers are high but not radially elongate or clavate, with tubulospines which are striate proximally and pointed distally. The surfaces of adult tests are smooth, but neanic chambers may be muricate. Eocene (Early Lutetian, P11b, to Late Priabonian, P17) (Plate 5.2, Figs. 8 and 9).

- *Clavigerinella* Bolli, Loeblich, and Tappan, 1957 (Type species: *Clavigerinella akersi* Bolli, Loeblich, and Tappan, 1957). Early chambers are globular and followed by later radial elongate and clavate chambers. The aperture is high with a porticus which broadens laterally. Umbilici are small, with no relict apertures. The test surface is smooth with no tubulospines and no accessory apertures. Eocene (Late Ypresian, P9b, to Middle Priabonian, P15a) (Plate 5.2, Figs. 16 and 17).

Superfamily EOGLOBIGERINOIDEA new superfamily

Members of this superfamily have a macroperforate, calcitic, trochospiral test with a rounded periphery or one that is angled with a muricocarina, and a smooth surface. The primary aperture is intraumbilical or intra-extraumbilical with no bullae. Chambers may be radially clavate and supplementary, sutural, spiral apertures may be present. The periphery may be angular or rounded in peripheral view and may be smooth or with imperforate carina. Paleocene to Oligocene.

Family Eoglobigerinidae Blow, 1979

The test is smooth macroperforate, or pitted by fine or macroperforations. The wall is weakly to strongly cancellate and spinose. Pustules are confined to umbilical areas if present at all. The periphery is rounded not carinate. The aperture is intraumbilical to intra-extraumbilical. Paleocene to Oligocene.

- *Eoclavatorella* Cremades Campos, 1980 (Type species: *Eoclavatorella benidormensis* Cremades Campos, 1980). The test is macroperforate, smooth with four chambers in the last whorl, and is radially elongate. The periphery is broadly rounded. Eocene (Ypresian, P6a–P9b) (Plate 5.3, Fig. 6).
- *Eoglobigerina* Morozova, 1959 (Type species: *Globigerina (Eoglobigerina) eobulloides* Morozova, 1959). The test is strongly lobulate with a broadly rounded periphery, but not carinate. The wall is macroperforate, smooth, or poorly cancellate. Aperture: a low intraumbilical arch. Paleocene (Danian, Pa to Selandian, P3a) (Plate 5.3, Figs. 7 and 8).
- *Parasubbotina* Olsson, Hemleben, Berggren, and Liu, 1992 (Type species: *Globigerina pseudobulloides* Plummer, 1926). The test is planoconvex with a flattened spiral side and a broadly rounded periphery. The wall is weakly to strongly cancellate and spinose. Paleocene (Late Pa)to Early Eocene (P6) (Plate 5.3, Figs. 11 and 12; Plate 5.11, Fig. 16).
- *Pseudoglobigerinella* Olsson and Pearson, 2006 (Type species: *Globigerina wilsoni bolivariana* Petters, 1954). The test is asymmetrically planispiral in the adult stage with a very low trochospiral juvenile stage and an asymmetrical equatorial aperture. Chambers are inflated, nearly involute. The wall is reticulate with a thickened crust. Eocene (Ypresian, P9 to Lutetian P11) (Fig. 5.8).
- *Subbotina* Brotzen and Pożaryska, 1961 (Type species: *Globigerina triloculinoides* Plummer, 1927). The test is trochospiral with globular chambers. No supplementary apertures are present. The periphery is broadly rounded, and the wall surface is pitted, forming medium to strong cancelation. The aperture is umbilical to extraumbilical, a low slit, with a lip of uniform breadth along its length. Paleocene (Danian, P1) to Late Oligocene (Chattian, P22) (Plate 5.3, Figs.13–16; Plate 5.11, Fig. 19; Plate 5.12, Fig. 1).
- *Turbeogloborotalia* new genus (Type species: *Globorotalia griffinae* Blow, 1979). The test is macroperforate, calcitic, and trochospiral. The surface is smooth, and the axial periphery is rounded to bluntly subangular. The last whorl of the test has four to five rapidly enlarging, globular chambers, but they are not radially elongate or carinate. The penultimate whorl is small. The umbilical side of test is flat to concave. The surface is smooth, without perforation pits, or with broad and widely separated perforation pits on the early whorls. The primary aperture is umbilical, extraumbilical with no bullae. The genus evolved from *Parasubbotina* and into *Pseudoglobigerinella bolivariana* (Fig. 5.8). Eocene (Danian, P1b, to Lutetian, P11) (Plate 5.2, Fig. 1; Plate 5.3, Figs. 17–20; Plate 5.11, Fig. 4).

Superfamily GLOBIGERINOIDEA Carpenter, Parker and Jones, 1862

Members of this superfamily have trochospiral calcitic tests, with chambers that are rounded or angular with a peripheral keel or an imperforate band surrounded by a double keel. When the portici are fused, accessory apertures or supplementary sutural apertures are formed. The wall is micro-perforate or macroperforate, and the surface may be smooth, with or without perforation cones, muricate, or may be spinose. The aperture is interiomarginal, umbilical, or intra-extraumbilical and bordered by a lip, and may have portici, or can be covered by tegilla with accessory apertures. Cretaceous to Holocene.

Family Globigerinidae Carpenter, Parker and Jones, 1862

The test is trochospiral. The wall is smooth, macroperforate with large perforations (up to 5 mm) regularly and geometrically arranged (when not obscured by encrustation), or with perforation pits which may coalesce to form medium cancelation, and with delicate radiating spines from the surface of the adult test, or spine-bases. The primary aperture is a simple arch and umbilical, which may have bullae or a conspicuous portical structure, to extraumbilical, and rimless or with a narrow lip. Eocene to Holocene.

- *Catapsydrax* Bolli, Loeblich, and Tappan, 1957 (Type species: *Globigerina dissimilis* Cushman and Bermudez, 1937). The test is macroperforate and may have spine-bases and muricae near the aperture. The primary aperture is intraumbilical and covered by a regularly developed bulla with few (one to four) large, arched infralaminal accessory apertures at smooth sutural margins of the bulla. No dorsal supplementary apertures and no teeth are present. Eocene (Ypresian, P7) to Pleistocene (N22) (Plate 5.4, Figs. 1–4).
- *Dentoglobigerina* Blow, 1979 (Type species: *Globigerina galavisi* Bermudez, 1961). The test is trochospiral with a wall-lacking strong cancelation. Surfaces are smooth, or with perforation pits, but may have spine-bases and muricae concentrated near the aperture. The aperture is an axio-intraumbilical arch with a tooth-like, subtriangular, symmetrical porticus projecting into the umbilicus. Eocene (Ypresian, P9a) to Pliocene (Zanclean, N20a) (Plate 5.5, Figs. 13–16; Plate 5.6, Figs. 1–5; Plate 5.12, Fig. 2).
- *Globigerina* d'Orbigny, 1826 (Type species: *Globigerina bulloides* d'Orbigny, 1826). A glob-igerinid with subglobular adult chambers, but not radially elongate. The wall is smooth with spines or spine-bases. The aperture is intraumbilical throughout ontogeny, a high umbilical arch which may be bordered by a thin lip, but is without an apertural tooth or bulla. *Globigerina* is distinguished from *Subbotina* and *Dentoglobigerina* by the absence of apertural structures such as a porticus or teeth and in possessing noncancellate walls. *Globigerina* differs from *Globoturborotalita* by the absence of coalescent perforation pits which produce a favose surface. Early Eocene (Ypresian, P6b) to Holocene (N23) (Plate 5.4, Figs. 5–7 and 9–15; Plate5.5, Figs. 1–6 and 16–19).
- *Globigerinoides* Cushman, 1927 (Type species: *Globigerina rubra* d'Orbigny, 1839). The test is a low trochospiral with globular to ovate chambers, with a spiral sutural supplementary aperture or apertures. The wall is coarselyperforate and spinose, sometimes with deep perforation pits. The primary aperture is intraumbilical without bulla or an apertural tooth. Late Oligocene (Late Chattian, P22b) to Holocene (N23) (Plate 5.5, Figs. 9–11).
- *Globoquadrina* Finlay, 1947 (Type species: *Globorotalia dehiscens* Chapman, Parr, and Collins, 1934). The test is macroperforate, trochospiral. The equatorial outline is rounded, subquadrate. The apertural face is flattened, sometimes bullate. The principal apertural extent changes ontogenetically to intraumbilical from early intra-extraumbilical. The aperture usually has an apertural tooth. The surface has spine-bases often concentrated at the distal limits of the apertural face, which may be convex or may be flattened and high. No supplementary apertures are present. Oligocene (Chattian, P22) to Pliocene (Zanclean, N19) (Plate 5.6, Figs. 6 and 7).
- *Globorotaloides* Bolli, 1957 (Type species: *Globorotaloides variabilis* Bolli, 1957). The test is trochospiral, macroperforate. Early ontogeny has the primary aperture umbilical–extraumbilical, but in late ontogeny, the aperture becomes intraumbilical, and may be bullate at any growth stage, but with no apertural tooth. The equatorial outline is rounded, and subcircular. Surfaces have few or no spine bases. Perforation pits may be present. Theapertural face is convexly rounded with no supplementary apertures. Eocene (Ypresian, P5b) to Pliocene(Early Zanclean, N19)(Plate 5.6, Fig. 8).
- *Globoturborotalita* Hofker, 1976 (Type species: *Globigerina rubescens* Hofker, 1956). The test has globular, closely appressed chambers. The wall is cancellate with simple spines. Perforation pits are large, coalescent, and prominent. Eocene (Ypresian, P5b) to Holocene (N23b) (Plate 5.4, Figs. 7 and 16–17; Plate 5.5, Figs. 20 and 21; Plate 5.11, Fig. 6).
- *Guembelitrioides* El Naggar, 1971 (Type species: "*Globigerinoides*" *higginsi* Bolli, 1957) = *Zeaglobigerina* Kennett and Srinivasan, 1983. The test is high spired, loosely coiled, micro-perforate, smooth, or with small spine-bases. The primary aperture is intraumbilical, with spiral, supplementary sutural apertures. The periphery is rounded with four to five chambers in the adult. Eocene (P8–P12a)(Plate 5.3, Figs. 9 and 10).

Family Globorotaliidae Cushman, 1927

The test is macroperforate, trochospiral, convex, acute or has a rounded lens-shape. Only one umbilicus is open in the umbilical side, and the principal aperture is intra-extraumbilical throughout ontogeny or may become intraumbilical in the last few chambers. Chambers may be radially elongate or ampullate. The periphery may be rounded, or carinate, or acutely angled in axial view. The primary aperture is never bullate and never has supplementary apertures. Surfaces are smooth or with spine-bases between perforation pits. Late Eocene to Holocene.

- *Paragloborotalia* Cifelli, 1982 (Type species: *Globorotalia opima* Bolli, 1957) = *Jenkinsella* Kennett and Srinivasan, 1983. The test has an axial periphery that is usually rounded, not carinate. Chambers are not radially elongate. Surfaces usually have spine-bases between perforation pits, or are smooth. The principal aperture is umbilical–extraumbilical throughout adultgrowth. Eocene (Late Bartonian, P15a) to Holocene (N23) (Plate 5.7, Figs. 1–14).

- *Protentelloides* Zhang and Scott, 1995 (Type species: *Protentelloides dalhousiei* Zhang and Scott, 1995). The test is low trochospiral in early stage, becoming planispiral in final whorl. Chambers are globular in early stages, but clavate in final whorl. It differs from *Protentella* (Plate 6.8, Figs. 9-10) in its large prolonged aperture that is axially symmetrical with distinct rim and dendritic apertural tooth, and cancellate surface structure. Late Oligocene (Chattian, P22) (Fig. 5.8).

Family Porticulasphaeridae Banner, 1982

The test is subglobular, spinose, macroperforate, and streptospiral at least in the adult, with the last chamber enveloping the initial trochospire. Spine-bases and/or spines may be present. Chambers in the final whorl cover the umbilicus. Supplementary apertures are only in sutures of ephebic whorls and never areal. Supplementary apertures are present in the last sutures, often furnished with bullae or thick lips. Accessory apertures on the margin of each bulla may be present. Eocene.

- *Globigerinatheka* Brönnimann, 1952 (Type species: *Globigerinatheka barri* Brönnimann, 1952). The test has four to five chambers in the final whorl covering the umbilicus. Many large supplementary apertures are present in the last sutures, furnished with small, separated bullae or broad lips. Few accessory apertures are present on the margin of each bulla. Eocene (Lutetian, P10b to Middle Priabonian, P16a) (Plate 5.5, Fig. 7; Plate 5.6, Figs. 15–17; Plate 5.11, Figs. 11 and 12; Plate 5.12, Fig. 8C).
- *Inordinatosphaera* Mohan and Soodan, 1967 (Type species: *Inordinatosphaera indica* Mohan and Soodan, 1967). The test has four to five chambers in the final whorl, which cover the umbilicus so that extraumbilical slit-like or pore-like apertures become developed in chamber sutures. Sutural apertures are covered by long, often closely adjacent bullae, each bulla has many, small accessory apertures in its margin. Spine-bases or spines are present. Eocene (Lutetian, P11, to Bartonian, P13) (Fig. 5.9).
- *Orbulinoides* Cordey, 1968 (Type species: *Porticulasphaera beckmanni* Saito, 1962). The test is almost globular. Early chambers are trochospirally coiled, with the last six to seven chambers in the final whorl streptospirally coiled, increasing gradually in size with a very large final chamber covering the umbilicus. Spine-bases or spines are present. Supplementary, extraumbilical apertures have no lips and no bullae. Eocene (Bartonian, P13) (Plate 5.6, Fig. 19; Plate 5.11, Fig. 10).
- *Porticulasphaera* Bolli, Loeblich, and Tappan, 1957 (Type species: *Globigerina mexicana* Cushman, 1925). The test has four to five chambers in the final whorl, which cover the umbilicus. Few, arched, supplementary, extraumbilical apertures are present in the last intercameral sutures. Apertures often have thick lips but are never bullate. Spine-bases or spines are present. Eocene (Middle Lutetian, P11a, to Middle Priabonian, P16b) (Plate 5.6, Figs. 10–14; Plate 5.11, Fig. 9).

Family Turborotalitidae Hofker, 1976

The species of this family are trochospiral to planispiral, microperforate to macroperforate, small, but the numbers of whorls and their calcitic crusts readily distinguish them from juvenile tests of taxa of other families. Eocene to Holocene.

- *Turborotalita* Blow and Banner, 1962 (Type species: *Truncatulina humilis* Brady, 1884). Test is macroperforate, trochospiral. The surface is spinose or with spine-bases between perforation pits, but maybe smoothed by added surface lamellae. The primary aperture is umbilical–extraumbilical. The apertural lip is broad, making the final chamber ampullate, with small, accessory infralaminal apertures at the margin of the ampulla, which may cover the umbilicus. Eocene (Lutetian, P10) to Holocene (N23b) (Fig. 5.8).

Superfamily TRUNCOROTALOIDINOIDEA new superfamily

Members of this superfamily have a calcitic, macroperforate, muricate test. Peripheral muricocarina may be present. The axial periphery may be rounded to acutely angular. The primary aperture is intraumbilical, or intra-extraumbilical with a thin lip, but with no porticus or tegillum. Supplementary dorsal apertures may be present. Walls are muricate, but without spines, spine-bases, spinules, or pustules. Paleocene (Danian) to Eocene (Bartonian).

Family Truncorotaloididae Loeblich and Tappan, 1961

The test is macroperforate, trochospiral, muricate, or pustulose, which may fuse at the periphery to form muricocarina. The aperture is intra-extraumbilical, with a thin lip. Supplementary spiral, sutural apertures may be present. Paleocene to Eocene.

- *Acarinina* Subbotina, 1953 (Type species: *Acarinina acarinata* Subbotina, 1953) = *Muricoglobigerina* Blow, 1979 = *Pseudogloboquadrina* Jenkins, 1965. The test is planoconvex, with a broadly rounded or subangular periphery, but with no enlarged peripheral muricae and no muricocarinae. Paleocene (Late Danian, P2) to Eocene (Late Bartonian, P15a) (Plate 5.7, Figs. 15–20; Plate 5.8, Figs. 1–6; Plate 5.11, Figs. 20 and 22).
- *Globigerapsis* Bolli, Loeblich, and Tappan,1957 (Type species:*Globigerapsis kugleri* Bolli, Loeblich, and Tappan, 1957). The test is macroperforate, muricate, initially trochospiral becoming ventrally streptospiral, with the closure

of the umbilicus by embracing ventral chambers. Multiple, supplementary, archedapertures occur in the last sutures. No spines, spine-bases, or pustules are present. Eocene (Early Lutetian, P10, to Priabonian, P16b) (Plate5.6, Figs. 18 and 19; Plate5.11, Figs. 7 and 8).

- *Igorina* Davidzon, 1978 (Type species: *Globorotalia tadjikistanensis* Bykova, 1953). The test is biconvex, cancellate, pustulose with a somewhat elevated spiral side. The axial periphery is subacute and noncarinate. The umbilicus is narrow and shallow. Paleocene (Selandian, P3) to Eocene (Bartonian, Early P15) (Plate 5.8, Figs. 16–20; Plate 5.11, Figs. 14 and 21).
- *Morozovella* McGoran, 1964 (Type species: *Pulvinulina velascoensis* Cushman, 1925). The test is planoconvex, with an axial periphery that is acutely angular with muricocarina. Dorsal intercameral sutures are usually muricate and limbate with no supplementary sutural aperture. Paleocene (Danian, P2) to Eocene (Middle Bartonian, P14b) (Plate 5.8, Figs. 7–15; Plate 5.9, Figs. 1–5 and 7–9; Plate 5.11, Figs. 18 and 23–26; Plate 5.12, Figs. 5 and 7).
- *Morozovelloides* Pearson and Berggren, 2006 (Type species: *Globorotalia lehneri* Cushman and Jarvis, 1929). The test is low trochospiral, generally lobulate or petaloid in outline, with concentrationof bladed muricae around theperiphery, forminga discontinuous muricocarina. The final whorl contains four to eight chambers. The aperture is umbilical–extraumbilical, or wholly extraumbilical. Eocene (Ypresian, P9b) to Bartonian (P15a). (Plate 5.12, Fig. 6).
- *Praemurica* Olsson, Hemleben, Berggren, and Liu 1992 (Type species: *Globigerina* (*Eoglobigerina*) *taurica* Morozova, 1961). The test is low trochospiral with globular to slightly ovoid chambers. The wall is weakly to strongly cancellate and nonspinose. The aperture is bordered by a narrow lip. Paleocene (Danian, Late Pa to P2) (Plate 5.9, Figs. 10–15; Plate 5.11, Fig. 17).
- *Testacarinata* Jenkins, 1951 (Type species: Globorotalia inconspicua Howe, 1939). The test is planoconvex with a subangular to broadly rounded periphery, with peripheral muricae enlarged but remaining separated, and unfused. Eocene (Middle Lutetian, P11,to Late Bartonian, P15a) (Plate 5.9, Fig. 16).
- *Truncorotaloides* Brönnimann and Bermudez, 1953 (Type species: *Truncorotaloides rohri* Brönnimann and Bermudez, 1953). The test is planoconvex with an initially rounded axial periphery but may become angular. Supplementary apertures occur in the spiral sutures. Spiral intercameral sutures are not limbate but may have a ring of enlarged muricae on the peripheral to umbilical faces of the later chambers. Eocene (Late Ypresian, P8, to Bartonian, P14) (Plate 5.9, Figs. 18–22; Plate 5.11, Fig. 15).

Superfamily GLOBOCONUSOIDEA new family

Members of this superfamily have trochospiral, calcitic, microperforate tests, with pustules. The primary aperture is intraumbilical or intra-extraumbilical, without a porticus or tegillum. No muricocarina, no spine-bases, or spines are present and perforation pits are always absent. Early Paleocene (Danian, Pa to P2).

Family Globoconusidae new family

The test is trochospiral, calcitic, microperforate, muricate with sometimes spiral, supplementary, sutural apertures. Four chambers are found in the last whorl. The primary aperture is intraumbilical, and lacking a porticus, but sometimes has bullae. These forms are always without tegillum, costellae, muricocarinae, perforation pits, or spines. Paleocene (Danian, Pa to P2).

- *Globoconusa* Khalilov, 1956 (Type species: *Globigerina daubjergensis* Brönnimann, 1953) = *Globastica* Blow, 1979. The test is minute with three to four inflated chambers in the whorl and a strongly convex spiral side.The wall has fine sharp pustules.The umbilicusis closed. The aperture is a small, low to rounded opening near the umbilicus that may be covered by a bulla and with one or more tiny secondary openings. Paleocene (Early Danian, Pa to P2)(Plate5.10, Figs.1 and2).
- *Postrugoglobigerina* Salaj, 1986 (Type species: *Postrugoglobigerina hariana* Salaj, 1986). The test has five or more chambers in the last whorl. The aperture is intra-extraumbilical. Paleocene (Early Danian, Pa)(Fig. 5.5).

ORDER HETEROHELICIDA FURSENKO, 1958

Tests are biserial or triserial, at least in the early stage, but they may be reduced to uniserial in later stages. They are microperforate or macroperforate, and smooth or muricate. Apertures are terminal, with a low to high arch. Walls are calcitic. Cretaceous (Aptian) to Holocene.

Superfamily HETEROHELICOIDEA Cushman, 1927

The test is mainly planispiral in the early stage, then biserial to triserial and possibly multiserial, rarely becoming uniserial in the adult stage. Apertures have a high to low arch at the base of the final stage or are terminal in the uniserial stage. Walls are calcitic, smooth, or muricate. Cretaceous (Aptian) to Holocene.

Family Cassigerinellidae Bolli, Loeblich and Tappan, 1957

The test is biserial or triserial in early stages, later becoming enrolled biserial, but with biseries coiled into a tight, involute trochospire. Chambers may be inflated or compressed. The primary aperture ranges from being asymmetric to being in the equatorial plane. Eocene to Miocene.

- *Cassigerinella* Pokorný, 1955 (Type species: *Cassigerinella boudecensis* Pokorný, 1955). The test has inflated chambers, which are reniform, or subglobular, with an anterior marginal aperture as a high, broad arch. Late Eocene (Priabonian, P14) to Middle Miocene (Serravallian, N13) (Plate 5.10, Figs. 3–6).
- *Cassigerinelloita* Stolk, 1965 (Type species: *Cassigerinelloita amekiensis* Stolk, 1965). The test is biserial, but the biseries is coiled into a tight, involute trochospire. Chambers are inflated. The aperture is interiomarginal and situated lateroposteriorly, with a low, broad arch. Middle Eocene (Lutetian, P10, to Bartonian, P13) (Plate 5.10, Fig. 5.7).

Family Guembelitriidae Montanaro Gallitelli, 1957

The test is triserial throughout with a straight axis of triseriality, or becoming multiserial in the adult, with globular inflated chambers. The aperture is a simple arch bordered by a rim, symmetrical about equatorial plane, or can have more than one aperture per chamber in the multiserial stage. Walls may be muricate. Cretaceous (Late Albian) to Holocene.

- *Chiloguembelitria* Hofker, 1978 (Type species: *Chiloguembelitria danica* Hofker, 1978) = *Jenkinsina* Haynes, 1981. The test is microperforate and triserial throughout. The wall is muricate. Cretaceous (Late Campanian) to Late Oligocene (Chattian, P22) (Plate 4.19, Figs. 37 and 38; Plate 5.10, Figs. 8–14).
- *Guembelitria* Cushman, 1940 (Type species: *G. cretacea* Cushman, 1933). The test is micro-perforate with pore mounds and is triserial throughout. The aperture is a simple, comma-shaped arch at the base of the final chamber, bordered with a lip. The wall is smooth or has perforation cones, or both. Cretaceous (Late Albian) to Paleocene (Early Danian, Pa) (Plate 4.19, Figs. 1–3).
- *Woodringina* Loeblich and Tappan, 1957 (Type species: *Woodringina claytonensis* Loeblich and Tappan, 1957). The test is initially triserial and terminally biserial. The wall is finely micro-perforate with pore mounds, the later part is smooth. The aperture is terminal with an apertural rim and is interiomarginal and asymmetric to the plane of biseries. Paleocene (Pa to P1c) (Plate 5.10, Fig. 20).

Family Heterohelicidae Cushman, 1927

The test is planispiral in the early stages, then biserial, or the biserial stage may be followed by a multiserial stage in the adult. Walls are microperforate to macroperforate. The primary aperture is asymmetrical about the equatorial plane. Cretaceous (Late Albian) to Eocene (Priabonian).

- *Bifarina* Parker and Jones, 1872 (Type species: *Dimorphina saxipara* Ehrenberg, 1854). The test is macroperforate, initially biserial, becoming uniserial, with the terminal aperture becoming areal. Chambers have no lateral extensions but remain subglobular, reniform, or ovoid. Cretaceous (Late Cenomanian) to Middle Eocene (Bartonian, P13) (Plate 4.18, Figs. 27–31; Plate 5.10, Fig. 21).
- *Zeauvigerina* Finlay, 1939 emend. Huber and Boersma, 1994 (Type species: *Zeauvigerina* zelandica Finlay, 1939). The test is microperforate. The surface is covered by irregularly scattered pustules. Chambers are biserially arranged in the early part, with a tendency to become uniserial in the latter part, increasing gradually in size. The aperture is terminally positioned, circular, or oval-shaped and may be produced on a short neck. Late Cretaceous (Maastrichtian, 3b) to Eocene (Priabonian, P17a) (Plate 5.10, Fig. 22).

Family Chiloguembelinidae Loeblich and Tappan, 1956

The test is biserial throughout. The aperture is interiomarginal, asymmetrical, extending up the face of the final chamber, bordered by an apertural rim, in the plane of biseriality, invaginated to make an internal plate, or not infolded or invaginated. Paleocene (Danian) to Holocene.

- *Chiloguembelina* Loeblich and Tappan, 1956 (Type species: *Guembelina midwayensis* Cushman, 1940). The test is compressed with chambers of each pair appressed, with no median plate or supplementary sutural apertures. Walls have poreless fine pustules. An interiomarginal aperture is asymmetric to the plane of the biseries, with the asymmetry alternating from chamber to successive chamber. The apertural margin in the equatorial plane is not infolded to form an internal plate. Paleocene (Danian, P1a) to Early Oligocene (Rupelian, P19b) (Plate 5.10, Figs. 15–18).
- *Streptochilus* Brönnimann and Resig, 1971 (Type species: *Bolivina tokelauae* Boersma, 1969). The test is compressed with chambers of each pair appressed, with no median plate or supplementary sutural apertures. An interiomarginal aperture is asymmetric to the plane of the biseries, with the asymmetry alternating from chamber to successive chamber, with the apertural margin in the equatorial plane infolded to form an invaginated, internal plate. Eocene (Ypresian, P8) to Holocene (N23) (Plate 5.10, Fig. 19).

Diagnostic First Occurrence of Planktonic Foraminifera

24.0 *Paragloborotalia mayeri, Globigerina brazieri*
26.8 *Globoturborotalia woodi, Paragloborotalia pseudokugleri*
28.1 *Globigerinita praestainforthi, Turborotalita praequinqueloba*
29.2 *Tenuitellinata juvenilis*
29.7 *Paragloborotalia opima*
30.3 *Tenuitella neoclemenciae*
31.0 *Tenuitella clemenciae*
32.0 *Globoturborotalia fariasi*
32.5 *Globigerina sellii*
33.9 *Tenuitella gemma*
Globigerina ampliapertura
Cribrohantkenina inflata
35.1 *Turborotalia pseudoampliapertura*
37.8 *Globorotaloides eovariabilis*
38.0 *Praetenuitella impariapertura*
38.5 *Subbotina angiporoides*
39.2 *Carinoturborotalia cocoaensis*
Orbulinoides beckmanni
42.1 *Hantkenina alabamensis/Truncorotaloides topilensis*
43.2 *Hantkenina mexicana*
44.5 *Acarinina spinuloinflata, Turborotalia centralis*
47.0 *Truncorotaloides haynesi*
47.8 *Igorina anapetes, Aragonella nuttalli*
49.0 *Globigerina hagni, Morozovella caucasica*
50.1 *Dentoglobigerina galavisi, Acarinina collactea*
50.5 *Guembelitrioides higginsi*
51.0 *Acarinina decepta*
51.7 *Acarinina pentacamerata*
52.3 *Morozovella aragonensis*
54.0 *Morozovella formosa, M. crater*
54.3 *Morozovella quetra*
56.0 *Morozovella subbotinae, Igorina broedermanni*
56.4 *Globanomalina australiformis, Acarinina wilcoxensis*
57.0 *Acarinina esnaensis, A. primitiva*
59.2 *Acarinina mckannai, Pseudomenardella pseudomenardii*
60.0 *Morozovella acuta, Morozovella velascoensis*
60.3 *Morozovella conicotruncata, M. angulata, Igorina albeari, Acarinina praesphaerica*
61.6 *Igorina pusilla, Morozovella angulata*
62.0 *Subbotina triangularis, Praemurica uncinata*
63.5 *Subbotina trinidadensis, S. triloculinoides, Parasubbotina varianta*
64.7 *Trochoguembelitria polycamera, Praemurica inconstans*
65.2 *Subbotina simplicissima, Parasubbotina pseudobulloides*
65.46 *Parvularugoglobigerina eugubina, Globanomalina planocompressa*

Diagnostic Last Occurrence of Planktonic Foraminifera

24.0 *Globigerina sellii*
26.8 *Tenuitella gemma/Paragloborotalia opima*
28.1 *Globoturborotalita angulisuturalis*
33.9 *Hantkenina alabamensis*
38.0 *Morozovella spinulosa/Truncorotaloides topilensis*
39.2 *Orbulinoides beckmanni/Morozovella lehneri*
41.2 *Globigerapsis kugleri/Guembelitrioides higginsi*
42.3 *Igorina anapetes*
44.9 *Acarinina decepta*
47.0 *Morozovella caucasica*
51.0 *Acarinina wilcoxensis*
54.0 *Morozovella occlusa*
56.0 *Subbotina velascoensis*
59.2 *Zeauvigerina teuria*
60.0 *Parasubbotina variospira/Igorina pusilla*
60.3 *Eoglobigerina spiralis*
61.0 *Eoglobigerina edita/Globoconusa daubjergensis*
64.7 *Hedbergella holmdelensis/H. monmouthensis*
65.2 *Parvularugoglobigerina eugubina*
65.46 *Guembelitria cretacea*

Figure 5.6. Diagnostic first and last occurrences of Paleogene zones compared with the earlier zonations of Berggren and Pearson (2005) and Wade *et al.* (2011).

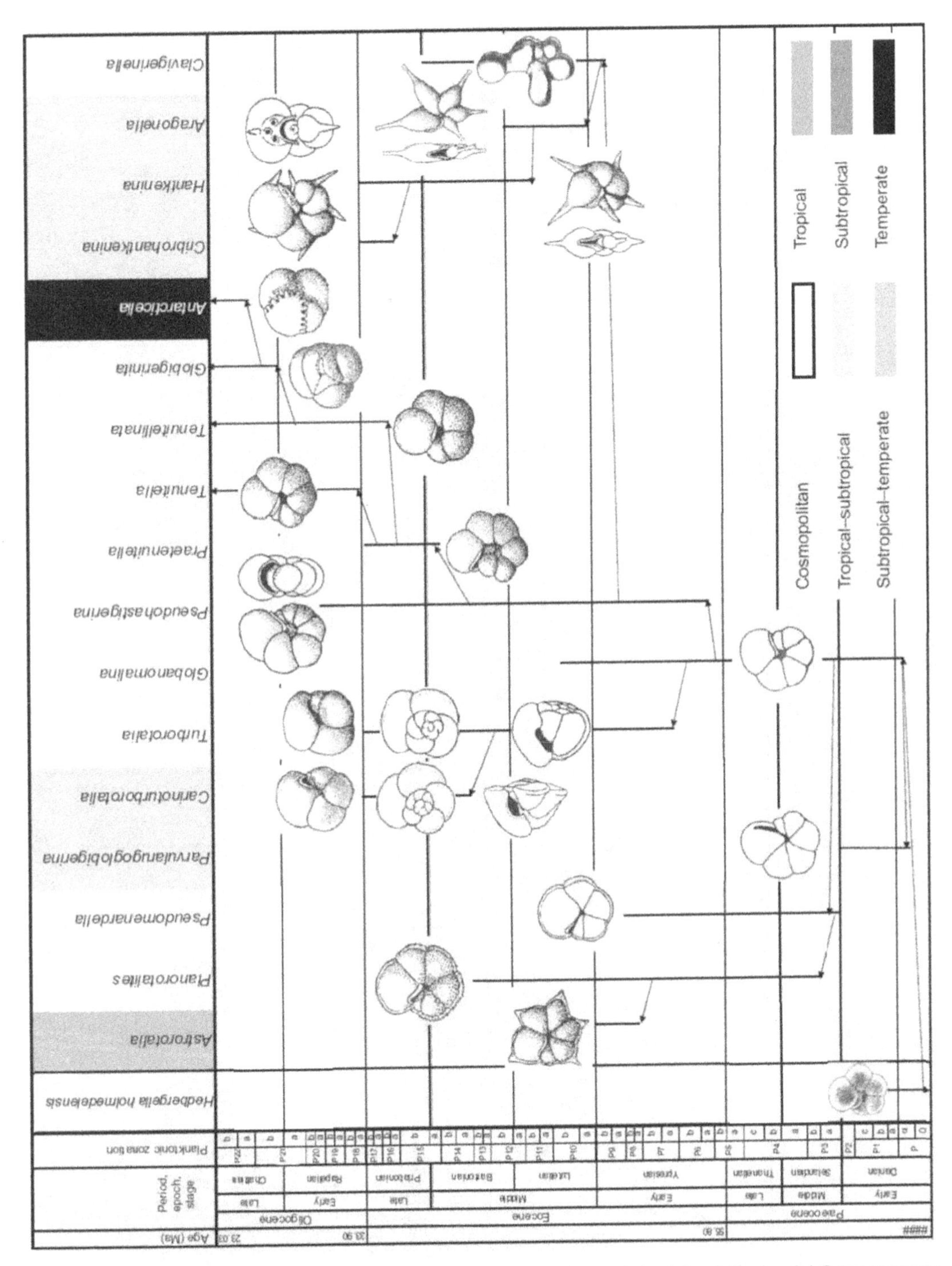

Figure 5.7. Phylogenetic evolution of the main nonspinose, trochospiral, and planispiral Paleogene planktonic foraminifera from their Cretaceous ancestor *Hedbergella holmdelensis*. The shading behind the genera names indicates their latitudinal range as defined in the embedded legend.

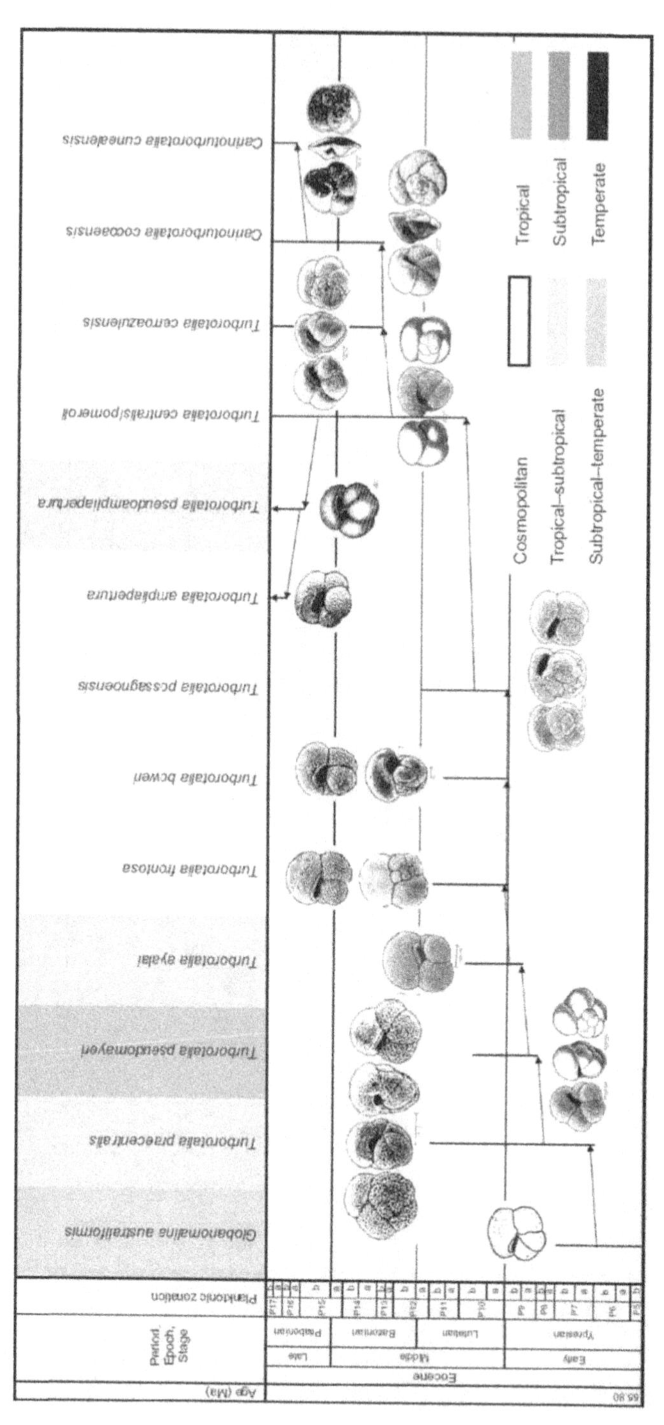

Figure 5.8. Phylogenetic evolution of the main species of *Turborotalia*. The shading behind the genera names indicates their latitudinal range as defined in the embedded legend.Images taken from original sources.

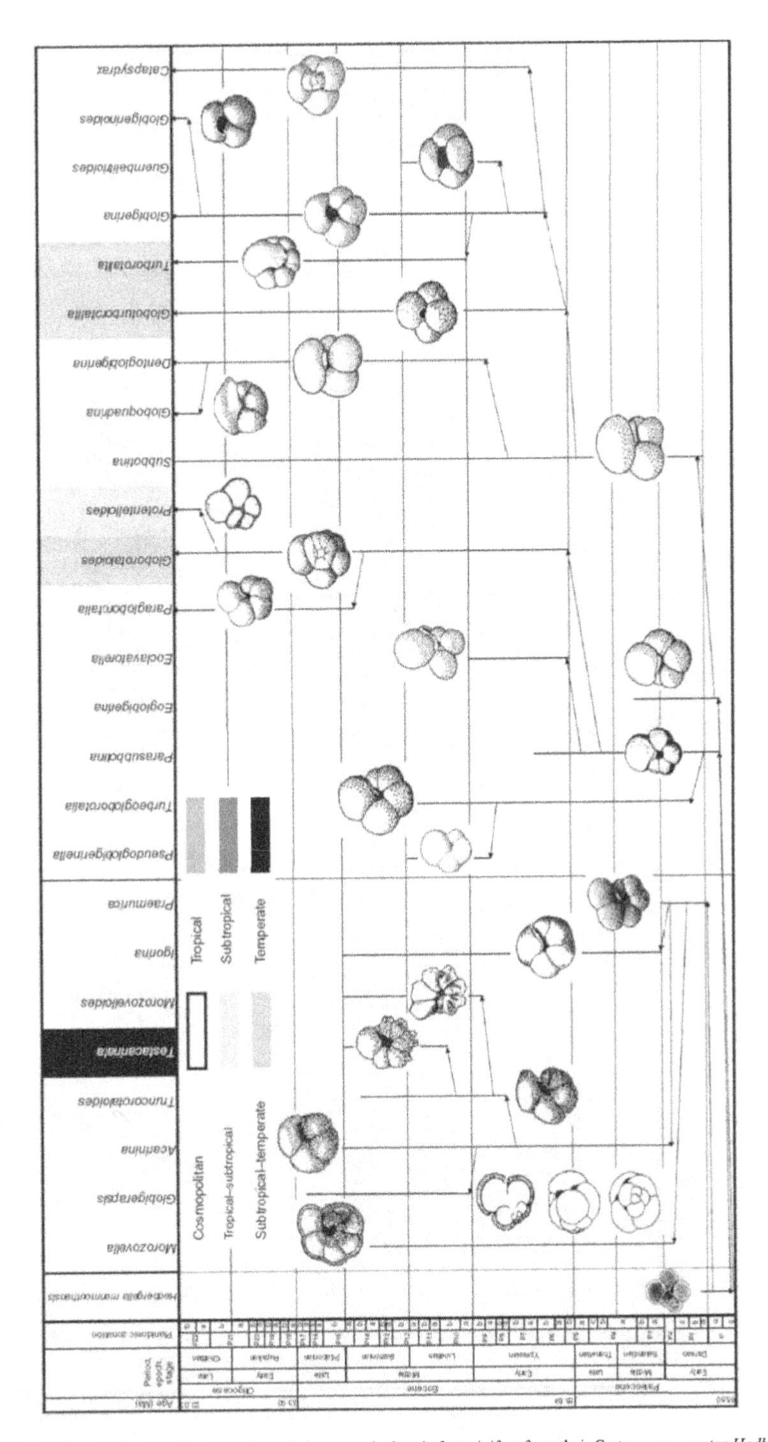

Figure 5.9. Phylogenetic evolution of the muricate or spinose Paleogene planktonic foraminifera from their Cretaceous ancestor *Hedbergella monmouthensis*. The shading behind the genera names indicates their latitudinal range as defined in the embedded legend.

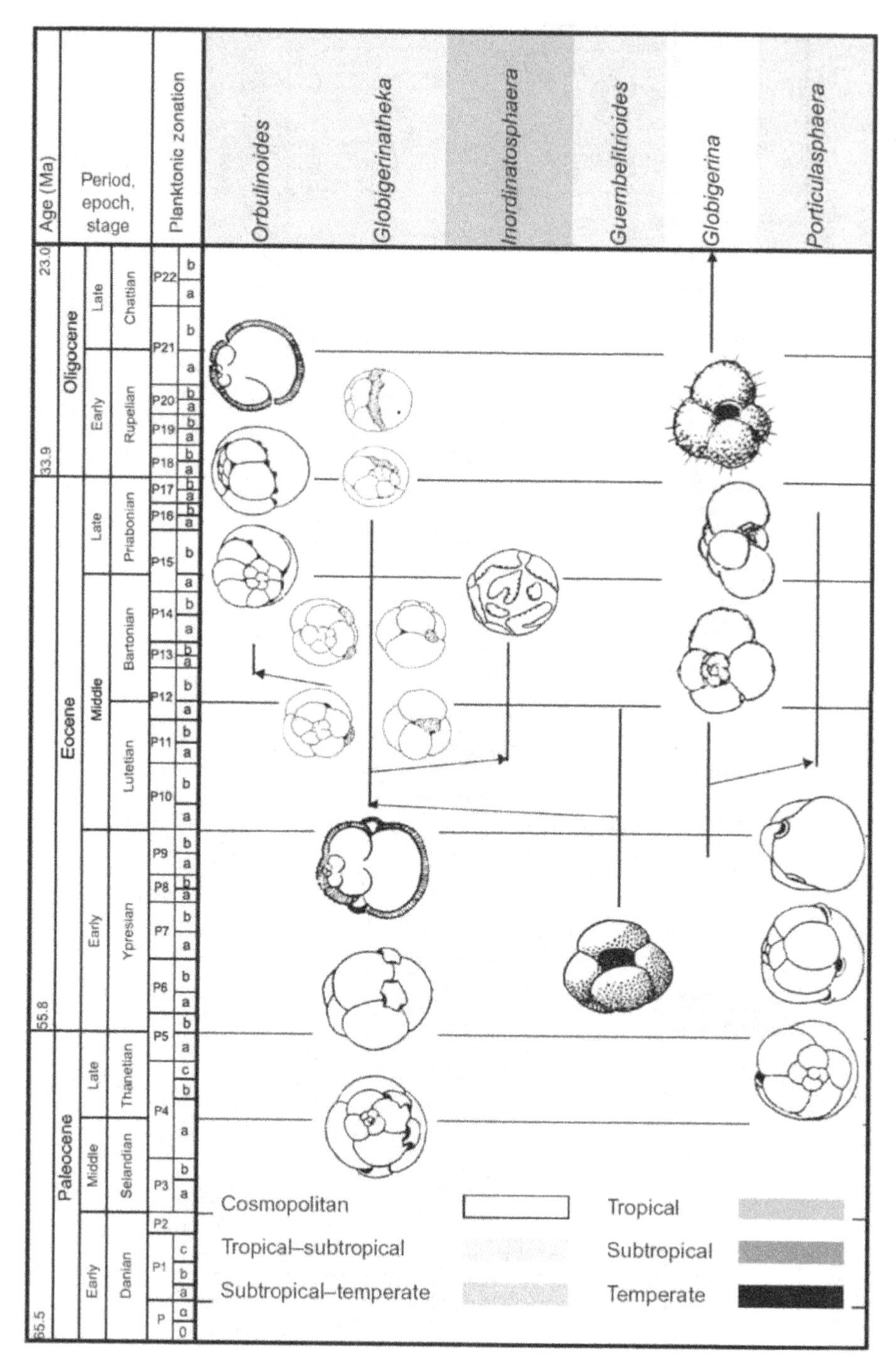

Figure 5.10. Phylogenetic evolution of the streptospiral porticulasphaerids. The shading behind the genera names indicates their latitudinal range as defined in the embedded legend.

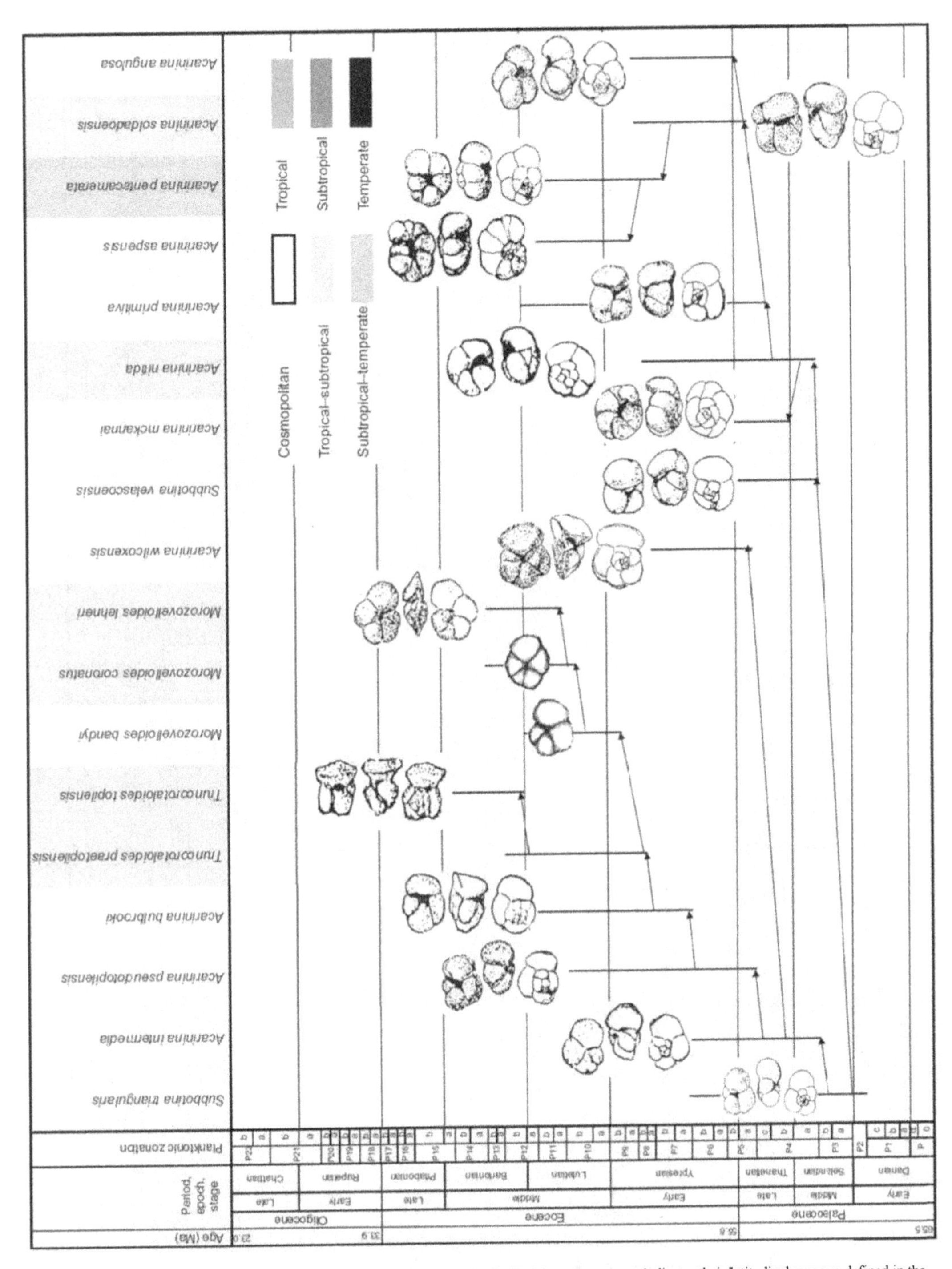

Figure 5.11. Phylogenetic evolution of the main acarinid species. The shading behind the genera names indicates their latitudinal range as defined in the embedded legend.

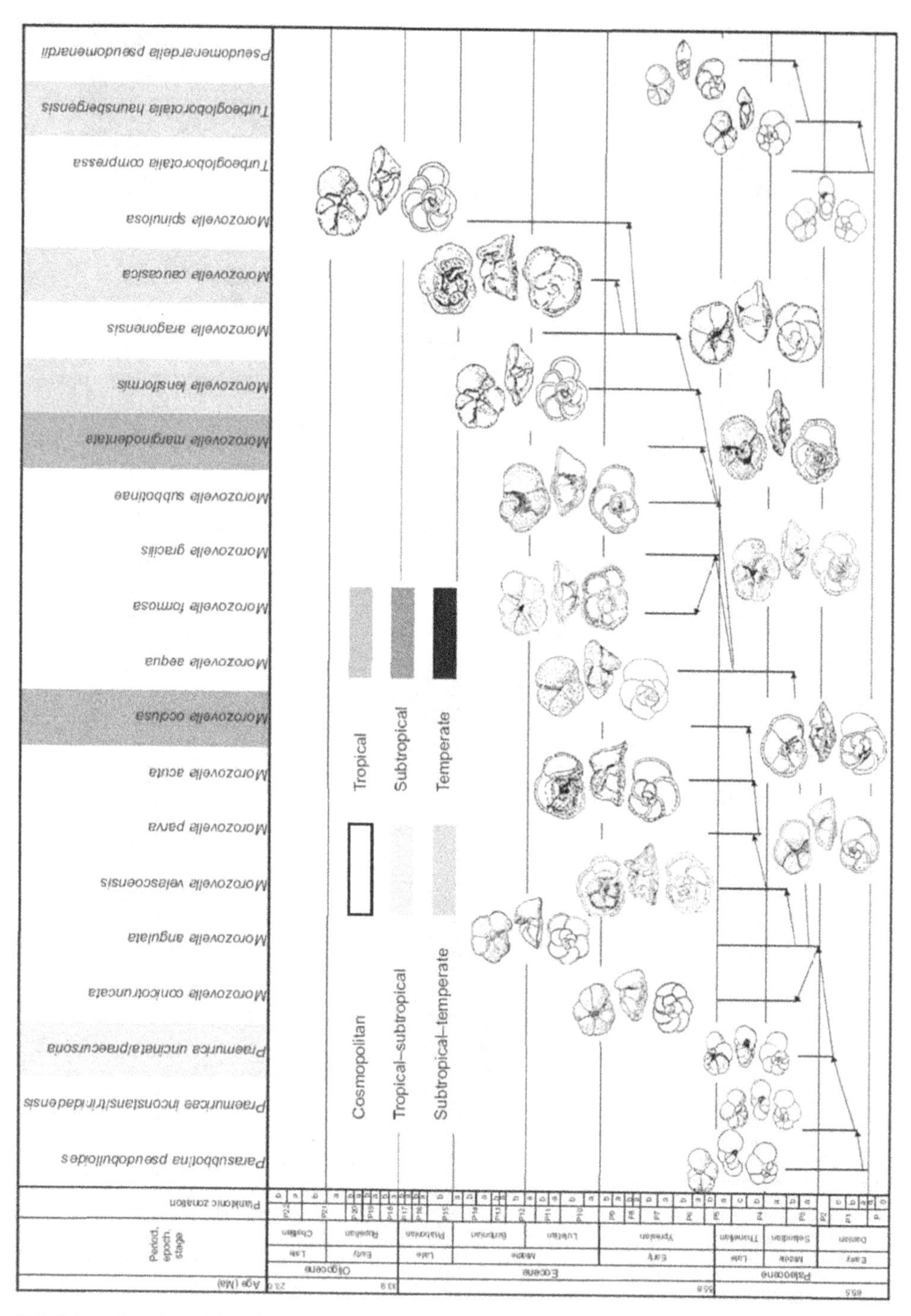

Figure 5.12. Phylogenetic evolution of the main morozovellid species. The shading behind the genera names indicates their latitudinal range as defined in the embedded legend.

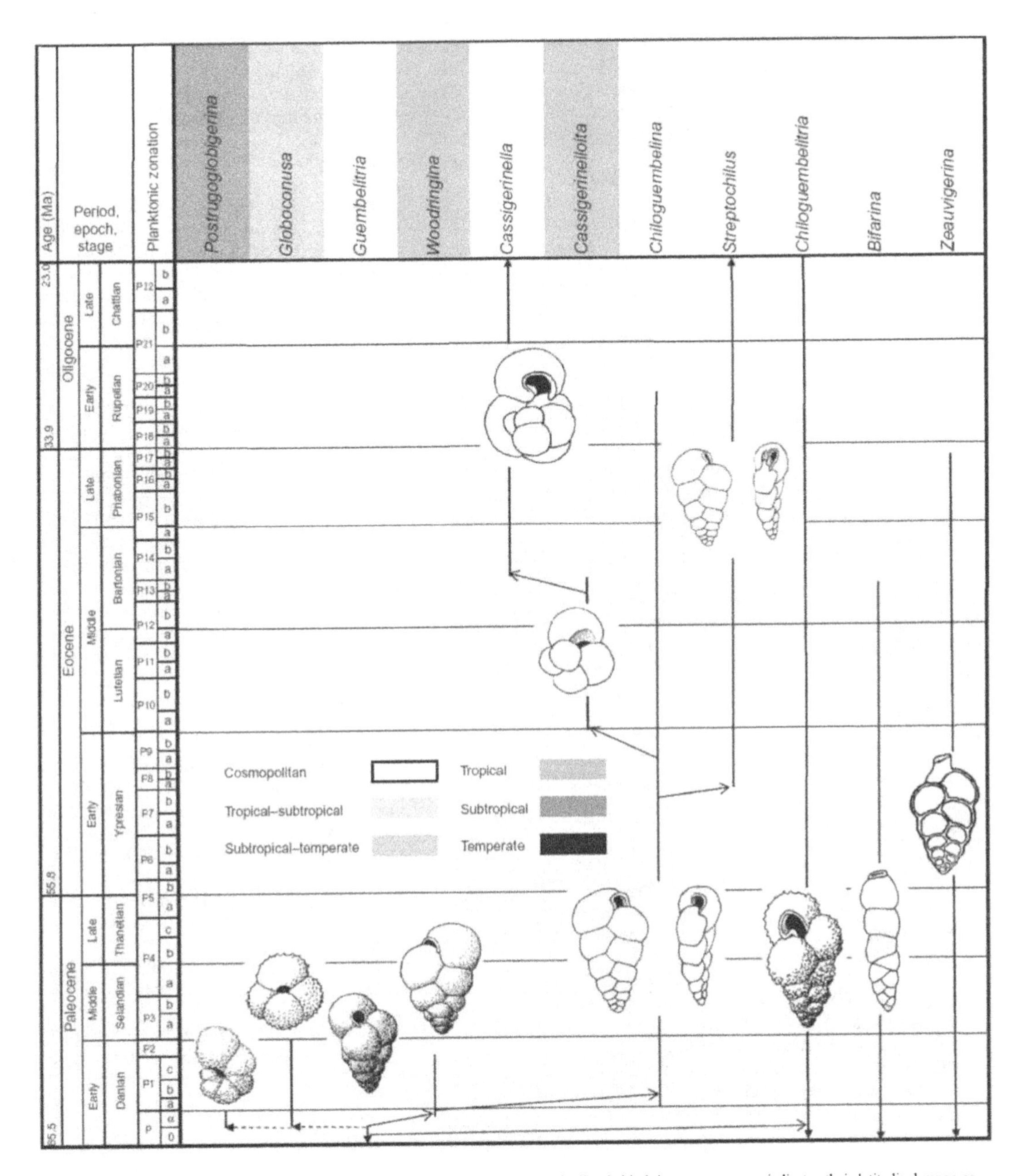

Figure 5.13. Phylogenetic evolution of the globoconusids and heterohelicids. The shading behind the genera names indicates their latitudinal range as defined in the embedded legend.

5.3 Biostratigraphy and phylogenetic evolution

During the Paleogene, planktonic foraminiferal assemblages became highly diverse. This, combined with their short biostratigraphic ranges, enables relatively high-resolution stratigraphic correlations to be made. They evolved gradually, forming a succession of morphological distinctive populations within important phylogenetic lineages, and have been essential in the development of biostratigraphical zonations for the Paleogene worldwide. Their origin and evolution has long been debated because of the extent of the extinction event at the

Cretaceous-Paleocene boundary, and because of overall significant morphologic differences between the Paleocene and Late Cretaceous assemblages. Reflecting this uncertainty, many variations of the phylogenetic relationships have been proposed for the Late Cretaceous and Paleogene planktonic foraminifera (see e.g., Berggren, 1962, 1969, 1977; Blow, 1979; Brinkhuish and Zachariasswe, 1988; D'Hondt, 1991; Liu and Olsson, 1994; Olsson, 1963, 1970, 1982; Orue-Etxebarria, 1985; Pearson, 1993; Smit, 1977, 1982; among others). The difficulty in resolving their phylogeny is due to several factors, including the sudden extinction of the Maastrichtian foraminifera, the relative rarity of continuous Cretaceous–Paleocene boundary sections containing well-preserved assemblages, and conflicting taxonomic attributions.

The updated and revised zonation presented here is a modified and refined version of the set of zones originally developed by Banner and Blow (1965) and Blow (1969), where they developed an alphanumeric shorthand (i.e., "Px," where "P" is for Paleogene, and "x" is a numerical index) to define stratigraphic zones during their extensive work on Cenozoic planktonic foraminifera from the Caribbean and Tanzania. This scheme was later revised and expanded uponby Blow (1979).A variant of this zonation was published by Berggren (1969), and numerous amendments to these zones have been suggested over the years. Berggren and Pearson (2005) produced a new revision for the Paleogene zones following extensive taxonomic work on the Paleocene and Eocene planktonic foraminifera (Olsson *et al.*, 1999; Pearson *et al.*, 2006a,b), extending the alphanumeric shorthand to include "P" for Paleocene, "E" for Eocene, and "O" for Oligocene. In an attempt to rationalize and harmonize the issues that have been debated in the literature, an updated taxonomy of these forms has been suggested in the previous section, and based on our recent work, revised species-level biostratigraphic ranges are presented in Charts 5.1–5.4 (http://dx.doi.org/10.14324/111.9781910634257). In the following, we describe an inferred phylogenetic relationship of the Paleogene forms, which is self-consistent and compatible with the taxonomic relationships defined in Section 5.2. We choose to define Paleogene planktonic zones (i.e., "P" stands for Paleogene), rather than use the letter descriptions (P, E, and O) adopted by Olsson *et al.* (1999) and Pearson *et al.* (2006a,b) (see Fig. 5.6).

About half of the few survivors of the Cretaceous–Paleocene event were small globular, opportunistic, shallow-dwelling hedbergellids that lived in eutrophic conditions (*Hedbergella monmouthensis*, Plate 4.16, Figs. 22–24 and *H. holmdelensis*, Plate 4.16, Figs. 14–16), and the remainder were small-sized, disaster/opportunists, triserial guembelitriids (e.g., *Guembelitria cretacea*, Plate 4.19, Figs. 1–3) and biserial to uniserial forms (e.g., *Bifarina*, Plate 4.18, Figs. 27–31; Plate 5.10, Fig. 21). As will be discussed below, however, only three survivor species went on to provide the stock from which the subsequent Cenozoic planktonic foraminifera derived.

Previously in Section 5.2, we defined seven Paleogene superfamilies. Members of these superfamilies flourished atdifferent stages and included 16 families. The morphological features of the families overlapped to some extent over time, but they can be summarized as follows:

- The *microperforate trochospiral, smooth or pustulose-walled, nonspinose* forms, which evolved from forms originating in the Cretaceous (Figs. 5.7 and 5.8), and are represented by member of the superfamily **Globigerinitoidea**, and include:
    - the smooth, microperforate **globanomalinids**, which evolved from *Hedbergella holmdelensis* in the Danian;
    - the keeled **planorotalids**, which evolved from the globanomalinids in the Selandian;
    - the microperforate, pustulose, low trochospiral **tenuitellids**, which evolved from the globanomalinids in the Late Bartonian; and
    - the microperforate, trochospiral **globigerinitids**, which evolved from the tenuitellids in the Late Priabonian.
- The *macroperforate planispiral smooth or pustulose-walled nonspinose* forms, which are represented by members of the superfamily **Hantkeninoidea** that evolved from the globanomalinids in the Late Ypresian (Fig. 5.7).
- The *macroperforate smooth, more or less cancellate, spinose, trochospiral* forms, which evolved from forms originating in the Cretaceous (Fig. 5.9), and are represented by member of the super-family **Eoglobigerinoidea**, and include:
    - the high trochospiral **eoglobigerinids** which first evolved from *H. monmouthensis* in the Danian (see Fig. 5.9).
- The *macroperforate trochospiral, planispiral, or streptospiral, smooth or punctate, spinose or nonspinose* forms, which are represented by members of the superfamily **Globigerinoidea** that evolved from the eoglobigerinids in the Early Eocene. They include
    - the **globigerinids**, which evolved from the eoglobigerinids in the Ypresian (see Fig. 5.9);
    - the **globorotaliids**, which evolved from the globigerinids in the Late Bartonian (see Fig. 5.9);
    - the **porticulasphaerids**, with a streptospiral coiling and a last chamber embracing the umbilicus. They evolved from the globigerinids in the Late Lutetian (see Fig. 5.10);
    - the **turborotalitids** with a small trochospiral test evolved from the globigerinids in the Lutetian (see Fig. 5.9).
- The *trochospiral muricate-walled, nonspinose* forms, which are represented by members of the superfamily **Truncorotaloidinoidea** that also evolved from the Cretaceous survivor *H. monmouthensis* (Figs. 5.9, 5.11, and 5.12).
- The *trochospiral, microperforate, pustulose, nonspinose* forms, which evolved from the Cretaceous survivors in the Danian. They are represented by members of the superfamily

Globoconusoidea (see Fig. 5.13). The **globoconusids** were initially triserial with a closed umbilicus and evolved from the guembelitriids. In the discussion below, this group is included in the discussion of the Heterohelicoidea.

- The *microperforate, pustulose* forms, which survived the K–P event or evolved from them (see Fig. 5.13). They are represented by members of the superfamily **Heterohelicoidea** and include
  - the triserial **guembelitriids**, which seem to have died out at the top of the Oligocene only to reappear in the Late Pliocene;
  - the biserial **chiloguembelinids**, which evolved from the guembelitriids in the Danian;
  - the enroled, biserial **cassigerinellids**, which evolved from the chiloguembelinids in the Early Lutetian; and
  - the **heterohelicids**, which survived the Cretaceous–Paleocene extinction event.

5.3.1 The microperforate trochospiral smooth or pustulose-walled, nonspinose Paleogene planktonic foraminifera

Many of the major morphological patterns of the Paleogene planktonic foraminifera became established during the Early Paleocene. The Maastrichtian survivors were rare and represented by two cosmopolitan species of *Hedbergella* (*H. holmdelensis*, Plate 4.16, Figs. 14–16 and *H. monmouthensis*, Plate 4.16, Figs. 22–24). However, these two species were very short lived, but before their extinction in the Danian, they gave rise to a number of new lineages (Liu and Olsson, 1994).

Hedbergella holmdelensis was a small planktonic foraminifera with a smooth test, except for a very few muricae restricted to the umbilical area and an imperforate peripheral marginal band. It gave rise in the Danian (Pa) to the cosmopolitan, flat, smooth-walled finely perforate *Globanomalina* (Plate 5.1, Figs. 1–3), which in turn developed a narrow elongate aperture in the Late Danian (P1) to give rise to *Parvularugoglobigerina* (Plate 5.1, Figs. 4 and 5), found in tropical and subtropical realms. The origin of the latter form, however, has been debated, as some authors have suggested that this form might have evolved from a benthic form, *Caucasina*, if mutation gave rise to an increase in the number of chambers thereby enhancing buoyancy (Brinkhuish and Zachariasswe, 1988). Others, it should be noted, have considered it to have originated from the triserial Cretaceous survivor, *Guembelitria*, on the grounds that early representatives of this genus are similar to this form (Li *et al.*, 1995, Olsson *et al.*, 1992). However, it is here considered as a globanomalinid because its wall lacks any pustules or perforation cones. After a short period of proliferation, *Parvularugoglobigerina* became extinct at the end of the Danian (P2), leaving no direct descendant.

In the Early Selandian (P1a), *Globanomalina* developed peripheral carina on its last chambers to give rise to *Pseudomenardella* (see Fig. 5.6), which in turn evolved in the Middle Selandian (P3b) into a compressed form, with a sharply angled, keeled periphery in axial view, *Planorotalites* (Plate 5.2, Figs. 2–4). Both forms occur in Paleocene and Eocene strata from various climatic belts, but in the Ypresian (P8 and P9) *Planorotalites*, living in the tropical and subtropical realms, developed radially orientated tubulospines (see Fig. 5.7) to give *Astrorotalia* (Fig. 5.6).

Globanomalina evolved separately into the cosmopolitan, planispiral *Pseudohastigerina* (Plate 5.2, Figs. 10–15) in the Ypresian (P6). *Pseudohastigerina* gave rise to *Praetenuitella* (Plate 5.2, Figs. 18 and 19) in the Middle Eocene, by developing a low trochospire and mimicking the high aperture of the Danian *Parvularugoglobigerina* (Li *et al.*, 1995). This represents an example of convergent morphological development among Paleogene planktonic foraminifera. The evolution of a planispiral test into a trochospiral test exemplifies a "morphological reversal" trend rather than a unidirectional one (Qianyu, 1987).

In the Late Eocene, *Praetenuitella* evolved into the microperforate globigerinitid, *Tenuitellinata* (Plate5.3, Figs.1-5), by developing an intra-umbilical aperture that in turn gave rise in the Late Oligocene (P21b) to *Globigerinita* (Plate5.3, Fig.1),which is characterized by obligate bullae in the adult stage. *Praetenuitella* was replaced at the Eocene–Oligocene by *Tenuitella* (Plate 5.2, Figs. 20–21), which has a simple aperture. These forms are found in sediments from tropical to temperate realms. The relationship between *Tenuitella* and *Globigerinita* was first suggested by Li (1987) and then confirmed by Pearson and Wade (2009), who described a common type of radial crystalline wall for both forms. Huber *et al.* (2006) suggested that the first tenuitellid, *T. insolita*, may have descended from the Eocene *Cassigerinella* (Plate 5.10, Figs. 3–6). However, it is here suggested that in fact the two genera are not related. They have different types of walls, chamber shape, and coiling arrangement. In the temperate realm, *Globigerinita* developed in the Late Oligocene (P22) numerous accessory apertures in its sutures between the many, closed, tunnel-like marginal extensions of the bulla that resulted in *Antarcticella* (Plate 5.1, Figs. 19 and 20). *Antarcticella* currently includes, from oldest to youngest, *pauciloculata, ceccionii, zeocenica,* and *anctarctica*. All species of this genus have restricted biogeographic distributions in the southern middle to high latitudes and occur in shallow marine clastic sedimentary facies. The species referred to this genus, *pauciloculata, ceccionii,* and *zeocenica* are considered as having a benthic paleohabitat because of their dominant occurrence in shallow facies of marginal basins in the southern high latitudes (Liu *et al.*, 1998). However, because of the gap in the stratigraphical occurrence between these forms, *Antarcticella zeozenica* is not considered here to be plausible the ancestor of the planktonic *Antarcticella anctartica* (Plate 5.1, Figs. 19 and 20).

In the Early Eocene (Ypresian, P7), *Globanomalina* gave rise to *Turborotalia*, which includes many species commonly used in biostratigraphic zonations. Since the 1950s, taxonomists have speculated that there was a phylogenetic development from relatively globular forms in the early Middle Eocene (*Turborotalia frontosa*, Plate

5.1, Figs. 13 and 14) to keeled, angular, and compressed forms (*Turborotalia cerroazulensis*, Plate 5.1, Figs. 16–18). The latter species gave rise to carinate forms (*Carinoturborotalia cocoaensis*, Plate 5.12, Figs. 12–13; *C. cunialensis*) in the late Eocene (see Figs. 5.6 and 5.7). These species are linked by overlapping morphospecies (Samuel and Salaj, 1968; Toumarkine and Bolli, 1970). Pearson *et al.* (2006a,b) hypothesize that *Turborotalia frontosa* was the first true member of the genus and was descended from *Globanomalina australiformis*. However, it is more likely that the fairly tight, compressed *Turborotalia praecentralis* is the ancestor stock of the *Turborotalia* lineages. It evolved in the tropical and subtropical realms from *Globanomalina australiformis* in the Middle Ypresian (P7) to give rise to *Turborotalia pseudomayeri* in the tropical realm. The latter had a less angular periphery and is more rounded than the advanced *Turborotalia cerroazulensis*. It subsequently gave rise to the cosmopolitan *Turborotalia frontosa* stock (*Turborotalia ayalai, boweri*, and *frontosa*) by retaining the high arch-like aperture with a faint lip, which separates this stock from the rest of the turborotalids (see Fig. 5.8). In *T. ayalai*, the chambers become more globular and more rapidly enlarging, with the last whorl of chambers more than doubling the test diameter. In the Lutetian, the chambers become more depressed with a very narrow umbilicus, but the aperture is still high and broad, *T. boweri*. The latter give rise to *T. frontosa* with subglobular chambers enlarging relatively slowly and having a lower aperture. *T. frontosa* developed a dorsally flattened test with a broadly rounded to broadly and very bluntly subangular periphery, *Turborotalia possagnoensis*. The latter gave rise to *T. centralis/pomeroli* in the Lutetian (P11) with a broadly rounded to bluntly subangular periphery. The test is weakly convex dorsally and strongly convex ventrally. It subsequently evolved into *T. cerroazulensis* in the Bartonian (P14) with the gradual flattening of the spiral side, giving the test a bluntly subangular or angular shape in axial view. The aperture is still high, but it tends to migrate toward the umbilicus, and does not reach the periphery. *T. cerroazulensis* in turn evolved in the Bartonian (P14) into forms with a more acutely angular periphery in axial view, but with imperforate carina in part of the test, as in *Carinoturborotalia cocoaensis*, to forms with prominent carina, as in *C. cunealensis*. Quite independently and in the Late Eocene, *Turborotalia centralis/pomeroli* gave rise to *T. peudoampliapertura* (Plate 5.1, Fig. 12) with a small umbilicus and a high-arched antero-intraumbilical aperture. This species survived the Late Eocene–Oligocene boundary only to disappear completely within the Early Oligocene. It evolved in the Late Priabonian (P17) into *T. ampliapertura* with reniform, depressed chambers, and a small umbilicus. It has a rough surface, with prominent perforation pits (see Plate 5.4, Fig. 8).

5.3.2 The macroperforate planispiral smooth or pustulose-walled nonspinose Paleogene planktonic foraminifera

Various hypotheses have been put forward concerning the phylogeny of the Hantkeninidae. *Hantkenina* (Plate 5.2, Figs. 8 and 9) has been suggested to be a monophyletic form that evolved from the Cretaceous tubulospinose genus *Schackoina* (Plate 4.15, Figs. 7–10) in the early Middle Eocene (e.g., Bolli *et al.*, 1957; Brönnimann, 1950; Cushman, 1933; Cushman and Wickenden, 1930; Rey, 1939; Thalmann, 1932, 1942). Others have traced its phylogeny to the genus *Pseudohastigerina* (Plate 5.2, Figs. 5–10) at the base of the Middle Eocene. However, others argued that *Hantkenina* is polyphyletic, and that Middle and Late Eocene groups evolved independently from different pseudohastigerinid ancestors (Berggren *et al.*, 1967; Blow, 1979; Blow and Banner, 1962), and yet others believed that it evolved gradually from the clavate species *Clavigerinella eocaenica* (Plate 5.2, Fig. 17) in the earliest Middle Eocene and is unrelated to the genus *Pseudohastigerina* (Benjamini and Reiss, 1979; Coxall *et al.*, 2003). This confusion results from the fact that no elongation of chambers into tubulospine-like structures has been observed in *Pseudohastigerina* (Berggren *et al.*, 1967). A clavigerinellid ancestry was traced to the low trochospiral genus *Parasubbotina* (Coxall *et al.*, 2003). However, as *Pseudohastigerina*, *Clavigerinella*, *Aragonella*, and *Hantkenina* all share many morphologic features, such as having smooth-walled tests with normal perforation, and showing similar developmental patterns of planispirality, they are considered here as being phylogenetically related. Thus, we suggest that *Pseudohastigerina* species developed radial elongate and clavate chambers in the Early Eocene (Late Ypresian, P9), leading to the cosmopolitan *Clavigerinella* (Plate 5.2, Figs. 16 and 17). The latter evolved in the subtropical and tropical realms into forms with hollow tubulospines, *Aragonella* (Lutetian, P10a–P12a) that in turn gave rise to *Hantkenina* (Fig. 5.6). The first appearance of the latter had been used to recognize the base of both the Middle Eocene and the Lutetian Stage. However, Rögl and Egger (2010) have correlated its first appearance to P11 within the Lutetian. *Aragonella* has been considered as a synonym of *Hantkenina* by many authors (e.g., Rögl and Egger, 2010); however, the tubulospines of *Aragonella* (Plate 5.2, Figs.5 and6) are smooth proximally and distally are bluntly digitate, while those of *Hantkenina* are striate proximally and pointed distally.

In the Priabonian (P16), the aperture of *Hantkenina* acquired a broad porticus, penetrated by areal accessory apertures, to give *Cribrohantkenina* (Plate 5.2, Figs. 5 and 6). Coxall *et al.*(2003) have observed asymmetry in the coiling of the test (in edge and side view) in many individuals of *C. eocanica* and *A. nuttalli*, indicating very low trochospiral coiling, and later *Hantkenina* morphotypes appear to be more fully planispiral (cf. the symmetrical apertural systems in *Cribrohantkenina inflata*, Plate 5.2, Figs. 5 and 6). This we suggest confirms that they were descended from the trochospiral *Globanomalina* (Plate 5.1, Figs. 1–3). Similar developments can be observed in the *Pseudohastigerina* lineage, with the earliest forms having an asymmetrical equatorial aperture and low trochospire (e.g., *P. wilcoxensis*, Plate 5.2, Figs. 10–11), and later morphotypes developing more perfect planispiral coiling (Berggren *et al.*, 1967; Blow, 1979). Complete shell dissections also reveal that the initial whorl morphologies, representing the onset of the neanic stage, as identified in fossil planktonic foraminifera of *H. nuttalli, C. eocanica* (Plate 5.6, Fig. 16), and *P. micra* by Brummer *et al.*(1987) and by Huber(1994), are broadly similar (Coxall *et al.*, 2003); therefore, they all belong to the same phylogenetic lineage (see Fig. 5.7).

5.3.3 The macroperforate trochospiral smooth, more or less cancellate, spinose Paleogene planktonic foraminifera

The eoglobigerinids (see Fig. 5.9) are the second lineage that evolved from *H. monmouthensis* and included the cancellate spinose genera, *Eoglobigerina* (Plate 5.3, Figs. 7 and 8), *Subbotina* (Plate 5.3, Figs. 13–16; Plate 5.11, Fig. 19), and *Parasubbotina* (Plate 5.3, Figs. 11 and 12). The eoglobigerinids reached their maximum diversity in the Late Danian and Lutetian (see Fig. 5.2). Spines in planktonic foraminifera evolved within the genus *Eoglobigerina* in less than 100,000 years after the Cretaceous–Tertiary extinction (Olsson *et al.*, 1992, 1999). As the spines are fragile and not always preserved, the presence of spine-holes and spine-bases are the primary means for recognizing species belonging to this phylogenetic group. Liu and Olsson (1994) suggested that the high trochospiral Paleocene eoglobigerinid species derived from the slightly elevated morphotype of *H. monmouthensis*,while the low trochospiral, planoconvex *Parasubbotina* species (Plate 5.3, Figs. 11 and12;Plate 5.11, Fig. 16) are related to the normal *Hedbergella monmouthensis*. All subsequent lineages with bilamellar spinose walls can be linked to this common ancestor (Aurahs *et al.*, 2009; Kennett and Srinivasan, 1983; Pearson *et al.*, 2006 a, b). The transition from the nonspinose hedbergellids to spinose eoglobigerinid was a rapid event associated with the filling of planktonic niches vacated after the mass extinction without any conclusive signal in the genes of modern descendants (Darling *et al.*, 2009; Whitfield and Lockhart, 2007).

The *Parasubbotina* genus was short lived (Paleocene to Early Eocene), but in the Danian, it quickly evolved into *Turbeogloborotalia* (Plate 5.2, Fig. 1; Plate 5.3, Figs. 17–20; Plate 5.11, Fig. 4), with a rounded (e.g., *Turbeogloborotalia compressa*, Plate 5.3, Figs. 18 and 19) to bluntly subangular axial periphery (e.g., *Turbeogloborotalia pseudomenardii*, see Fig. 5.12). *Parasubbotina* developed clavate chambers to give *Eoclavatorella* (Plate 5.3, Fig. 6) in the Early Eocene. *Turbeogloborotalia* in turn evolved in the Early Eocene (Late Ypresian, P9) into a form with an asymmetrically planispiral test in the adult stage, but with a very low trochospiral juvenile stage, *Pseudoglobigerinella* (e.g., *P. bolivariana*, Fig. 5.8). However, prior to its demise, *Parasubbotina* gave rise in the Ypresian to the compressed globigerinid, *Globorotaloides*, which in turn gave rise to the Neogene forms of Globorotaliidae (see Section 5.3.4). Simultaneously, the phylogenetic lineage of the high trochospiral *Eoglobigerina* gave rise in the Danian to *Subbotina*, with globular chambers and a broadly rounded periphery. *Subbotina* survived into the Oligocene where it was gradually replaced by the spinose globigerinids, which became common in the Miocene to Holocene (see Section 5.3.4).

5.3.4 The macroperforate trochospiral, planispiral, or streptospiral, smooth or punctate, spinose, or nonspinose Paleogene planktonic foraminifera

In the Early Eocene (Ypresian, P5b), the cancellate *Subbotina* evolved into the first cancellate spinose globigerinid, *Globoturborotalita*, which has globular and slightly embracing chambers (see Fig. 5.8). *Globoturborotalita bassriverensis* is the first species in a generic line which evolved from *Subbotina* (e.g., *S. hornibrooki*) that extends to the Holocene (Pearson *et al.*, 2006a,b). *Globoturborotalita* gave rise rapidly in the Ypresian (P6b) to the non-cancellate, spinose, long-ranging, extant genera, *Globigerina* (Plate 5.4, Figs. 5–17; Plate 5.5, Figs. 1–6 and 16–20; Plate 5.11, Fig. 6), with sub-globular chambers. *Globigerina*, in turn, developed regular bulla covering the aperture leading to *Catapsydrax* (Plate 5.4, Figs. 1–4). These forms became extinct in the Pleistocene (N22). Later, in the Late Oligocene (P22b), *Globigerina* evolved spiral sutural supplementary apertures leading to *Globigerinoides* (Plate 5.5, Figs. 9–11).

In the Eocene (Lutetian, P10), *Turborotalita* (n.b. not to be confused with *Turborotalia* discussed above), a long ranging genus with a final ampullate chamber, evolved from an uncertain ancestor. The number of whorls and calcitic crusts readily distinguish this form from juvenile tests of taxa of other families. Kennett and Srinivasan (1983) speculated that the minute size of the test and the surface ultrastructure are similar to features of *Tenuitella*, but Pearson *et al.* (2006) suggested that morphologically and in size *Turborotalita* appears closest to *Globoturborotalita*, in the arrangement of chambers around a small umbilicus and in the umbilical aperture. However, the test of *Turborotalita* lacks the cancelation and large pits of *Globoturborotalita* (Plate 5.5, Fig. 12), and it differs from *Turborotalia* and *Tenuitella* in being spinose and by possessing an ampulla (Holmes, 1984). We believe that it evolved from the spinose *Globigerina* by developing a broad apertural lip, making the final chamber ampullate.

In the Ypresian (P9), *Subbotina* gave rise to the globigerinid *Dentoglobigerina* (Plate 5.5, Figs. 13–16; Plate 5.6, Figs. 1–5) by loosing most of its cancelation. This form has a tooth-like, sub-triangular, symmetrical porticus projecting into the umbilicus (see Fig. 5.9) and had a long range as it survived into the Pliocene (Zanclean, N20a). *Dentoglobigerina* became subquadrate in equatorial outline to evolve into *Globoquadrina* (Plate 5.6, Figs. 6 and 7) in the Late Oligicene (P22). *Globoquadrina* became extinct in the Pliocene (Zanclean, N19).

The planoconvex subbotinid, *Parasubbotina* gave rise to *Globorotaloides* (Plate 5.6, Fig. 8) in the Late Lutetian that has a compressed test, which is biconvex with a lobate equatorial periphery. *Paragloborotalia* (Plate 5.7, Figs. 1–14) with a rounded to carinate periphery evolved from *Globorotaloides* in the latest Middle Eocene (Late Bartonian, P15a). *Paragloborotalia* species are long ranging and were the root stocks of many Neogene species of *Globorotalia*. *Paragloborotalia* opima (Plate 5.7, Figs. 2–5) extinction in the Chattian marks the top of the P21 zone. Quite separately, *Globorotaloides suteri* gave rise in the Chattian to *Protentelloides* with a compressed test, which was cancellate and more coarsely perforate (Zhang and Scott, 1995).

Quite separately, *Globigerina* gave rise to the high spired, loosely coiled *Guembelitrioides* (Plate 5.3, Figs. 9 and 10) in the Early to Middle Eocene (Ypresian, P8–P12a). *Guembelitrioides* became globular and closed the umbilicus by developing streptospirality (see Fig. 5.10) in the encrusted porticulasphaerid *Globigerinatheka* (Plate 5.5, Fig. 7; Plate 5.6, Figs. 15–19; Plate 5.11, Figs. 11 and 12) and *Orbulinoides* (Plate 5.6, Fig. 9; Plate 5.11, Fig. 10). *Globigerinatheka* possesses many large supplementary apertures in the last sutures, furnished with small, separated bullae, or broad lips. On the other hand, *Orbulinoides* is monotypic and differs from *Globigerinatheka* in possessing spiral areal apertures in the large globular end-chamber, which connect via vestibules with apertures of the covering thick wall. *Orbulinoides beckmanni* (Plate 5.6, Fig. 9; Plate 5.11, Fig. 10) has a restricted stratigraphic range, confined to the middle Eocene Zone (P13).

Globigerina in the Middle Eocene also developed globular forms, *Porticulasphaera* (P11–P16) in which the early globigerine whorl is still preserved, but the last chamber closes the umbilicus by umbilically directed streptospirality (Fig. 5.10). The apertures in *Porticulasphaera* (Plate 5.6, Figs. 10–14; Plate5.11, Fig. 9) often have thick lips but are never bullate. *Porticulasphaera* developed multiple meandriform bullae with very many small, sutural accessory apertures in Inordintosphaera (P11–P13).

5.3.5 The trochospiral muricate-walled, nonspinose Paleogene planktonic foraminifera

Populations of the Late Cretaceous muricate survivor *H. monmouthensis* display significant morphological variation (Liu and Olsson, 1994). These morphological differences gave rise to all Paleocene and Eocene cancellate spinose and nonspinose forms as virtually parallel lineages.

The first lineage, the truncorotaloidid *Praemurica* (Plate 5.9, Figs. 10–15; Plate 5.11, Fig. 17), evolving from *Hedbergella monmouthensis* was a weakly cancellate, praemuricate forms (see Fig. 5.9) that first appeared within a hundred thousand years of the K–P boundary (Berggren and Norris, 1997; Hemleben *et al.*, 1991; Olsson *et al.*, 1999). Around 3.5 Ma later, at the Danian–Selandian boundary, *Praemurica* gave rise to the tropical and subtropical muricate forms, namely, *Acarinina* (Plate 5.7, Figs. 15–20; Plate 5.8, Figs. 1–6; Plate 5.11, Figs. 20 and 22) and *Igorina* (Plate 5.8, Figs. 16–20; Plate 5.11, Figs. 14 and 21, see Fig. 5.9).

The cancellate nonspinose *Igorina* evolved through different species until its extinction at the end of the Middle Eocene (P15a). *I. pusilla* (Plate 5.11, Fig. 21) was the first representative of the *Igorina* lineage and is characterized by a thicker layer of pustules that covers the praemuricate wall of the inner spire (Olsson *et al.*, 1999; Soldan *et al.*, 2011). Quillévéré *et al.* (2001) used isotopic and biogeographic data to suggest that *Acarinina* species (Plate 5.7, Figs. 15–20; Plate 5.8, Figs. 1–6; Plate 5.11, Figs. 20 and 22) evolved from a photosymbiotic ancestor, which they identified as *Praemurica inconstans*, or early representatives of *P. uncinata*. *Acarinina* rapidly became cosmopolitan, dominating the upper Paleocene to Middle Eocene assemblages. *Acarinina* species (see Fig. 5.11) were initially small and globular, but they increased in size during the Paleocene and developed robust muricae on the umbilical surface. In the Late Ypresian and through to the Middle Eocene, *Acarinina* evolved supplementary apertures in spiral sutures, giving rise to *Truncorotaloides* (Plate 5.9, Figs. 18–22; Plate 5.11, Fig. 15) and *Globigerapsis* (P10–P16). *Globigerapsis* (Plate 5.6, Figs. 18 and 19; Plate 5.11, Figs. 7 and 8) had multiple supplementary arched apertures in its last sutures. *Truncorotaloides* gave rise in the Lutetian (P11) to *Testacarinata* (Plate 5.9, Fig. 16), with enlarged peripheral muricae, a genus preferring temperate climates (see Fig. 5.8). The evolution from *Truncorotaloides* to *Morozovelloides* group (*crassata, spinulosa, lehneri, coronata,* and *bandyi*) occurred in the Late Ypresian (P9b). *Truncorotaloides praetopilensis* developed bladed muricae along the periphery of each chamber in *Morozovelloides* (see Fig. 5.10). These developments may have been functionally linked to a discoidal arrangement (Pearson and Berggren, 2006). *Acarinina* died out completely at the end of the Middle Eocene.

The evolution of *Praemurica* (Plate 5.9, Figs. 10–15; Plate 5.11, Fig. 17) into *Morozovella* (Plate 5.8, Figs. 7–15; Plate 5.9, Figs. 1–5 and 7–9; Plate 5.11, Figs. 18 and 23–26) during the late Danian entailed a transformation from a trochospiral muricate ancestral texture to a smooth-walled, heavily pustulose (muricate) texture in its descendants. This event represents the beginning of the *Morozovella* species radiation (Plate 5.8, Figs. 7–15; Plate 5.9, Figs. 1–5 and 7–9; Plate 5.11, Figs. 18 and 23–26) and the first appearance of keeled foraminifera after the K–P extinction. During this evolution, the group diversified into three lineages: the *angulata* (Plate 5.8, Figs. 10–11) – *velascoensis* lineage which became extinct at the Paleocene–Eocene boundary (Berggren and Miller, 1988a,b); and the main Eocene lineages which are rooted in *Morozovella aequa* (see Fig. 5.12). There are two main lineages as seen in Fig. 5.12: one lineage evolved from *M. aequa* (Plate 5.8, Fig. 9; Plate 5.11, Fig. 24) to *M. formosa* (Plate 5.11, Fig. 23), while the second one developed from *Morozovella aequa* to M. *subbotinae* (Plate 5.9, Figs. 1–3), which in turn evolved into the *aragonensis* (Plate 5.11, Fig. 25) and *spinulosa* (Plate 5.8, Fig. 15) groups. The *aequa* group is characterized by predominantly dextral coiling, and the *velascoensis* (Plate 5.8, Fig. 14) group, by predominantly sinistral coiling (Pearson, 1993), while the *aequa* group gave rise to the remaining Eocene *Morozovella* and *Morozovelloides* groups (see Figs. 5.9 and 5.11). All *Morozovella* species became extinct at the top of the Middle Miocene (Bartonian, P14) (see Fig. 5.9 and Chart 5.3 online).

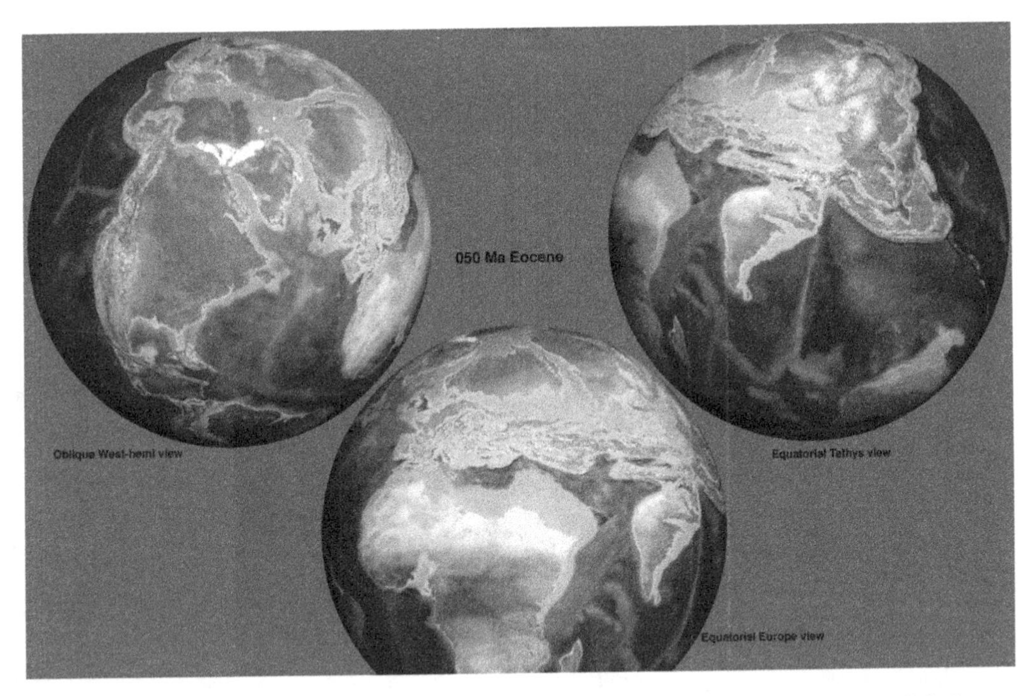

Figure 5.14. Paleogeographic and tectonic reconstruction of the Late Eocene (by R. Blakey, http://jan.ucc.nau.edu/~rcb7/paleogeographic.html).

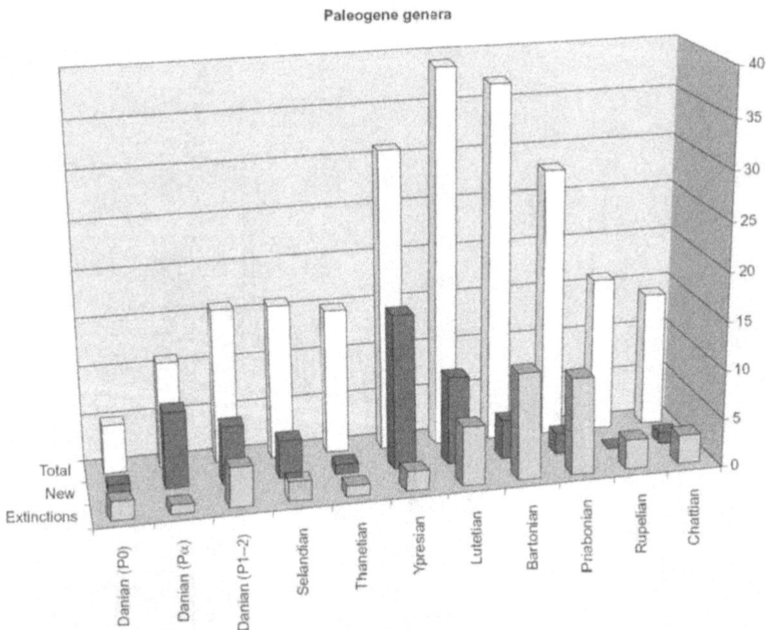

Figure 5.15. The total number of genera, extinctions, and new appearances of planktonic foraminifera in each stage of the Paleogene. The extinctions are defined relative to the end of each stage and the appearances with the beginning of the stage.

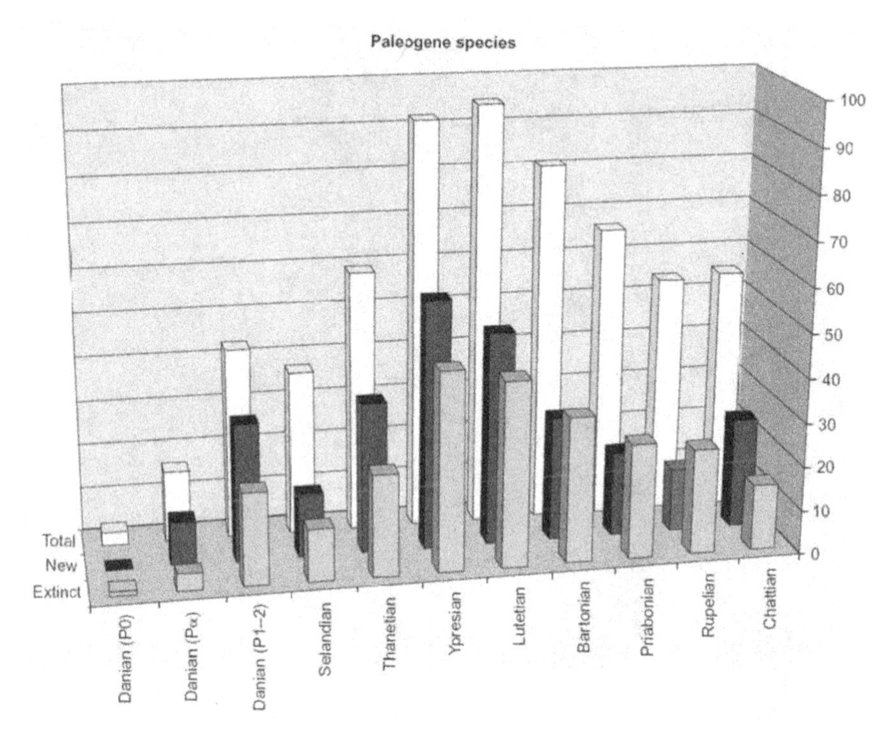

Figure 5.16. The total number of species, extinctions, and new appearances of planktonic foraminifera in each stage of the Paleogene. The extinctions coincide with the end of each stage and the appearances with the beginning of the stage.

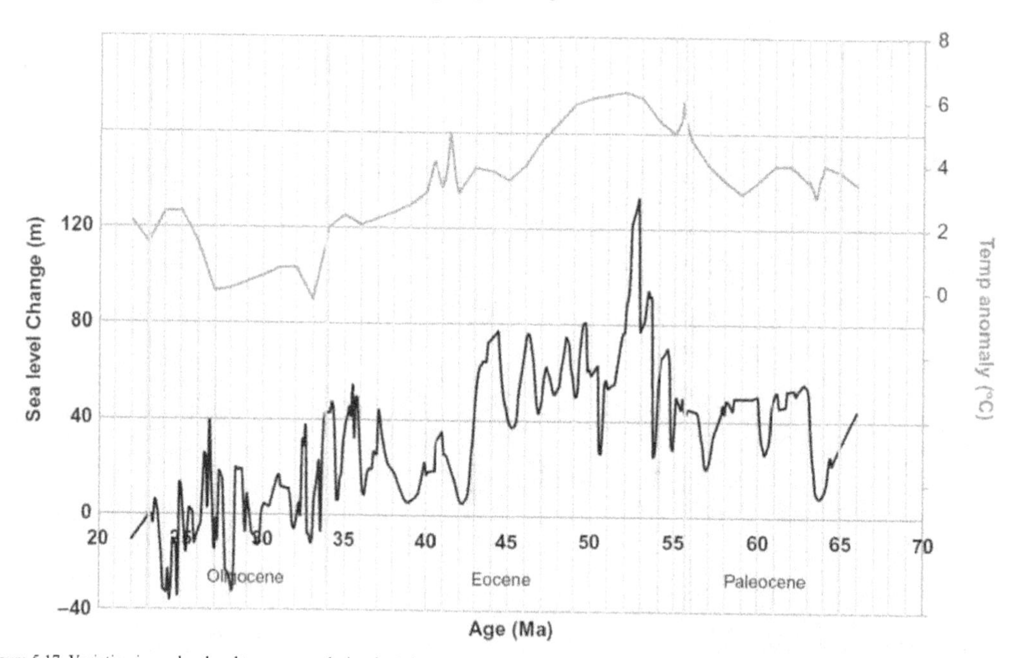

Figure 5.17. Variation in sea level and temperature during the Paleogene based on Miller *et al.* (2011) and Zachos *et al.* (2001).

5.3.6 The microperforate, pustulose Paleogene trochospiral globoconusids, and heterohelicids

Among the few Cretaceous mass extinction survivors, the heterohelicoids were the most successful as they crossed the boundary with four surviving genera, *Guembelitria*, *Chiloguembelitria* (Plate 5.10, Figs. 8–14), *Bifarina* (Plate 5.10, Fig. 21), and *Zeauvigerina* (Plate 5.10, Fig. 22). Liu and Olsson (1992) suggested that all Early Paleocene microperforate planktonic foraminifera are derived from the Cretaceous-surviving species *Guembelitria cretacea*. Four evolutionary lineages are recognized as deriving from *Guembelitria* (see Fig. 5.13):

- a lineage that developed globigeriniform trochospiral coiling along, with poreless minute pustules, *Globoconusa* (Pa–P2);
- another leading to the final whorl becoming biserial, *Woodringina* (Pa–P1);
- a lineage with an intra-extraumbilical aperture in *Postrugoglobigerina* (Pa);
- and the biserial *Chiloguembelina* (Plate 5.10, Figs. 15–18), that developed independently from *Woodringina* in the Danian (P1) and ranged into the Oligocene (P19).

The *Chiloguembelina* lineage gave rise to compressed biserial tests, *Streptochilus* (Plate 5.10, Fig. 19) in the Early Eocene (P8). It also gave rise to forms with the biseries coiled in a tight, involute trochospire in the Middle Eocene (P10–P13); *Cassigerinelloita*, which in turn evolved into biserially–planispirally enrolled forms; and *Cassigerinella* (Plate 5.10, Figs. 3–6) in the Late Eocene (Priabonian, P14) to Middle Miocene.

The other three Cretaceous survivors, *Bifarina*, *Zeauvigerina*, and *Chiloguembelitria*, died out, respectively, in the Middle Eocene (P13), Late Eocene (P17a), and Late Oligocene (P22) without leaving any descendants. *Zeauvigerina* resemble benthic taxa, but they have stable isotopic values that are equivocal to a planktonic mode of life (Huber *et al.*, 2006).

Recently, geochemical evidence have confirmed that these biserial and triserial groups are not monophyletic, descending from a single Jurassic–Cretaceous ancestry, but must have been polyphyletic, evolving many times in their history from benthic foraminifera (Darling *et al.*, 2009). Therefore, although evolutionary lineages are traced throughout the geological records for these forms (see Fig. 5.13), there is still a possibility that at any stage of their life and under stressful conditions they swapped their mode of life from benthic to planktonic or possibly vice versa. Any phylogeny based simply on morphology and biostratigraphic continuity must therefore be viewed with a degree of caution.

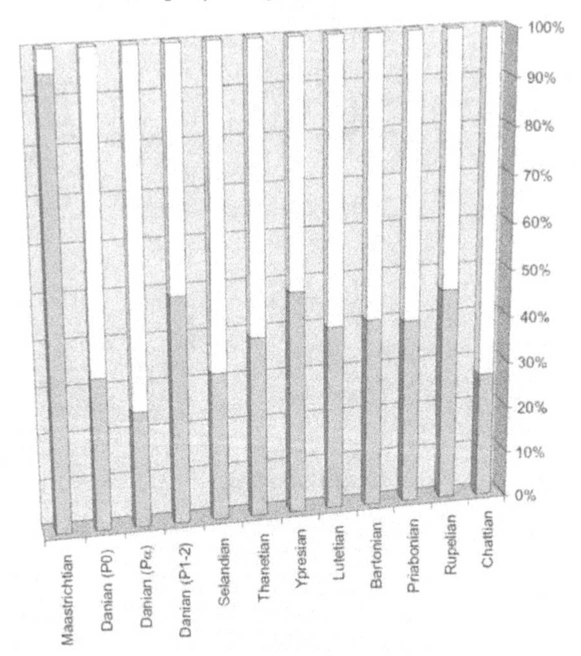

Figure 5.18. Percentage of planktonic foraminifera species extinctions during the Paleogene.

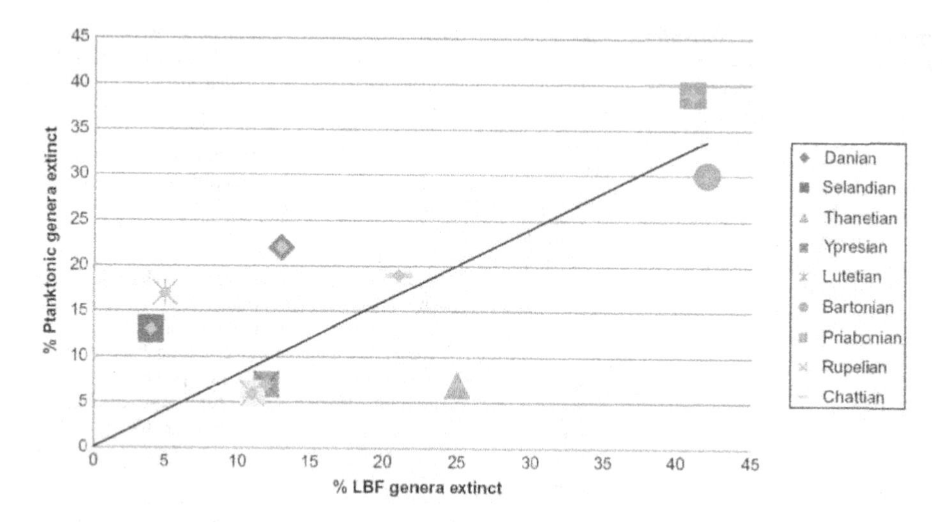

Figure 5.19. The total number of generic extinctions of planktonic foraminifera compared with those of the larger benthic foraminifera in each stage of the Paleogene.

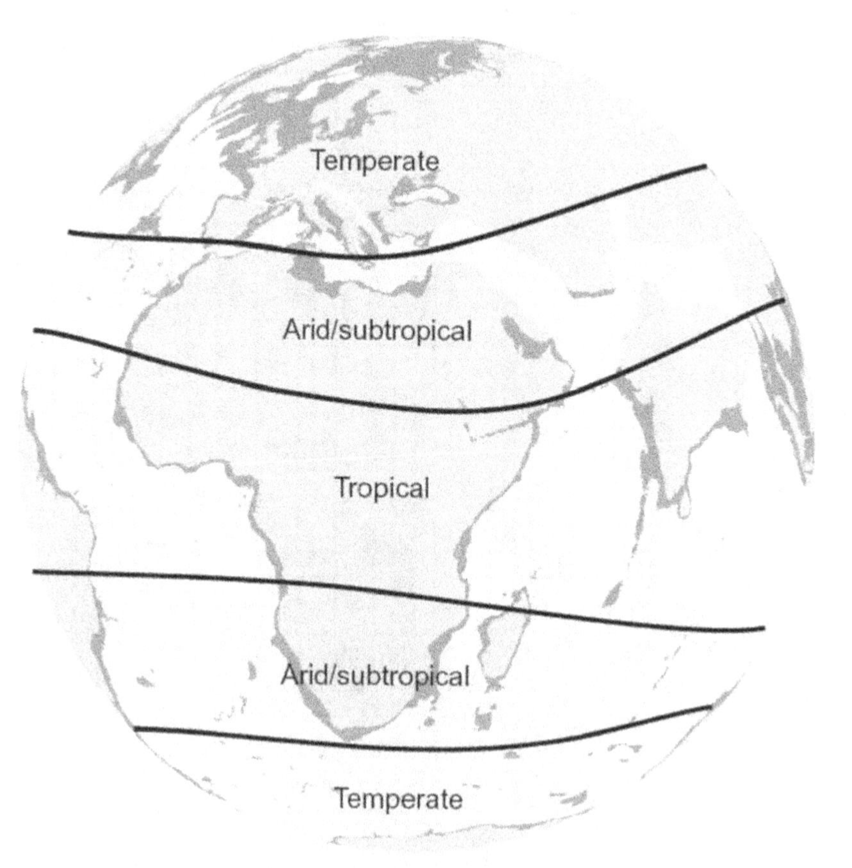

Figure 5.20. Climate zones in the Paleogene.

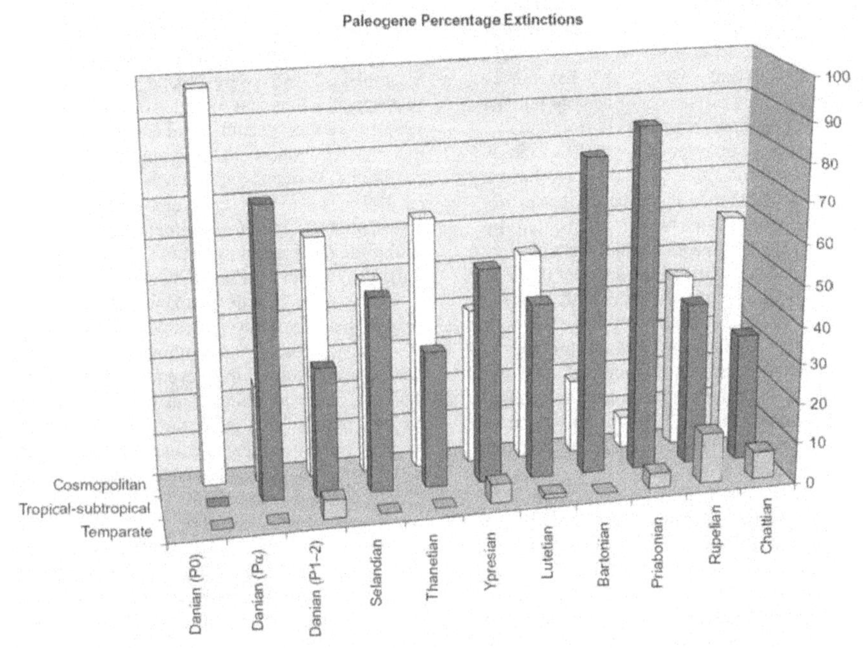

Figure 5.21. The total number of planktonic foraminifera species extinctions in the three climatic zones of the Paleogene.

5.4 Paleogeography and paleoecology of the Paleogene planktonic foraminifera

The Paleogene period starts with the Paleocene (ancient recent life) epoch, and together they mark the beginning of the Cenozoic era. Tectonically, during this period, the continents continued to drift toward their present positions, with the development of the Alpine–Himalayan orogeny, the opening of the circum-Antarctic seaway, but with South and North America still remaining separated (Fig. 5.14). The vast inland seas of the Cretaceous period dried up, exposing large land areas in North America and Eurasia.

In the wake of the End-Cretaceous crisis, when about 90% of the planktonic foraminifera (see Chapter 4) and 83% of the Maastrichtian shallow-water larger benthic became extinct (see BouDagher-Fadel, 2008), the Early Paleocene was a period of recovery for both groups of foraminifera. The extinction of the relatively large Late Cretaceous globotruncanids enabled the small planktonic Cretaceous survivors to evolve into, and rapidly fill the vacant niches, resulting in the appearance of the new planktonic foraminiferal forms that occurred early within the Danian (Figs. 5.15 and 5.16). In contrast to the planktonic realm, it took longer for the complex, interdependent reef-ecosystems to recover (Fig. 5.16), and it was not until the Late Paleocene (Thanetian) that many of the new larger benthic foraminiferal forms made their first appearance (BouDagher-Fadel, 2008).

Two ecological models have been proposed for recovery from the End-Cretaceous mass extinctions. The first assumes that evolutionary recoveries are structured by trophic interactions alone, resulting in a predictable recovery of species richness and abundance earlier in lower trophic levels (Sole *et al.*, 2002). The second, the so-called contingent model, maintains that both ecology and chance are the basis for the structure of recoveries and, therefore, are inherently unpredictable (Yedid *et al.*, 2009). In a recent study of the marine recovery after the K–P extinction, however, Hull *et al.* (2011) concluded that data favored the contingent recovery model. They suggested that ecological and environmental mechanisms may explain the transition from early recovery to late recovery communities, but chance may explain local differences in the timing and recovery path observed.

Many of the planktonic forams that survived the K–P event had restricted environmental distributions. *Bifarina* and *Zeauvigerina* occurred in epicontinental seas, but *Guembelitria* was more successful in the pelagic realm. However, based on their predominant occurrence in sediments from slope and continental shelf environments, all species surviving the Maastrichtian crisis mass extinction are best described as "nerito-plankton," that is living predominantly in coastal, relatively shallow waters (Olsson *et al.*, 1999). This is a mode of occurrence that is similar to that of the still extant form, *Streptochilus* (Kroon and Nederbragt, 1990).

Interestingly, Darling *et al.* (2009) have presented genetic evidence that the living, biserial, planktonic *Streptochilus globigerus* (Schwager) is in fact the same biological species as the benthic *Bolivina variabilis* (Williamson). However, geochemical evidence also suggests that *S. globigerus* actively grows within the open-ocean surface waters. Thus, it must be inferred that this form is ecologically flexible and is capable of occupying both planktonic and benthic domains. Darling *et al.* (2009) argued that at least some of the assumedly nerito-planktonic species may very well have been "tychopelagic," that is they are normally benthic but can exist and grow in the planktonic state as well. Additionally, available isotopic evidence on some *Heterohelix* species is also ambiguous and indicates that they too could have a benthic as well as a planktonic existence (Huber *et al.*, 2006). These observations are further augmented by findings that the living triserial planktonic foraminifera *Gallitellia vivans* had a Miocene benthic ancestor (Ujiié *et al.*, 2008). It would seem, therefore, that some "planktonic" forms have the ability to thrive in both planktonic and benthic habitats. This would provide an ideal ecological adaptation strategy for survival, enabling tychopelagic species to rapidly recolonize the pelagic realm from their benthic stock, which one must assume would have had a greater chance of surviving a major impact event, such as that at Chicxulub.

During the recovery period of the earliest Danian (P0), therefore, the oceans were still occupied worldwide by the small Cretaceous survivors, the microperforate triserial opportunistic *Guembelitria* and the small macroperforate muricate hedbergellids. These survivors are reported from both Northern Hemisphere sediments (New Jersey) and tropical–subtropical regions (Liu and Olsson, 1994). Although their occurrence is rare, and not yet fully studied, they do not seem to have responded to the K–P event in the same way as the nannoplankton species studied by Jiang *et al.* (2010). These authors found that the extent of nannoplankton extinction was higher in the Northern Hemisphere oceans and diversity remained low there for 310,000 years. In contrast, Southern Hemisphere oceans showed lower degrees of extinction and a nearly immediate recovery to normal levels of nannoplankton populations. Jiang *et al.* (2010) and Wignall (2010) suggest that for some calcareous nannoplankton, residence in the relatively protected Southern Hemisphere oceans may have been the key to surviving the mass extinction caused by the K–P impact. The fact that the planktonic foraminifera do not seem to have responded in the same way as the nannoplankton to the K–P event may reflect the survival strategy of some of the "planktonic" forms noted above, which may have involved being able to exchange between living in the planktonic and benthic realms. Indeed, it seems likely that invasions of the planktonic realm by benthic foraminifera, typically during times of ecological crisis, may have occurred throughout the geological past. Such invasions may be expected to have been initiated by perturbations in the food supply, or by oxygen stress in the benthos, or, as in the case of the post K–P recovery, by the creation of empty planktonic niches caused by the extinction of previously existing planktonic forms (Leckie, 2009).

Of the Danian planktonic foraminifera, 69% rapidly became cosmopolitan, and the bulk of the residue (25%) were restricted to tropical–subtropical waters, while only 6% were found solely in temperate conditions. A million years after the K–P boundary forming event, the Cretaceous survivors are found to be associated with small, microperforate, trochospiral globoconusids, which rapidly became dominant in the earliest Danian, Pa (see Figs. 5.5 and 5.13). They were mainly tiny forms of *Parvularugoglobigerina*, which often constituted 50% of assemblages and were found in mainly tropical–subtropical waters and in regions of upwelling (Li *et al.*, 1995). Like their guembelitriid ancestors, they sought out nutrient-rich waters in eutrophic to highly eutrophic conditions (Boersma and Premoli Silva, 1991; Hallock *et al.*, 1991). However, it seems that gradual climate warming contributed to *Parvularugoglobigerina* extinction at the end of the Danian and to that of *Globoconusa* (Plate 5.10, Figs. 1–2) within the Early Selandian.

In the Late Danian, planktonic assemblages became progressively more diversified with an explosive radiation of nonspinose, smooth-wall globanomalinids, praemuricate forms that eventually gave rise to the muricate truncorotallids, and the spinose, cancellate eoglobigerinids (see Figs. 5.2 and 5.6). Ecologically, each of these groups occupied distinct niches, as inferred from their morphology and stable isotope geochemistry (Berggren and Norris, 1997). The hedbergellids had been cosmopolitan and probably lived within thermocline depths, and their first Paleogene descendant, *Globanomalina* (Plate 5.1, Figs. 1–3), continued to grow in the thermocline throughout the remainder of the Paleocene. The earliest truncorotalid, *Praemurica*, appears to have become a near-surface dweller by the Danian (Pa). On the other hand, the cancellate, spinose eoglobigerinids began as a deep-water group in the Late Danian (P1) and remained in this habitat throughout the Paleogene (Berggren and Norris, 1997).

Besides the probable invasion of benthic forms in the aftermath of the K–P event, another driver of evolutionary change for Paleogene planktonic foraminifera was the development and change of the whole phytoplankton community structure of the upper water column. The diversification and rise in dominance of dinoflagellates and coccolithophorids during the Cenozoic (Erba, 2006; Falkowski, *et al.*, 2004; Leckie, 2009) is likely also to have facilitated or driven planktonic foraminiferal evolution, culminating in the development of the photosymbiotic truncorotalids, which became very abundant during the long-term warming of the early Eocene. The development of photosymbiosis in these extinct forms has been inferred from the study of living forms, in which the development of the cancellate, spinose morphologies is linked to the hosting of algalphotosymbionts (Quillévéré *et al.*, 2001).

The end of the Danian witnessed a sharp drop in sea level (see Fig. 5.17), which may have contributed to the extinction of the planktonic forms seen at this time. All of the small microperforate *Parvularugoglobigerina* and praemuricate forms died out at the end of the Danian. However, a steady increase in temperature contributed to the expansion of oligotrophic conditions in the global ocean, which would have facilitated the diversification of the muricate planorotalids and truncorotalids, including *Morozovella* and *Acarinina* (see Fig. 5.3). They reached their maximum diversity toward the end of the Thanetian.

Global warming events occurred during the Late Paleocene and Early Eocene, from about 59 to 50 Ma (Zachos *et al.*, 2008), resulting in an anomalously warm global climate, optimum spanning some 4–5Ma during the

Early Eocene (Berggren and Prothero, 1992). This climate optimum was preceded by a brief period of extreme warming at the onset of the Eocene (see Fig. 5.17), called the Paleocene–Eocene Thermal Maximum (PETM) (Cramer and Kent, 2005). Intense flood basalt magmatism, accompanying the opening of the North Atlantic, dated at 58–55 Ma, gave rise to volcanic fields that covered 1.3 million km^2 (Courtillot and Renne, 2003). This period also saw a major perturbation of the carbon cycle as indicated by a sharp negative $\delta^{13}C$ excursion (Dickens, 2001; Lunt *et al.*, 2011), which may have been caused by metamorphically generated methane from sill intrusion into basin-filling carbon-rich sediments (Storey *et al.*, 2007), or orbitally induced changes in ocean circulation that triggered the destabilization of methane hydrates (Lunt *et al.*, 2011). On the other hand, the reported presence of an iridium anomaly at the P–E boundary could be indicative of meteorite collision with Earth (Dolenec *et al.*, 2000; Schmitz *et al.*, 1996), which might in turn have triggered a rapid change in the climate (Higgins and Schrag, 2006). Cramer and Kent (2005) argue that the very rapid onset of the PETM is best explained by such an impact mechanism and refer to the after effects of this proposed impact as the "Bolide Summer." However, whatever the origin of the PETM, the Paleocene–Eocene boundary saw the extinction of 39% of planktonic foraminifera species (Fig. 5.18), represented by 7% of the Thanetian genera (see Fig. 5.15) and 25% of larger benthic foraminiferal species (Fig. 5.19), but no family extinctions.

The warm climatic period persisted through the Early Eocene, and the continents continued to drift toward their present positions. The Eocene global climate was perhaps the most homogeneous of any in the Cenozoic; the temperature gradient from equator to pole was only half of that found today, and deep-ocean currents were exceptionally warm (Fig. 5.20). A pole-ward expansion of coral reefs occurred, together with a broader latitudinal distributionof temperature-sensitive organisms such as larger benthic foraminifera, mangroves, palms, and reptiles (Adams *et al.*, 1990; BouDagher-Fadel, 2008; Kiessling, 2002; Pearson *et al.*, 2007). These warmer conditions were, it seems, ripe for enhanced speciation, and seven new families of planktonic foraminifera came into existence in this epoch.

The Ypresian witnessed the appearance of most new species (58%) in the tropics and subtropics (see Fig. 5.16), and saw the beginning of the worldwide radiation of spinose globigerinids. The photo-symbiotic *Acarinina–Morozovella* species, which lived in the surface ocean habitat, expanded their geographic range to become mainly cosmopolitan (55% of the new species). In the tropics, the hantkeninids evolved broad tubulospines in the Late Ypresian, which significantly increased the surface area of their chambers and therefore the potential proportion of their extrathalamous cytoplasm available to enable the cross-membrane diffusion needed to survive in oxygen-depleted waters. On the other hand, the number of the globanomalinids, which lived primarily in the thermocline but that lacked photosymbiosis, remained constant at the beginning of the Ypresian but began to dwindle toward the end of that stage.

Gradual cooling at the end of the Ypresian was accompanied by drops in sea level and saw the extinction of 55% of the planktonic foraminifera species in the tropical–subtropical waters (see Fig. 5.21). By the end of the Ypresian, 88% of the deep-water globanomalinids had become extinct (see Fig. 5.2). Species of *Morozovelloides*, which were cosmopolitan (e.g., *Morozovelloides bandyi*) in the Ypresian, survived into the Middle Eocene but evolved into species confined to the tropics and subtropics (e.g., *Morozovelloides lehneri*, Plate 5.12, Fig. 6) before they disappeared completely at the end of the Middle Eocene. The Middle Eocene was a time of sea-level fluctuation and a continued gradual decrease in global temperature. During this time, the truncorotalids persisted in being dominant in the world oceans, and with the acquisition of strong muricae in *Testacarinata* (Plate 5.9, Fig. 16), they were able to move into deeper oceanic layers than those inhabited by nonmuricate or weakly muricate forms (BouDagher-Fadel *et al.*, 1997). The spinose trochospiral globigerinids and eoglobigerinids continued slowly to diversify, evolving into new lineages (Fig. 5.9). The spines of the latter and the muricae of *Morozovella*, presumably covered by extrathalamous extensions of the cytoplasm, were capable of carrying and transporting vacuoles, which would have given them a competitive advantage, and might explain why they were able to spread faster than the smooth globanomalids, which died out in the early part of the Middle Eocene (see Fig. 5.2). The nonmuricate, nonspinose globanomalinids would have had no skeletal control on the distribution of sites for pseudopodial extensions into the surrounding seawater. The pseudopodia would have extended either irregularly or weakly from the approximately spherical test. If the same volume of extrathalamous cytoplasm, however, were to be compressed into a disk, its surface area for food collection would have been very greatly increased. Therefore, species which could extrude pseudopodia into a disk must have had a considerable advantage in the collection of available suspended particulate nutrients. This could be the reason for the repeated success of spinose or muricate taxa, and for the success of the globigerinids in supplanting the globanomalinids from their dominant position in the Eocene biota (see Fig. 5.2). Living globigerinids (and by inference fossil forms) include species which possess dinoflagellate symbiont and those which do not. The former are restricted to the photic zone of the water column, while the latter can live both in deeper waters (down to 400 m) and in the photic zone (see Fig. 5.22). Even those which have symbionts may also be carnivorous (e.g., feeding on calanoid copecopds), but this is not universally the case (BouDagher-Fadel *et al.*, 1997). Living species of spinose *Turborotalita* (e.g., *Turborotalita quinqueloba*) feed mostly on algae and small zooplankton (Asano *et al.* 1968; Kroon *et al.*, 1988) and are especially abundant during spring bloom conditions. Bullate forms, such as *Catapsydrax*, are mainly found in subthermocline water (from 600 to 800 m), while those quadrate forms, such as *Dentoglobigerina*, can inhabit the upper and deep thermocline (Pearson and Wade, 2009a,b).

In the Middle to Late Eocene, the globigerinids developed streptospiral coiling, giving rise to the porticulasphaerids including *Globigerinatheka* and *Orbulinoides*, which diversified mainly in the inner neritic waters of the tropics and subtropics. Stable isotope evidence suggests that all globigerinathids were mixed layer-dwelling species which probably sank in the water column at the end of their life cycle and added a thick calcite crust over the cancellate spinose wall (Boersma *et al.*, 1987; Pearson *et al.*, 1993, 2001; Premoli Silva *et al.*, 2006). This was also a period when there was an increase in digitate hantkeninids, with elongate chambers, that were mainly found in the tropics and subtropics.

During the Middle Eocene, digitate species were common in deep habitats, characterized by lower temperatures, reduced oxygen, and enrichment of dissolved inorganic carbon. The occurrence of modern digitates in deep subthermocline(>150 m) habitats supports this interpretation. BouDagher-Fadel *et al.* (1997) suggested that the broad, radial elongation of the adult chambers increase the surface area of the chambers and therefore the potential proportion of extrathalamous cytoplasm and the ability of the foraminifera to survive in waters lower in oxygen. Coxall *et al.* (2007) propose that the primary function of digitate chambers was as a feeding specialization that increased effective shell size and food gathering efficiency, for survival in a usually food-poor environment, close to the oxygen minimum zone. In either case, episodes of increased digitate abundance and diversity generally indicate expansion into deep-water conditions that are unfavorable to other planktonic species (BouDagher-Fadel *et al.*, 1997).

By the late Middle Eocene, the globigerinids developed flattened tests to give *Paragloborotalia*, the ancestral form of the keeled Neogene *Globorotalia*. *Paragloborotalia* species (Plate 5.7, Figs. 1–14) were wide spread latitudinally (being quite common in temperate seas as well as the tropics).

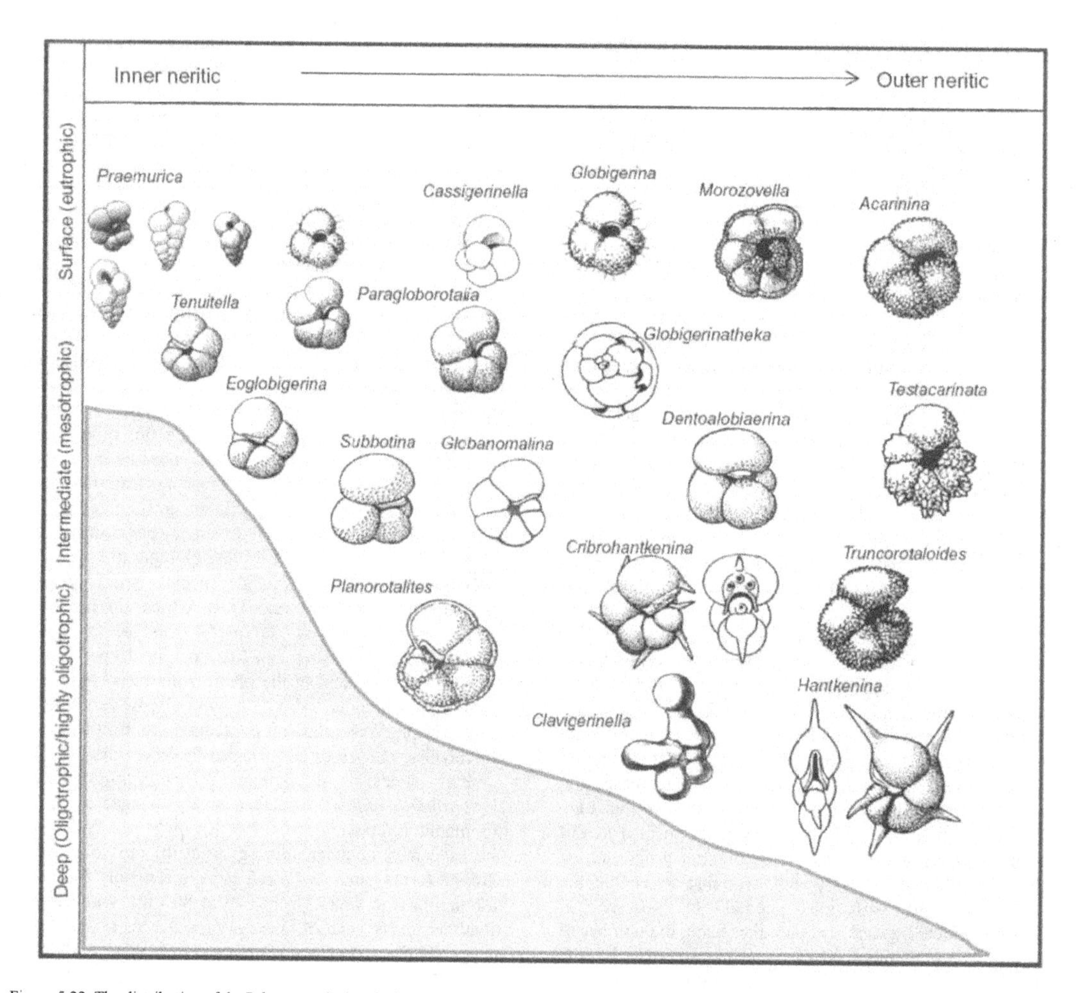

Figure 5.22. The distribution of the Paleogene planktonic foraminiferal in the neritic environment.

The Late Eocene and the transition from the Eocene to the Oligocene witnessed several major pulses of extinctions, which were probably driven by the significant changes in global climate and ocean circulation that occurred at this time (Berggren and Prothero, 1992). The changes from the warm "greenhouse" climate of the Early Eocene to a much cooler, more temperate climate might have been triggered by tectonic events (see Fig. 5.23), such as the opening of the Tasmanian gateway which led to the expansion of the Antarctic ice cap following its isolation from other continental masses (Smith and Pickering, 2003). In addition, multiple impacts (e.g., at Chesapeake Bay, and Popigai) at ~35 Ma, when associated with the later large flood basalt event in Ethiopia and Yemen around 30 Ma (Courtillot and Renne, 2003), may have also contributed significantly to the environmental stress during these stages. These environmental stresses and global cooling certainly contributed to the planktonic extinction rate exceeding the speciation rate, and hence the global reduction in the diversity of planktonic foraminifera species (see Fig. 5.16).

Most planktonic foraminifera extinctions were concentrated in the late (but not end) Priabonian and not at the Eocene–Oligocene boundary. The extinctions of two major groups, the truncorotalids (at about 35.4 Ma) and the porticulasphaerids (at about 34.2 Ma) occurred within the last 3.3 Ma of the Eocene. Many of the taxa disappeared as tropical–subtropical faunas (see Fig. 5.21). Others, such as the spinose globigerinids, migrated toward temperate climates. The truncorotalids which included the morozovellids and acarininids were highly successful shallow-dwelling groups occupying the near-surface photic zone (Boersma *et al.*, 1987; Pearson *et al.* 1993, 2001; Wade and Kroon, 2002) and dominated subtropical and tropical oceans of the early Paleogene (see Chart 5.3 online). In the Late Bartonian, ~37 Ma, a faunal turnover in planktonic foraminifera occurred which saw the extinction of this group. Carbon isotope results indicate an ontogenetic change in the ecology and life strategy of the truncorotalids in the interval preceding their extinction in the Late Priabonian. They had a strong reliance of photosymbionts, and the destruction of this ecological partnership was probably a cause of their extinction (Wade *et al.*, 2008). These extinctions within the Priabonian coincided with the onset of Late Eocene rapid sea-level changes, and the move into an "ice house" climate (see Fig. 5.23).

In contrast to these events within the Late Eocene, the Eocene–Oligocene boundary (at about 33.9 Ma) is a relatively minor event, marked only by the extinction of the spiny hantkeninids and the smooth turborotalids. The former had had their habitats restricted to the tropical–subtropical realm, which became under stress in a cooling world. The latter (e.g., *Turborotalia cerroazulensis*) were asymbiotic and cosmopolitan, except species of *Carinoturborotalia* which were confined to the tropics and subtropics. *Turborotalia* may have occupied a deeper niche in the water column (thermocline) compared with the morozovellids (Boersma *et al.*, 1987; Pearson *et al.*, 2001; Wade and Kroon, 2002; Wade *et al.*, 2008) which died out at the end of the Middle Eocene. Further extinctions and climatic changes occurred 4.6Ma after the Eocene–Oligocene boundary.

In the Early Oligocene (from about 33.5 Ma), the Drake Passage opened and there was a sustained cool period (Berggren and Prothero, 1992). This, when added to further stress from the gradual global sea-level falls, and the extensive volcanism of the Ethiopian Flood Basalts (31–29 Ma), almost certainly contributed to the complete demise of an important Paleogene group, the globanomalinids. This group had not changed their deep-water habitat throughout the Paleogene, and had suffered reductions in diversity during the Middle to Late Eocene, but did not disappear completely till the Early Oligocene (about 29.3 Ma). They were replaced immediately in the Late Eocene and Oligocene by opportunists (e.g., the replacement of *Praetenuitella* by typical tenuitellids; see Fig. 5.7).

In summary, therefore, the main events in the Paleogene are presented in Fig. 5.23. Overall, species diversity increased gradually from the Paleocene and up to the Lutetian during a time of globally warm conditions. From the Bartonian to the Chattian in the Late Oligocene, there was a notable decrease in diversity (see Fig. 5.16), as extinction rates exceeded speciation rates as global conditions cooled. Throughout the Paleogene, extinction rates remained fairly constant at an average of ~ 35–40% (see Fig. 5.18). However, extinction events were more notable on the family levels (see Fig. 5.1), as seven major families disappeared at different times during the Paleogene. During the Thanetian, cosmopolitan forms suffered relatively high extinction rates relative to tropical and sub-tropical forms (see Fig. 5.21) as increases in global temperature at this time (see Fig. 5.17) favored the diversification of the truncorotalids and the planorotalids. On the other hand, tropical and subtrop¬ical species extinctions were more notable in the Bartonian and Priabonian as the onset of the cooling period of the Late Eocene began, but preceded maximum glacial conditions in the early Oligocene by ~3 Ma.

The deep-water globigerinids diversified in the Oligocene, dominating most of the assemblages and evolving into compressed forms, the globorotaliids which became very successful in the Neogene. Very few extinctions occurred in the Middle Oligocene, perhaps because the surviving globigerinids and globigerinitids were already adapted to cooler environments from the earlier attrition of war¬m-climate foraminifera. Temperate planktonic foraminifera were more common in the Oligocene, but they never really flourished when compared with the tropical and subtropical forms. Most of them became extinct at the P21/P22 boundary (e.g., Globigerinita boweni, Plate 5.3, Fig. 1; Tenuitella neoclemenciae, Tenuitella munda, Tenuitella gemma). The eoglobigerinids dwindled gradually until they disappeared, with the last species of the genus Subbotina, at the end of the Oligocene which also saw the final extinction of the Guembelitriidae. This stratigraphic boundary is not marked by any major discontinuity, and as we will see in the following chapter, the planktonic foraminifera of the Neogene show a gradual, continuous development from those forms of the seven families of Paleo¬gene planktonic foraminifera which survived uninterrupted into the Miocene.

Age (Ma)	Period, Epoch, Stage			Planktonic zonation	Planktonic foraminifera events	Tectonic and oceanic events	Climatic events
23.0	Oligocene	Late	Chattian	P22 b, a; P21 b	Extinction of **eoglobigerinids** and the **guembelitrids**	Alpine-Himalayan orogeny	Late Oligocene global warming (26 Ma)
33.9		Early	Rupelian	P21 a; P20 b, a; P19 b, a; P18 b, a	Extinction of the **globanomalinids**	Gradual global sea-level falls Ethiopian Flood Basalts (31–29 Ma)	Rapid global cooling and major Antarctic ice cap initiation (34 Ma)
	Eocene	Late	Priabonian	P17 b, a; P16 b, a; P15 b	Extinction of the **hantkeninids** Extinction of the **porticulasphaerids** and appearance of **globigerinitids** **Truncorotalids** and **heterohelicids** extinct	Opening of Drake and Tasmanian Passages (35–33 Ma). Chesapeake and Popigai impacts (35.5 Ma)	
		Middle	Bartonian	P15 a; P14 b, a; P13 b, a; P12 b	Diversification of **globigerinids**	Rapid sea-level changes	Mountain glaciers in antarctica
			Lutetian	P12 a; P11 b, a; P10 b, a	First appearance of **globigerinitids** High diversity of **eoglobigerinids** and extinction of **planorotalids** First appearance of **porticulasphaerid**	Major global sea-level fall (43 Ma)	Gradual global cooling
55.8		Early	Ypresian	P9 b, a; P8 b, a; P7 b, a; P6 b, a; P5 b	First appearance of the **hantkeninds** Worldwide radiation of spinose **globigerinids** First appearance of **globigerinids** and **cassigerinellids**	India- Asian collision (50 Ma) Major global sea-level high stand (54–51 Ma)	Eocene climatic optimum (54–50 Ma)
	Paleocene	Late	Thanetian	P5 a; P4 c, b	Diversification of the muricate **planorotalids** and **truncorotalids**	North Atlantic rifting and volcanism (58–55 Ma)	PETM (55.8 Ma)
		Middle	Selandian	P4 a; P3 b, a	**Gloconusids** extinction	Expansion of oligotrophic Sharp drop in sea level	Gradual global warming
65.5		Early	Danian	P2; P1 c, b, a; Pα	First appeamce of major Paleogene groups, **globanomalinids,** truncorotalids, and **eoglobigerinids** Small Cretaceous survivors (*Guembelitria, Hedbergella*)	Post-crisis oceanic recovery period	

Figure 5.23. The major tectonic, oceanic, and climatic events affecting planktonic foraminifera evolution and extinction during the Paleogene.

Plates 5.1–5.12

Plate 5.1. Figure 1. *Globanomalina archeocompressa* (Blow). Figured by Blow (1979), PacificOcean, DSDP 47, Paleocene, Early Pa, x450. Figure 2. *Globanomalina planocompressa* (Shutskaya). Figured by Blow (1979), Pacific Ocean, DSDP 47, Paleocene, P1, x400. Figure 3. *Globanomalina ovalis* (Haque). Paratype figured by Banner (1989), Salt Range, Pakistan, NHMP42400a, Paleocene, P5a, x400. Figures 4–5. *Parvularugoglobigerina longiapertura* Blow. (4) Holotype, (5) paratype, both figured by Blow (1979), Pacific Ocean, DSDP 47, Paleocene, Pa, x400. Figures 6–7. *Pseudomenardella ehrenbergi* (Bolli). Holotype figured by Bolli (1957) from Trinidad, Paleocene, P3, x200. Figures 8–9. *Pseudomenardella pseudomargaritae* new species, (8) holotype (from Blow, 1979, pl. 94, fig. 1), (9) paratype (from Blow, 1979, pl. 94, fig. 4 reversed image), with spiral sutures limbate and meeting spiral suture perpendicularly, strong carina throughout, Tanzania, Paleocene, P5, x133. Figures 10–11. *Pseudomenardella quinquecamerata* new species. Holotype, called *Globorotalia troelseni* by Blow (1979, pl. 90, figs 1–2), with spiral sutures notlimbate, carinate throughout, five chambers in the last whorl, spiral side flattened, aperture a high arch, DSDP 21A, S. Atlantic, Paleocene, P4, x127. Figure 12. *Turborotalia pseudoampliapertura* (Blow and Banner). Figured by Blow (1979), Tanzania, Eocene, P16, x133. Figures 13–14. *Turborotalia frontosa* (Subbotina). Figured by Blow (1979), Kane 9 core, Atlantic, Eocene, P10, x133. Figure 15. *Turborotalia centralis* (Cushman andBermudez). Figured by Blow (1979) from Tanzania, East Africa, Eocene, P15, x60. Figures 16–18. *Turborotalia cerroazulensis* (Cole). Hypotypes figured by Blow (1979), Tanzania, P15, x60. Figures 19-20. *Antarcticella antarctica* (Leckie and Webb). Figured by Leckie and Webb (1985), DSDP 270, Southern Ross Sea, Antarctica, Late Oligocene–Early Miocene, x200.

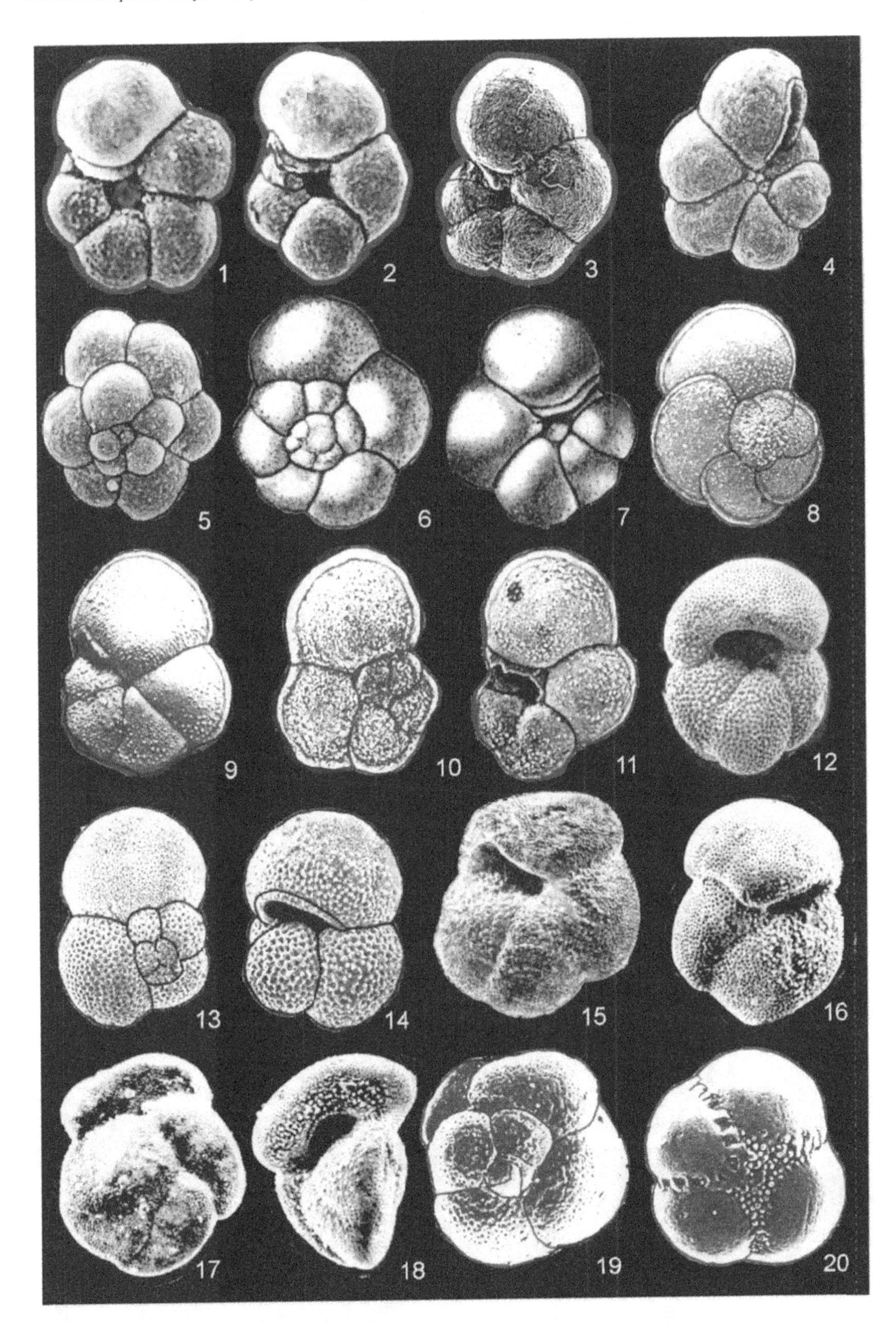
1
2
3
4
5
6
7
8
9
10
11
12
13
14
15
16
17
18
19
20

Plate 5.2. Figure 1. *Turbeogloborotalia rainwateri* (Blow). Paratype, figured by Blow (1979), DSDP 20C, S. Atlantic, Paleocene, P2, x135. Figures 2-4. (2-3) *Planorotalites pseudomenardii* (Bolli), (4) *Planorotalites* sp., DSDP 86, UCL Collection, Paleocene, P4a, x400. Figures 5-6. *Aragonella dumblei* (Weinzierl and Applin). Trinidad, UCL Collection, Eocene, P13, x400. Figure 7. *Cribrohantkenina inflata* (Howe). Figured by Blow (1979), Tanzania, Eocene, P16, x30. Figure 8. *Hantkenina mexicana* (Cushman). Figured by Tourmakine and Luterbacher (1985), Venezuela, Eocene (*suconglobata* to *lehneri* Zone), x133. Figure 9. *Hantkenina primitiva* (Cushman and Jarvis). Figured by Blow (1979), Tanzania, Eocene, P16, x200. Figures 10-11. *Pseudohastigerina wilcoxensis* (Cushman and Ponton). Figured by Blow (1979), Kane 9 core, Atlantic, Eocene, P10, x400. Figure 12. *Pseudohastigerina danvillensis* (Howe and Wallace). Figured by Blow (1979) as *P. micra* , Tanzania, Eocene, P16, x200. Figure 13. *Pseudohastigerina micra* (Cole). Figured by Blow (1979), Mexico, Eocene, P12, x350. Figure 14 *Pseudohastigerina sharkriverensis* (Berggren and Olsson). Figured by Blow (1979), Belgium, P12-14, Eocene, x300. Figure 15. *Pseudohastigerina naguewichiensis* (Myatliuk). Figured by Blow (1979), Red Bluff Clay, Mississippi, Oligocene, P18, x400. Figure 16. *Clavigerinella akersi* (Bolli, Loeblich, and Tappan). Figured by Blow (1979), Kane 9 core, Atlantic, Eocene, P10, x130. Figure 17. *Clavigerinella eocaenica* (Nuttall). Figured by Blow (1979), Kane 9 core, Atlantic, Eocene, P10, x133. Figures 18-19. *Praetenuitella praegemma* (LiQianyu). Holotype figured by Li (1987), Alabama, Eocene, P16-P17, x250. Figures 20-22. *Tenuitella postcretacea* (Myatliuk). UCL Collection, Trinidad, Cipero Formation, Oligocene, P21, x250.

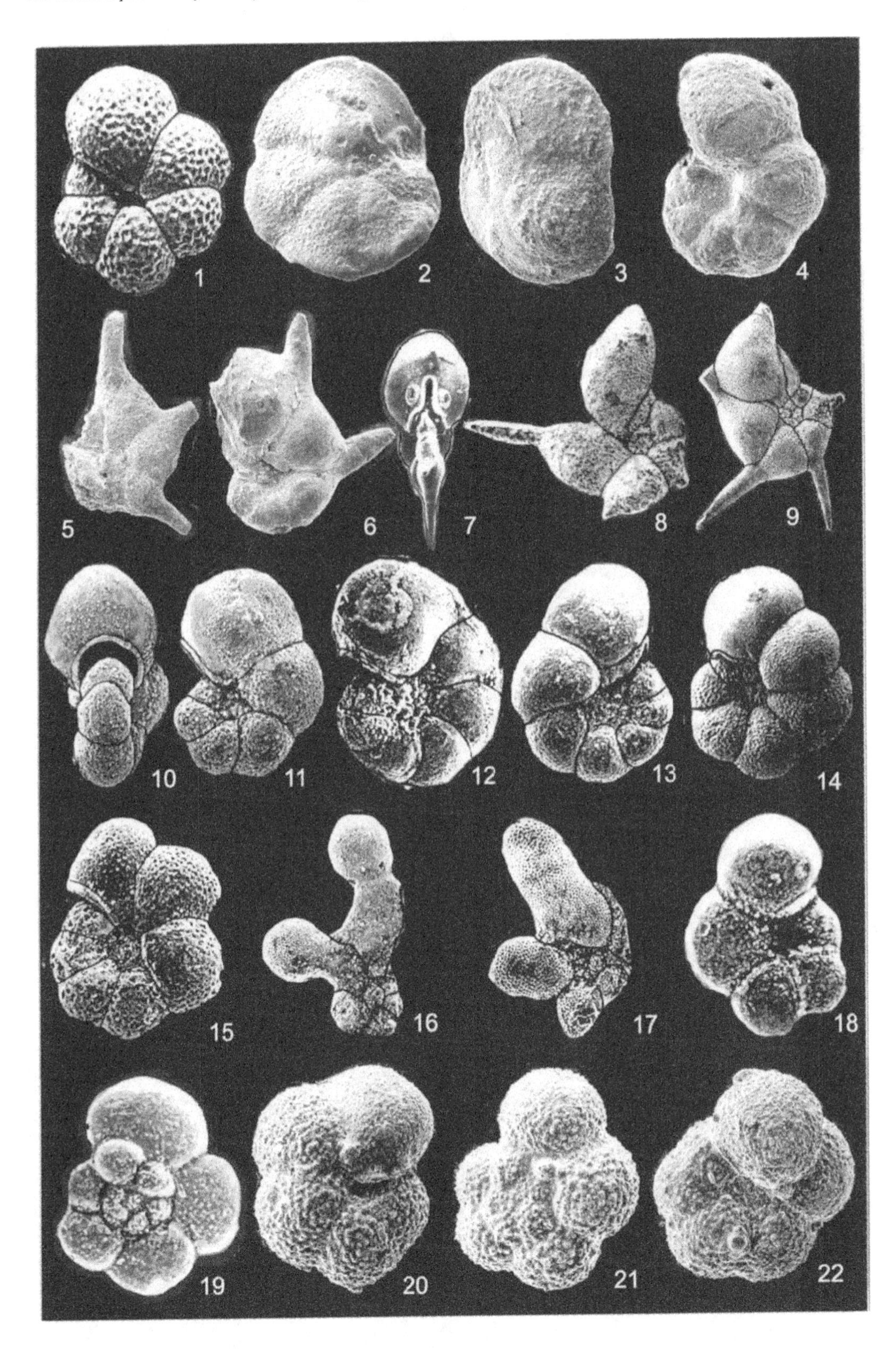
1
2
3
4
5
6
7
8
9
10
11
12
13
14
15
16
17
18
19
20
21
22

Plate 5.3. Figure 1. *Globigerinita boweni* (Brönnimann and Resig). Holotype, figured by Brönnimann and Resig (1971), DSDP 64, Oligocene, P21, x650. Figure 2. *Tenuitellinata juvenilis* (Bolli). Figured by Kennett and Srinivasan (1983) as *Globigerinita glutinata,* DSDP Site 284, Late Pliocene, x200. Figures 3-4. *Tenuitellinata praepseudoedita* new species. Figured by Li (1987, pl. 3, figs 1-2) as *Tenuitellinata,* cf. *T. pseudoedita* (Subbotina), with five chambers in the last whorl, aperture antero-intraumbilical, Trinidad, *C. dissimilis* Zone, x60. Figure 5. *Tenuitellinata angustiumbilicata* (Bolli). Topotype, figured by Li (1987), Trinidad, *G. ciperoensis* Zone, x133. Figure 6. *Eoclavatorella benidormensis* (Cremades Campos). Holotype, figured by Cremades Campos (1980), Spain, Early Eocene, x200. Figure 7. *Eoglobigerina edita* (Subbotina). Hypotype, figured by Blow (1979), DSDP 6, Station 47/2, Paleocene, P1, x200. Figure 8. *Eoglobigerina eobulloides* (Morozova). Hypotype, figured by Blow (1979), DSDP 6, Paleocene, Pa, x200. Figures 9-10. *Guembelitrioides higginsi* (Bolli). Trinidad, UCL Collection, Eocene, P13, x600. Figures 11-12. *Parasubbotina pseudobulloides* (Plummer). Texas, Mexia Clay, Wills Point Formation, UCL Collection, Paleocene, P3, x133. Figure 13. *Subbotina simplissima* (Bolli). Paratype, figured by Blow (1979), DSDP 6, Station 47/2, Paleocene, P1, x150. Figures 14–16. *Subbotina triangularis* (White). Texas, Mexia Clay, Wills Point Formation, UCL Collection, Paleocene, P3, x150. Figure 17. *Subbotina triloculinoides* (Plummer). DSDP 86, UCL Collection, Paleocene, P4, x150. Figures 18-19. *Turbeogloborotalia compressa* (Plummer). Texas, Mexia Clay, Wills Point Formation, UCL Collection, Paleocene, P2, x80. Figure 20. *Turbeogloborotalia haunsbergensis* (Gohrbandt). Figured by Blow (1979) as *Globorotalia compressa,* DSDP Site 20C, Paleocene, P2, x200. Figure 21. *Turbeogloborotalia griffinae* (Blow). Holotype, figured by Blow (1979), Kane 9 core, Atlantic, Eocene, P9, x133.

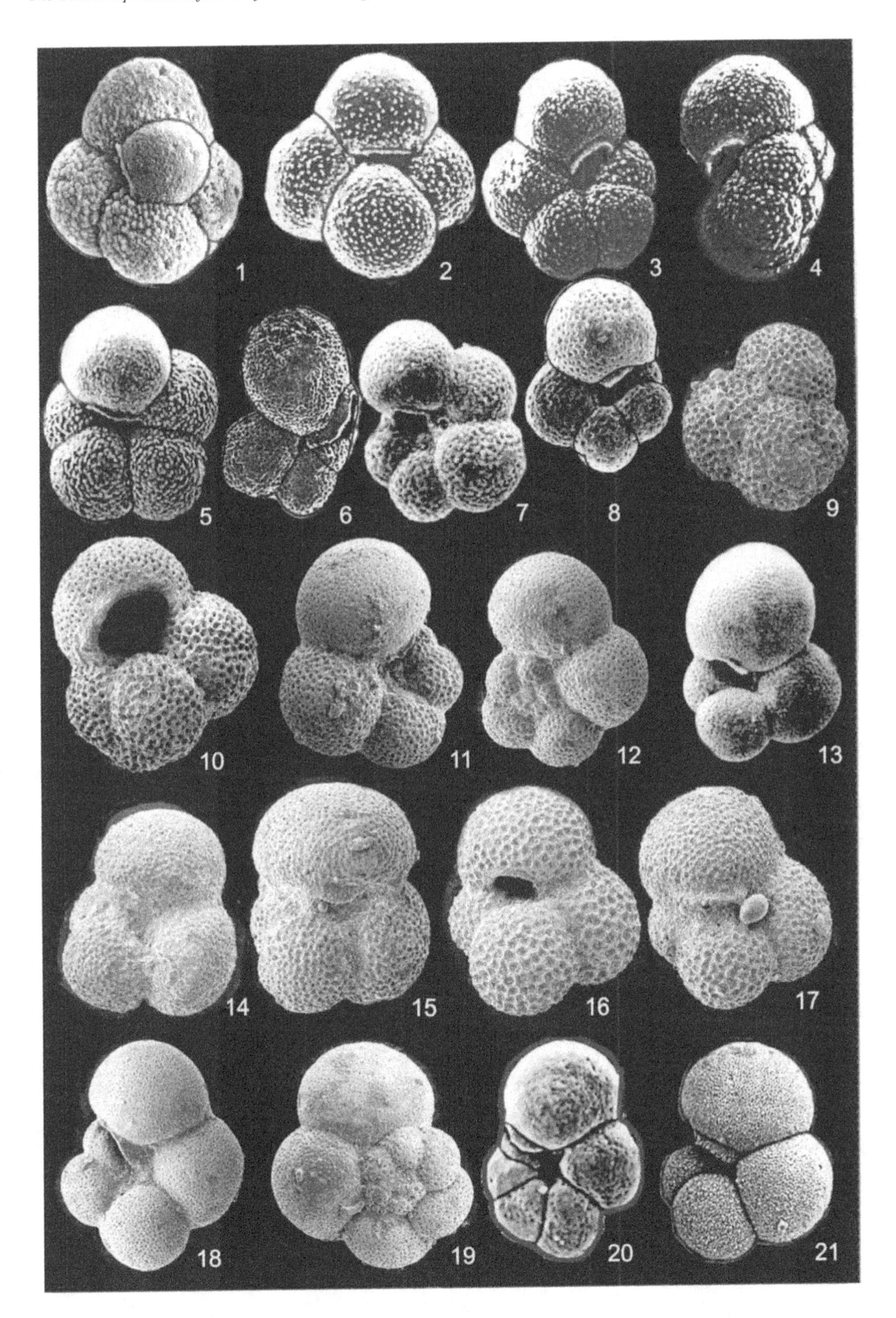
1
2
3
4
5
6
7
8
9
10
11
12
13
14
15
16
17
18
19
20
21

Plate 5.4. Figures 1–3. *Catapsydrax unicavus* (Bolli, Loeblich, and Tappan). Trinidad, UCL Collection, Early Oligocene, P18, x37. Figure 4. *Catapsydrax dissimilis* (Cushman and Bermudez). Cuba, UCL Collection, Early Oligocene, P19, x37. Figures 5- 6. *Globigerina angulisuturalis* (Bolli). Cipero Formation, Bo 291A, Trinidad, UCL Collection, Early Oligocene, P18, x96. Figure 7. *Globoturborotalita anguliofficinalis* (Blow). Holotype, figured by Blow (1969), North of San Fernando, Trinidad, West Indies, Oligocene, P19, x72. Figure 8.*Turborotalia ampliapertura* (Bolli). As all globerinids, the surface is rough with prominent perforation pits, figured by Blow (1969), Trinidad, N1 type locality of Blow 1969, x100. Figure 9. *Globigerina sellii* (Borsetti). Figured by Blow (1969), Tanzania, Early Oligocene, P19, x100. Figure 10. *Globigerina tapuriensis* (Blow and Banner). Figured by Blow (1969), Lindi Creek, southern Tanzania, Early Oligocene, P19, x100. Figures 11-12. *Globigerina prasaepis* (Blow). Cipero Formation, Trinidad, UCL Colelction, Oligocene, P21, x67. Figures 13-14. *Globigerina ciperoensis* (Bolli). Trinidad, Early Oligocene, P18, x70. Figure 15. *Globigerina officinalis* (Subbotina). Cocoa Sands formation, Alabama, UCL Collection, Late Eocene, x82. Figures 16-17. *Globoturborotalia woodi* (Jenkins). Figured by Kennett and Srinivasan (1983), DSDP 207, Middle Miocene, x100. Figure 18. *Subbotina angiporoides* (Hornibrook). Figured by Blow (1979), Tanzania, x100. Figure 19. *Globigerina hagni* (Gohrbandt). Figured by Toumarkine and Luterbacher (1985), Possagno, Italy, *H. nutalli* Zone (i.e., P10), x100. Figure 20. *Globigerina senilis* (Bandy). Figured by Blow (1969), Jackson Formation, Alabama, Late Eocene, P16, x133.

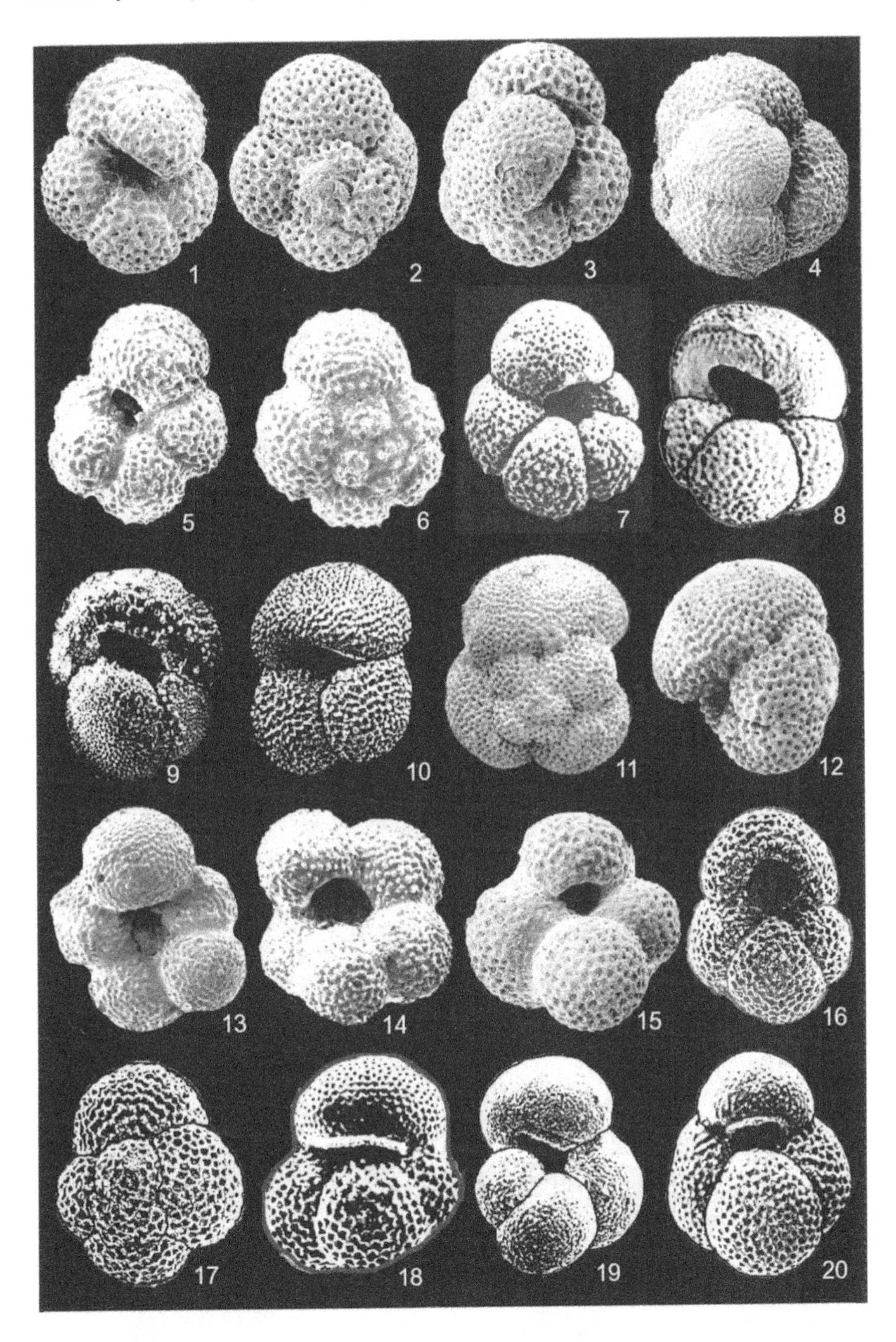
1
2
3
4
5
6
7
8
9
10
11
12
13
14
15
16
17
18
19
20

Plate 5.5. Figure 1. *Globigerina occlusa* (Blow andBanner). Figured by Kennett and Srinivasan (1983) as *Globigerina praebulloides*, DSDP Site 289, Ontong Java Plateau, West Pacific, Zone N4A, x210. Figure 2. *Globigerina praebulloides* (Blow). Figured by Blow and Banner (1962), Tanzania, Oligocene, P20, x210. Figures 3–6. *Globigerina prolata* Bolli. Figured by Blow (1979) as *G. lozanoi* (Colom), Jamaica, Eocene, P9, x130. Figure 7. *Globigerinatheka senni* (Beckmann). Trinidad, UCL Collection, Eocene, P15a, x100. Figure 8. *Turborotalia ampliapertura* (Blow and Banner). Ideotype, figured by Blow (1979), Trinidad, N1, x70. Figures 9–11. *Globigerinoides primordius* (Blow and Banner). Figured by Kennett and Srinivasan (1983), Zone N4a, DSDP 289, x67. Figure 12. *Globoturborotalia gnaucki* (Blow and Banner). Figured by Blow and Banner (1962), Tanzania, Oligocene, x80. Figures 13-14. *Dentoglobigerina tripartita* (Koch). Cipero Formation, Trinidad, UCL Collection, Oligocene, P21, x200. Figures 15-16. *Dentoglobigerina* sp. Cipero Formation, Trinidad, UCL Collection, Oligocene, P21, x200. Figure 17. *Dentoglobigerina galavisi* (Bermudez). Figured by Blow (1969), Mississippi, P16, x100. Figure 18. *Globigerina brazieri* (Jenkins). Figured by Kennett and Srinivasan (1983), DSDP Site 208, Early Miocene, x60. Figure 19. *Globigerina gortanii* (Borsetti). Hypotype, figured by Blow (1969), Tanzania, Oligocene, P19, x61. Figures 20-21. *Globoturborotalita ouachitaensis* (Howe and Wallace). Hypotype, figured by Blow (1969), Trinidad, *T. ampliapertura* Zone, x226. Figure 22. *Globigerina eamesi* (Blow). Figured by Kennett and Srinivasan (1983), DSDP Site 281, *dissimilis* Zone, x112.

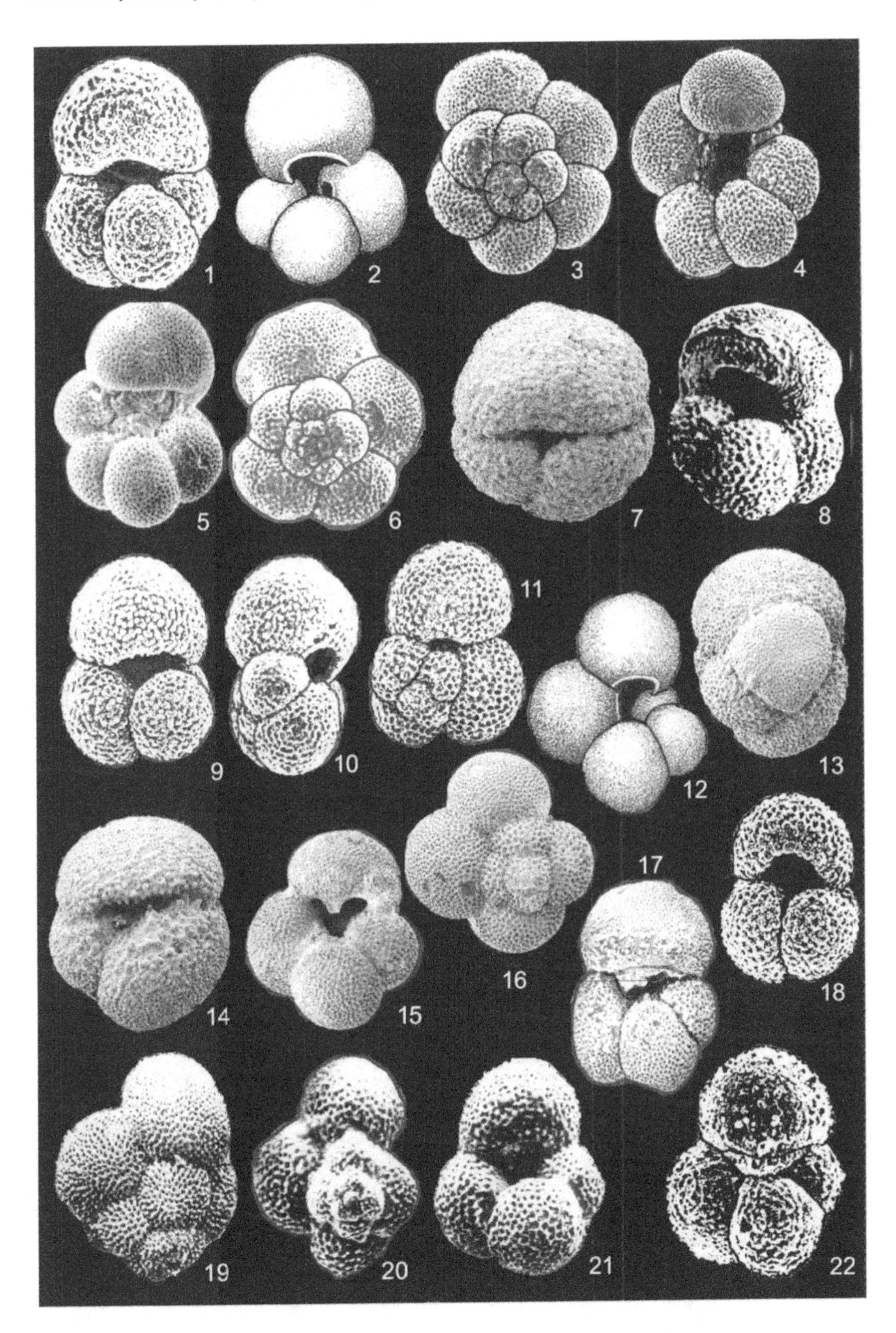
1
2
3
4
5
6
7
8
9
10
11
12
13
14
15
16
17
18
19
20
21
22

Plate 5.6. Figure 1. *Dentoglobigerina globularis* (Bermudez). Figured by Blow (1969), Cuba, level not given, x100. Figure 2. *Dentoglobigerina baroemoenensis* (Le Roy). Figured by Kennett and Srinivasan (1983), DSDP 289, N16, x100. Figure 3. *Dentoglobigerina venezuelana* (Hedberg). Figured by Kennett and Srinivasan (1983), DSDP 289, N10, x100. Figure 4. *Dentoglobigerina pseudovenezuelana* (Blow and Banner). Tanzania, Oligocene, P19, Banner's Collection UCL, x200. Figure 5. *Dentoglobigerina praedehiscens* (Blow and Banner). Tanzania, Oligocene, P22, Banner's Collection UCL, x200. Figures 6-7. *Globoquadrina binaiensis* (Koch). Figured by Kennett and Srinivasan (1983), DSDP 289, N5, x100. Figure 8. *Globorotaloides praestainforthi* (Blow). Figured by Blow (1969), Trinidad, N4a, x200. Figure 9. *Orbulinoides beckmanni* (Saito). Figured by Toumarkine (1975), DSDP Leg 32, Eocene, *O. beckmanni* Zone, x100. Figure 10. *Porticulasphaera howei* (Blow and Banner). Figured by Blow (1979), Tanzania, Eocene, P15, x200. Figure 11. *Porticulasphaera mexicana* (Cushman). Hypotype, figured by Blow (1979), Tanzania, Eocene, P13, x108. Figure 12. *Porticulasphaera rubriformis* (Subbotina). Figured by Toumarkine (1978), DSDP Site 363, Walvis Ridge, *M. lehneri to G. subconglobata* Zone, x108. Figures 13-14. *Porticulasphaera semiinvoluta* (Keijzer). Hypotypes figured by Blow (1979), Tanzania, Eocene, P15, x108. Figure 15. *Globigerinatheka curryi* (ProtoDecima and Bolli). Figured by Toumarkine and Luterbacher (1985), Beluchistan, *M. lehneri* Zone, x75. Figure 16. *Globigerinatheka barri* (Brönnimann). Trinidad, UCL collection, Eocene, P14, x75. Figure 17. *Globigerinatheka euganea* (ProtoDecima and Bolli). Figured by ProtoDecima and Bolli (1970), DSDP Site 313, Eocene, *O. beckmanni* Zone, x105. Figure 18. *Globigerapsis index* (Finlay). Figured by Blow (1979), Tanzania, Eocene, P11, x133. Figure 19. *Globigerapsis kugleri* (Bolli, Loeblich, and Tappan). Figured by Bolli (1972), Trinidad, Eocene, *O. beckmanni* Zone, x200.

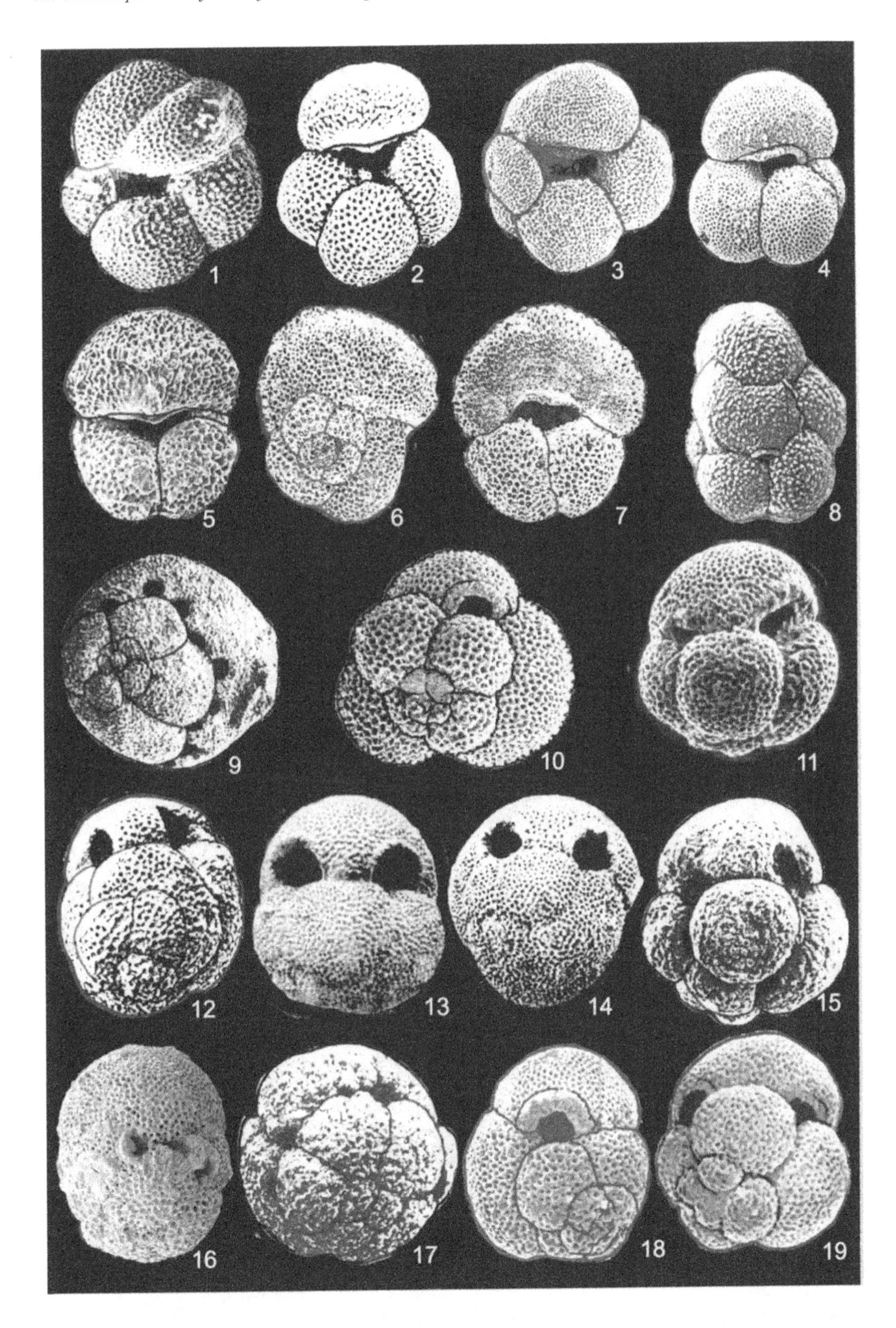
1
2
3
4
5
6
7
8
9
10
11
12
13
14
15
16
17
18
19

Plate 5.7. Figure 1. *Paragloborotalia nana* (Bolli). Trinidad, UCL Collection, Oligocene, P21, x570. Figures 2–5. *Paragloborotalia opima* (Bolli). Trinidad, UCL Collection, Oligocene, P21, x570. Figure 6. *Paragloborotalia obesa* (Bolli). Figured by Kennett and Srinivasan (1983), DSDP Site 281, *G. puncticulata* Zone, Early Pliocene, x100. Figure 7. *Paragloborotalia increbescens* (Bandy). Figured by Blow (1969), Tanzania, Eocene, P17, x200. Figure 8. *Paragloborotalia siakensis* (Le Roy). Figured by Blow (1969), Trinidad, *G. mayeri* Zone, x133. Figure 9. *Paragloborotalia semivera* (Hornibrook). Figured by Kennett and Srinivasan (1983), DSDP 206, *G. miozea* Zone, x133. Figure 10. *Paragloborotalia mendacis* (Blow). Figured by Blow (1969), JOIDES 3, Blake Plateau, Oligocene, P22, x200. Figures 11-12. *Paragloborotalia acrostoma* (Wezel). Figured by Kennett and Srinivasan (1983), DSDP Site 408, N8, x200. Figures 13-14. *Paragloborotalia mayeri* (Cushman and Ellisor). Figured by Kennett and Srinivasan (1983), DSDP 206-33, *G. miozea* Zone, x200. Figure 15. *Acarinina bulbrooki* (Bolli). Trinidad, UCL Collection, Eocene, P10, x100. Figure 16. *Acarinina praeaequa* (Blow). Trinidad, UCL Collection, Paleocene, P2, x100. Figure 17. *Acarinina strabocella* (Loeblich and Tappan). Figured by Blow (1979), Tanzania, Eocene, P7, x133. Figure 18. *Acarinina pentacamerata* (Subbotina). Figured by Blow (1979), DSDP 47, Eocene P8, x100. Figure 19. *Acarinina spinuloinflata* (Bandy). Figured by Blow (1979), Alabama, Eocene, P10, x100. Figure 20. *Acarinina nicoli* (Martin). Figured by Blow (1979), Tanzania, Eocene, P5, x200.

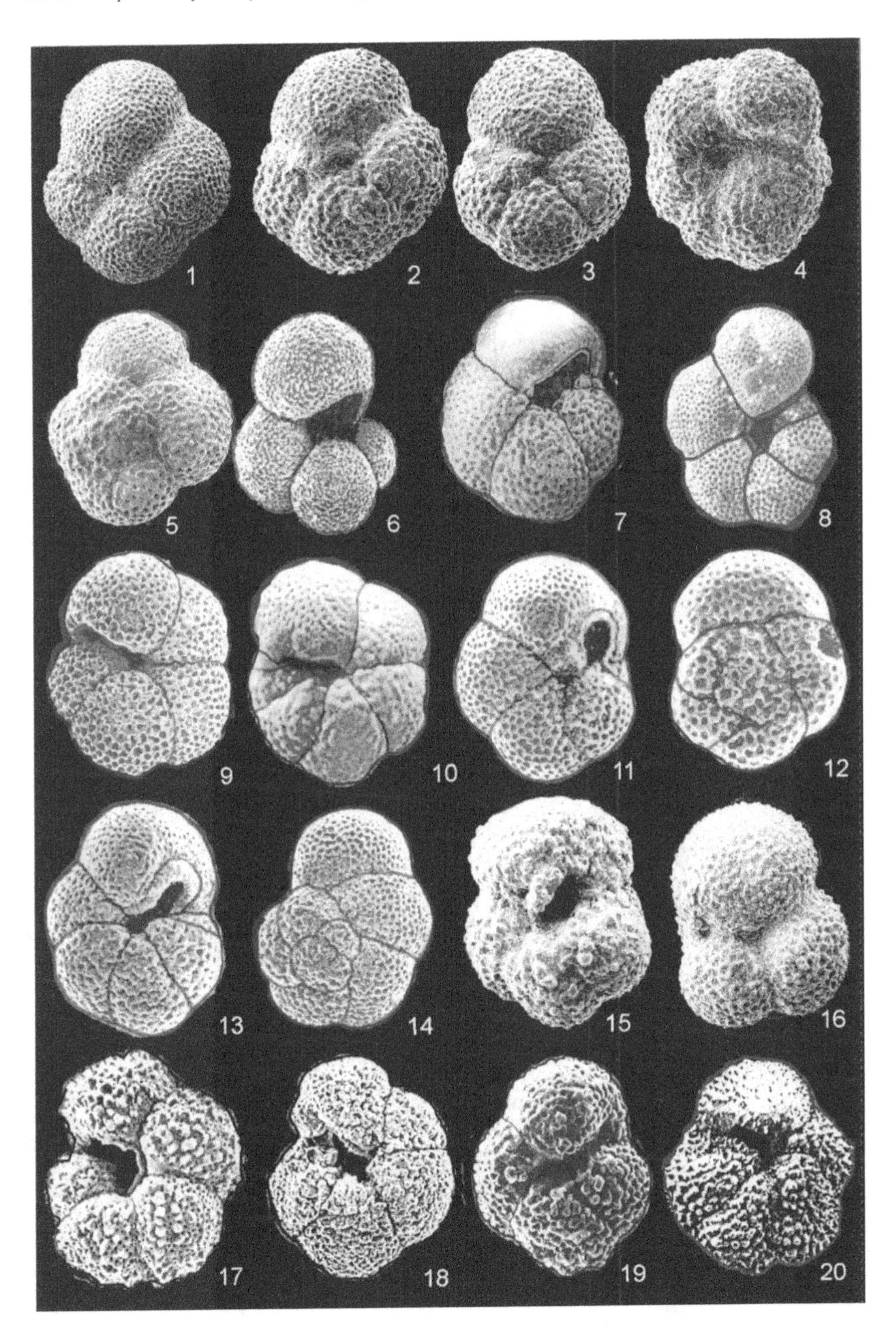
1
2
3
4
5
6
7
8
9
10
11
12
13
14
15
16
17
18
19
20

Plate 5.8. Figure 1. *Acarinina pentacamerata* (Subbotina). Figured by Blow (1979), DSDP47, Leg 6, Eocene, P8b, x133. Figure 2. *Acarinina cuneicamerata* (Blow). Figured by Blow (1979) as *A. camerata* Khalilov, DSDP 47, Eocene, P8b, x133. Figures 3-4. *Acarinina pseudotopilensis* (Subbotina). Figured by Blow (1979), Tanzania, Eocene, P7, x133. Figures 5-6. *Acarinina collactea* (Finlay). Figured by Blow (1979), Tanzania, Eocene, P11, x200. Figures 7-8. *Morozovella acuta* (Toulmin). (7) Trinidad, x67; (8) DSDP 86, x67, both from UCL Collection, Paleocene, P5a. Figure 9. *Morozovella aequa* (Cushman and Renz). Alabama, Coal Bluff Marl member of Naheola Formation, UCL Collection, Paleocene, P5a, x67. Figures 10-11. *Morozovella angulata* (White). Alabama, Coal Bluff Marl member of Naheola Formation, UCL Collection, Paleocene, P4, x65. Figure 12. *Morozovella marginodentata* (Subbotina). DSDP 86, UCL Collection, Eocene, P6, x80. Figure 13. *Morozovella occlusa* (Loeblich and Tappan). DSDP 94, UCL Collection, Paleocene, P4, x65. Figure 14. *Morozovella velascoensis* (Cushman). DSDP 86, UCL Collection, Paleocene, P4, x67. Figure 15. *Morozovella spinulosa* (Cushman). Trinidad, UCL Collection, Eocene, P13, x115. Figures 16-17. *Igorina broedermanni* (Cushman and Bermudez). Figured by Blow (1979), DSDP 21A, S. Atlantic, Eocene, P11, x150. Figures 18-19. *Igorina anapetes* (Blow). Figured by Blow (1979), Tanzania, Eocene, P11, x180. Figure 20. *Igorina convexa* (Subbotina). DSDP 86,UCL Collection, Paleocene, P4, x80.

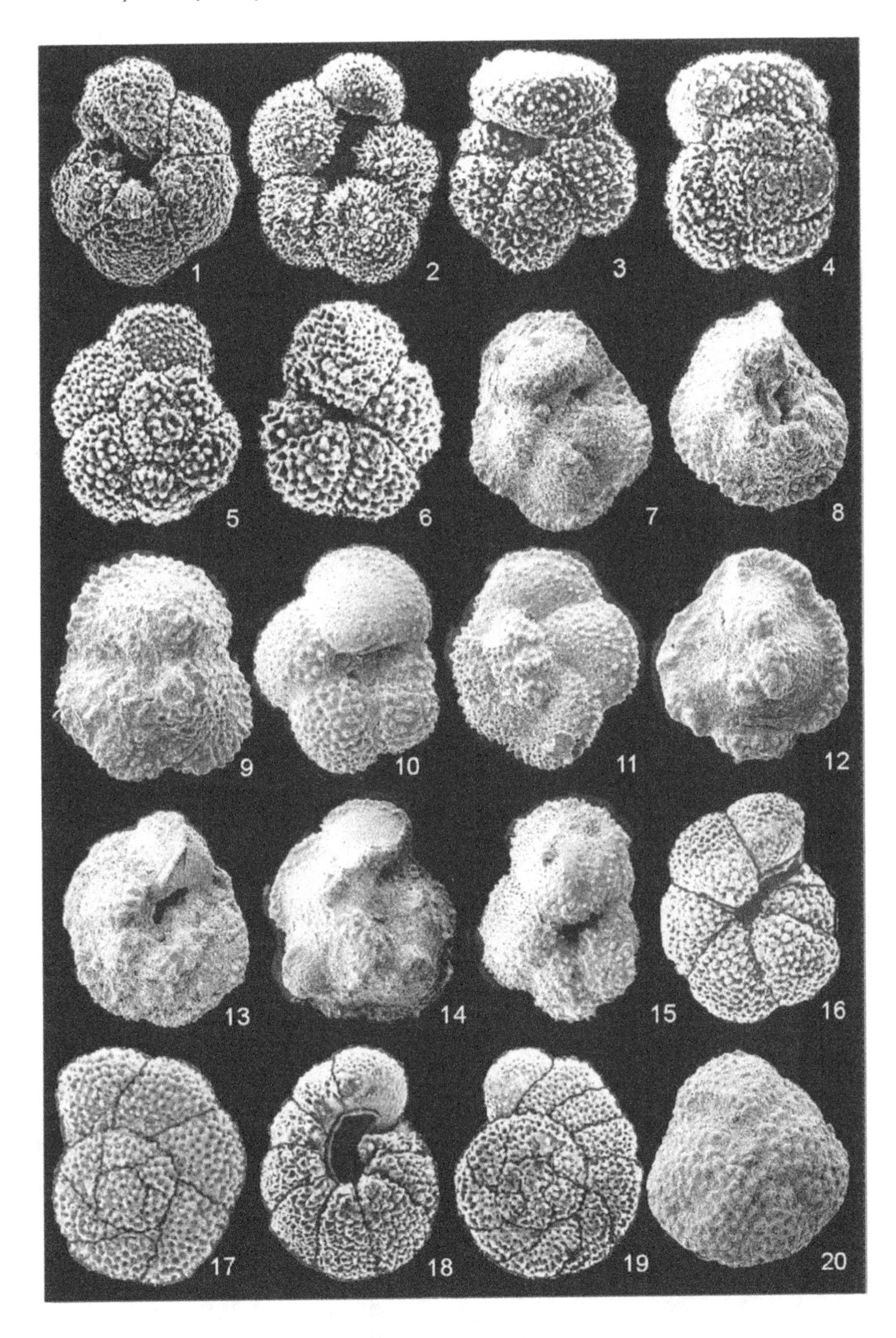
1
2
3
4
5
6
7
8
9
10
11
12
13
14
15
16
17
18
19
20

Plate 5.9. Figures 1–3. *Morozovella Subbotinae* (Morozova). DSDP 33-2 60–61, UCL Collection, Eocene, P6, x100. Figure 4. *Morozovella gracilis* (Bolli). Figured by Blow (1979), DSDP 47, Pacific, Eocene, P7, x100. Figure 5. *Morozovella quetra* (Bolli). Figured by Blow (1979), DSDP 47, Pacific, Eocene, P7, x100. Figure 6. *Praemurica praecursoria* (Morozova). Texas, Mexia Clay, Wills Point Formation, UCL Collection, Paleocene, P3, x125. Figures 7–9. *Morozovella trinidadensis* (Bolli). Texas, Mexia Clay, Wills Point Formation, UCL Collection, Paleocene, P2, x80. Figure 10. *Praemurica carinata* (El Naggar). Figured by Blow (1979), DSDP 47/2, Paleocene, P3, x200. Figures 11–13. *Praemurica uncinata* (Bolli). Reproduced from Banner's Collection UCL, Paleocene, P2, x200. Figures 14-15. *Praemurica pseudoinconstans* (Bolli). Figured by Blow (1979), DSDP 47, Paleocene, Pa, x200. Figure 16. *Testacarinata inconspicua* (Howe). Figured by Blow (1979), Louisiana, Eocene, P11-P12, x170. Figure 17. *Truncorotaloides rohri* (Brönnimann and Bermudez). Figured by Blow (1979), Tanzania, Eocene, P13, x133. Figure 18. *Truncorotaloides topilensis* (Cushman). Figured by Blow (1979), Tanzania, Eocene, P13, x100. Figures 19-20. *Truncorotaloides mayoensis* (Brönnimann and Bermudez). Figured by Blow (1979), Tanzania, Eocene, P13, x70. Figures 21- 22. *Truncaorotaloides piparoensis* (Brönnimann and Bermudez). Figured by Blow (1979), Tanzania, Eocene, P13, x80.

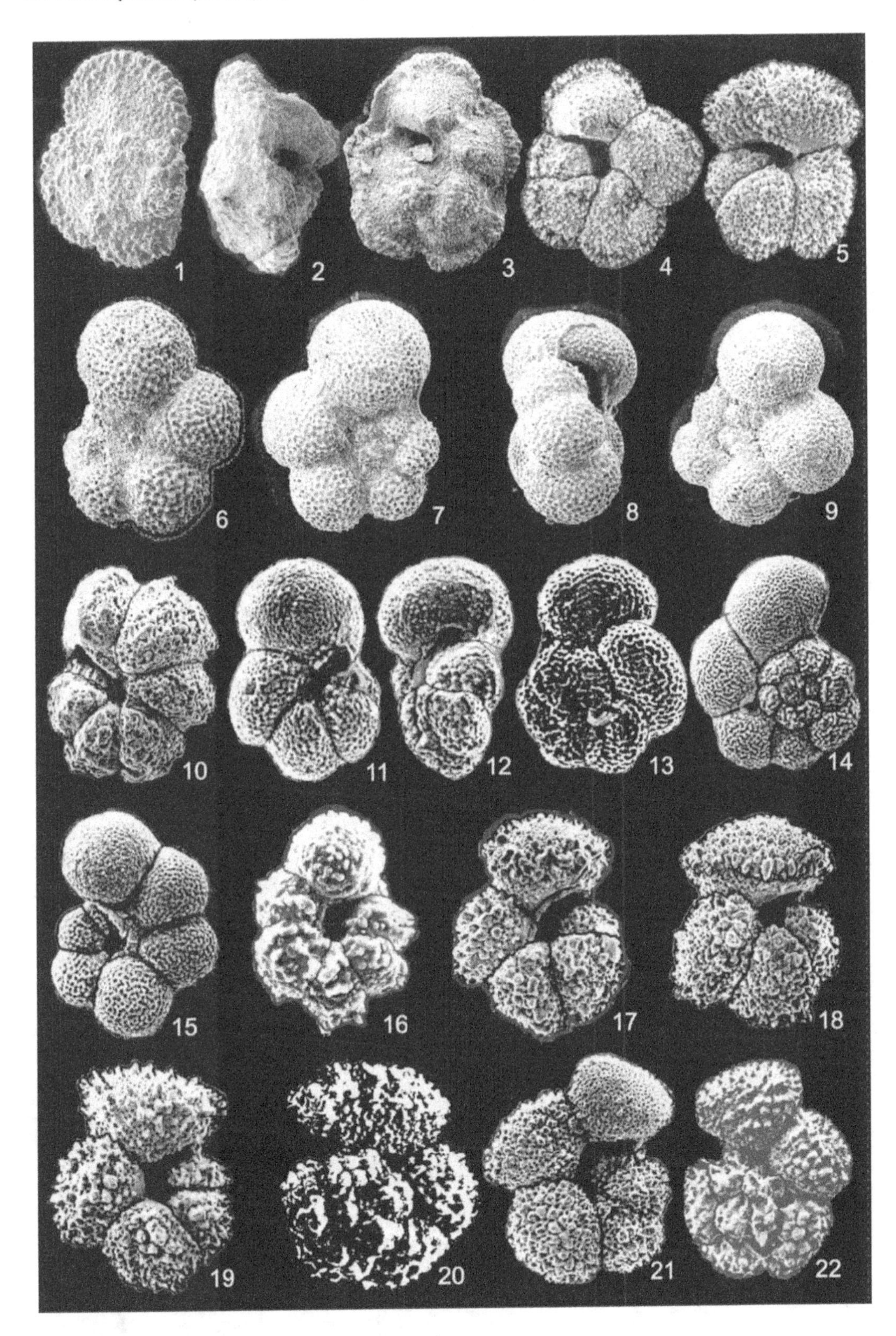
1
2
3
4
5
6
7
8
9
10
11
12
13
14
15
16
17
18
19
20
21
22

Plate 5.10. Figure 1. *Globoconusa daubjergensis* (Brönnimann). Texas, Mexia Clay, Wills Point Formation, UCL Collection, Paleocene, Pa, x355. Figure 2. *Globoconusa kozlowskii* (Brotzen and Pozaryska). Figured by Blow (1979), Pugu Well 2, East Africa, Early Danian, P1, x350. Figures 3-4. *Cassigerinella chipolensis* (Cushamn and Ponton). Trinidad, UCL Collection, Oligocene, P22, x133. Figure 5. *Cassigerinella winniana* (Howe). Figured by Blow (1979), DSDP 14, P18, x250. Figure 6. *Cassigerinella boudecensis* (Pokorny). Figured by Li (1986), Trinidad, Oligocene, P21, x250. Figure 7. *Cassigerinelloita amekiensis* (Stolk). Reproduced from Banner's Collection UCL., Nigeria, Eocene, P10, x400. Figures 8-9. *Chiloguembelitria danica* (Hofker). Reproduced from Hofker (1978), DSDP Hole 47.2, sample depth unknown, Shatsky Rise, NW Pacific Ocean Paleocene, Danian, P1, x320. Figure 10. *Chiloguembelitria oveyi* (Ansary). Reproduced from Banner's Collection UCL, Eocene, P6, x133. Figure 11. *Chiloguembelitria stavensis* (Bandy). Reproduced from Banner's Collection UCL, Eocene, P10, x500. Figures 12–14. *Chiloguembelitria* samwelli (Jenkins). Holotypes, figured by Jenkins (1978), off coast New Zealand, Oligocene, *G. euapertura* Zone, x400. Figure 15. *Chiloguembelina subcylindrica* (Beckmann). Holotype, figured by Beckmann (1957), Trinidad, Eocene, *M. rex* and *M. formosa* Zones, x200. Figure 16. *Chiloguembelina wilcoxensis* (Cushman and Ponton). Figured by Beckmann (1957), Trinidad, Eocene, *M. rex* and *M. formosa* Zones, x133. Figures 17-18. *Chiloguembelina parallela* (Beckmann). Holotype, figured by Beckmann (1957), Trinidad, Eocene, *M. rex* and *M. formosa* Zones, x180. Figure 19. *Streptochilus martini* (Pijpers). Reproduced from Banner's Collection UCL, Eocene, P14, x200. Figure 20. *Woodringina claytonensis* (Loeblich and Tappan). Figured by Loeblich and Tappan (1957), Alabama, Early Paleocene, Danian, x400. Figure 21. *Bifarina laevigata* (Loeblich and Tappan). Holotype, figured by Loeblich and Tappan (1957), Early Paleocene, Upper Danian, x105. Figure 22. *Zeauvigerina aegyptiaca* (Said and Kenawy). Reproduced from Banner's Collection UCL, Egypt, Paleocene, P5a, x95.

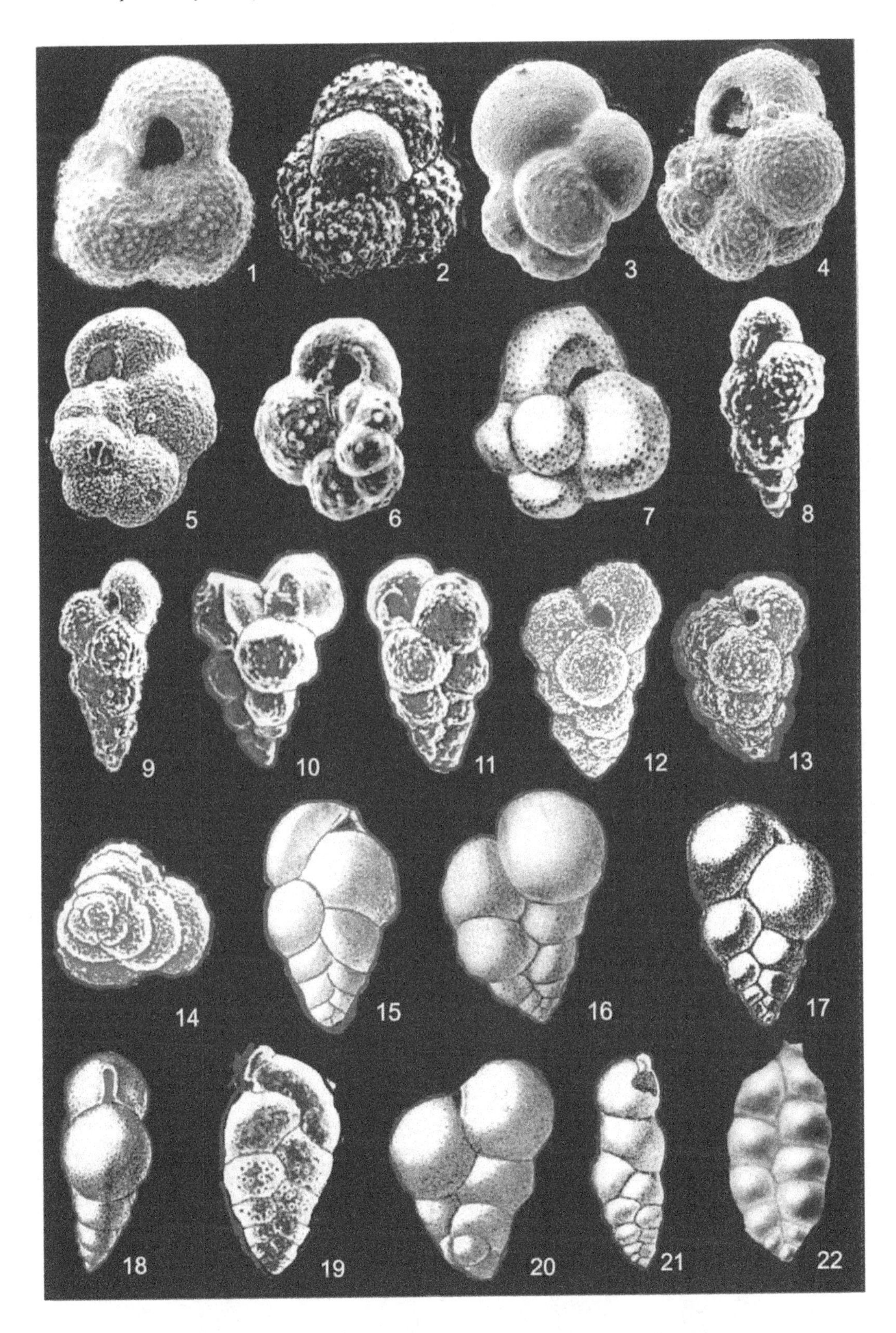
1
2
3
4
5
6
7
8
9
10
11
12
13
14
15
16
17
18
19
20
21
22

Plate 5.11. Figure 1. *Turborotalia centralis* (Cushman and Bermudez). Figured by Postuma (1971), Trinidad, Eocene, P13, x110. Figure 2. *Turborotalia cerroazulensis* (Cole). Figured by Postuma (1971), Alabama, Eocene, P16, x88. Figure 3. *Pseudomenardella ehrenbergi* (Bolli). Figured by Postuma (1971). Tunisia, Paleocene, P4, x126. Figure 4. *Turbeogloborotalia compressa* (Plummer). Figured by Postuma (1971), Texas, Paleocene, P2, x165. Figure 5. *Pseudohastigerina micra* (Cole). Figured by Postuma (1971), Mexico, Eocene, P11, x55. Figure 6. *Parasubbotina pseudobulloides* (Plummer). Figured by Postuma (1971), Navarro County, Texas, USA, Paleocene, x113. Fig. 7. *Globoturborotalita ouachitaensis* (Howe and Wallace). Figured by Postuma (1971), Ouchita River, Louisiana, USA, Eocene, P17, x150. Figure 8. *Globigerapsis index* (Finlay). Figured by Postuma (1971), Kakaho Creek, New Zealand, Eocene, P14, x105. Figure 9. *Globigerapsis kugleri* (Bolli, Loeblich, and Tappan). Figured by Postuma (1971), Navet Formation, Trinidad, Eocene, P12, x70. Figure 10. *Porticulasphaera mexicana* (Cushman). Figured by Postuma (1971), DB229, Trinidad, Eocene, P13, x100. Figure 11. *Orbulinoides beckmanni* (Saito). Figured by Postuma (1971), Trinidad, Eocene, P13, x56. Figure 12. *Globigerinatheka barri* (Brönnimann). Figured by Postuma (1971), DB276, Trinidad, Eocene, P14, x105. Figure 13. *Globigerinatheka senni* (Beckmann). Figured by Postuma (1971), DB129, Trinidad, Eocene, P15a, x105. Figure 14. *Globoconusa daubjergensis* (Brönnimann). Figured by Postuma (1971), Navarro County, Texas, Paleocene, Pa, x400. Figure 15. *Igorina broedermanni* (Cushman and Bermudez). Figured by Postuma (1971), Tschop, Cuba, Eocene, P6b, x150. Figure 16. *Truncorotaloides rohri* (Brönnimann and Bermudez). Figured by Postuma (1971), Trinidad, Eocene, P13, x88. Figure 17. *Subbotina triloculinoides* (Plummer). Figured by Postuma (1971), Texas, Paleocene, P4, x108. Figure 18. *Acarinina soldadoensis* (Brönnimann). Figured by Postuma (1971), Trinidad, Eocene, P7, x100. Figure 19. *Praemurica uncinata* (Bolli). Figured by Postuma (1971), Trinidad, Paleocene, P2, x133. Figure 20. *Morozovella trinidadensis* (Bolli). Figured by Postuma (1971), Tunisia, Paleocene, P2, x120. Figure 21. *Igorina pusilla* (Bolli). Figured by Postuma (1971), Tunisia, Paleocene, P3, x172. Figure 22. *Acarinina bulbrooki* (Bolli). Figured by Postuma (1971), Cuba, Eocene, P10, x124. Figure 23. *Morozovella formosa* (Bolli). Figured by Postuma (1971), Cuba, Eocene, P10, x63. Figure 24. *Morozovella aequa* (Cushman and Renz). Figured by Postuma (1971), Trinidad, Paleocene, P5, x110. Figure 25. *Morozovella aragonensis* (Nuttall). Figured by Postuma (1971), La Antigua, Mexico, Eocene, P10, x72. Figure 26. *Morozovella velascoensis* (Cushman). Figured by Postuma (1971), Mexico, Paleocene, P4, x70. Figure 27. *Truncorotaloides topilensis* (Cushman). Figured by Postuma (1971), Trinidad, Eocene, P14, x95. Figure 28. *Hantkenina alabamensis* (Cushman). Figured by Postuma (1971), Trinidad, Eocene, P14, x35. Figure 29. *Cribrohantkenina bermudezi* (Thalmann) (= *Cribrohantkenina inflata* (Howe)). Figured by Postuma (1971), Palmer Station, Camaguay Province, Cuba, Eocene, P17, x90.

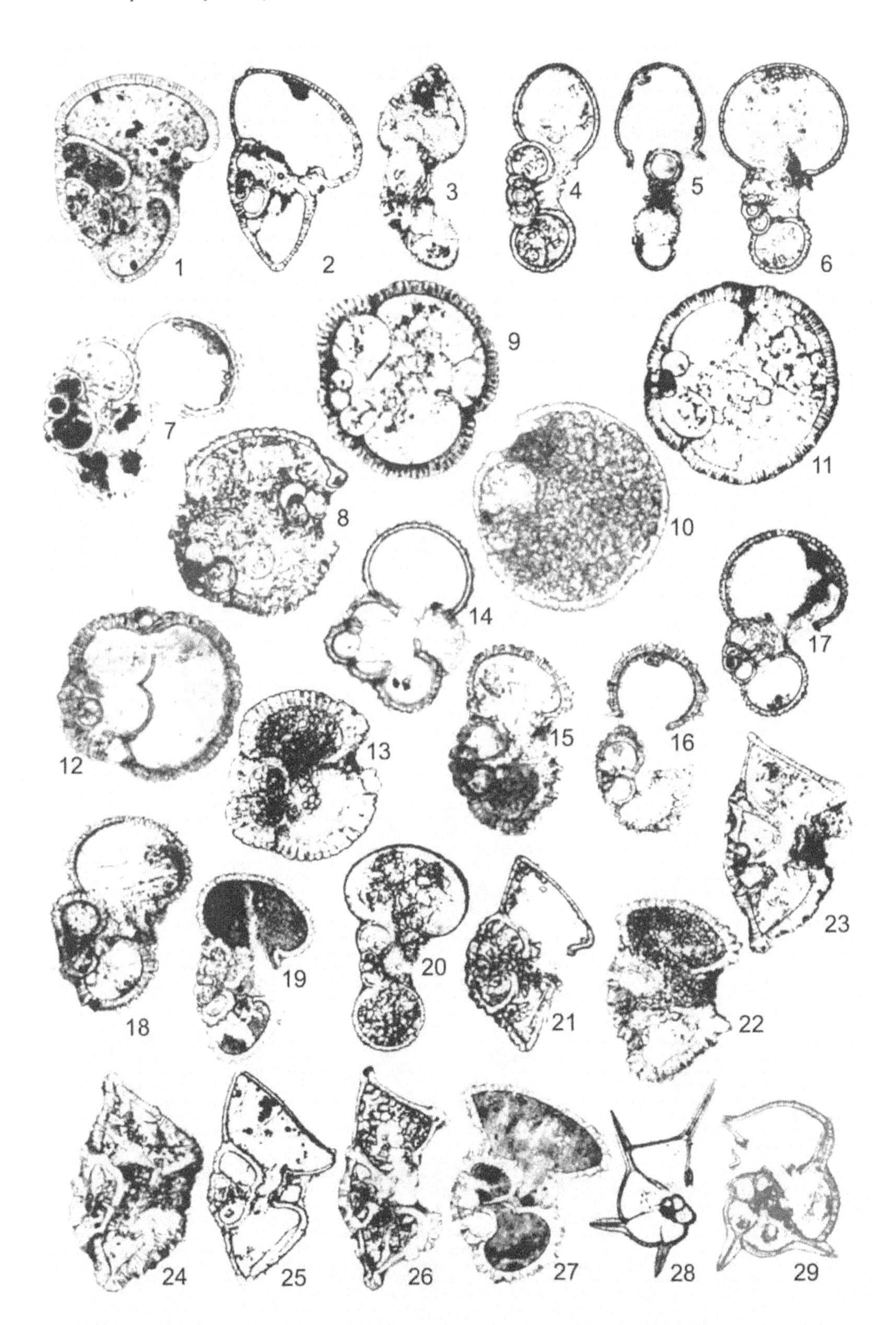
1
2
3
4
5
6
7
8
9
10
11
12
13
14
15
16
17
18
19
20
21
22
23
24
25
26
27
28
29

Plate 5.12. (All figures from the UCL Collection). Figure 1. *Subbotina eocaenica* (Terquem). Syria, Eocene, Lutetian, P10, x50. Figure 2. *Dentoglobigerina venezuelana* (Hedberg). Syria, Eocene, Lutetian, P10, x86. Figure 3. (A) *Globigerapsis index* (Finlay), (B) *Turborotalia possagnoensis* (Tourmakine and Bolli). Syria, Eocene, Lutetian, P11, x65. Figure 4. *Turborotalia frontosa* (Subbotina). Syria, Eocene, Lutetian, P11, x75. Figure 5. *Morozovella spinulosa* (Cushman). Syria, Eocene, P11, x60. Figure 6. *Morozovelloides lehneri* (Cushman and Jarvis). Syria, Eocene, P11, x60. Figure 7. *Morozovella lensiformis* (Subbotina). Syria, Eocene, P11, x60. Figure 8. (A) *Guembelitrioides higginsi* (Bolli), (B) *Discocyclina* sp., (C) *Globigerinatheka subconglobata* (Chalilov). Pakistan, Lower Kirthar Formation, Eocene, P11-P12, x53. Figure 9. (A) *Discocyclina* sp., (B) *Morozovella caucasica* (Glaessner). Pakistan, Lower Kirthar Formation, Eocene, P11-P12, x50. Figure 10. *Acarinina decepta* (Martin). Syria, Eocene, P11, x56. Figure 11. *Catapsydrax dissimilis* (Cushman and Bermudez). Syria, Oligocene, P18, x67. Figure 12. *Carinoturborotalia cocoaensis* (Cushman). On the Tyre-Abassieh Road, Tyre, Lebanon, Late Eocene, P16, x100. Figure 13. Micritic packstone of Rodophyte algae, Eocene planktonic foraminifera, *Carinoturborotalia cocoaensis* (Cushman), *Planorotalites pseudoscitula* (Glaessner), *Subbotina* spp., and larger benthic foraminifera, *Discocylina* sp. Pakistan, Lower Kirthar Formation, Eocene, P11- P12, x25.

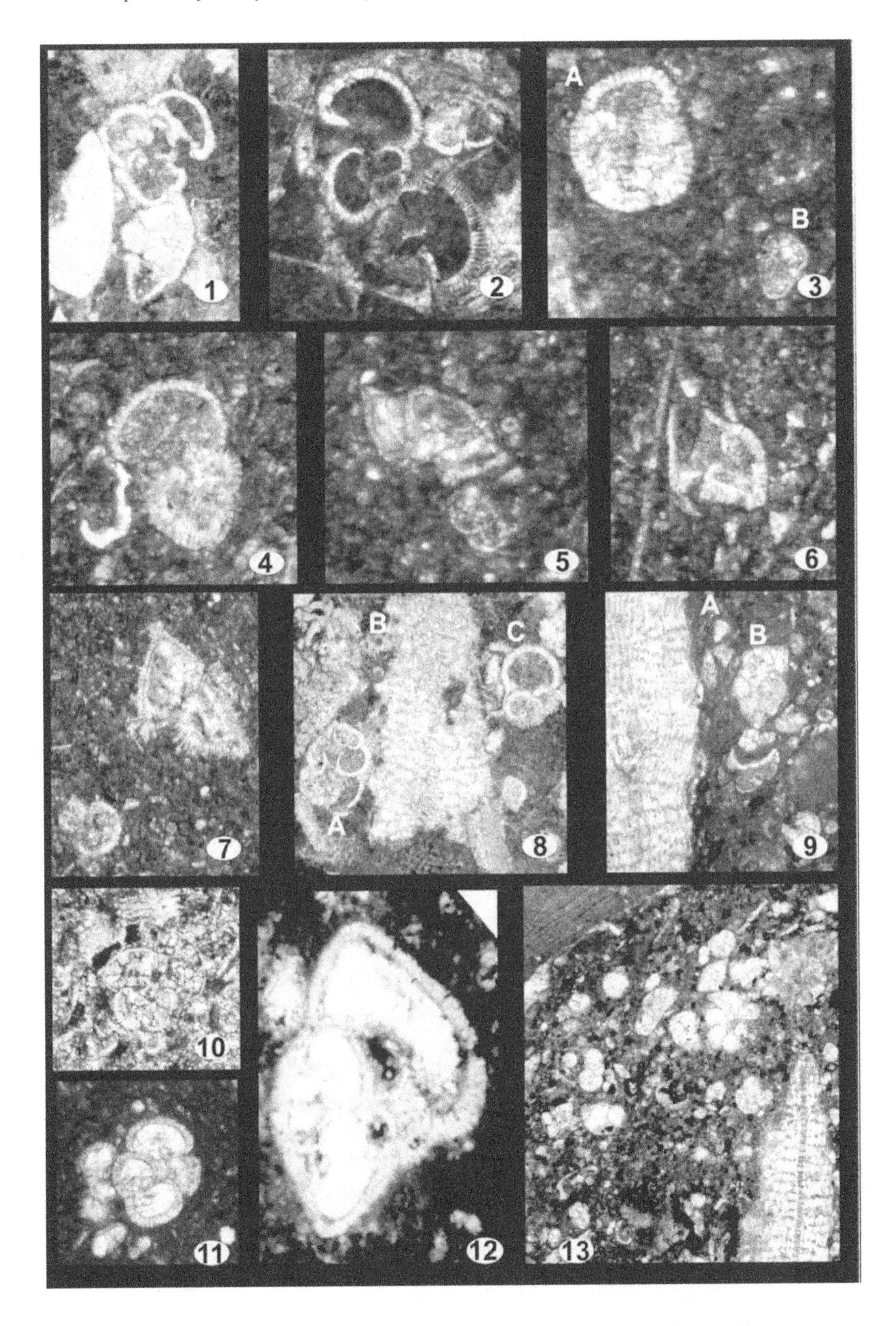
A
B
1
2
3
4
5
6
B
C
A
7
8
A
B
9
10
11
12
13

Chapter 6

The Cenozoic planktonic foraminifera: The Neogene

6.1 Introduction

As seen in the previous chapter, the Paleogene–Neogene boundary was not such a significant event for planktonic foraminifera as the sharply defined Cretaceous–Paleogene boundary. Indeed, eight out nine families (see Fig. 6.1) and almost 73% of species survived the boundary into the Miocene. These Paleogene forms dominated the first 3 Ma of the Miocene in the Aquitanian, and it was not before the Burdigalian that the first stage of Neogene diversification occurred.

In the Aquitanian, the small globular Globigerinidae that passed through the Paleogene–Neogene boundary thrived from the photic zones to the deeper waters of the tropics and subtropics. They continued to dominate the planktonic foraminiferal assemblages throughout the Neogene but were most abundant in the Burdigalian and Tortonian (see Fig. 6.2). The highly eurythermal Globigerinitidae and Tenuitellidae were to be found in surface and subsurface, uppermost waters from the tropics to temperate seas throughout the Neogene, but with a relatively low diversity. Other rare Paleogene forms that passed into the Neogene include the biserially–planispirally enrolled Cassigerinellidae, which died out in the Middle Miocene; the biserial Chiloguembelinidae, which lives in the thermocline of the tropics and subtropics; and the Turborotalitidae, which has a trans-global or tropical–subtropical distribution (Fig. 6.3).

An important event in the evolution of the Neogene planktonic foraminifera began in the Burdigalian, with the development of the keeled Globorotaliidae. At this time, the unkeeled Oligocene *Paragloborotalia* developed a keeled periphery, which allowed the globorotaliids to move deeper into the water column, and rapidly to spread in the tropics and subtropics, giving rise to several important Neogene lineages. The globorotaliids increased in diversity throughout the Neogene to reach a maximum in the Pliocene before dwindling slightly toward the Holocene (see Fig. 6.4).

Additionally, the Burdigalian saw the emergence of new families in the tropics and subtropics (see Fig. 6.1). Some of these new forms were short ranged, such as the Globigerinatellidae, while others are still extant, such as the Globigerinellidae and the Sphaeroidinellidae (see Fig. 6.5). The Globigerinatellidae were rare and did not proliferate beyond the Middle Miocene, while the last two families evolved gradually throughout the Neogene to form a succession of morphologically distinctive populations within important phylogenetic lineages. The Globigerinellidae were most abundant in the Late Pleistocene and the Holocene, while the Sphaeroidinellidae increased in diversity gradually until the Pliocene, when they were at their most abundant, but dwindled in number during the Pleistocene and Holocene.

The Late Burdigalian also saw the development of the Hastigerinidae, with massive calcitic triradiate spines, which evolved from the Globigerinellidae. They were rare throughout the Neogene, (Fig. 6.5) but their diversity increased slightly in the Pleistocene. There are only four extant genera that are deep-dwelling in the tropics and subtropics.

In the Middle Miocene (Langhian, N8a), the intensely streptospiral Orbulinidae made their first appearance in the tropics and subtropics but very rapidly spread to temperate waters to become cosmopolitan. Another peak of species diversification occurred in the Late Miocene and coincided with the appearance of the Candeinidae in the Tortonian, but today only one species survives in intermediate depth waters in the tropics (Fig. 6.5). In the Messinian, these forms were joined by the nonspinose Pulleniatinidae, which inhabited the same depths in the tropics and subtropics. The Pulleniatinidae were at their most diverse in the Late Pliocene and Pleistocene. The only new family appearing post the Miocene is the Neoacarininidae, a short range family with only one known species, which appeared to die out within the Late Pleistocene (Fig. 6.5).

Throughout the Cenozoic, the planktonic foraminifera were most abundant and diverse in the tropics and subtropics. However, following the Mid-Miocene Climatic Optimum (MMCO) (14–16.5 Ma), global cooling was established and climatic gradients became more pronounced, and many species adapted their development to populate temperate and sub-polar oceans.

Paleogene planktonic foraminifera from different localities around the world have been studied in detail by many workers within the past 20 years. Many new data have come from DSDP and ODP studies that have enabled the development of a more precise Neogene chronology, calibrated against magneto-and astro-chronological time scales (e.g., Aze *et al.*, 2011; Berggren, 1995; Wade *et al.*, 2011). However, despite, or because of, this increase in data, several taxonomic and phylogenetic issues persist. In this chapter, therefore, in order to resolve some of the outstanding problems with the existing classification of the Neogene planktonic foraminifera, a revised taxonomy is presented in which four new families are introduced. The chapter concludes with a discussion of the biostratigraphic, phylogenetic, paleoenvironmental, and paleogeographic significance of all the Neogene forms up to and including those of the present day.

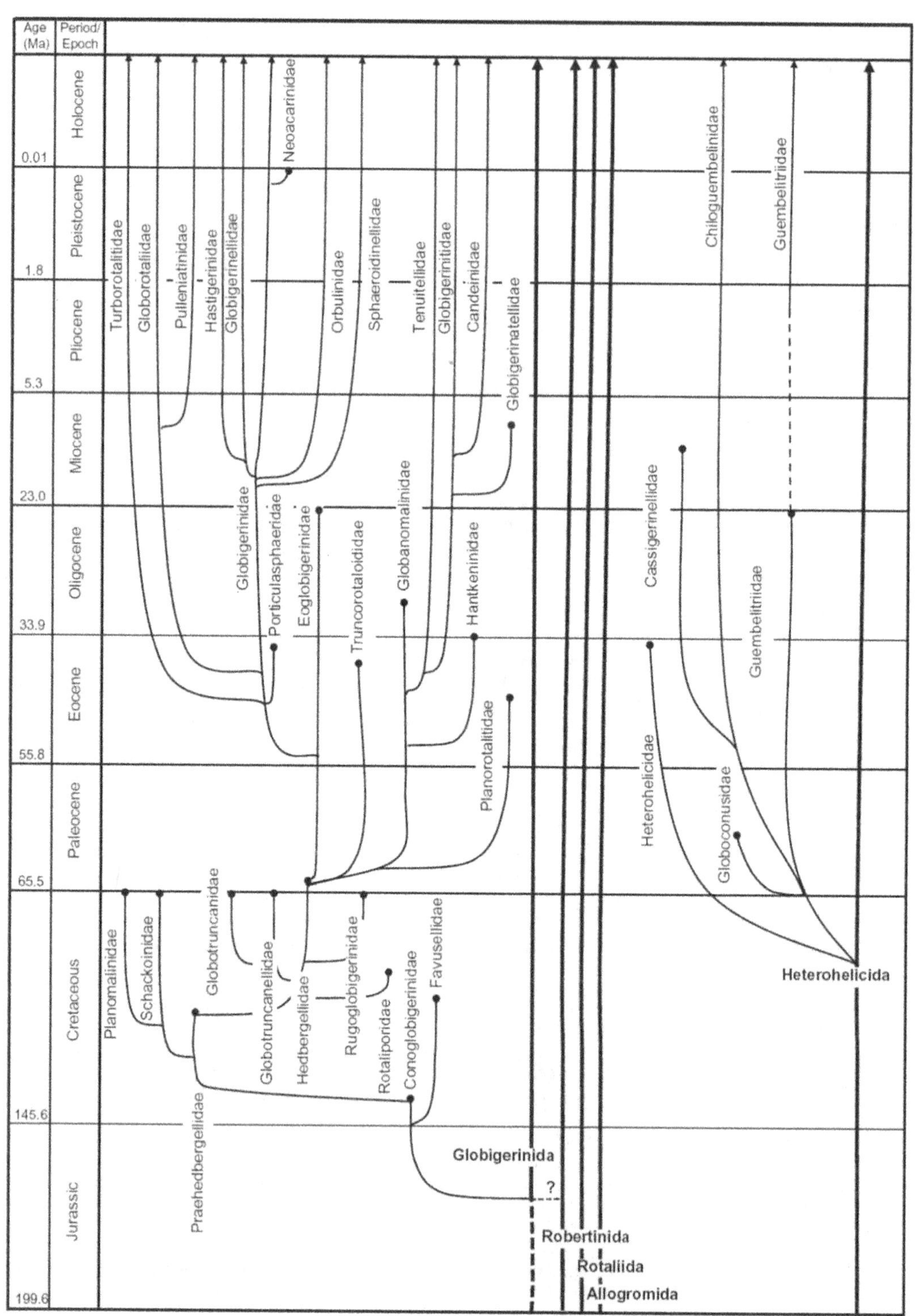

Figure 6.1. The evolution of the Neogene planktonic foraminiferal families (thin lines) from their Paleogene ancestors.

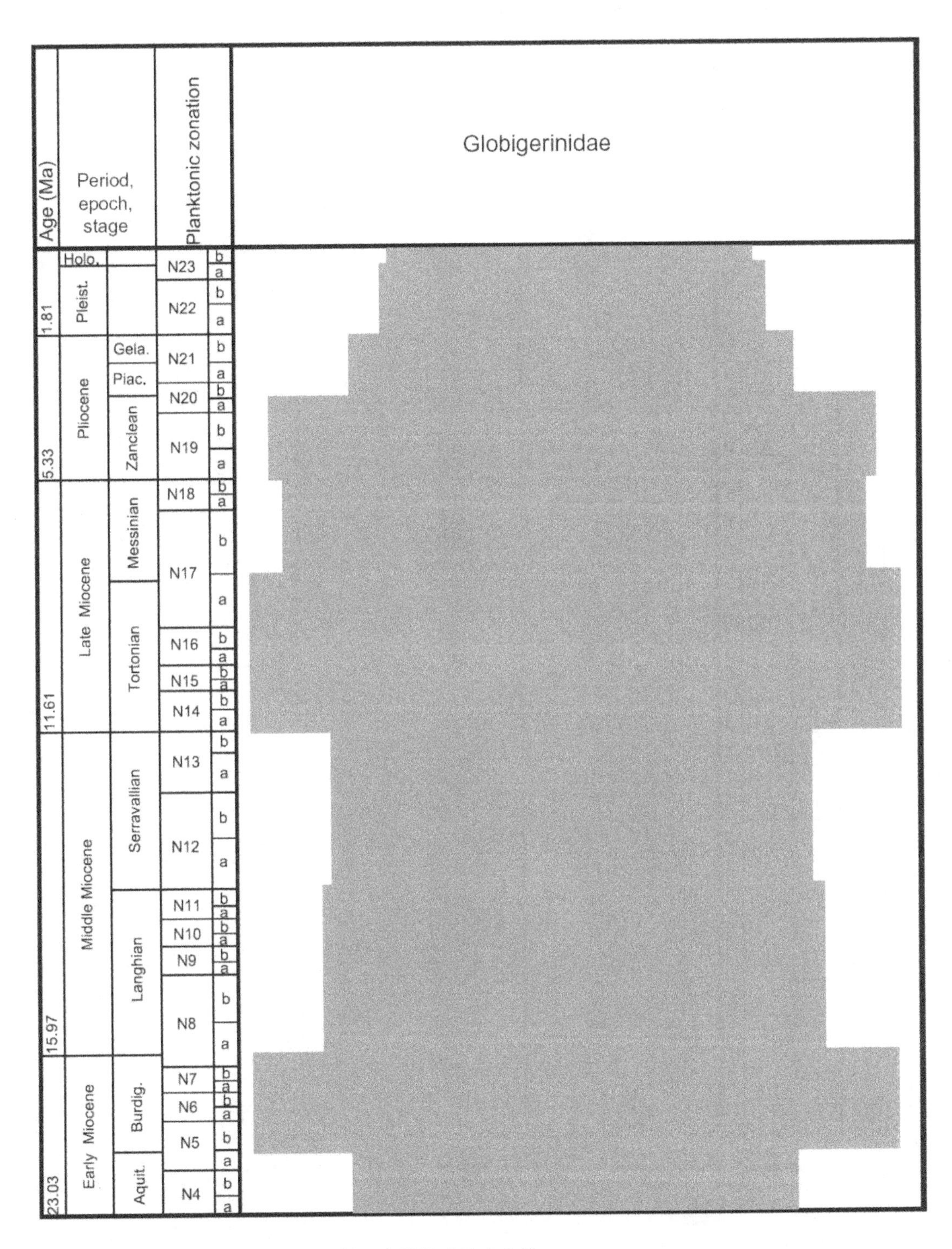

Figure 6.2. The biostratigraphic range and diversity of the main Globigerinidae in the Neogene.

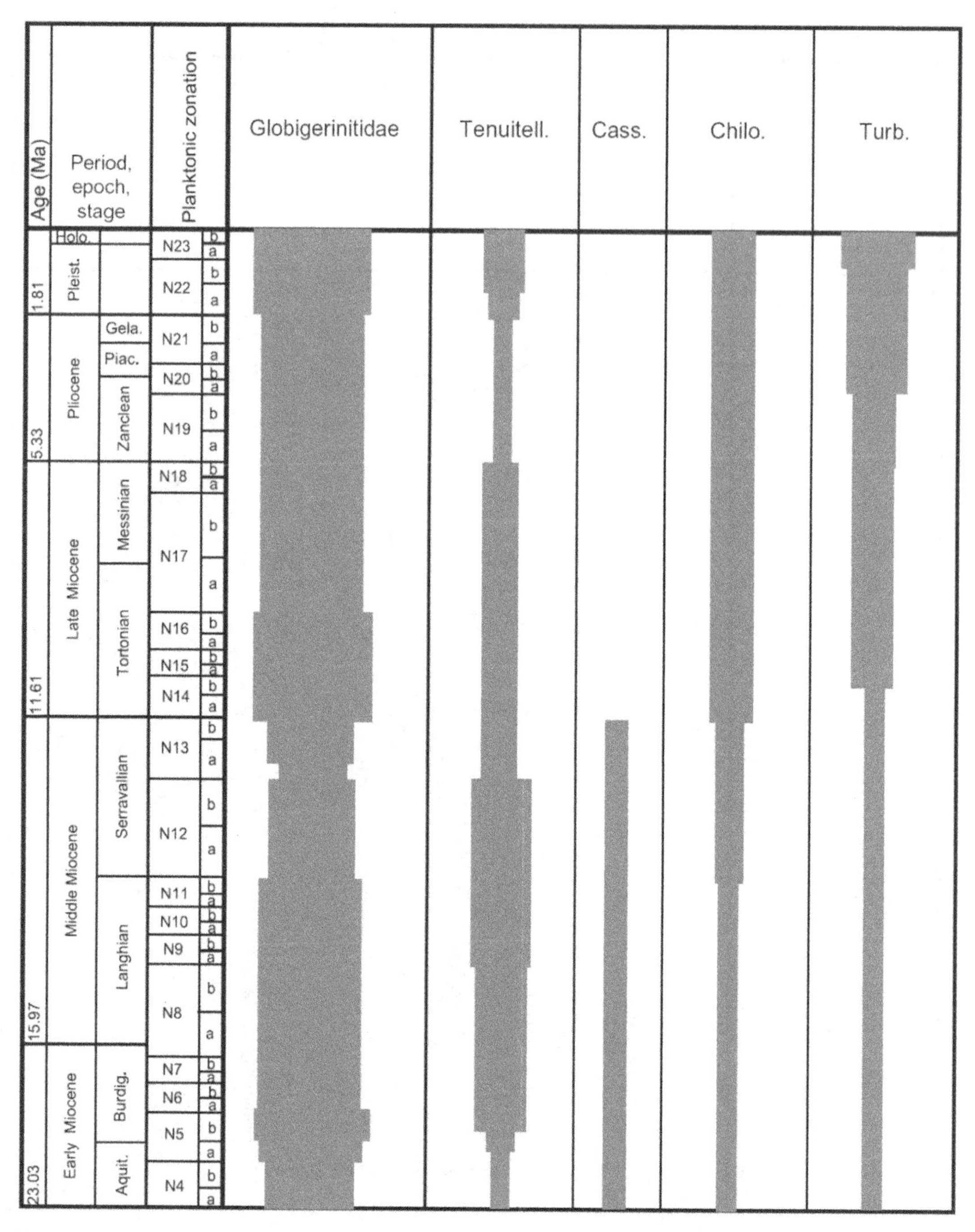

Figure 6.3. The biostratigraphic range and diversity of the main Globigerinitidae, Tenuitellidae, Cassigerinellidae, Chiloguembelinidae, and Turborotalitidae in the Neogene.

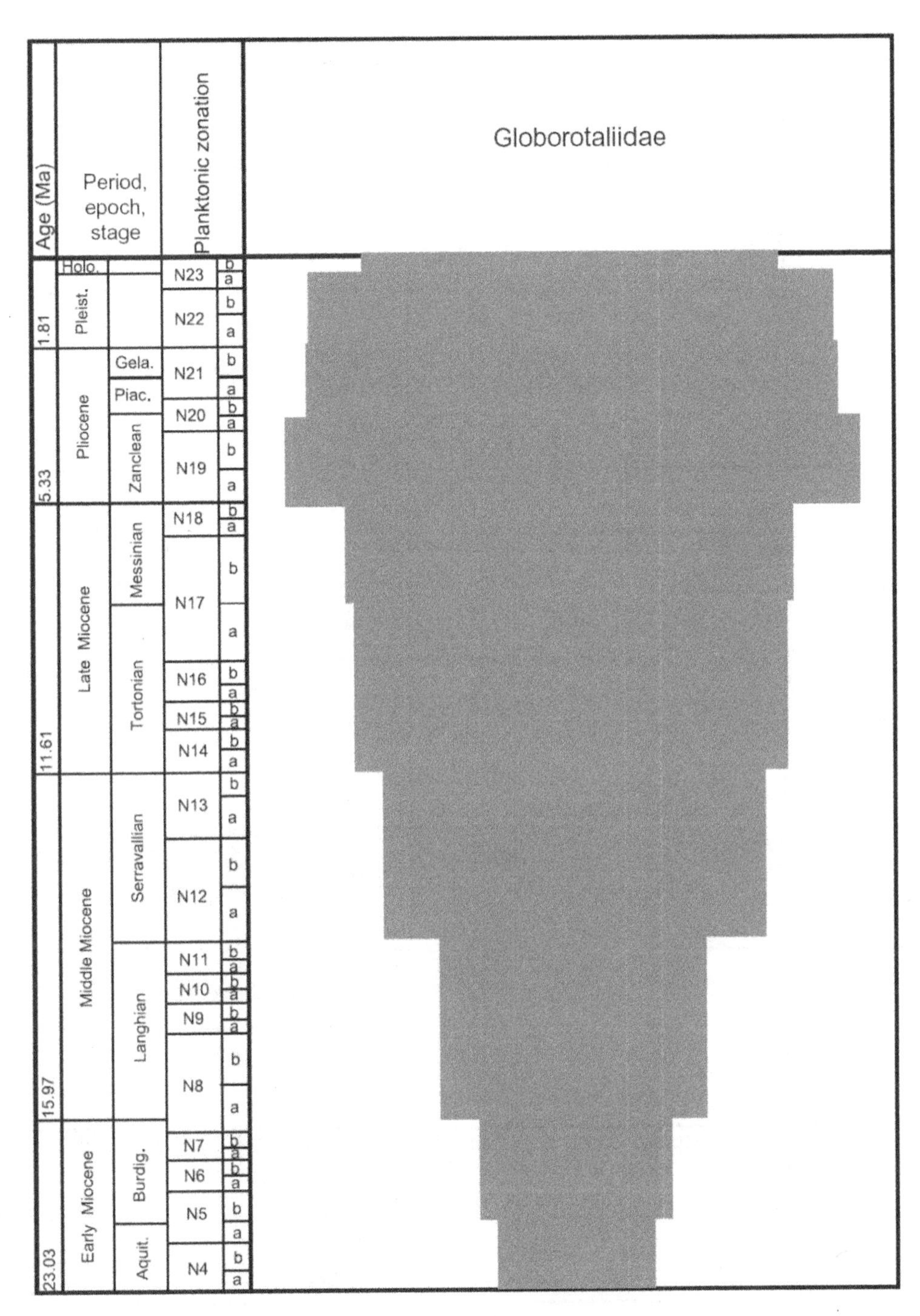

Figure 6.4. The biostratigraphic range and diversity of the Globorotaliidae in the Neogene.

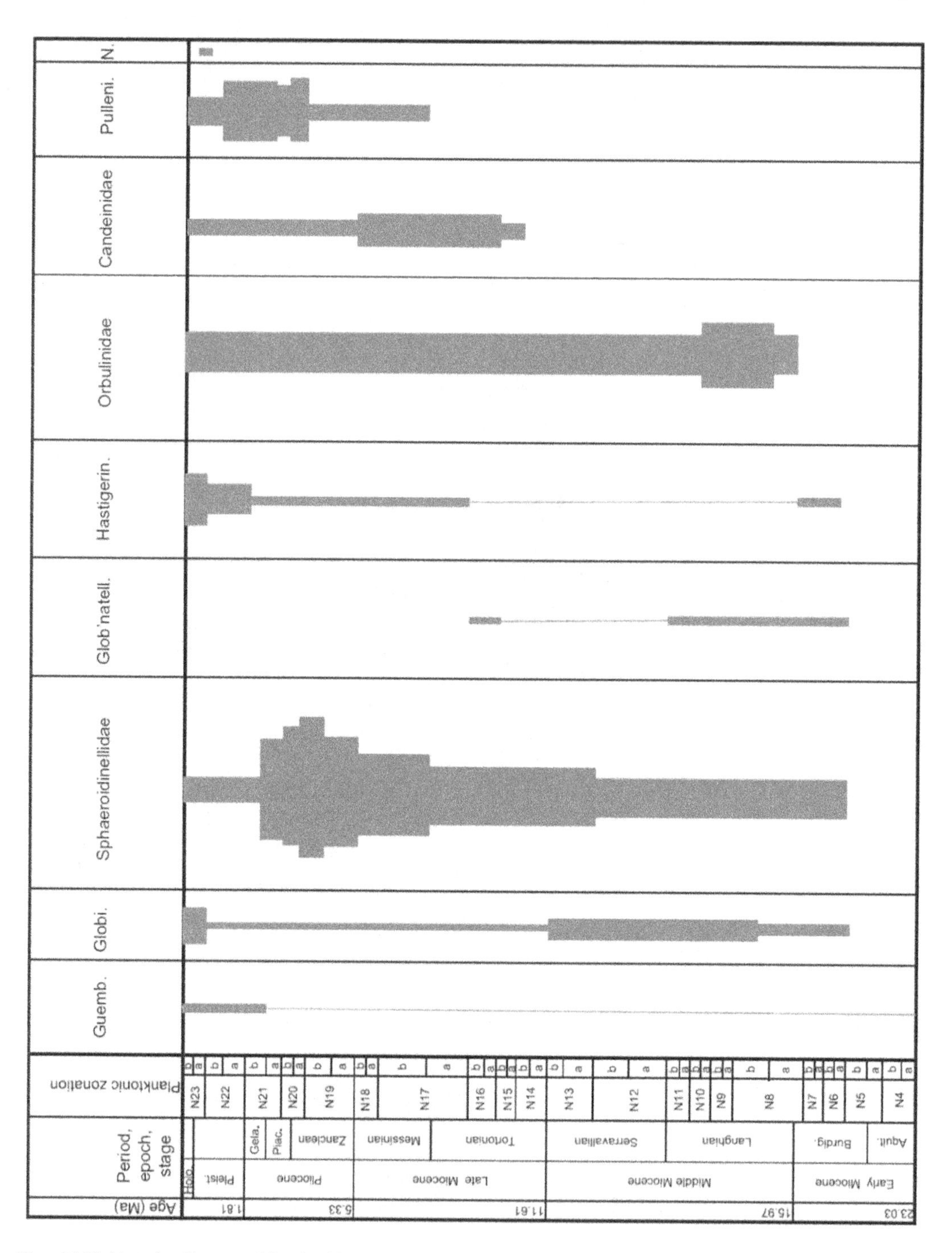

Figure 6.5. The biostratigraphic range and diversity of the main Guembelitriidae, Globigerinellidae, Sphaeroidinellidae, Globigerinatellidae, Hastigerinidae, Orbulinidae, Candeinidae, Pulleniatinidae, and Neoacarinidae in the Neogene.

6.2 Morphology and taxonomy of the Neogene planktonic foraminifera

The classification of the Neogene planktonic foraminifera was traditionally based on the morphology and characteristics of their test. Many Neogene forms are still extant, and the tests of most of these species have been illustrated by many authors (e.g., Saito *et al.*, 1981). Importantly, for the study of paleoenvironment, the biological characteristics of extant forms have been summarized by Hemleben *et al.* (1989). More recently, it has become possible to construct the molecular phylogenetic relationships among existing planktonic forms, by analysing, for example, the sequence divergence within their small subunit ribosomal RNA gene (SSU rDNA) (Aurahs *et al.*, 2009; Darling *et al.*, 2009; see Chapter 2). This has provided an unprecedented level of understanding of the evolutionary relationships of the planktonic foraminifera, and such recently determined information is used in this chapter to reassess the classification and phylogeny of the Neogene forms and to amend previous biostratigraphical zonal schemes.

In our analysis, we suggest that the Neogene planktonic foraminifera should be divided into three superfamilies:

- the Globigerinitoidea
- the Globigerinoidea
- the Heterohelicoidea

We divide these 3 superfamilies into 16 families (4 of which are newly defined) and recognize 44 genera, 26 of them still being extant (in the Holocene stage, N23b). There are about 100 species living in modern seas. As in Chapter 5, the plates in this chapter contain new images as well as those taken from original sources. In some cases, images from the late Prof. Banner's Collection, now in UCL, have had morphological features highlighted for pedagogical purposes.

CLASS FORAMINIFERA Lee, 1990

ORDER GLOBIGERINIDA LANKASTER, 1885

The test is planispiral or trochospiral, at least in the early stage, microperforate or macroperforate, smooth, muricate, or with spines. Apertures are terminal, umbilical, intra-extraumbilical, or peripheral. Walls are calcitic, but early forms may be aragonitic. Jurassic (Late Bajocian) to Holocene.

Superfamily GLOBIGERINITOIDEA BouDagher-Fadel, 2012

The test is calcitic, microperforate, nonspinose, without continuous encrustations and without perforation cones, but with blunt pustules, and is low trochospiral to planispiral. Spiral sutural apertures are absent or very few and sporadic. The umbilicus is open, except when it is covered by bulla. Sutural pores may be present. The primary aperture is single, intraumbilical, or intra-extraumbilical. Paleocene to Holocene.

Family Candeinidae Cushman, 1927

The test is trochospiral with an early growth stage similar to the Globigerinitidae, but adults have a closed umbilicus and the single primary aperture is replaced by many small, uniform apertures in each suture. Apertures are interiomarginal, with narrow a lip (or lips) of uniform breadth. The wall is microperforate in the adult, with perforations irregularly scattered (the perforations are about 1.5 μm (or less) in diameter). The surface is unencrusted and smooth or with scattered blunt pustules and has no spines or perforation pits. Miocene to Holocene.

- *Candeina* d'Orbigny, 1839 (Type species: *Candeina nitida* d'Orbigny, 1839). Adults have a closed umbilicus and the single primary aperture is replaced by many small, uniform apertures in each suture. The surface is smooth. Miocene (Tortonian, N15) to Holocene (N23b) (Plate 6.1, Figs. 1–3).

Family Globigerinatellidae new family

The test is calcitic, microperforate, and initially trochospiral, becoming streptospiral. The surface sometimes has pustules. The umbilicus is terminally embraced by the primary chamber. Supplementary sutural apertures are found on the dorsal and ventral sides or on the ventral side only. Primary apertures are multiple and may be areal or sutural. The aperture is extraumbilical, and always bullate. Bullae may cover all primary apertures in adult. Miocene.

- *Globigerinatella* Cushman and Stainforth, 1945 (Type species *Globigerinatella insueta* Cushman and Stainforth, 1945). The surface of the test is smooth or postulate. The chambers embrace and cover the umbilicus. The primary apertures are cribrate, areal, and sutural, with lips. The bullae do not cover all primary apertures. Miocene (Burdigalian, N6a to Langhian, N9a) (Fig. 6.7; Plate 6.11, Fig. 1).
- *Polyperibola* Liska, 1980 (Type species: *Polyperibola christiani* Liske, 1980). The surface of the test is smooth or postulate. The globular chambers embrace and cover the umbilicus. Primary and secondary apertures are cribrate pore-like, with thick and broad lips, and are always interiomarginal, in sutures between chambers and at the margin of numerous variable bullae. Miocene (Tortonian, N16a–N16b) (Fig. 6.7; Plate 6.1, Fig. 5).

Family Globigerinitidae Loeblich and Tappan, 1984

The test is microperforate and trochospiral. The surface is smooth or pustulate. There is a single primary aperture, that is intra or intra-extraumbilical, and may have a bulla, with or without accessory apertures. The umbilicus is open except when covered by bulla, and few or no sutural pores are present. The surface sometimes has pustules. Eocene to Holocene.

- *Antarcticella* Loeblich and Tappan, 1988 (Type species: *Candeina antarctica* Leckie and Webb, 1985). The test is smooth, microperforate, with pustules only in the intercameral sutures. The aperture is covered by a large bulla with small accessory apertures occurring in its suture between the many, closed, tunnel-like marginal extensions of the bulla. Oligocene (Chattian, P22) to Miocene (Serravallian, P12) (Plate 5.1, Figs. 19 and 20).
- *Globigerinita* Brönnimann, 1951 (Type species: *Globigerinita naparimaensis* Brönnimann, 1951). The test is trochospiral and microperforate. The primary aperture is intraumbilical, but bullate. Accessory apertures occur in the untunnelled, smooth bulla marginal sutures. Pustules are scattered over wall surfaces, but there are no spines or muricae. Oligocene (Chattian, P21b) to Holocene (N23b) (Plate 5.3, Fig. 1; Plate 6.1, Figs. 6 and 7; Plate 6.11, Fig. 2).
- *Tenuitellinata* Li, 1987 (Type species: *Globigerinita angustiumbilicata* Bolli, 1957). Holotype (Plate 5.3, Figs. 3 and 4). The wall is smooth or pustulate. The aperture is simple, intraumbilical, with no bulla. Eocene (Priabonian, P16b) to Holocene (N23b) (Plate 5.3, Figs. 1–5; Plate 6.1, Figs. 8–13).
- *Tinophodella* Loeblich and Tappan, 1957 (Type species: *Tinophodella ambitacrena* Loeblich and Tappan, 1957). The test is trochospiral and microperforate. Pustules are scattered over wall surface. The primary aperture is intraumbilical and bullate. The accessory apertures are at the ends of the extended marginal tunnels of the bulla. Miocene (Burdigalian, N5b) to Holocene (N23b) (Fig. 6.7; Plate 6.1, Fig. 14).

Family Tenuitellidae new family

The test is low trochospiral to planispiral. The wall is microperforate in the adult, with perforations irregularly scattered. The surface is unencrusted, but with scattered pustules over the wall surface or only in intercameral sutures, but has no spines. The aperture is interiomarginal, intra-extraumbilical, with a narrow lip (or lips) of uniform breadth. Few or no sutural pores are present. The umbilicus is open except when it is covered by bulla. Eocene to Holocene.

- *Tenuitella* Fleisher, 1974 (Type species: *Globorotalia gemma* Jenkins, 1966). The test is low trochospiral, with a postulate wall. The aperture is intra-extraumbilical, lacking a bulla. Oligocene (Rupelian, P18) to Holocene (N23b) (Plate 5.2, Figs. 20 and 21; Plate 6.1, Figs. 15 and 16).
- *Tenuitellita* Li, 1987 (Type species: *Globigerinita iota* Parker, 1962). The test is trochospiral. The surface is smooth or pustulate. The primary aperture is intra-extraumbilical and in the adult is covered by a bulla which has accessory interiomarginal apertures situated above intercameral sutures. Pleistocene (N22) to Holocene (N23b) (Fig. 6.7).

Superfamily GLOBIGERINOIDEA Carpenter, Parker and Jones, 1862

Members of this superfamily have trochospiral calcitic tests, with chambers that are rounded or angular with a peripheral keel or an imperforate band surrounded by a double keel. When the portici are fused, accessory apertures or supplementary sutural apertures are formed. The wall is micro-perforate or macroperforate, and the surface may be smooth, with or without perforation cones, muricate, or may be spinose, sometimes may be encrusted or coated with smooth cortex. The aperture is interiomarginal, umbilical, or intra-extraumbilical and bordered by a lip, and may have portici, or can be covered by tegilla with accessory apertures. Cretaceous to Holocene.

Family Globigerinidae Carpenter, Parker and Jones, 1862

The test is trochospiral. The chambers may be subglobular or radially elongate. The wall is smooth, macroperforate with large perforations (up to 10 mm in diameter) regularly and geometrically arranged (when not obscured by encrustation), or with perforation pits which may coalesce to form medium cancelation, and with delicate radiating spines from the surface of the adult test, or spine bases. Dorsal supplementary apertures may be present. The primary aperture is a simple arch and umbilical, which may have bullae or conspicuous portical structures, to extraumbilical, and rimless or with a narrow lip. Eocene to Holocene.

- *Beella* Banner and Blow, 1960 (Type species: *Globigerina digitata* Brady, 1879). The test is trochospiral, mostly or wholly with spine bases and spines. The adult chambers are radially elongate and higher than long. The aperture is intraumbilical. Miocene (Tortonian, N17a) to Holocene (N23b) (Fig. 6.8).
- *Catapsydrax* Bolli, Loeblich and Tappan, 1957 (Type species: *Globigerina dissimilis* Cushman and Bermudez, 1937). The test is macroperforate and may have spine bases and muricae near the aperture. The primary aperture is intraumbilical and covered by a regularly developed bulla with few (one to four) large, arched infralaminal accessory apertures at smooth sutural margins of the bulla. No dorsal supplementary apertures and no teeth are present. Eocene (Ypresian, P7) to Pleistocene (N22) (Plate 5.4, Figs. 1–4; Plate 6.1, Figs. 17 and 18).
- *Dentoglobigerina* Blow, 1979 (Type species: *Globigerina galavisi* Bermudez, 1961). The test is trochospiral with a wall lacking strong cancelation. Surfaces are smooth, or with perforation pits, but may have spine bases and

muricae concentrated near the aperture. The aperture is an axio-intraumbilical arch with a tooth-like, sub-triangular, symmetrical porticus projecting into the umbilicus. Eocene (Ypresian, P9a) to Pliocene (Zanclean, N20a) (Plate 5.5, Figs. 13–16; Plate 5.6, Figs. 1–5; Plate 5.12, Fig. 2; Plate 6.2, Figs. 1–7; Plate 6.14, Fig. 5B; Plate 6.15, Fig. 3).

- *Globigerina* d'Orbigny, 1826 (Type species: *Globigerina bulloides* d'Orbigny, 1826). A globigerinid with subglobular adult chambers, but not radially elongate. The wall is smooth with spines or spine bases. The aperture is intraumbilical throughout ontogeny, a high umbilical arch which may bordered by a thin lip, but is without an apertural tooth or bulla. *Globigerina* is distinguished from *Subbotina* and *Dentoglobigerina* by the absence of apertural structures such as a porticus or teeth and in possessing noncancellate walls. *Globigerina* differs from *Globoturborotalita* by the absence of coalescent perforation pits which produce a favose surface. Early Eocene (Ypresian, P6b) to Holocene (N23b) (Plate 5.4, Figs. 5–15; Plate 5.5, Figs. 1–6, 16–19; Plate 6.2, Figs. 8–16).
- *Globigerinoides* Cushman, 1927 (Type species: *Globigerina rubra* d'Orbigny, 1839) = *Globicuniculus* Saito and Thompson, 1976 = *Globigerinoidesella* El Naggar, 1971. The test is a low trochospiral with globular to ovate chambers, with a spiral sutural supplementary aperture or apertures. The wall is coarsely perforate and spinose, sometimes with deep perforation pits. The primary aperture is intraumbilical without bulla or an apertural tooth. Late Oligocene (Late Chattian, P22b) to Holocene (N23b) (Plate 5.5, Figs. 9–11; Plate 6.2, 17–20; Plate 6.3, Figs. 1–10; Plate 6.11, 3–8, 10; Plate 6.12, Figs. 1, 7A; Plate 6.13, Fig. 7; Plate 6.12, Figs. 4 and 8; Plate 6.13, Figs. 5A and 7; Plate 6.14, Fig. 6; Plate 6.15, Fig. 5).
- *Globigerinoita* Brönnimann, 1952 (Type species: *Globigerinoita morugaensis* Brönnimann, 1952) = *Velapertina* Popescu, 1969. The surface is spinose, sometimes with deep perforation pits. The primary aperture is intraumbilical. Supplementary spiral sutures are also present. Bullae are present on primary or supplementary apertures or both. Miocene (Burdigalian, N6 to Tortonian, N15) (Fig. 6.8).
- *Globoquadrina* Finlay, 1947 (Type species: *Globorotalia dehiscens* Chapman, Parr, and Collins, 1934). The test is macroperforate, trochospiral. The equatorial outline is rounded, subquadrate. The apertural face is flattened, sometimes bullate. The principal apertural extent changes ontogenetically to intraumbilical from early intra-extraumbilical. The aperture is usually has an apertural tooth. The surface has spine bases often concentrated at the distal limits of apertural face, which may be convex or may be flattened and high. No supplementary apertures are present. Oligocene (Chattian, P22) to Pliocene (Zanclean, N19) (Plate 5.6, Figs. 6 and 7; Plate 6.3, Figs. 14–17; Plate 6.13, Fig. 2; Plate 6.13, Figs. 1A and 2; Plate 6.15, Fig. 12).
- *Globorotaloides* Bolli, 1957 (Type species: *Globorotaloides variabilis* Bolli, 1957). The test is trochospiral, macroperforate. Early ontogeny has the primary aperture umbilical–extraumbilical, but in late ontogeny, the aperture becomes intraumbilical, and may be bullate at any growth stage, but with no apertural tooth. The equatorial outline is rounded, and subcircular. Surfaces have few or no spine bases. Perforation pits may be present. The apertural face is convexly rounded with no supplementary apertures. Eocene (Ypresian, P5b) to Pliocene (Early Zanclean, N19) (Plate 5.6, Figs. 8; Fig. 6.3, Figs. 18–20).
- *Globoturborotalita* Hofker, 1976 (Type species: *Globigerina rubescens* Hofker, 1956). The test has globular, closely appressed chambers. The wall is cancellate with simple spines. Perforation pits are large and coalescent. Eocene (Ypresian, P5b) to Holocene (N23b) (Plate 5.4, Figs. 16 and 17; Plate 5.5, Fig. 12; Plate 6.4, Figs. 1–12).

Family Globigerinellidae new family

The test is calcitic, macroperforate, biumbilicate, with a biconcave, thick-disc shape. Adult coiling is planispiral and adult chambers are subglobular or radially elongate. The surface is covered with spines or spine bases. The aperture is umbilical–extraumbilical throughout ontogeny. Miocene to Holocene.

- *Bolliella* Banner and Blow, 1959 (Type species: *Hastigerina* (*Bolliella*) *adamsi* Banner and Blow, 1959). The adult test is planispiral. The surface is covered with spine bases. The chambers are initially subglobular, with broadly rounded periphery, later become high and radially elongate. The primary aperture is umbilical–extraumbilical throughout ontogeny. Holocene (N23b) (Plate 6.4, Figs. 13–15).
- *Globigerinella* Cushman, 1927 (Type species: *Globigerina aequilateralis* Brady, 1879) = *Globigerina siphonifera* d'Orbigny, 1939. The adult test is planispiral with subglobular chambers. The test is biumbilicate with a broadly rounded periphery. The surface is covered with spine bases. The primary aperture is umbilical–extraumbilical throughout ontogeny. Miocene (Burdigalian, N7b) to Holocene (N23b) (Plate 6.4, Fig. 16).

Family Globorotaliidae Cushman, 1927

The test is macroperforate, trochospiral, convex, acute or has a rounded lens shape. Only one umbilicus is open in the umbilical side, and the principal aperture is intra-extraumbilical throughout ontogeny, or may become intraumbilical in the last few chambers. Chambers may be radially elongate or ampullate. The periphery may be rounded, or carinate, or acutely angled in axial view. The primary aperture is never bullate and never has supplementary apertures. Surfaces are smooth or with spine bases between perforation pits. Late Eocene to Holocene.

- *Clavatorella* Blow, 1965 (Type species: *Hastigerinella bermudezi* Bolli, 1957). The chambers are radially elongate, often clavate. The surface is smooth or with spine bases often prominent between perforation pits. The aperture is umbilical–extraumbilical. Miocene (Langhian, N8b) to Pliocene (Zanclean, N19) (Plate 6.4, Figs. 17 and 18).
- *Globorotalia* Cushman, 1927 (Type species: *Pulvinulina menardii* (d'Orbigny) var. *tumida* Brady, 1877) = *Fohsella* Bandy, 1972 = *Globoconella* Bandy, 1975 = *Hirsutella* Bandy, 1972 = *Obandyella* Haman, Huddleston and Donahue, 1981 = *Menardella* Blow, 1972. The test is carinate or acutely angled at the periphery. The surface is mostly smooth, with spine bases rarely prominent, and often obscure. The aperture is umbilical–extraumbilical, never bullate or ampullate. Miocene (Burdigalian, N5b) to Holocene (N23b) (Plate 6.4, Figs. 19–21; Plate 6.5, Figs. 1–18; Plate 6.6, Figs. 6–22; Plate 6.7, Figs. 1–11; Plate 6.14, Figs. 3, 4A, 6; Plate 6.15, Fig. 5).
- *Neogloboquadrina* Bandy, Frerichs, and Vincent, 1967 (Type species: *Globigerina dutertrei* d'Orbigny, 1839). The test is trochospiral and macroperforate. The surface of the test is covered with spine bases or with perforation pits, or both. The primary aperture is antero-intraumbilical. The apertural lip sometimes is a sub-triangular tooth. Pliocene (N19) to Holocene (N23b) (Plate 6.7, Figs. 14 and 15).
- *Truncorotalia* Cushman and Bermudez, 1949 (Type species: *Rotalia truncatulinoides* d'Orbigny). The test is conicotruncate, keeled. The surface is mostly smooth, with spine bases rarely prominent, and often obscure. The aperture is umbilical–extraumbilical, never bullate or ampullate. Miocene (Tortonian, 16b) to Holocene (N23b) (Plate 6.7, Figs. 16–20; Plate 6.11, Fig. 13).
- *Paragloborotalia* Cifelli, 1982 (Type species: *Globorotalia opima* Bolli, 1957) = *Jenkinsella* Kennett and Srinivasan, 1983. The test has an axial periphery that is usually rounded, not carinate. Chambers are not radially elongate. Surfaces usually have spine bases between perforation pits or are smooth. The principal aperture is umbilical–extraumbilical throughout adult growth. Eocene (Late Bartonian, P15a) to Holocene (N23b) (Plate 5.7, Figs. 1–14; Plate 6.5, Figs. 19–20; Plate 6.6, Figs. 1–5; Plate 6.8, Figs. 1–8; Plate 6.15, Figs. 6 and 7A).
- *Protentella* Lipps, 1964 (Type species: *Protentella prolixa* Lipps, 1964). Test is macroperforate and planispiral, with spines or spine bases widely separated or absent. Chambers are high, becoming radially elongate. Miocene (Serravallian, N12 and N13) (Plate 6.8, Figs. 9 and 10).

Family Hastigerinidae Bolli, Loeblich and Tappan, 1957

The test is initially trochospiral or planispiral, may be becoming streptospiral in the adult. Chambers are globular to clavate. The surface of the test is smooth except for distal, peripheral ends of chambers where thick spines, triadiate in section, are situated. The aperture is interiomarginal and equatorial or may become spiroumbilical. Miocene to Holocene.

- *Hastigerina* Thomson (Type species: *Hastigerina murrayi* Thomson, 1876) = *Nonionina pelagica* d'Orbigny, 1839. The test is microperforate, planispiral, spinose, and triadiate in section and barbed. Chambers are not radially elongate. Miocene (Tortonian N17a) to Holocene (N23b) (Plate 6.8, Fig. 11).
- *Hastigerinella* Cushman, 1927, emended Banner, 1960 (Type species: *Hastigerina digitata* Thumbler, 1911). The test is macroperforate, planispiral to streptospiral in the adult, and biumbilicate. The aperture is high, umbilical–extraumbilical-equatorial, with a porticus which broadens laterally. The umbilici are small, with no relict apertures. The test surface is smooth. The chambers are radially elongate, and sometimes clavate. Pleistocene (N22b) to Holocene (N23b) (Fig. 6.11).
- *Hastigerinopsis* Saito and Thompson, 1976 (Type species: *Hastigerinopsis digitiformans* Saito and Thompson, 1976). The test is macroperforate, often pseudoplanispiral, becoming streptospiral in the adult, with radially elongate, clavate adult chambers, often partly distally divided into two or more lobes. The aperture is equatorial becoming spiroumbilical in the adult. Miocene (Burdigalian, N8a) to Holocene (N23b) (Fig. 6.11).
- *Orcadia* Boltovskoy and Watanabe, 1982 (Type species: *Hatigerinella riedeli* Rögl and Bolli, 1973). The test is macroperforate, trochospiral with a tendency toward streptospiral coiling, and spinose with spines widely separated that are confined to distal parts of chambers. Chambers are not radially elongate. The aperture is intraumbilical, never bullate. Pleistocene (N22) to Holocene (N23b) (Plate 6.8, Fig. 12).

Family Neoacarininidae new family

The test is calcitic, macroperforate, and trochospiral. The surface is covered with clumps of spinules, but no true spines or muricae. The aperture is intraumbilical, with a thin lip. There is no porticus, and no supplementary or accessory apertures are present. Pleistocene (N23a).

- *Neoacarinina* Thompson, 1973 (Type species: *Neoacarinina blowi* Thompson, 1973). The chambers are globular increasing rapidly in size with strongly embracing and flattened spiral side. The aperture is bordered by a narrow lip. Pleistocene (N23a) (Plate 6.8, Figs. 13–15).

Family Orbulinidae Schultze, 1854

The coiling is intensely streptospiral, with the last globular chamber completely embracing the umbilical side (or all) of the juvenile test. The interiomarginal principal aperture of the nepionic and neanic stages is replaced by a multitude of rounded pores in the area of the last chamber (and also in the last intercameral sutures, if the nepionic test is not completely enclosed by the ephebic, adult chamber). The surface of the adult test is covered with perforation pits and spines. Miocene to Holocene.

- *Orbulina* d'Orbigny, 1829 (Type species: *Orbulina universa* d'Orbigny, 1839) = *Candorbulina* Jedlitschka, 1934. The test is macroperforate. Coiling becomes ventrally embracing so that chambers cover the umbilicus until the test becomes spherical (or bispherical), with the last chamber covering much or all of the earlier, coiled test. There are apertural pores in both the last sutures (if any are exposed) and in the area of the walls of the last chamber. It is never bullate. Miocene (Langhian, N9) to Holocene (N23b) (Plate 6.8, Figs. 16–19; Plate 6.9, Figs. 1–3; Plate 6.13, Fig. 4A).
- *Praeorbulina* Olsson, 1964 (Type species: *Globigerinoides glomerosa* subsp. *glomerosa* Blow, 1956). The coiling is embracing so that the chambers cover the umbilicus, and the apertures become wholly extraumbilical. Subglobular later chambers cover much of earlier test. Apertures are multiple but are confined to the suture of the last chamber and are never areal and never bullate. Miocene (Langhian, N8 to N9) (Plate 6.9, Figs. 4–7; Plate 6.11, Fig. 9).

Family Pulleniatinidae Cushman, 1927

The test is macroperforate. The initial coils, like the adult ones, are streptospiral. The chambers increasingly embrace ventrally so that the umbilicus becomes closed and the aperture becomes wholly extraumbilical. In the morphologically most advanced forms, the chambers begin to embrace dorsally so that the last whorl covers much of the preceding whorls. The surface of the neanic chambers is punctate, perforation pitted, and spinose, but the surface of the adult chambers is encrusted and smooth (except for the tubercles around the aperture). There is only one extant genus. Miocene to Holocene.

- *Pulleniatina* Cushman, 1927 (Type species: *Pullenia sphaeroides* (d'Orbigny) var. *obliquiloculata* Parker and Jones, 1862). The test is pseudotrochospiral, but actually streptospiral. Chambers cover the umbilicus so that the aperture becomes wholly extraumbilical even though it is ventrally interiomarginal. There is no cortex, but the surface of the test is smooth with no spine bases or perforation pits on the adult surface. It is never bullate nor has radially elongate chambers, but it may be tuberculate near the aperture. Miocene (Messinian, N17b) to Holocene (N23b) (Plate 6.9, Figs. 8–13; Plate 6.14, Fig. 2; Plate 6.15, Fig. 9A).

Family Sphaeroidinellidae Banner and Blow, 1959

The family includes genera with trochospiral, macroperforate tests. Adult tests are coated in a thick, smooth cortex of calcite. The cortex diminishes or closes the perforations and greatly thickens the apertural lips. Tests that are broken show the younger whorls, below the cortex, to be punctuate and spinose. There is a single principal primal aperture that is interiomarginal and intraumbilical in position and extent. Miocene to Holocene.

- *Prosphaeroidinella* Ujii, 1976 (Type species: *Sphaeroidinella disjuncta* Finlay, 1940). No spiral supplementary apertures are present. The cortex covers only part of the adult test, and remainder shows perforation pits and spine bases. The primary aperture is intraumbilical. Miocene (Burdigalian, N6) to Pleistocene (N22a) (Plate 6.9, Figs. 14 and 15; Plate 6.10, Fig. 1–3).
- *Sphaeroidinella* Cushman, 1927 (Type species: *Sphaeroidina dehiscens* Parker and Jones, 1865). The cortex covers all of the adult test, reducing or sealing perforations externally. The primary aperture is intraumbilical. Spiral sutural supplementary aperture may be equally big. Pliocene (N19b) to Holocene (N23b) (Plate 6.10, Figs. 9 and 10).
- *Sphaeroidinellopsis* Banner and Blow, 1959. (Type species: *Sphaeroidinella dehicens* Parker and Jones subsp. *subdehiscens* Blow, 1959). The cortex covers the entire test, reducing the perforation size externally and covering spine bases. The primary aperture is intraumbilical. No spiral supplementary apertures are present. Miocene (Burdigalian, N6) to Pliocene (Piacenzian, N21a) (Plate 6.9, Figs. 12–14; Plate 6.10, Figs. 4–8; Plate 6.13, Fig. 3A; Plate 6.14, Fig. 5A).

Family Turborotalitidae Hofker, 1976

The species of this family are trochospiral to planispiral, microperforate to macroperforate, spinose, small, but the numbers of their whorls and their calcitic crusts readily distinguish them from juvenile tests of taxa of other families. Eocene to Holocene.

- *Berggrenia* Parker, 1976 (Type species: *Globanomalina praepumilio* Parker, 1967). The test is microperforate, planispiral; macroperforations sometimes are enlarged in sutural depressions, where pustules may form as in umbilical areas. It has no bullae or peripheral carina. The apertural lip is weak or absent. Pliocene (Piacenzian, N20b and N21b) (Fig. 6.7).

- *Turborotalita* Blow and Banner, 1962 (Type species: *Truncatulina humilis* Brady, 1884). The test is macroperforate, trochospiral. The surface is spinose or with spine bases between perforation pits but may be smoothed by added surface lamellae. The primary aperture is umbilical–extraumbilical. The apertural lip is broad, making the final chamber ampullate, with small, accessory infralaminal apertures at the margin of the ampulla, which may cover the umbilicus. Eocene (Lutetian, P10) to Holocene (N23b) (Plate 6.10, Figs. 11–12).

ORDER HETEROHELICIDA FURSENKO, 1958

Tests are biserial or triserial, at least in the early stage. They may be reduced to being uniserial in later stages and are microperforate or macroperforate, and smooth or muricate. Apertures are terminal, with a low to high arches. Walls are calcitic. Cretaceous (Aptian) to Holocene.

Superfamily HETEROHELICOIDEA Cushman, 1927

The test is calcitic, mainly planispiral in the early stage, then biserial to triserial and possibly multiserial, rarely becoming uniserial in the adult stage. Apertures have a high to low arch at the base of the final stage or are terminal in the uniserial stage. Walls are calcitic, smooth, or muricate. Cretaceous (Aptian) to Holocene.

Family Cassigerinellidae Bolli, Loeblich and Tappan, 1957

The test is biserial or triserial in early stages, later becoming enrolled biserial, but with the biseries coiled in tight, involute trochospire. Chambers may be inflated or compressed. The primary aperture ranges from being asymmetric to being in the equatorial plane. Eocene to Miocene.

- *Cassigerinella* Pokorný, 1955 (Type species: *Cassigerinella boudecensis* Pokorný, 1955). The test has inflated chambers, which are reniform, or subglobular, with an anterior marginal aperture as a high, broad arch. Late Eocene (Priabonian, P14) to Middle Miocene (Serravallian, N13) (Plate 5.10, Figs. 3–6; Plate 6.10, Fig. 13).
- *Riveroinella* Bermudez and Seiglie, 1967 (Type species: *Riveroinella martinezpicoi* Bermudez and Seiglie, 1967). The wall is smooth. The test is biserial, but biseries are loosely coiled trochospirally. Chambers are compressed. The aperture is a high, interiomarginal arched slit at the anterior end of the chamber. Miocene (Burdigalian, N6 and N7) (Plate 6.10, Figs. 14 and 15).

Family Chiloguembelinidae Loeblich and Tappan, 1956

The test is biserial throughout. The aperture is interiomarginal, asymmetrical, extending up the face of the final chamber, bordered by an apertural rim, in the plane of biseriality, invaginated to make an internal plate or not infolded or invaginated. Paleocene (Danian) to Holocene.

- *Streptochilus* Brönnimann and Resig, 1971 (Type species: *Bolivina tokelauae* Boersma, 1969). The test is compressed with chambers of each pair appressed, with no median plate or supplementary sutural apertures. An interiomarginal aperture is asymmetric to the plane of the biseries, with the asymmetry alternating from chamber to successive chamber, with the apertural margin in the equatorial plane infolded to form an invaginated, internal plate. Eocene (Ypresian, P8) to Holocene (N23b) (Plate 5.10, Fig. 19; Plate 6.10, Figs. 16–20).

Family Guembelitriidae Montanaro Gallitelli, 1957

The test is triserial throughout with a straight axis of triseriality, or becoming multiserial in the adult, with globular inflated chambers. The aperture is a simple arch bordered by a rim, symmetrical about equatorial plane, or can have more than one aperture per chamber in the multiserial stage. Walls may be muricate. Cretaceous (Late Albian) to Holocene.

- *Gallitellia* Loeblich and Tappan, 1986 (Type species: *Guembelitria? vivans* Cushman, 1935). The test is triserial in the adult and may sometimes proliferate its terminal chambers irregularly. The wall is smooth and microperforate. The aperture is a simple rounded arch. Pliocene (Piacenzian, N21) to Holocene (N23b) (Plate 6.10, Figs. 21–23).

Age (Ma)	Period, Epoch, Stage	Berggren and Pearson, 2005	Wade *et al.*, 2011	Planktonic Zonation (this study)
	Holo.			N23 b, a
1.81	Pleist.	PT1	PT1	N22 b, a
	Pliocene – Gela., Piac.	PL3-PL6	PL3-PL6	N21 b, a; N20 b, a
	Pliocene – Zanclean	PL2	PL2	N19 b, a
5.33	Pliocene – Zanclean	PL1	PL1	N18 b, a
	Late Miocene – Messinian, Tortonian	M13-M14	M12-M14	N17 b, a; N16 b, a
	Late Miocene – Tortonian	M12	M11	N15 b, a
11.63	Late Miocene – Tortonian	M11		N14 b, a
	Middle Miocene – Serravallian	M10	M10	N13 b, a
	Middle Miocene – Serravallian	M8-M9	M7-M9	N12 b, a
	Middle Miocene – Langhian	M7		N11 b, a; N10 b, a
	Middle Miocene – Langhian	M6	M6	N9 b, a
15.97	Middle Miocene – Langhian	M5	M5	N8 b, a
	Early Miocene – Burdigalian	M4	M4	N7 b, a
	Early Miocene – Burdigalian	M3	M3	N6 b, a
	Early Miocene – Burdigalian	M2	M2	N5 b, a
23.03	Early Miocene – Aquitanian	M1	M1	N4 b, a

Diagnostic First Occurrence of Planktonic Foraminifera

Age (Ma)	Species
0.01	*Globorotalia neoflexuosa*
0.12	*Sphaeroidinella excavata*
1.0	*Hastigerinella digitata*
1.8	*Beella digitata*
2.5	*Truncorotalia tenuitheca*
3.4	*Tenuitella anfracta*
3.6	*Globoturborotalita rubescens*
3.8	*Neogloboquadrina eggeri, Pulleniatina obliquiloculata*
4.2	*Globorotalia ungulata*
5.3	*Globorotalia margaritae*
5.8	*Globorotalia tumida*
7.2	*Globorotalia miocenica*
8.6	*Globorotalia plesiotumida*
9.0	*Truncorotalia oceanica*
9.6	*Globoturborotalita apertura, Globorotalia merotumida*
10.2	*Globigerinoides ruber, Turborotalita neominutissima*
11.6	*Globigerinoides ruber, Globorotalia paralenguaensis*
11.8	*Globoturborotalia nepenthes*
12.0	*Globorotalia lenguaensis, Sphaeroidinellopsis subdehiscens*
12.8	*Globorotalia lobata, Globigerinoides bollii*
13.6	*Globorotalia praemenardii, , Globorotalia fohsi*
13.7	*Globorotalia praefohsi, Globigerina druryi*
14.0	*Paragloborotalia peripheroacuta, Globorotalia rifensis*
14.5	*Orbulina universa*
15.0	*Orbulina suturalis*
15.4	*Praeorbulina glomerosa*
15.9	*Paragloborotalia peripheroronda, Praeorbulina sicana*
17.0	*Globigerinoides bisphericus*
17.1	*Paragloborotalia birnageae, Globigerina diplostoma*
17.2	*Globorotalia miozea, Catapsydrax parvulus*
17.3	*Globorotalia archeomenardii, Sphaeroidinellopsis seminulina*
18.0	*Prosphaeroidinella disjuncta*
20.4	*Globorotalia praescitula, Paragloborotalia zealandica*
21.0	*Tenuitella minutissima, Globigerinita incrusta*
22.0	*Turborotalita quinqueloba, Paragloborotalia continuosa*
23.03	*Paragloborotalia kugleri, Globoturborotalita brazieri, Globigerinoides parawoodi*

Diagnostic Last Occurrence of Planktonic Foraminifera

Age (Ma)	Species
0.01	*Globorotalia flexuosa*
0.12	*Truncorotalia hessi/Truncorotalia tosaensis*
1.0	*Globorotalia pertenuis*
2.5	*Neogloboquadrina acostaensis*
3.6	*Globorotalia cibaoensis*
3.8	*Globorotalia juanai*
4.2	*Globorotalia conomiozea*
6.3	*Globorotalia merotumida*
8.6	*Paragloborotalia paralenguaensis*
11.6	*Cassigerinella chipolensis*
12.0	*Paragloborotalia peripheroacuta*
12.8	*Globorotalia praefohsi*
13.6	*Paragloborotalia peripheroronda*
[illegible]	*Globorotalia archeomenardii*
14.5	*Globigerinoides diminutus*
16.0	*Paragloborotalia semivera, Globoturborotalita brazieri*
15.9	*Dentoglobigerina praedehiscens*
17.0	*Globigerina connecta*
18.0	*Globigerinita praestainforthi*
21.0	*Globigerina ciperoensis*
22.0	*Globigerina angulisuturalis*

Figure 6.6. Diagnostic first and last occurrences of Neogene Planktonic Foraminifera defined in this study compared with the earlier zonations of Berggren and Pearson (2005) and Wade *et al.* (2011).

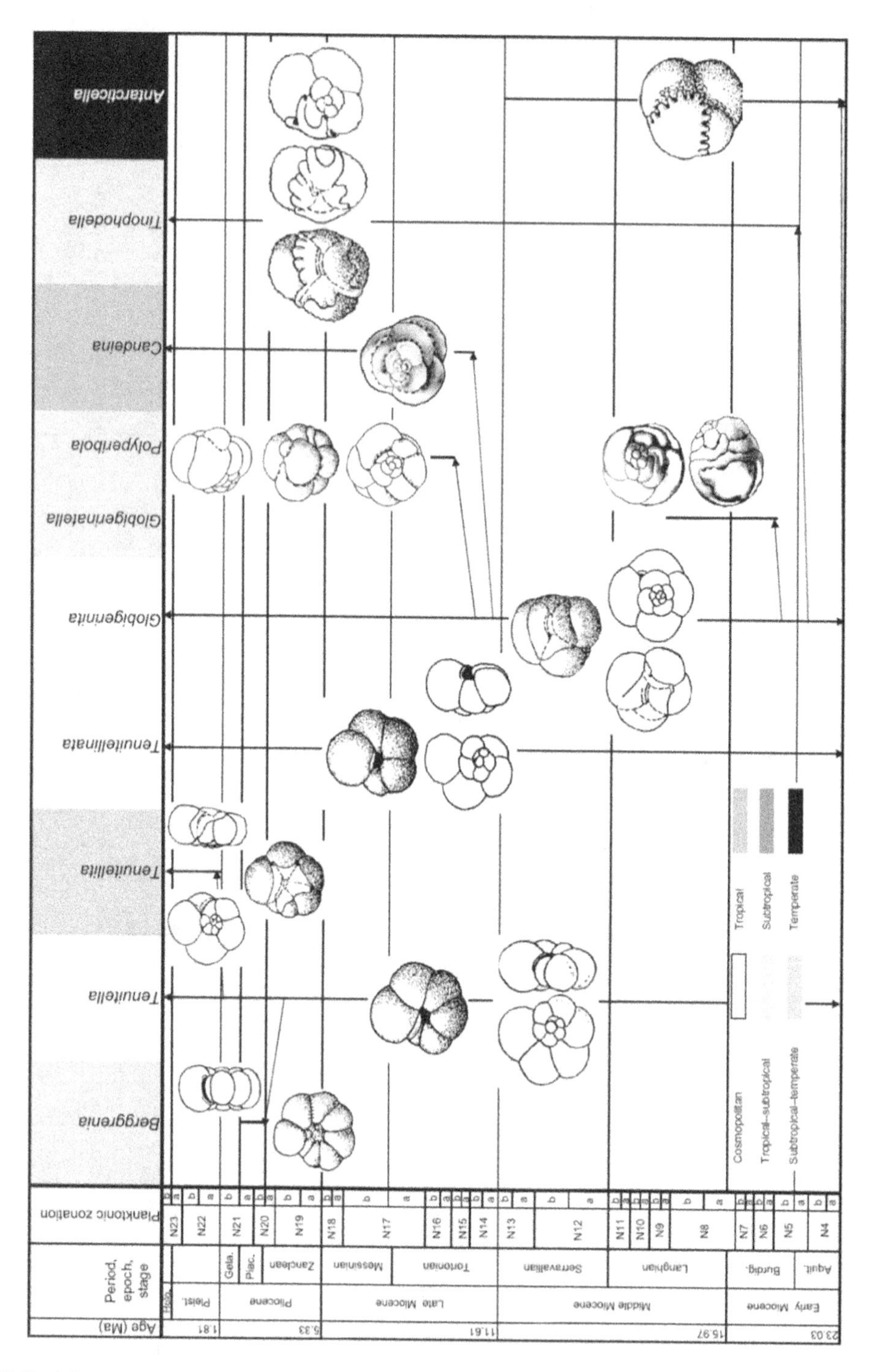

Figure 6.7. The phylogenetic evolution of the main nonspinose, microperforate, Neogene planktonic foraminifera. The shading behind the genera names indicates their latitudinal range as defined in the embedded legend.

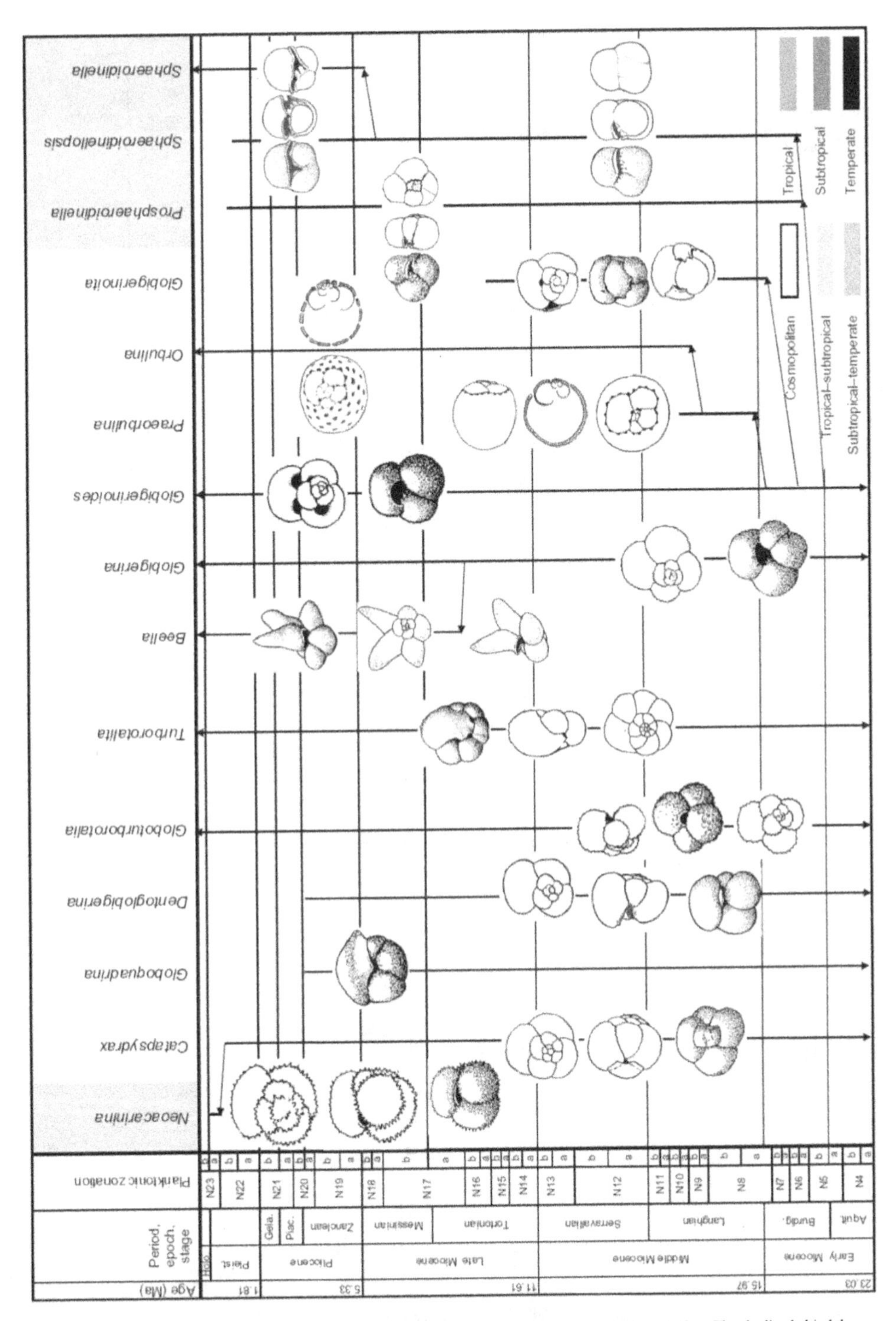

Figure 6.8. The phylogenetic evolution of the main spinose, macroperforate, Neogene planktonic foraminifera. The shading behind the genera names indicates their latitudinal range as defined in the embedded legend.

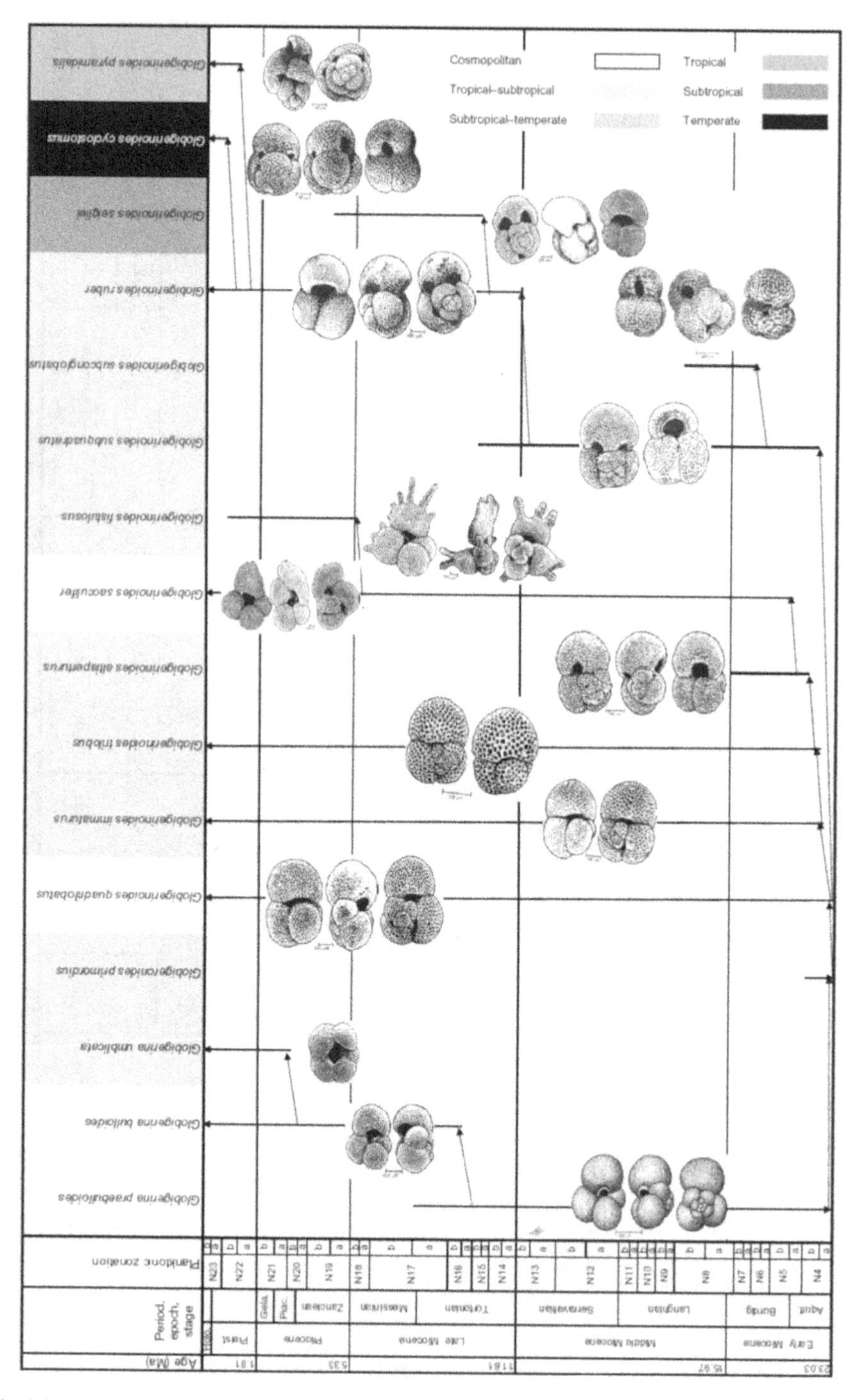

Figure 6.9. The phylogenetic evolution of some species of Globigerinoides. The shading behind the genera names indicates their latitudinal range as defined in the embedded legend. Images from original sources.

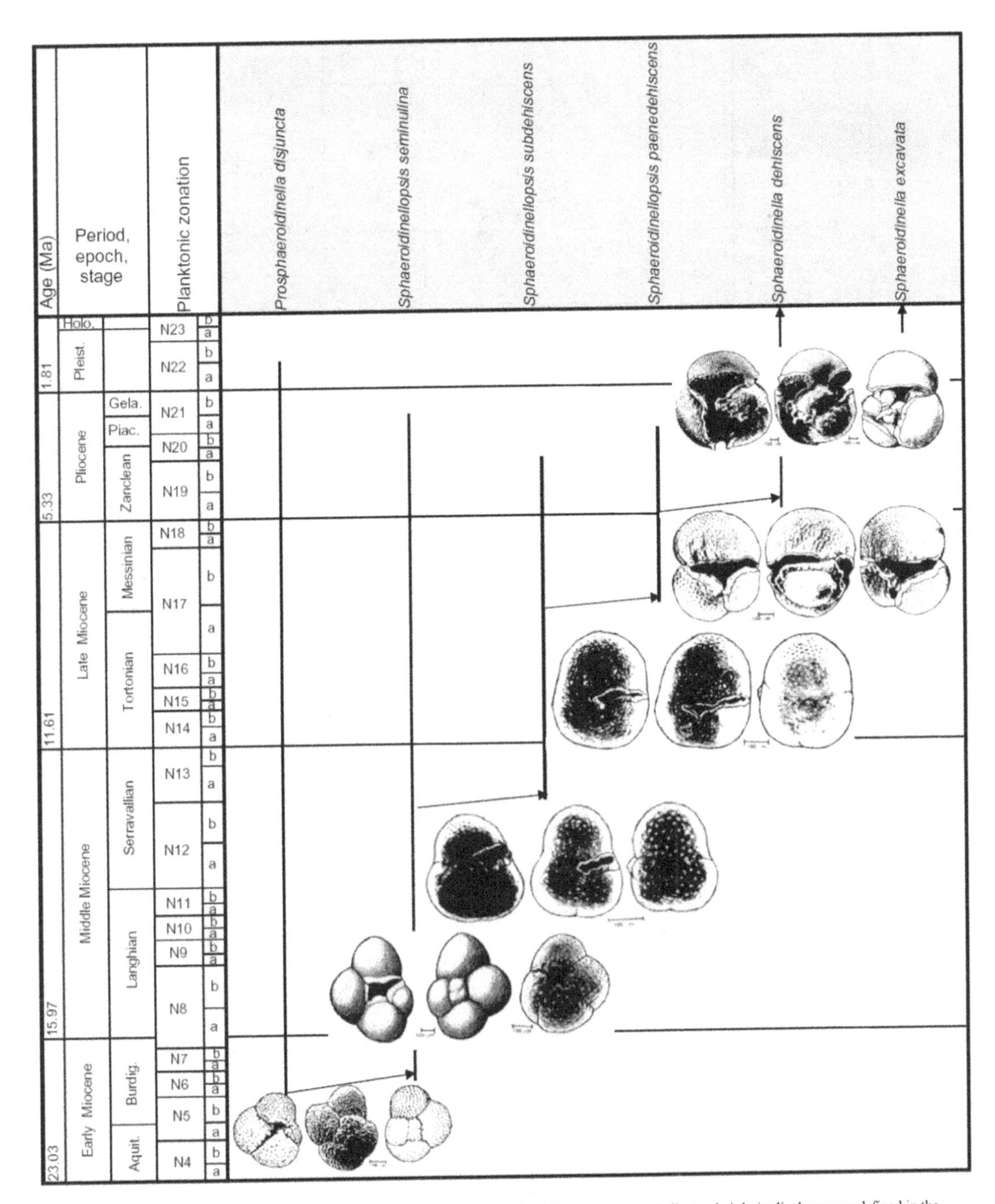

Figure 6.10. The phylogenetic evolution of the sphaeroidinellids. The shading behind the genera names indicates their latitudinal range as defined in the embedded legend. Images from original sources.

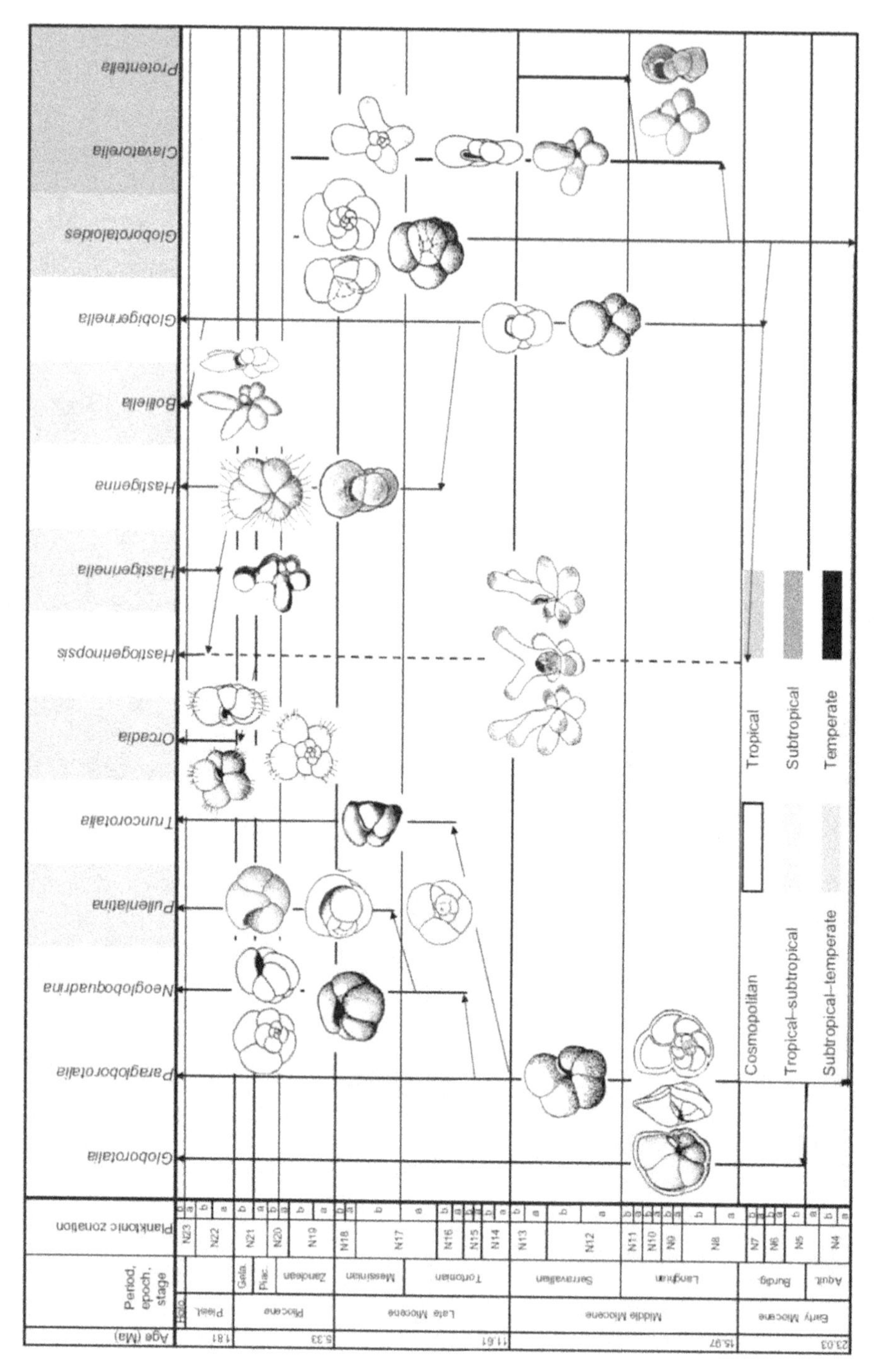

Figure 6.11. The phylogenetic evolution of main macroperforate, trochospiral, streptospiral, or planispiral Neogene planktonic foraminifera. The shading behind the genera names indicates their latitudinal range as defined in the embedded legend.

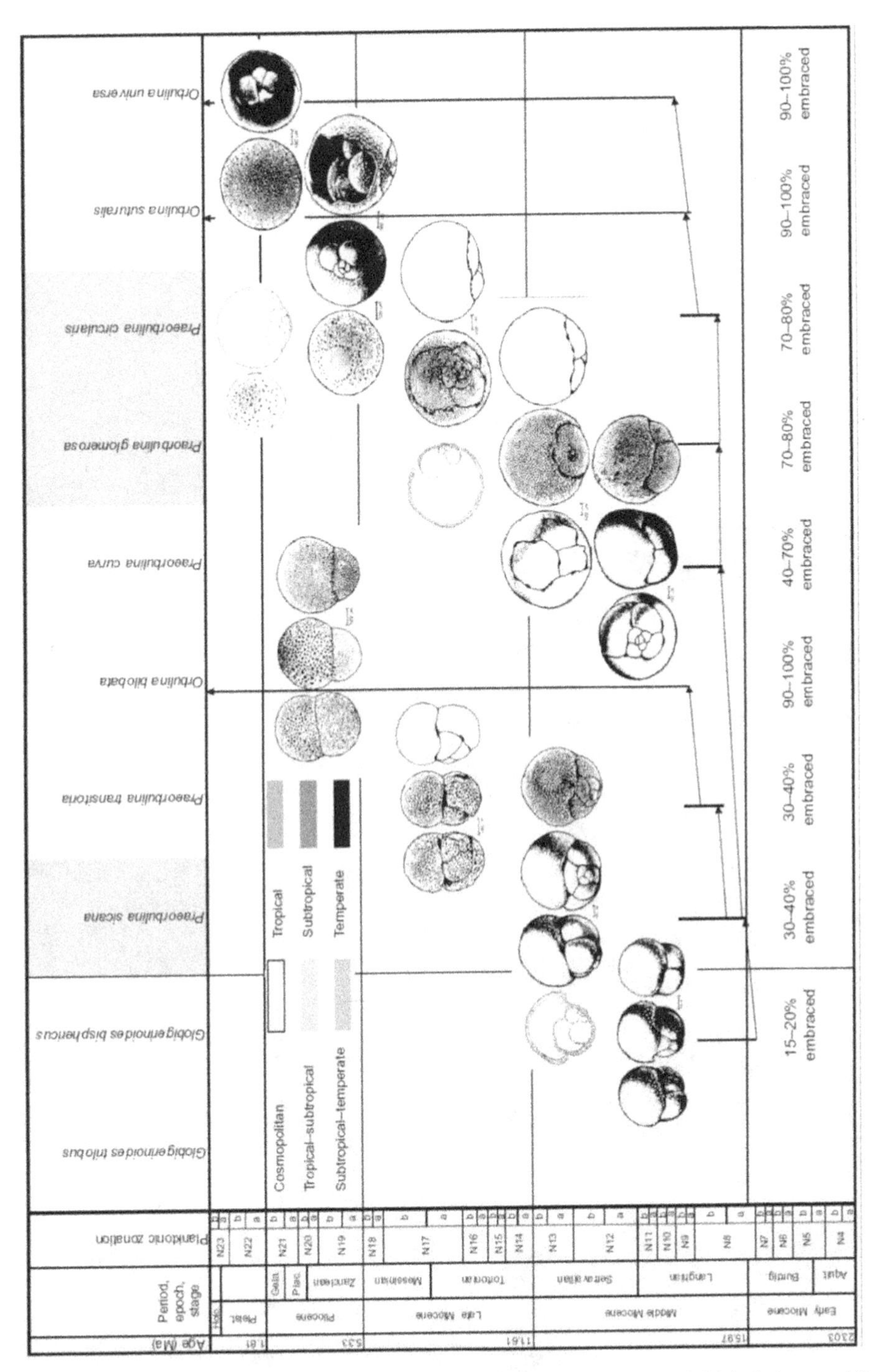

Figure 6.12. The phylogenetic evolution of main orbulinid species. The shading behind the genera names indicates their latitudinal range as defined in the embedded legend. Images from original sources.

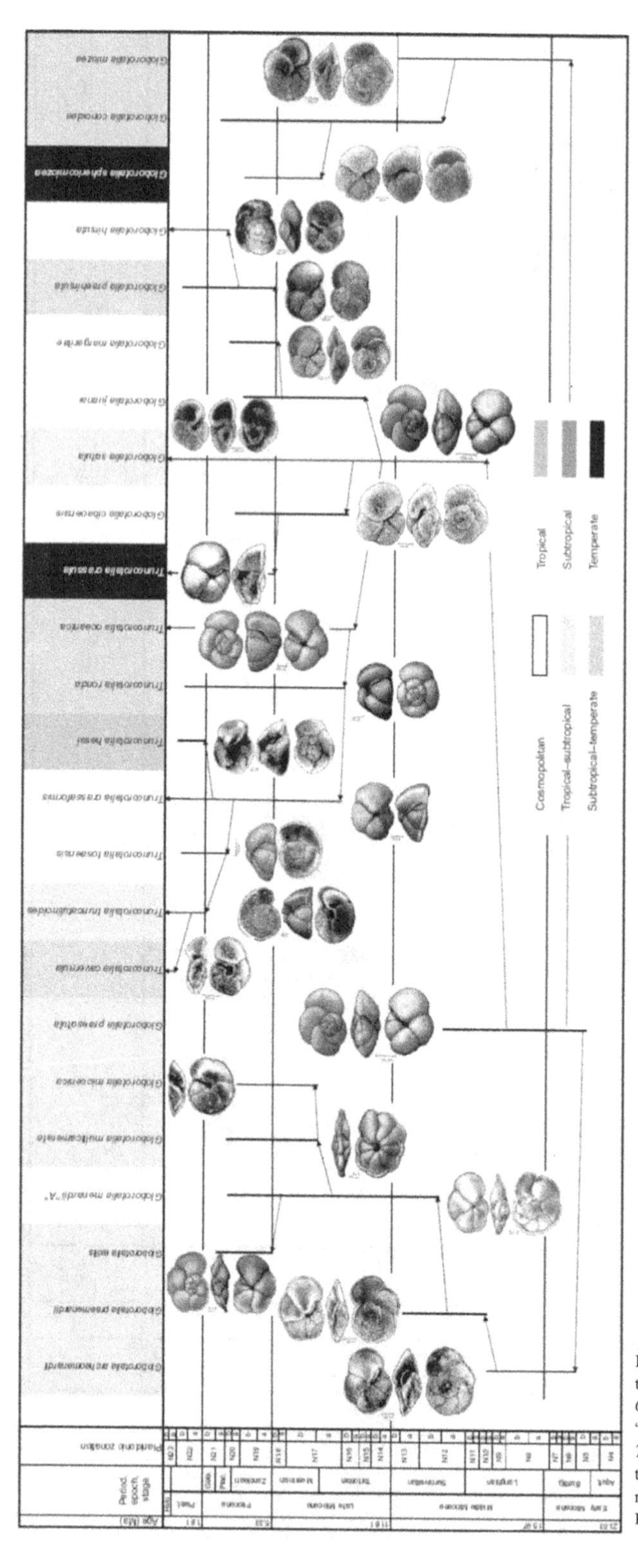

Figure 6.13. The phylogenetic evolution of the globorotaliids which evolved from *Globorotalia praescitula*, including the "*Menardella*," "*Hirsutella*" types, and *Truncorotalia* species. The shading behind the genera names indicates their latitudinal range as defined in the embedded legend. Images from original sources.

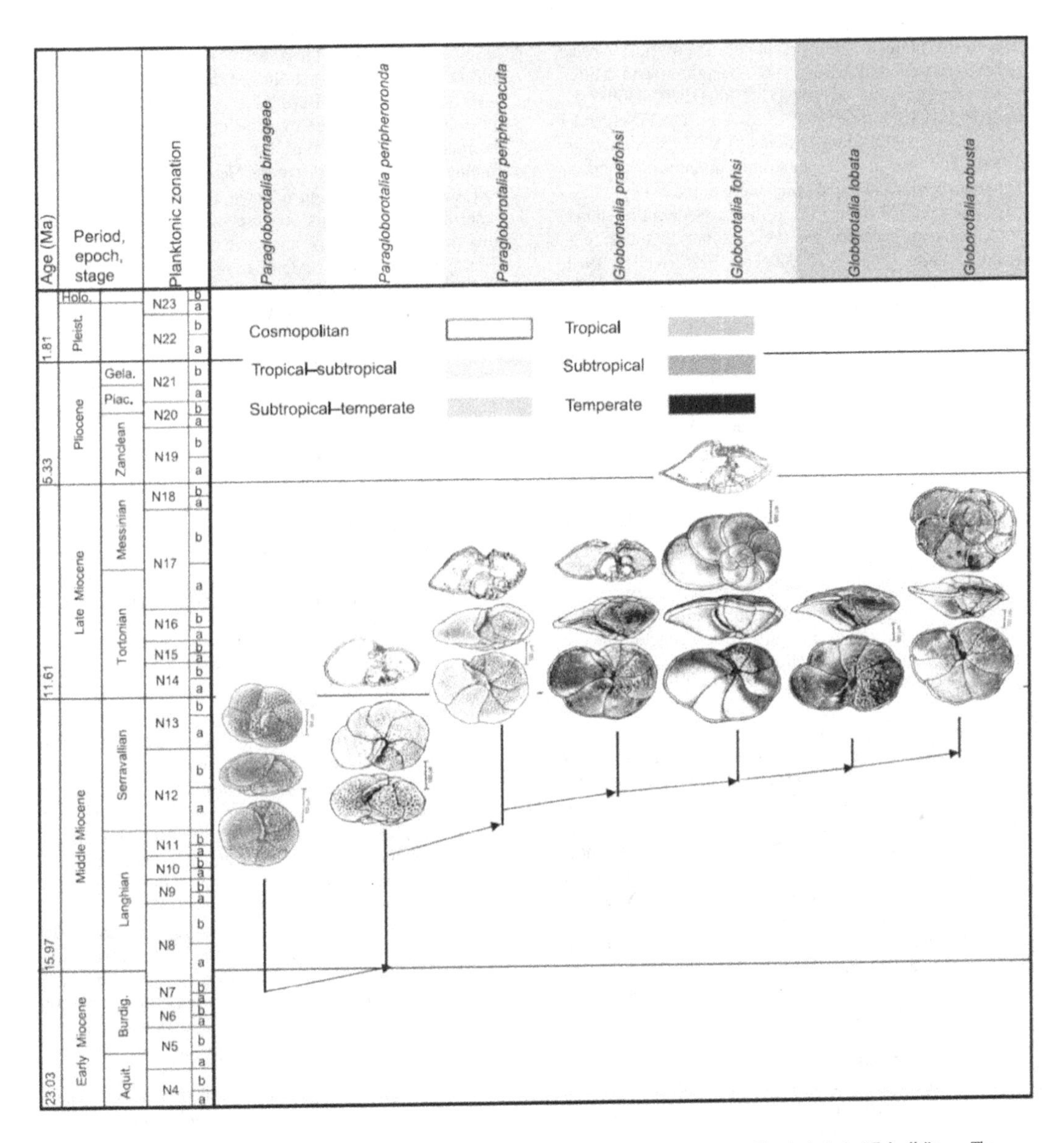

Figure 6.14. The phylogenetic evolution of the globorotaliids which evolved from *Paragloborotalia birnageae*. They include the "*Fohsella*" type. The shading behind the genera names indicates their latitudinal range as defined in the embedded legend. Images from original sources.

6.3 Biostratigraphy and phylogenetic evolution

The International Commission on Stratigraphy recommended the lowering of the base of the Quaternary Period and the Pleistocene epoch to 2.58 Ma, at the same time capping the Neogene Period at that age (Gibbard *et al.*, 2009). However, this recommendation is not universally accepted (Van Couvering *et al.*, 2009), and for simplicity, the Neogene period in this chapter is taken to be, as it was formerly defined, as consisting of the Miocene, Pliocene, Pleistocene, and Holocene (Gradstein *et al.*, 2004; McGowran *et al.*, 2009).

As noted above, during the first 3 Ma of the Neogene, the older Paleogene forms were still dominant, and it was not before the Burdigalian that truly new Neogene forms of planktonic foraminifera began to diversify. The origin and evolution of the Neogene planktonic foraminifera has long been debated because of their overall significance and their wide distribution in the tropical and subtropical marine realms (Banner and Blow, 1967; Berggren and Miller, 1988; Berggren *et al.*, 1985; Blow,1979; Bolli and Saunders, 1985; Kennett and Srinivasan, 1983; Schneider and Kennett, 1999; Stainforth *et al.*, 1975; Stanley *et al.*, 1988). The scarcity of planktonic foraminifera in the temperate realm and the absence there of many key species (see Charts 6.1–6.3 online) led to the development of entirely separate zonal schemes for these higher latitudes (e.g., Jenkins, 1966, 1967, 1971; Kaneps, 1975; Keller, 1981; Kennett and Srinivasan, 1983; Poore and Berggren, 1975; Stott and Kennett, 1990). Recently, an increasing number of key biostratigraphic events have been accurately calibrated against magneto-, radiometric, and astrochronological time scales (Agnini *et al.*, 2009; Berggren *et al.*, 1995a,b; Norris, 1991; Sinha and Singh, 2008; Wade *et al.*, 2011 among others). Detailed biostratigraphic investigations of the Tropical Atlantic Middle Miocene–Pleistocene (Chaisson and Pearson, 1997; Norris, 1998; Pearson and Chaisson, 1997; Turco *et al.*, 2002), the Tropical Pacific (Chaisson and Leckie, 1993; Shackleton *et al.*, 1995), and the Mediterranean Miocene–Pleistocene (Borsetti *et al.*, 1979; Iaccarino, 1985; Iaccarino and Salvatorini, 1982; Sprovieri *et al.*, 2006) have resulted in revisions of the calibrations of numerous bioevents. In addition, the presence of and genetic study of numerous extant genera has allowed the proposal of a more natural phylogeny based on inferred biological lineages (Aze *et al.*, 2011).

Banner and Blow (1965) and Blow (1969) described a set of zones using an alphanumeric shorthand ("P" for Paleogene and "N" for Neogene, with the Neogene zones being divided from N1 to N22). Numerous revisions and amendments to these zones have been suggested over the years, and Berggren *et al.* (1995a, b) introduced a new development to the alphanumeric notation for epoch-level intervals, namely, the introduction of "M" for Miocene, "PL" for Pliocene, and "PT" for Pleistocene in place of the earlier "N" for Neogene zones. In an attempt to rationalize and harmonize the issues that have been debated of late in the literature, an updated taxonomy of these forms was suggested in the previous section of this chapter, and species-level biostratigraphic ranges are presented in Charts 6.1–6.3 (http://dx.doi.org/10.14324/111.9781910634257). In the following, we describe an inferred phylogenetic relationship between the Neogene forms, which is self-consistent and compatible with the taxonomic relationships defined in Section 6.2 above and informed by recent genetic studies of living forms. For simplicity, we follow Banner and Blow (1965) and choose to define Neogene planktonic zones (i.e., "N" stands for Neogene), rather than using the letter descriptions (M, PL, and PT) adopted by Olsson *et al.* (1999) and Pearson *et al.* (2006a,b) (see Fig. 6.6).

Many of the major morphological patterns of the Neogene planktonic foraminifera became established during the Eocene, and most of the Paleogene families survived the Oligocene–Miocene boundary. Among the trochospiral families that survived the minor extinction event at that boundary were the Tenuitellidae, the Globigerinitidae, the Globigerinidae, and the Globorotaliidae. The Aquitanian in the Early Miocene saw little development for these families, but the low diversity during this time was followed by a period of high diversification in the Burdigalian and Langhian. In fact, most Neogene families evolved during this period, and extinctions occurred only at genera or species level throughout the remainder of the Neogene.

Previously in Section 6.2, we defined three Neogene superfamilies. Members of these superfamilies flourished throughout the Neogene and their morphological features overlapped to some extent over time. They can be summarized as follows:

- The *microperforate trochospiral, smooth, or pustulose-walled, nonspinose* forms, which evolved from forms originating in the Cretaceous (Figs. 5.6 and 6.7), and are represented by member of the superfamily **Globigerinitoidea** and include:
  - the microperforate, pustulose, low trochospiral **tenuitellids**, which evolved from the globanomalinids in the Late Bartonian,
  - the microperforate, trochospiral **globigerinitids**, which evolved from the tenuitellids in the Late Priabonian,
  - the microperforate, trochospiral to streptospiral **globigerinatellids**, which evolved from the globigerinitids in the Burdigalian, and
  - the microperforate, pustulose, trochospiral **candeinids**, which evolved from the globigerinitids in the Tortonian.
- The *macroperforate trochospiral, planispiral, or streptospiral, smooth or punctate, spinose or nonspinose* forms, which are represented by members of the superfamily **Globigerinoidea** that evolved from the eoglobigerinids in the Early Eocene. They include:
  - the globular trochospiral **globigerinids**, which evolved from the eoglobigerinids in the Ypresian (see Figs. 5.7, 6.8, and 6.9),
  - the trochospiral **sphaeroidinellids**, coated by a cortical crust, which evolved from the globigerinids in the Burdigalian (see Figs. 6.8 and 6.10),
  - the planispiral **globigerinellids**, which evolved from the globigerinids in the Burdigalian (see Fig. 6.11),
  - the pseudoplanispiral to streptospiral **hastigerinids**, which evolved from the globigerinellids in the Burdigalian (see Fig. 6.11),
  - the streptospiral **orbulinids**, with the last globular chamber completely embracing the umbilical side, which evolved from the globigerinids in the Langhian (Figs. 6.8 and 6.12),
  - the trochospiral **neoacarininids**, which evolved from the globigerinids in the Pleistocene (see Fig. 6.8),

  - the small, extant **turborotalitids**, which evolved in the Lutetian from the globigerinids (see Figs. 5.8 and 6.8),
  - the compressed trochospiral **globorotaliids**, which evolved from the globigerinids in the Late Bartonian (see Figs. 5.7, 6.11, and 6.13–6.15),
  - the trochospiral **neogloboquadrinids**, with an antero-intraumbilical aperture, which evolved from the globorotaliids in the Tortonian (Figs. 6.11 and 6.16), and
  - the streptospiral **pulleniatinids**, which evolved from the neogloboquadrinids in the Messinian (Figs. 6.16).
- The *microperforate, pustulose, nonspinose* forms that survived the K-P event or evolved from those survivors (see Figs. 5.10 and 6.17). They are represented by members of the superfamily **Heterohelicoidea** and include
  - the triserial **guembelitriids**, which survived the Cretaceous–Paleocene extinction event,
  - the biserial **chiloguembelinids**, which evolved from the guembelitriids in the Danian, and
  - the enrolled, biserial **cassigerinellids**, which evolved from the chiloguembelinids in the Late Ypresian.

6.3.1 The microperforate trochospiral, smooth or pustulose-walled, nonspinose Neogene planktonic foraminifera

Among the Oligocene forms that passed into the Miocene were the microperforate, trochospiral modern *Tenuitella* (Plate 5.2, Figs. 20 and 21; Plate 6.1, Figs. 16–18), which had their roots at the Eocene–Oligocene boundary. *Tenuitella* evolved into the planispiral form *Berggrenia*, in the Middle Pliocene (N20b and N21a), and developed a bulla in the Pleistocene to Holocene with *Tenuitellita* (see Fig. 6.7).

The globigerinitids, which evolved from the tenuitellids in the Paleogene (see Fig. 5.6), are represented in the early Neogene by three forms that originated in the Oligocene, namely, the temperate form *Antarcticella*, which disappeared at the top of the Middle Miocene (N13), and the still extant *Tenuitellinata* (Plate 6.1, Figs. 8–13) and *Globigerinita* (Plate 6.1, Figs. 6–7, 15; Plate 6.11, Fig. 2). The latter gave rise to the streptospiral globigerinatellids, in which the primary chamber terminally embraces and covers the umbilicus in the Burdigalian (N6). *Globigerinatella* (Plate 6.11, Fig. 1) has areal and sutural cribrate primary apertures, while *Polyperibola* (Plate 6.1, Fig. 5), which has interiomarginal cribrate apertures, developed in the Tortonian (N16a). Both genera have short stratigraphic ranges; *Globigerinatella* disappeared in the Langhian (9a), while *Polyperibola* became extinct in the Tortonian (N16b). Quite independently, *Globigerinita* (Plate 6.1, Figs. 6, 7, 15; Plate 6.11, Fig. 2) evolved the extant *Candeina* (Plate 6.1, Figs. 1–4) in the Tortonian by closing the umbilicus and replacing the single primary aperture by many small, uniform sutural apertures.

6.3.2 The macroperforate trochospiral, planispiral or streptospiral, smooth or punctate, spinose or nonspinose Neogene planktonic foraminifera

The spinose planktonic foraminifera first evolved, as seen in Chapter 5, with *Eoglobigerina*, less than 100,000 years after the Late Cretaceous extinction event. This was followed by the extensive radiation of spinose forms, but all subsequent lineages of spinose planktonic foraminifera with bilamellar shells (Fig. 5.8) can be linked to this one common ancestor (Aurahs *et al.*, 2009). Many of the Aquitanian spinose globigerinids (83%) are pre-existing Paleogene forms, and indeed, they are still living in present day oceans (Fig. 6.8). The extant globular *Globoturborotalita* (Plate 6.4, Figs. 1–12), with closely appressed chambers, originated in the Ypresian and passed into the Neogene. However, others such as, *Dentoglobigerina* (Plate 5.5, Figs. 13–16; Plate 5.6, Figs. 1–5; Plate 6.2, Figs. 1–7; Plate 6.11, Fig. 11; Plate 6.12, Figs. 3 and 5A; Plate 6.14, Fig. 5B; Plate 6.15, Fig. 3), a form with a tooth-like, sub-triangular, symmetrical porticus projecting into the umbilicus (see Fig. 5.7) originated in the Ypresian (P9a) and passed through the Paleogene–Neogene boundary only to go extinct in the Pliocene (Zanclean, N20a).

Although the first *Globigerinoides* evolved in the Oligocene (P22b) from *Globigerina* (Plate 6.2, Figs. 8–16; Plate 6.15, Fig. 1A), by developing supplementary dorsal apertures, the radiation of species of this genus only occurred in the Early Miocene (Fig. 6.8). The polyphyletic genus *Globigerinoides* exhibits many phylogenetic lineages that seem to have evolved at different times from *Globigerina* during the Neogene (see Kennett and Srinivasan, 1983). Their diversity makes them useful biostratigraphic markers for this period (Keller, 1981). For instance, the base of the Burdigalian is recognized by the appearance of the cosmopolitan *Globigerinoides sacculifer* (Plate 6.3, Figs. 9 and 10; Plate 6.11, Fig. 6) and *Globigerinoides obliquus* (Plate 6.3, Figs. 1 and 2; Plate 6.11, Fig. 8) in the tropics and subtropics, while the appearance of *Globigerinoides bulloideus* marks the beginning of the Messinian in the tropics and subtropics, and that of *Globigerinoides emeisi* the start of the Zanclean in the tropics. *Globoturborotalita brazieri* (Plate 6.4, Figs. 1 and 2) marks the beginning of the Aquitanian in the tropics and subtropics, while *Globigerinoides parawoodi* is diagnostic in the temperate latitudes (see Chart 6.1 online and Fig. 6.6).

Takayanagi and Saito (1962) and Kennett and Srinivasan (1983) broadly classified the genus *Globigerinoides* into two groups based on the position of the primary aperture (Fig. 6.9). "Group A" exhibit an ultrastructure with spines and spine bases (e.g., *Globigerinoides primordius* (Plate 6.3, Figs. 3 and 4), *Gdes obliquus*), while "Group B" exhibit a cancellate structure (e.g., *Globigerinoides trilobus* (Plate 6.11, Fig. 10; Plate 6.13, Fig. 3; Plate 6.15, Fig. 8), *Gdes subquadratus*). Molecular biology based on SSU rDNA analysis has proved the close relationships between *Globigerina bulloides* (Plate 6.2, Fig. 9) *Globoturborotalita falconensis*, *Globigerinoides ruber* (Plate 6.3, Figs. 7 and 8)–*Gdes conglobatus* (Plate 6.2, Fig.

20; Plate 6.11, Fig. 4), and *Globigerinoides sacculifer* (Aurahs *et al.*, 2009; de Vargas *et al.*, 1997).

The Sphaeroidinellidae made their first appearance in the Burdigalian (N6) (Figs. 6.8 and 6.10). Members of this family had adult tests coated in a thick, smooth cortex of calcite; the cortex diminished or closed the perforations and greatly thickened the apertural lips. Tests which are broken show the surface of the younger whorls, below the cortex, to be punctuate and spinose. *Prosphaeroidinella disjuncta* (Plate 6.9, Fig. 14; Plate 6.10, Figs. 1–3), the stratigraphically earliest species of the lineage, evolved in the Burdigalian (N6) from a globigerinid form. Kennett and Srinivasan (1983) proposed *Globoturborotalita woodi* (Plate 6.4, Figs. 2–4) as a possible ancestor, while Chaisson and Leckie (1993) described it as gradational with *Globigerina druryi* (Plate 6.4, Figs. 7–8), from which it is distinguished by its distinctly thickened test wall. In *P. disjuncta*, the wall is coarsely cancellate and the cortex is weakly developed, covering only part of adult test, and the remainder shows its perforation pits and spine bases. *Prosphaeroidinella* evolved gradually into *Sphaeroidinellopsis* in the Burdigalian (Figs. 6.8 and 6.10) as the cortex expanded to cover the entire test; however, the primary aperture is still intraumbilical (e.g., *Sphaeroidinellopsis seminulina* (Plate 6.10, Figs. 6 and 7) and *Sphaeroidinellopsis kochi*). In the Zanclean, *Sphaeroidinellopsis paenedehiscens* (Plate 6.10, Fig. 8) evolved into *Sphaeroidinella* (Plate 6.10, Figs. 9 and 10), which possesses additional, dorsal, supplementary apertures. Only two species of *Sphaeroidinella* are extant, and they are found rarely in subsurface, tropical, and subtropical waters (see Chart 6.2 online).

The Paleogene *Globorotaloides suteri* (Plate 6.3, Figs. 18–20) survived into the Miocene (Early Langhian (N8a), see Chart 6.1 online) where it evolved into many species. In lateral outline, *Globorotaloides suteri* may vary from a somewhat compressed to a subglobular test, and umbilical bulla may be absent or extremely small in size in the immature stage. In the Langhian (N8), the compressed, trochospiral *Globorotaloides* developed radially elongate, often clavate chambers, *Clavatorella* (see Fig. 6.11, Plate 6.4, Figs. 17 and 18), which in turn developed planispiral forms with high, radially elongate chambers, *Protentella* (Kennett and Srinivasan, 1983). Aze *et al.* (2011) draw a phylogenetic chart with *Protentella* evolving directly from "*Globigerinella*" *obesa*. However, *Protentella* is closer in morphology to *Clavatorella* (Plate 6.4, Figs. 17 and 18), which is here considered as its direct ancestor.

In parallel, during the Late Burdigalian (N8), *Globorotaloides* gave rise to *Globigerinella* (Plate 6.4, Fig. 16), by developing a planispiral, biumbilicate test with subglobular chambers (Fig. 6.11). Later chambers become high and radially elongate in the Holocene, giving rise to *Bolliella*. It has been proposed that *Globigerinella* (e.g., *Glla aequilateralis*, Plate 6.4, Fig. 16) evolved into the extant mono-lamellar hastigerinids (Kennett and Srinivasan, 1983) by developing thick spines, triadiate in section, mainly at ends of the chambers (Fig. 6.11). However, the origin of *Hastigerina* is still unknown, and given the position of *H. pelagica* (Plate 6.8, Fig. 11) in SSU rDNA trees, Aurahs *et al.* (2009) speculated that this species might represent the latest colonization of the planktonic realm from a completely new group of benthic foraminifera. *Hastigerinopsis*, only known from the Late Burdigalian and the Holocene, has radially elongate chambers, later becoming elongate and occasionally divided distally, while the chambers in *Hastigerina*, which appeared in the Late Tortonian, are never radially elongate. *Orcadia* (Plate 6.8, Fig. 12), an extant trochospiral form with spines confined to distal parts, evolved from *Hastigerinopsis* in the Pleistocene (Fig. 6.11). All hastigerinids pass through a trochospiral stage with an extraumbilical–umbilical aperture, before becoming planispiral or streptospiral (Holmes, 1984).

In the Middle Miocene (Langhian, N8a), the orbulinid lineages made their first appearance (Fig. 6.8). They evolved gradually from *Globigerinoides bisphericus* as the rate of enlargement of the last chamber increased, until it covered the umbilicus with the new streptospirality (see Fig. 6.12). *Globigerinoides* species (e.g., *Gdes trilobus*, Plate 6.11, Fig. 10) gave rise to *Gdes bisphericus* by increasing the embrace of the earlier test to just 20% (Jenkins *et al.*, 1981). It has only two apertures in the last intercameral suture. *Praeorbulina sicana* (Plate 6.9, Fig. 4; Plate 6.11, Fig. 9) evolved in the earliest Langhian (N8a) from *Gdes bisphericus* by increasing the embrace of the last chamber to 30–40% of the earlier test, and by the development of three to four apertures in the final intercameral suture. According to Pearson *et al.* (1997), *Praeorbulina sicana* fully intergrades with *Gdes bisphericus*, and via that form with *Gdes trilobus*. These three morphotypes apparently constitute a single very variable population. The appearance of *P. sicana* provides a useful biohorizon and it is used to denote the base of the Langhian (N8) (see Figs. 6.6 and 6.12). This level correlates globally at 15.97 Ma (Gradstein *et al.*, 2004).

Also at this stage, *P. sicana* gave rise to two distinct lineages which led to forms that are placed in the extant genus *Orbulina* (Plate 6.9, Figs. 1–3), and which are characterized by having a spherical shell with a small, *Globigerinoides*-like trochospiral test attached to the inner wall of the sphere (Vilks and Walker, 1974). The taxonomy of the orbulinid species was reviewed by Kennett and Srinivasan (1983) and Bolli and Saunders (1985). Genetic data reveal the presence of three species of *Orbulina*, namely, *O. universa* (Plate 6.9, Fig. 1; Plate 6.13, Fig. 4A), *O. suturalis* (Plate 6.9, Fig. 2), and *O. bilobata* (Plate 6.9, Fig. 3) (De Vargas *et al.*, 1999). The morphological developments, such as the evolution of supplementary apertures along the spiral suture and the modifications of the last chamber, are largely congruent with the SSU rDNA phylogenies (Aurahs *et al.*, 2009).

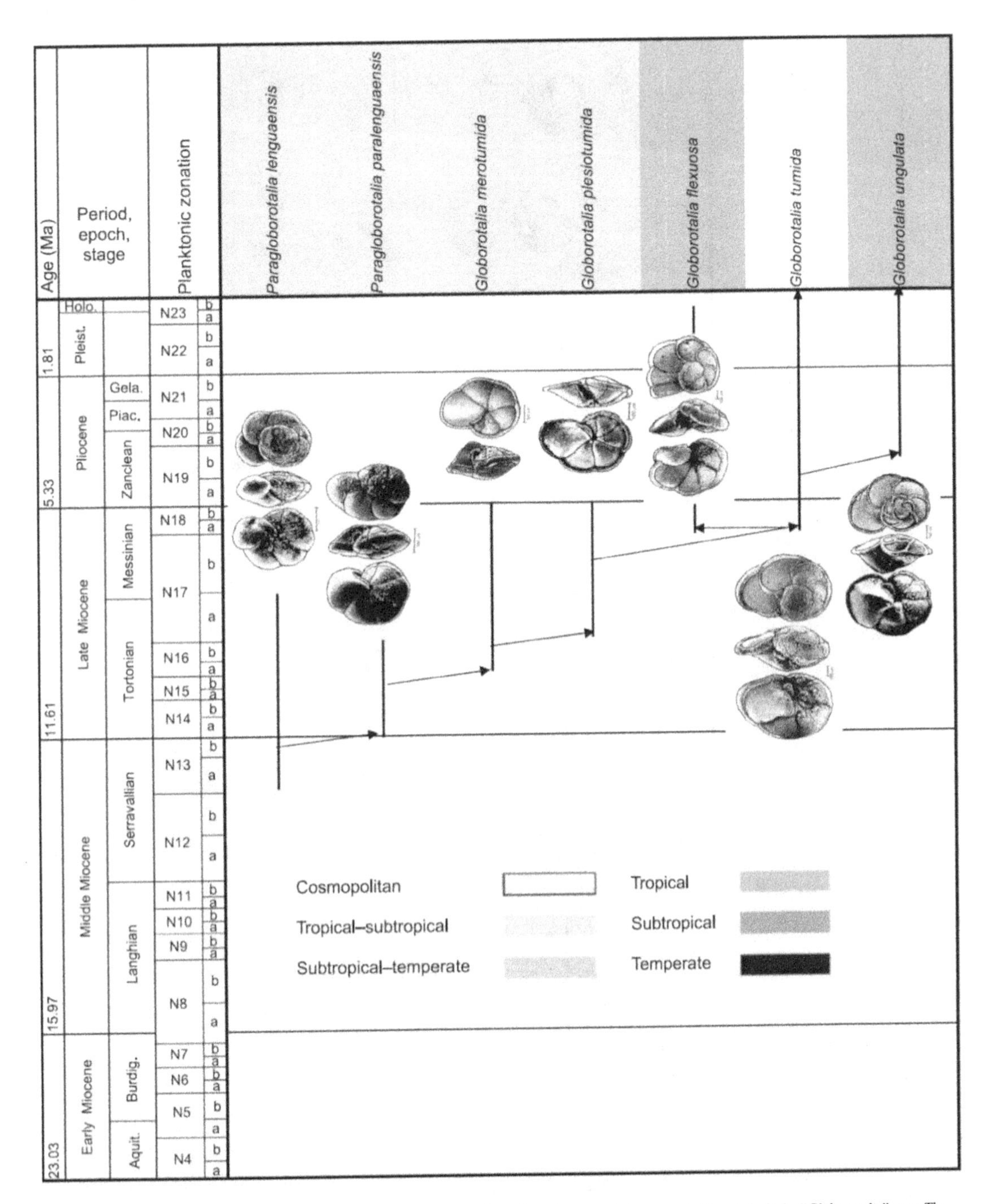

Figure 6.15. The phylogenetic evolution of the globorotaliids which evolved from *Paragloborotalia lenguaensis*. They include the "*Globorotalia*" type. The shading behind the genera names indicates their latitudinal range as defined in the embedded legend. Images from original sources.

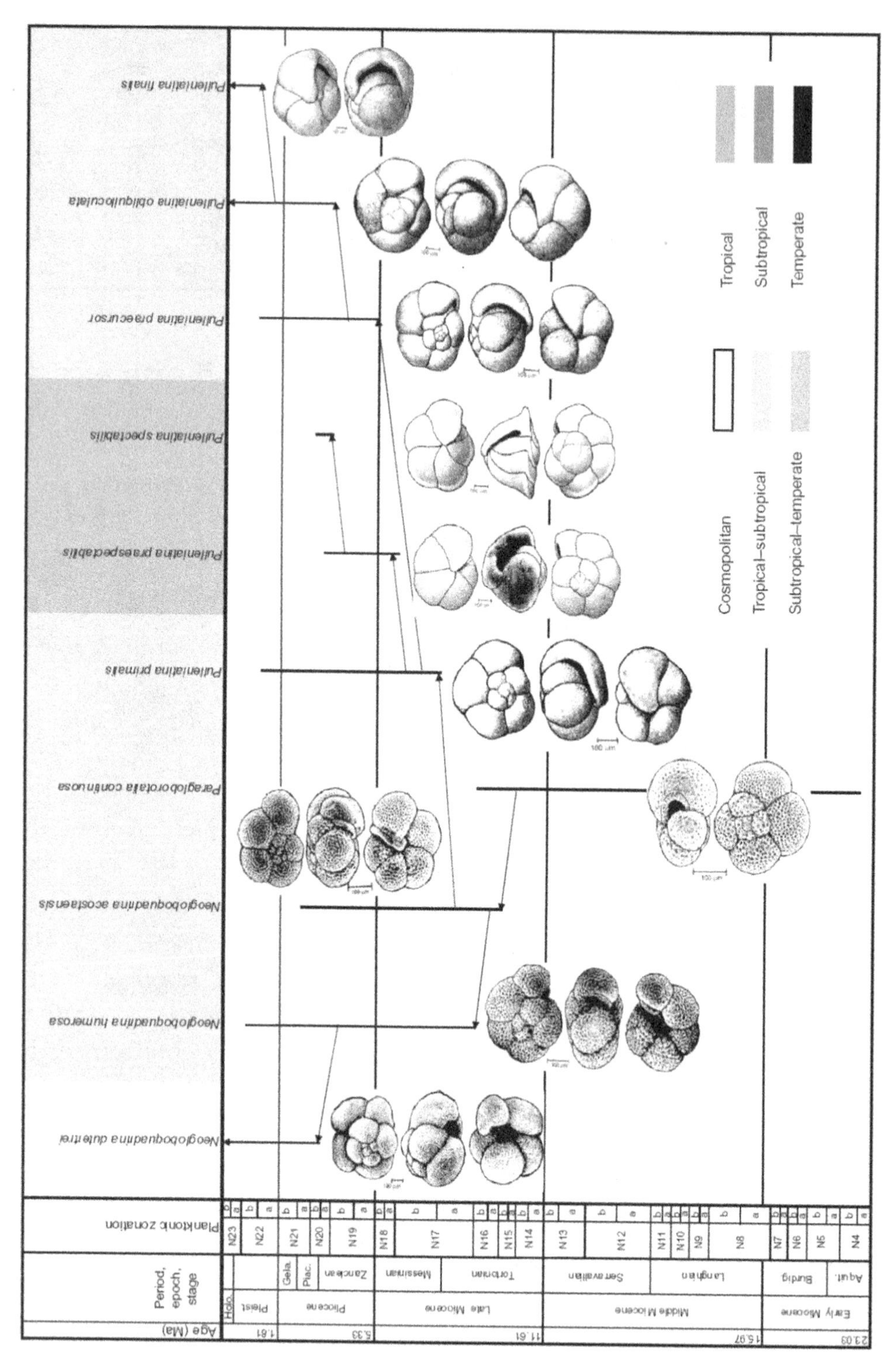

Figure 6.16. The phylogenetic evolution of *Neogloboquadrina* and *Pulleniatina* species which evolved from *Paragloborotalia continuosa*. The shading behind the genera names indicates their latitudinal range as defined in the embedded legend. Images from original sources.

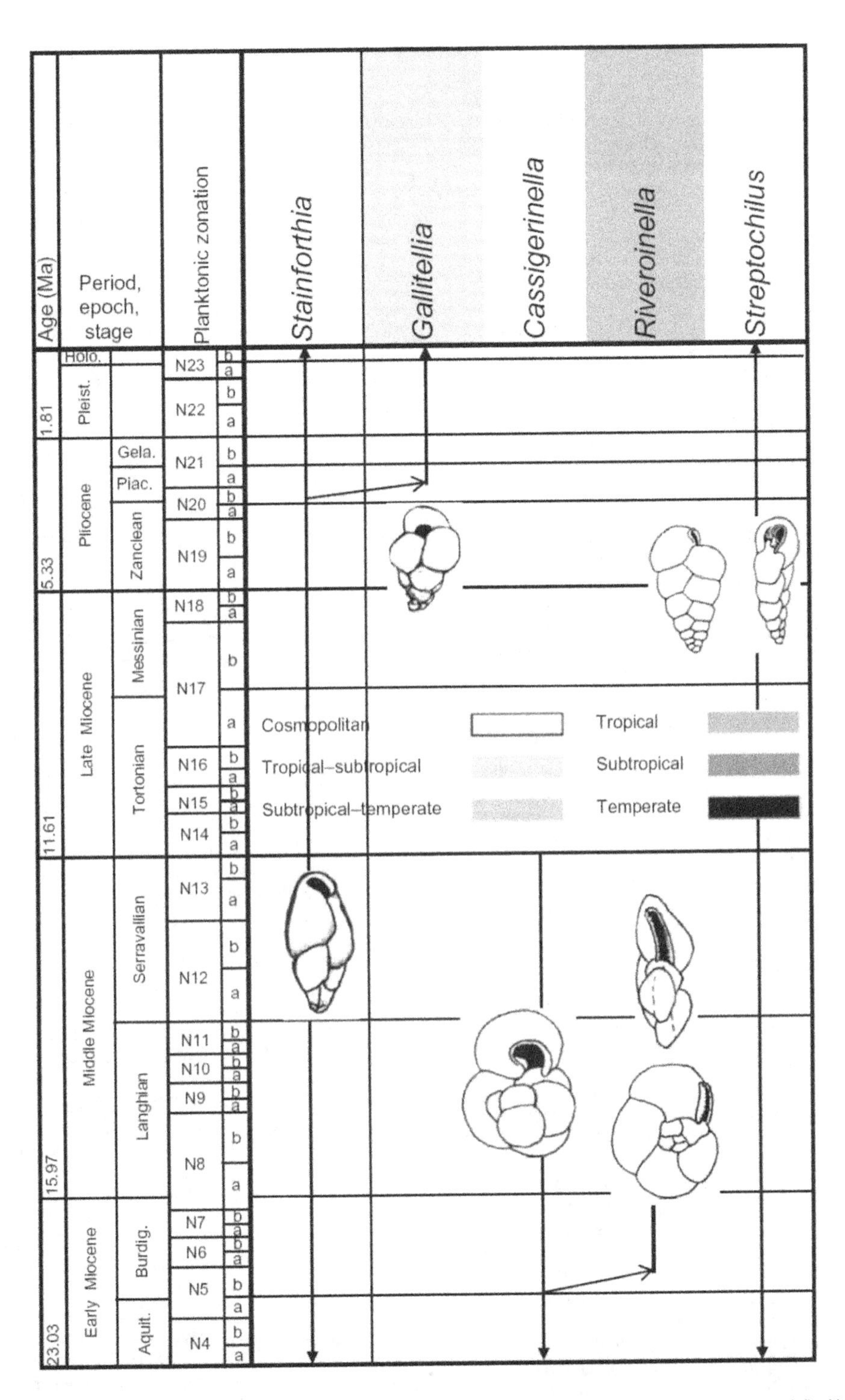

Figure 6.17. The phylogenetic evolution of the heterohelicids. The shading behind the genera names indicates their latitudinal range as defined in the embedded legend.

The first orbulinid lineage evolved directly from *Praeorbulina sicana*, in the Early Langhian, with the development of *Praeorbulina transitoria*. The latter has penultimate and final chambers of nearly equal size. The penultimate chamber embraces about 30% of earlier test. Many small, slit-like apertures were developed in the last and penultimate intercameral apertures. *P. transitoria* was replaced by *Orbulina bilobata* in the Middle Langhian (N9a), when the early coiled test became completely enclosed by the penultimate chamber. The penultimate and last chambers comprise all that is visible of the test in *O. bilobata*.

The second orbulinid lineage evolved from *P. sicana* in the Early Langhian (N8b) to give rise to the *Praeorbulina glomerosa* group in the Langhian, which in turn evolved into *Orbulina*. The *P. glomerosa* group originated from *Praeorbulina curva*. The latter evolved rapidly in the Early to Middle Langhian (N8b and N9a) into *Praeorbulina glomerosa* (Plate 6.9, Fig. 6), by developing so that the last chamber embraces 70–80% of the earlier test and by developing many (more than eight) small, slit-like apertures in the last intercameral suture. *Praeorbulina glomerosa* gave rise to *P. circularis* (Plate 6.9, Fig. 7), in which the last chamber embraces 70–90% of the earlier test, and by developing more numerous sub-circular apertures in last intercameral suture. *Praeorbulina circularis* evolved directly into *Orbulina suturalis* (Plate 6.9, Fig. 2), in which the coiling becomes intensely streptospiral, with the last, globular chamber completely embracing the coiled early test, which is still visible in *Orbulina universa*. The wall of last chamber in *O. universa* is penetrated by both perforations and larger circular pore-like apertures, which are present both in the last suture and in the area of the last chamber. The coiled early test is completely enclosed by the spherical final chamber (which may even be empty). According to Aurahs *et al.* (2009), the sister clade of *Globigerinoides conglobatus–ruber* comprises *O. universa* and *Gdes sacculifer*, which implies a common origin of these four morphospecies. In addition, strontium isotope evidence points to a shared common ancestor in the early Miocene for the two extant species of spinose planktonic foraminifera, *Globigerinoides trilobus* and *Orbulina universa* (Pearson *et al.*, 1997). Both *Orbulina universa* and *O. suturalis* are extant and cosmopolitan.

Quite independently, in the Late Burdigalian, *Globigerinoides* gave rise to a small form, *Globigerinoita*, with three chambers in the last whorl, and with one or two bullae developed over the apertures in the adult stage (Fig. 6.8). *Globigerinoita* remained monotypic and died out in the Tortonian (N15). Another monotypic genus, *Neoacarinina* (Plate 6.8, Figs. 13–15) had a short range in the Pleistocene (N23a). It evolved from *Catapsydrax* (Plate 6.1, Figs. 19 and 20; Plate 6.15, Fig. 11) as the surface of its test became covered by clumps of spinules. In the Tortonian, *Globigerina* gave rise to an extant genus, *Beella*, as the chambers of the last whorl became radially elongate (Fig. 6.8). *Beella* has a short range and differs from *Bolliella* (Plate 6.4, Figs. 13–15) in possessing a trochospiral test with a tendency toward streptospiral coiling in adult test, suggesting an affinity with *Hastigerinella*.

The macroperforate trochospiral globorotaliids made their first appearance in the Eocene (Fig. 5.9); however, it was not till the Neogene that species radiation occurred, when *Paragloborotalia*, which first evolved in the Eocene from the globigerinids, *Globorotaloides*, gave rise to acutely angled or peripherally carinate (keeled) species of *Globorotalia* (Fig. 6.11). They have smooth tests that may possess scattered, small tubercles but which are not cancellate or punctuate, unlike the unkeeled *Paragloborotalia*, which never have smooth adult tests (except when rarely encrusted). The flattened *Globorotalia* species, with acute-edged profiles, tend to be larger in size than the globular globigerinids but are less abundant in Miocene assemblages. However, they tend to have shorter biostratigraphic ranges as they evolved quickly through the Neogene and so form biostratigraphically important lineages.

Globorotalia lineages repeatedly evolved from species of *Paragloborotalia*. The base of the Aquitanian is defined here by the first appearance of *Paragloborotalia kugleri* (see Chart 6.3 online). The separate lineages have been given different subgeneric names (see Kennett and Srinivasan, 1983), including:

- The "*Menardella*" lineage (*Globorotalia praescitula – archaeomenardii – praemenardii–menardii* lineage). Species of this lineage are characterized by lenticular tests and a prominent keel. This lineage evolved from *Globorotalia praescitula* (Plate 6.7, Fig. 1) in the Early Miocene (N6) (see Fig. 6.13);
- The "*Hirsutella*" lineage (*Globorotalia margaritae–praehirsuta–hirsuta* lineage). This lineage includes sharp-edged to keeled and inflated to compressed forms. Spinosity is widespread ventrally and dorsally. Dorsal chambers are much broader than high (reniform). The evolution of *hirsuta* (Plate 6.6, Figs. 14–16) from *margaritae* (Plate 6.6, Figs. 17 and 18; Plate 6.13, Fig. 5B; Plate 6.14, Fig. 3B; Plate 6.15, Fig. 5) via *praehirsuta* (Plate 6.6, Figs. 19 and 20) is accompanied by a gradual tightening of the coil, which is eventually shown in the development of a small but open umbilical depression in *hirsuta* (Fig. 6.13);
- The "*Globoconella*" lineage with a high-arched aperture (*Globorotalia praescitula–miozea–conoidea– conomiozea–inflata* lineage). Dorsal chambers are much higher than broad. They evolved from *Globorotalia praescitula* as the test enlarged and inflated and showed increasing development of secondary thickening and rugosity of the umbilical side as in *Globorotalia inflata* (Plate 6.5, Figs. 19 and 20; Fig. 6.13);
- The "*Fohsella*" lineage (*Globorotalia praefohsi–fohsi* lineage, see Fig. 6.14) is characterized by an increase in the size of specimens and the development of a keeled periphery from a nonkeeled *Paragloborotalia peripheroacuta* to the partly keeled *Gt praefohsi* (Plate 6.6, Figs. 6 and 7) and the fully keeled *Globorotalia fohsi*

(Plate 6.6, Figs. 8 and 9; Plate 6.13, Fig. 6A). Spinosity is widespread over ventral side (see Fig. 6.14);

- The "*Globorotalia*" lineage (*Paragloborotalia lenguaensis-P. paralenguaensis–Globorotalia merotumida–tumida–ungulata*, Fig. 6.15) that evolved from nonkeeled forms to distinctly keeled ones, with biconvex tests and tending to be more convex ventrally than dorsally in the most advanced form *Globorotalia ungulata* (Plate 6.7, Figs. 9–11);
- The *Truncorotalia* lineage that includes nonkeeled and keeled planoconvex species (*Truncorotalia crassula–crassaformis–tosaensis*, Fig. 6.13). *Truncorotalia* species (Plate 6.7, Figs. 16–20) evolved gradually from biconvex *Globorotalia cibaoensis* into conical and strongly conical ventrally forms in the Pliocene to Holocene.

In addition to these *Paragloborotalia*-derived globorotaliids, *Paragloborotalia continuosa* gave rise in the Tortonian to the *Neogloboquadrina* lineage (see Figs. 6.11 and 6.16). The relationship between *Paragloborotalia incompta* and the neogloboquadrinids is confirmed by SSU rDNA trees (see Aurahs *et al.* 2009). According to Kennett and Srinivasan (1983), this evolution development involved a progressive change from an interiomargina-extraumbilical aperture in *P. continuosa* to an antero-intraumbilical one in *Neogloboquadrina* (Plate 6.7, Figs. 12–15; Plate 6.11, Fig. 14). However, different genetic studies have provided new information about the evolutionary history of the neogloboquadrinids (Aze *et al.*, 2011). For example, prior to genetic studies, it had been thought that the right-coiling *Neogloboquadrina pachyderma* and the left coiling *Paragloborotalia incompta* were one species, with coiling direction being an ecophenotypic response to temperature (Ericson, 1959). Genetic studies, however, have revealed a substantial divergence between the two forms that may have occurred during the late Miocene approximately 10 Ma ago. Their fluctuating abundance down particular ocean sediment cores is now thought to reflect fluctuations in the location of the polar front that serves as the range boundary between the two species (Darling *et al.*, 2004, 2006).

The pulleniatinellids evolved from *Neogloboquadrina* in the Messinian (see Fig. 6.16) by the initial coil, like the adult ones, becoming streptospiral. *Pulleniatina primalis* (Plate 6.9, Figs. 11–12; Plate 6.14, Fig. 2) evolved from *Neogloboquadrina acostaensis* through a series of morphological intermediates present as a subpopulation within the *N. acostaensis* population (Belyea and Thunell, 1984). The chambers increasingly embrace ventrally so that the umbilicus becomes closed and the aperture becomes wholly extraumbilical. In the morphologically most advanced forms (see Fig. 6.16), the chambers begin to embrace dorsally so that the last whorl covers much of the preceding whorls. The late chambers encroach onto the dorsal surface of earlier chambers so that the dorsal side begins to become partly involute, as in *Pulleniatina obliquiloculata* (Plate 6.9, Figs. 8–10), and the adult test becomes pseudoplanipiral, as in *P. finalis*.

6.3.3 The microperforate, pustulose Neogene trochospiral globoconusids, and heterohelicids

Among the Paleogene forms that continued into the Neogene were two genera from the heterohelicoids (see Fig. 6.17), namely, *Cassigerinella* (Plate 5.10, Figs. 3–6; Plate 6.10, Fig. 13), which died out in the Middle Miocene (top of the Serravallian, N13), and the still extant *Streptochilus* (Plate 5.10, Fig. 19; Plate 6.10, Figs. 16–20). However, of late, the relationship between the Paleocene *Streptochilus* species and those of the Neogene has been questioned by molecular and genetic evidence. As mentioned in Chapters 2 and 5, recent geochemical evidence has confirmed that the heterohelicoids groups are not monophyletic, descending from a single Jurassic–Cretaceous ancestor, but that they must have been polyphyletic, evolving many times throughout the geological record from benthic foraminifera (Darling *et al.*, 2009). Indeed, through analysis of the SSU-rRNA gene, Darling *et al.* (2009) demonstrated that the extant biserial planktonic *Streptochilus globigerus* belongs to the same biological species as the benthic *Bolivina variabilis*. In the light of these recent findings, it could be argued that the Neogene triserial forms should be given a different genus name than *Streptochilus*. However, for practicality, we keep them assigned to the same genus, *Streptochilus*, because of its well-established usage. Furthermore, any revision currently proposed may be superseded in the near future, as it is probable that further genetic research will indicate that more major and systemic taxonomic revisions of these forms are required.

The triserial guembelitriids seemingly became extinct in the Late Oligocene, only to reappear in the Pliocene. They are represented in the Pliocene and Holocene by small triserial forms, which, following the indications of polyphyletic behavior in this order, might not be related to the Cretaceous–Paleogene triserial species of *Gallitellia*. *G. vivans* (Plate 6.10, Figs. 21–23) is the only triserial coiling species among modern planktonic foraminifera. In molecular analyses, it evolved from the benthic rotaliids *Stainforthia* and *Virgulinella* (Ujiié *et al.*, 2008). Both of these benthic genera resemble *Gallitellia* in general morphological appearance, having elongate triserial tests at least in their early ontogenic stages. Ujiié *et al.* (2008), by analyzing sequences of the SSU rDNA, estimated that *Gallitellia vivans* diverged from the *Stainforthia* lineage in the Early Miocene, around 18 Ma and that this divergence is much earlier than the fossil record of *Gallitellia vivans* (see Chart 6.2 online). According to these authors, this overestimate may be due to the fact that the transitional lineages between *Gallitellia vivans* and *Stainforthia* have not been found. The *Gallitellia vivans* lineage might have acquired a planktonic mode of life by multiple transitions from the benthic mode to the planktonic mode from the Miocene to Pliocene. Despite therefore being genetically unrelated to the Paleogene guembelitriids, these Neogene forms are included in the Guembelitriidae as they are isomorphic.

6.4 Paleogeography and paleoecology of the Neogene planktonic foraminifera

Unlike the Eocene–Oligocene boundary, the end of the Oligocene witnessed relatively few planktonic foraminiferal extinctions. Worldwide, 27% of planktonic foraminifera when extinct (Fig. 5.18) compare to just 13% of the shallow water reef-forming larger benthic foraminifera (see BouDagher-Fadel, 2008). Most of the planktonic foraminifera extinctions occurred before the end of the Oligocene at the P21/P22 boundary (see Charts 5.1 and 5.2 online and Chapter 5). As explained in Chapter 5, the Oligocene–Miocene stratigraphic boundary is probably mainly associated with plate tectonic events, and/or gradual changes in climate (Berggren and Prothero, 1992), and because the surviving globigerinids and globigerinitids were already adapted to cooler environments (see Fig. 6.18) from the earlier attrition of warm-climate foraminifera, they passed largely unaffected into the Miocene. Most of the species which became extinct within the Late Oligocene were temperate surface and subsurface dwelling forms (e.g., *Globigerinita boweni* (Plate 5.3, Fig. 1), *Tenuitella neoclemenciae*, *T. munda*, *T. gemma*). The only important Paleogene genus to be eliminated during this transition was *Subbotina*, when the last species *Subbotina cryptomphala* died out.

During the Early Miocene, modern patterns of atmospheric and ocean circulation began to be established. The sea-way linking the proto-Mediterranean to the Indian Ocean was still open but was narrower than before. However, the isolation of the Antarctica from Australia and South America was more pronounced and led to the establishment of the circum-polar ocean current. This significantly reduced the mixing of warmer tropical water and cold polar water, thereby leading to the buildup of the Antarctic ice cap. Additionally, the Tibetan platform uplift (which began around the Eocene–Oligocene boundary, ~34 Ma) continued, and the Red Sea rifting accelerated (Aitchison *et al.*, 2007; Thomas *et al.*, 2006). It has been suggested by Jakobsson *et al.* (2007) that these tectonic changes affected oceanic circulation patterns and caused a flow reversal through the Panama seaway sometimes between the Late Oligocene and Early Miocene.

The global thermohaline circulations in Oligocene and Miocene oceanic models are significantly different from the present day pattern. In particular, in the Oligocene, the salinity contrast between the Atlantic and Pacific oceans was reduced because of water mass exchange through the low-latitude connections between the two oceans (von der Heydt and Dijkstra, 2006). Although that geochemical proxies suggest that the Drake Passage between South America and Antarctica was open earlier, in the Late Oligocene (32.8 Ma), to intermediate and deep-water circulation (Latimer and Filippelli, 2002; Lawver *et al.* 1992), the continued opening of this gateway eventually led to the thermal isolation of Antarctica in the Early Miocene and the creation of the clockwise strong Antarctic circumpolar current (Smith and Pickering, 2003).

The tectonic and climate changes during the Early Miocene inevitably influenced the biotic distributions within the oceans. The planktonic foraminifera, which colonized the oceans during the Early Miocene, were distributed mainly in the tropical–subtropical (58%) and tropical–temperate (7%) realms. However, some were confined just to the tropical (4%), subtropical (1%), or temperate waters (7%). There are also many cosmopolitan forms (23%) (see Charts 6.1–6.3 online). In the discussion below, we focus on the latitudinal ranges of planktonic foraminifera and discuss assemblages either considered cosmopolitan, or those limited to the tropical–subtropical, or temperate regions. Although the extinctions (Fig. 6.19) throughout the Neogene were not as pronounced as in other periods, each of these regions had its own patterns of evolutionary lineages and extinctions, and throughout the Neogene, the highest extinction rates we in tropical and subtropical realms.

The reef expansion (Mutti *et al.*, 2011) and the 26% increase in the diversity of the larger benthic foraminifera (BouDagher-Fadel, 2008) point to an increase in temperature during the Aquitanian (see Fig. 6.18). However, no new planktonic foraminiferal genera appeared at this time (see Fig. 6.20), but rather the preexisting Oligocene genera evolved into new species (40% of the Aquitanian planktonic foraminifera were new species; see Fig. 6.21) out of which 70% thrived in the subtropics and tropics, 22% were cosmopolitan, and just 8% were restricted to the temperate realm. The first appearance of *Paragloborotalia kugleri* is the main planktonic foraminifera bioevent at this stage.

The Aquitanian forms were mainly small trochospiral globigerinids, which are inferred to have included both species that possessed dinoflagellate photosymbionts and those that did not; the former were restricted to the photic zone of the water column, while the latter could live in deeper water (down to depths of ~400 m). By analogy with living forms, it seems that planktonic foraminifera that have symbionts may nevertheless also be carnivorous (e.g., feeding on calanoid cope-pods), although this is not universally the case. The globigerinids (e.g., *Globigerinoides primordius*, Plate 6.3, Figs. 3 and 4) occurred mainly in the tropical to subtropical seas, but some preferred living in temperate waters (e.g., *Globigerina euapertura*). Others (e.g., *Dentoglobigerina venezuelana*, Plate 6.2, Fig. 7) are non-symbiotic forms and show ontogenetic variations in their depths habitat, as they appeared to have shifted their depth of calcification from surface waters to deep-dwelling thermocline during the Early Miocene (Spezzaferri and Pearson, 2009; Stewart *et al.*, 2012). *Catapsydrax* species are also found in the tropics and subtropics. Their fossilized forms have the most enriched oxygen isotope values of those currently studied, and they are inferred to have been the deepest-dwelling of all the planktonic foraminifera (Spezzaferri and Pearson, 2009). They deposited calcitic crusts and may have lived within and below the thermocline (Boersma and Premoli Silva, 1991).

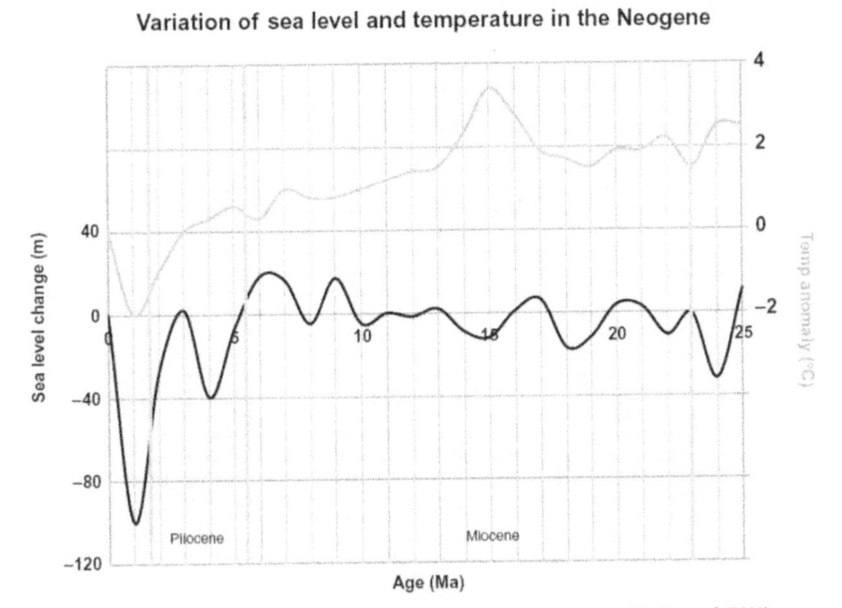

Figure 6.18. The variation in sea level and temperature during the Neogene based on Miller *et al.* (2011) and Zachos *et al.* (2001).

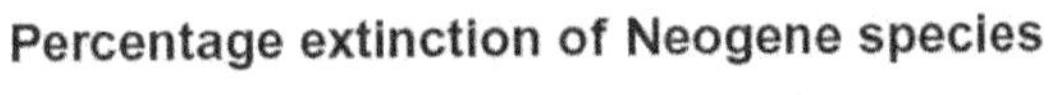

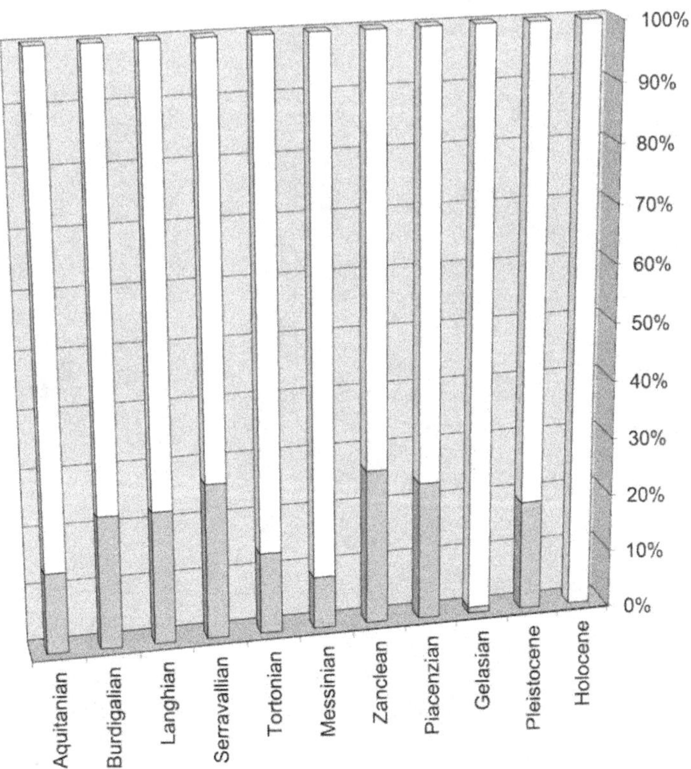

Figure 6.19. The percentage of planktonic foraminifera extinctions during the Neogene.

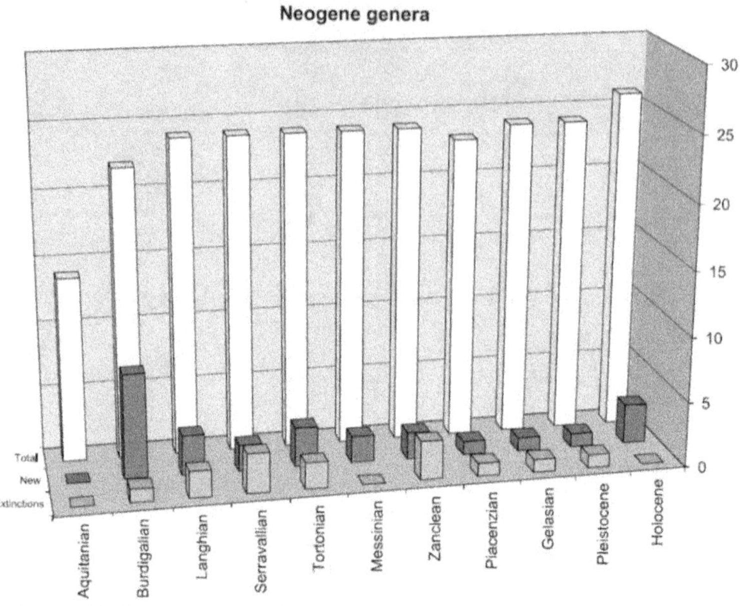

Figure 6.20. The total number of genera, extinctions, and new appearances of planktonic foraminifera in each stage of the Neogene. The extinctions are defined relative to the end of each stage and the appearances with the beginning of the stage.

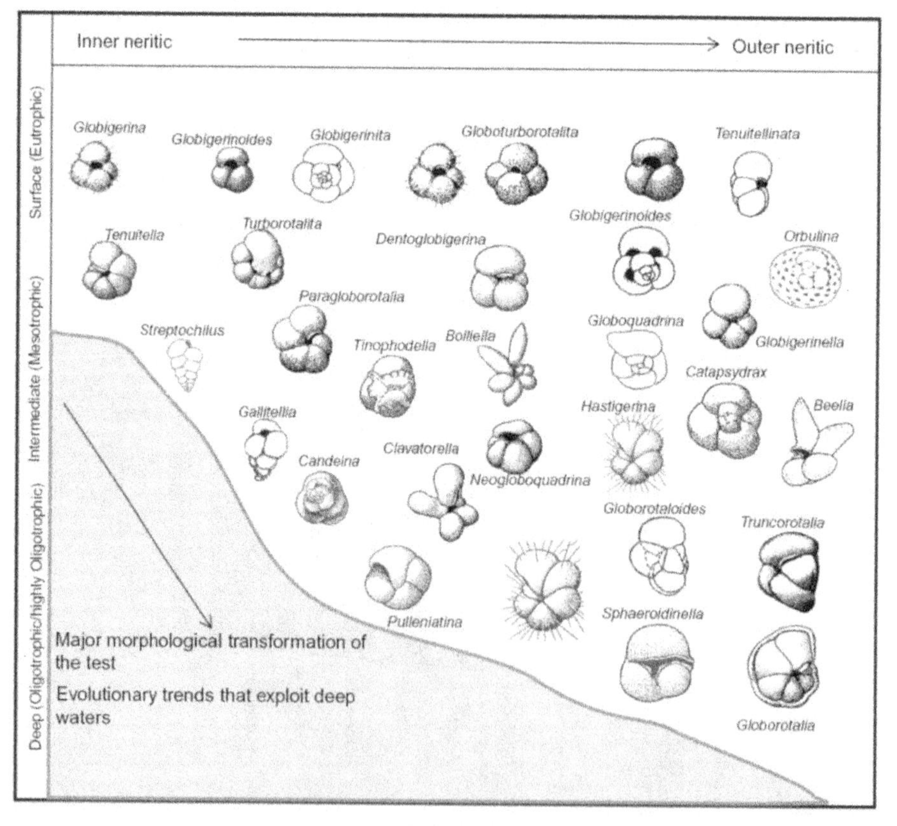

Figure 6.21. The distribution of the Neogene planktonic foraminiferal in the neritic environment.

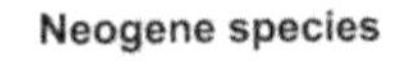

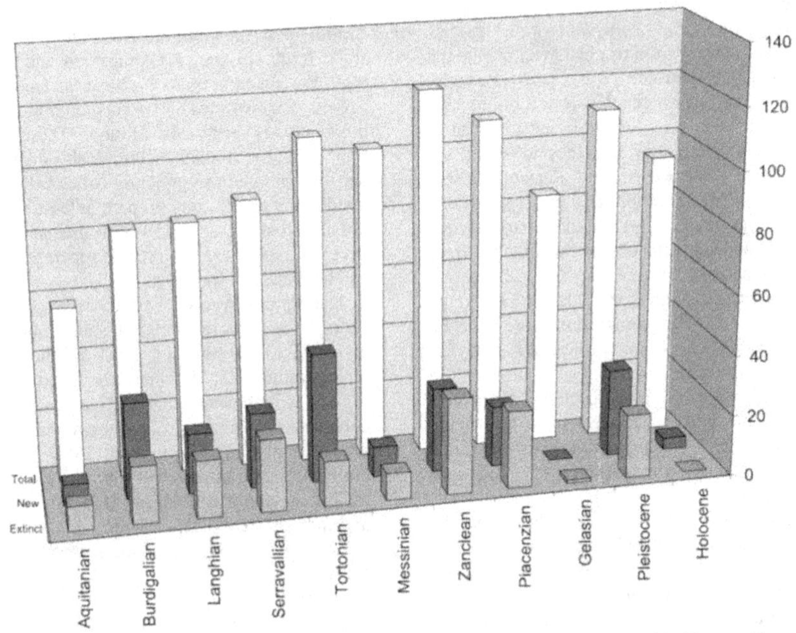

Figure 6.22. The total number of species, extinctions, and new appearances of planktonic foraminifera in each stage of the Neogene. The extinctions coincide with the end of each stage and the appearances with the beginning of the stage.

Neogene percentage extinctions

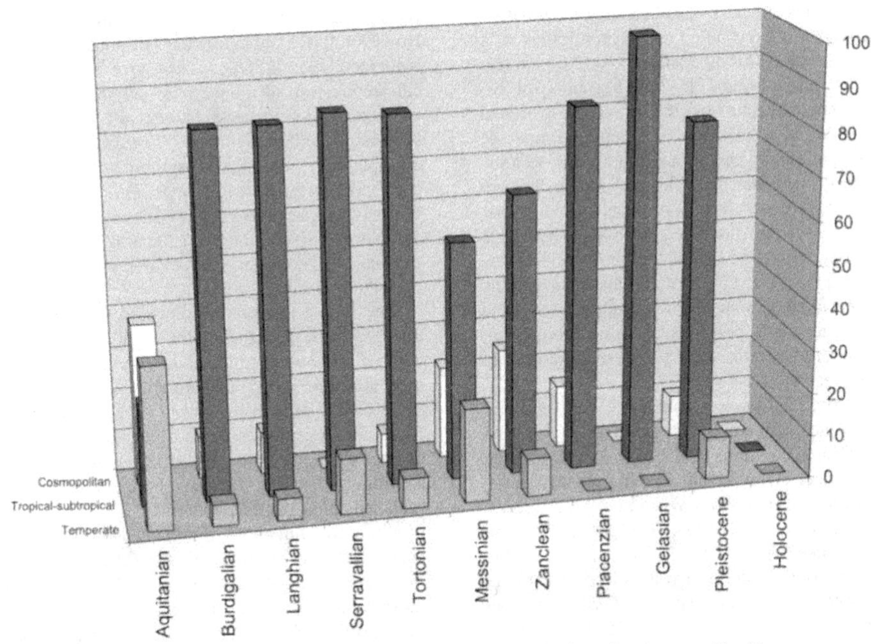

Figure 6.23. The percentage of planktonic foraminifera species extinctions at each stage in the three climatic zones of the. Neogene.

In the Aquitanian, second in abundance to the globigerinids are the globorotaliids (making up 17% of recorded species). These latter are still represented by the non-keeled *Paragloborotalia*. *Paragloborotalia* species are wide spread latitudinally from tropical to temperate but are most common in subtropical seas (e.g., *Paragloborotalia bella, P. incompta*, see Chart 6.3 online and Fig. 6.23). Species such as *P. kugleri* have the most negative $\delta^{18}O$ values, suggesting that they inhabited the surface mixed layer (Spezzaferri and Pearson, 2009). Globigerinitids make up 12% of the Aquitanian planktonic foraminifera. These species are highly eurythermal, thriving in surface and subsurface uppermost waters from the tropical to temperate seas (see Fig. 6.21). Some are restricted to the temperate subpolar seas (e.g., *Antarcticella antarctica*). The globigerinitids are believed to ingest fine vegetable debris, diatoms, and chrysophytes, but none is known to be carnivorous.

Toward the end of the Aquitanian, global extinctions were low. Planktonic genera survived the Aquitanian–Burdigalian boundary and only 14% of the species became extinct. They were mainly cosmopolitan (38%) and from temperate seas (38%).

The Burdigalian was an important stage for the evolution of the Neogene planktonic foraminifera. During this time, a significant climatic change occurred from a cooling trend to a warming one, which led into the MMCO. However, the first 2 Ma of the Burdigalian are characterized by a drop in the temperature resulting from tectonic and paleoceanographic conditions, involving the closure of the Tethyan Seaway link to the Indo-Pacific Ocean and the advancing of the Antarctic glaciation (see Fig. 6.18). This was followed, however, by a climatic amelioration toward subtropical conditions around 18 Ma heralding the MMCO. This warming led to the evolution and diversification of a large number of species. This diversification occurred in two phases: the first phase around 18 Ma (beginning of N6 planktonic foraminiferal zone) and the second phase around 17.2 Ma (beginning of N7 planktonic foraminiferal zone, see Charts 6.1–6.3 online). At this time, a number of tectonic events occurred, including the closure of the Eastern Mediterranean. Approximately 36% of the Neogene genera, represented by 39% of the species, made their first appearance during this period (the highest proportion of new forms seen in the Neogene, Fig. 6.22). Of these, 88% of the new species occurred in tropical to subtropical seas, and many of these genera and species have persisted throughout the Neogene to form the bulk of the planktonic foraminifera in the modern ocean.

In the Burdigalian, planktonic foraminiferal assemblages were dominated by globular globigerind species (see Chart 6.1 online and Fig. 6.2). The mixed layer, shallow-water fauna included predominantly small, microperforate species *Globigerina*, and *Globigerinoides*, as well as *Globigerinita* (Plate 6.11, Fig. 2) and *Tenuitellinata* (Plate 6.1, Figs. 8–13). The latter thrived in surface and subsurface uppermost waters from the tropics to the temperate seas. Species such as *Globigerinoides ruber* (Plate 6.3, Figs. 7 and 8; Plate 6.15, Fig. 4) lived permanently in the shallow waters of the tropics and subtropics, while others lived in intermediate waters between 50 and 100 m, at the base of the mixed layer and upper thermocline (Majewski, 2003; Pearson and Wade, 2009). These waters were occupied in the Burdigalian by heavily ornamented species of *Paragloborotalia, Catapsydrax, Dentoglobigerina*, and *Tenuitella*, as well as some large species of *Globigerina* and *Globigerinoides* (e.g., the cosmopolitan, *Globigerina bulloides*, Plate 6.2, Fig. 9). Rare species of *Turborotalita* (e.g., *Turborotalita quinqueloba*) were also found. These forms have a trans-global, tropical–subtropical distribution. They carried endophotosymbionts (often chrysophytes) and lived in the photic zone for part, at least, of their lives; it is possible that the adults lived at greater depths and there develop thicker calcitic crusts to increase the smoothness and heaviness of their tests.

During this warmer period, new species of unkeeled *Paragloborotalia* acquired a subangular periphery that give rise to the ancestor stock of the Neogene keeled globorotaliid (e.g., *Globorotalia praescitula*, Plate 6.7, Fig. 1), which were able to live deeper into the water columns. This taxon originally evolved in the temperate seas (Kennett and Srinivasan, 1983), but subsequently migrated to the tropics and subtropics to give rise to two important Neogene lineages (see Fig. 6.13), the "*Menardella*" lineage of the tropics and subtropics and the "*Globoconella*" lineage of the subtropical and temperate areas.

In the Burdigalian (N6, ~17.3 Ma), species of *Globorotalia* were joined by the first sphaeroidinelloinids, which throughout their evolution were restricted to the tropical and warm-subtropical waters. They lived, at least in the adult stage, in deep water exceeding 100 m (Murray, 1991). The hastigerinids made their first appearance in the subtropical and temperate areas toward the end of the Burdigalian, inhabiting deep or intermediate water layers. However, it was not before the Tortonian that hastigerinids colonized lower latitudes. They have been found to have a calcareous test surrounded by a bubble-like capsule of vacuolated cytoplasm, a feature not known to occur in species of other families. *Hastigerina* has no algal, dinoflagellate, or diatom symbionts, although it may contain *Pyrocystis robusta* or *P. notiluca*, and seems to be exclusively carnivorous using massive spines supporting cytoplasm to form rhizopoda that drag prey to the digestive capsule. Extant forms have a synodic lunar periodic reproduction cycle.

The Late Burdigalian coincided with a major global transgressions and the major eruption of the Columbia River Flood basalts and ended with a significant sea level drop accompanied by a sharp rise in the temperature (see Fig. 6.18). However, only 12% of the planktonic foraminifera genera and 23% of the species became extinct at the stage, 84% of them being in the tropics and subtropics (see Fig. 6.23). These extinct species were replaced rapidly in the earliest Langhian, a period of a pronounced climatic optimum (16–14 Ma), by species such as the cosmopolitan *Paragloborotalia peripheroronda* (Plate 6.6, Figs. 1–3; Plate 6.15, Fig. 6) and those which were restricted to the tropics, such as the deep-dwelling *Clavatorella* (Plate 6.4, Figs. 17 and 18).

In the Langhian, the diverse planktonic genera that survived the end of the Burdigalian thrived and diversified in the warmer global temperatures. Following the morphological variance of the orbulinid ancestors in the Burdigalian, *Praeorbulina* and its descendant *Orbulina* appeared in the Langhian. The entire evolutionary transition occurred within mixed-layer habitats, followed by subsequent habitat partitioning (Pearson *et al.*, 1997). *Praeorbulina* first evolved in the tropics and subtropics but very rapidly gave rise to the cosmopolitan *Orbulina* before disappearing completely in the Middle Langhian (N9) around 14 Ma. *Orbulina* species are distributed from tropical to temperate seas in the upper 100 m but are much more common in low latitudes. It is known to thrive at temperatures below 12 °C. It is carnivorous, preying upon several species of copecod and other organisms.

The End Langhian (13–14 Ma) saw the beginning of a slow decline in global temperatures (Fig. 6.18). In the Serravallian, plate collisions sealed off the eastern proto-Mediterranean and the connections between the proto-Mediterranean and the proto-Indian Ocean closed. Evaporitic deposits formed in the Red Sea, the Mesopotamian basin, and along the northern margin of Africa in the Sirt basin (Gvirtzman and Buchbinder, 1978; Jolivet *et al.*, 2006; Rögl and Steininger, 1983). During this period, the rapid expansion of the East Antarctic Ice Sheet (12 Ma) (Flower and Kennett, 1993, 1995; Holbourn *et al.*, 2005) marked the final step in the transition toward the current "icehouse" climate (Zachos *et al.*, 2001). These tectonic and paleoclimatic events mainly affected the provincial reef-forming larger benthic foraminifera as many of them (60%) became extinct in Tethys (see BouDagher-Fadel, 2008). However, only 27% of the planktonic foraminifera became extinct, but most of these extinctions (85%) were in the tropics (see Fig. 6.23).

During the Serravallian, the empty niches caused by these extinctions were replenished mainly by globorotaliids (28% of species are new compared with just 8% of genera). The flattened keeled forms evolved rapidly in the tropics and subtropics into different lineages (see Figs. 6.13–6.15) that inhabited deeper waters than their unkeeled ancestors (see Fig. 6.21). Similar forms in the modern ocean ingest diatoms or chrysophytes but are not known to possess photosymbionts.

In the Tortonian, a distinct speciation pulse occurred (see Fig. 6.22) associated with the onset of Northern Hemisphere glaciation. Although only 9% of the genera were newly evolved, 39% of species made their first appearance during this interval. The majority of planktonic foraminifera species that originated in the Tortonian were short-lived, keeled globorotaliids (38%) and globigerinids (36%), which do not contribute to the modern fauna as most of the globigerinids (74%) died out in the Pliocene. Many of them were restricted to the subtropical (27%, e.g., *Globigerinoides kennetti*) or to temperate waters (20%, e.g., *Globoquadrina langhiana*). The globotoraliids were mainly to be found in the tropical and subtropical realms. Only very few (6%) were restricted to temperate waters. They included, however, the ancestor stock of many important lineages that form the basis of Pliocene biostratigraphy. Among these was the *Truncorotalia* lineage, species of which are deep-dwelling (Chaisson and Ravelo, 1997), with many species, such as *T. crassula*, being restricted to temperate waters (see Chart 6.3 online).

Almost all species of *Neogloboquadrina* evolved in the Middle Tortonian (Fig. 6.16) and were the ancestor stock of many Pliocene and modern lineages. *Neogloboquadrina* species are deep-water dwellers, with 60% dwelling in the tropics and subtropics and smaller forms ranging into temperate realms. *N. pachyderma* coiling directions are used as a proxy for paleoclimate through this time, as left coiling is largely restricted to cold water masses. According to Darling *et al.* (2004), through time, the left coiling type of *N. pachyderma* has been transformed from a cosmopolitan form to a high-latitude specialist, leading to the isolation of the Northern and Southern Hemispheric stock. Cold-water specimens can withstand high salinities and may live in small channels within the ice (Spindler and Dieckmann, 1986). It seems that both allopatric or geographic speciation and ecological evolutionary processes have played an important role in the diversification of these high-latitude planktonic protists (Darling *et al.*, 2004).

In the Middle Tortonian, globular forms made their first appearances (see Fig. 6.7). *Polyperibola*, a short-range form, appeared in the tropics and subtropics, and *Candeina*, an extant form, is rare but when found occurs in the tropical waters. Only 14% of the species became extinct at the end of the Tortonian (Fig. 6.19). In the Messinian, only 10% of the species were new and are represented by just 9% of the new genera. These new genera were *Pulleniatina* and *Sphaeroidinella*. Both genera are thermocline dwellers (Chaisson and Ravelo, 1997). They occur in tropical and subtropical seas. *Pulleniatina* appears to be solely herbivorous, and it has been found to contain both diatoms and chrysophytes, which may have been ingested, or may be living endosymbionts. Over geologic time, *Pulleniatina* switch coiling direction from dominantly dextral to sinistral and back again. These morphological variations have been widely used for stratigraphic correlation as well as to infer changes in water mass conditions (Bandy, 1960; Ericson *et al.*, 1954; Lohmann, 1992; Norris and Nishi, 2001; Saito, 1976; Xu *et al.*, 1995).

The pre-closure interval of the Central American seaway (7.6–4.2 Ma) saw enhanced seasonal input of phytodetritus and a reduction of oceanic ventilation (Roth *et al.*, 2000; Jain and Collins, 2007). By the Late Messinian, a further 10% of the planktonic foraminifera species became extinct (Fig. 6.19), but all representative genera survived the Miocene–Pliocene boundary. In the Late Messinian–Early Zanclean, there may have been a short-lived warming period or at least a modest reduction in the rate of global cooling. The Early Zanclean (from 5.33 to 3.60 Ma, N19) saw the highest species diversity of the Neogene (Fig. 6.22) as 63% of the new Zanclean species appeared around this time. The new species were spread in the tropics and subtropics (42%), while some were cosmopolitan (19%). A similar number of species were restricted to the temperate realm (19%), while a smaller number (10%) were constrained just to the tropics. Very few planktonic foraminifera originated in the subtropics (5%) and an equal percentage (5%) had a subtropic to temperate distribution. Most of the Zanclean species were thermocline-dwelling taxa (e.g., the subtropical *Globorotaloides hexagonus*); others are mixed-layer dwellers (e.g., the cosmopolitan *Globigerina bulloides*). Many species of *Sphaeroidinella* evolved in the tropics and subtropics. They were mainly thermocline dwellers, but none survived the Late Pliocene, with the exception of the extant species *Sphaeroidinella dehiscens* (Plate 6.10, Fig. 9) and *Sphaeroidinella excavata* (Plate 6.10, Fig. 10).

The Middle and Late Pliocene saw a continuation in the reduction of global temperatures (see Fig. 6.18). By 3.60 Ma in the Middle Pliocene, a new pulse of speciation occurred (accounting for 24% of the stock), and more than half of the species originating at this stage are still living. Among the new species of globorotaliids were forms that thrived in the tropics (e.g., *Globorotalia ungulata* (Plate 6.7, Figs. 9–11), a mixed-layer dweller (Shackleton and Vincent, 1978)) and others that were cosmopolitan (e.g., *Globorotalia margaritae*, Plate 6.6, Figs. 17 and 18; Plate 6.14, Fig. 6) a thermocline dweller (Chaisson and Ravelo, 1997)). Many of the extant planktonic foraminifera that evolved in the Late Miocene and Pliocene are descendants of well-established evolving lineages (e.g., the cosmopolitan *Globorotalia tumida*, Plate 6.7, Figs. 5–8 and *Truncorotalia crassaformis*, Plate 6.7, Figs. 16 and 17) or of tropical–subtropical species (e.g., *S. dehiscens*, Plate 6.11, Fig. 19). Their development of new specific morphological features was doubtless driven by change in climates or the thermocline habitat. While temperate and subtropical faunas were significant in the Early Pliocene, tropical and subtropical planktonic foraminifera speciation increased in the Late Pliocene. Most thermocline dwellers, such as *Neogloboquadrina* and *Truncorotalia*, became cosmopolitan.

In the Caribbean, the total closure of the Central American seaway at 4.2 Ma caused the formation of the Isthmus of Panama and allowed direct land-to-land connection between North and South America. This closure blocked the Atlantic–Pacific water interchange, changing profoundly the ocean circulation, isolating the Arctic Ocean and initiating the northern polar ice (Jain and Collins, 2007). The closure of the seaway in the Caribbean produced a cascade of environmental consequences, including the reorganization of circulation in the Gulf of Mexico and much of the Caribbean, leading to reduced upwelling and paleoproductivity (Allmon, 2001). As a result of these changes in ocean circulation, 23% of the planktonic foraminifera species went extinct in the Middle Pliocene. Extinctions at the end of the Pliocene were largely restricted to the tropics and subtropics (see Fig. 6.23).

The Late Neogene was a time of exceptionally strong global cooling and oceanographic change (Zachos *et al.*, 2001). Global cooling after 3 Ma had for the first time a direct effect on tropical sea surface temperatures, resulting in high-amplitude fluctuations in global ice volume and sea levels (Fedorov *et al.*, 2006; Johnson *et al.*, 2007). The Pleistocene saw moderate species diversification (26%), and only 4% new genera. Of the forms that came into being in the Late Pleistocene, all are still extant, and many of them are subtropical and tropical forms. However, there were also some significant new temperate forms (e.g., *Globigerina cryophila, Paragloborotalia oscitans*). In addition, the present day polar affinity of temperate forms (e.g., *Neogloboquadrina pachyderma*) evolved in the Pleistocene (Huber *et al.*, 2000; Kucera and Kennett, 2002). In the surface water of the tropics and subtropics, the heterohelicids were joined by their modern representative *Gallitellia vivans* (Plate 6.10, Figs. 21–23). This form has its highest relative frequencies in south of India, where upwelling waters cause highly variable conditions on the outer shelf. They live inbetween the surface dwellers *Globigerinoides trilobus* (Plate 6.11, Fig. 10) and *Globigerina bulloides* (Plate 6.2, Fig. 9) and the deep-dwelling *Globorotalia menardii* (Plate 6.5, Figs. 2–6; Plate 6.13, Fig. 1B) (Kroon and Nederbragt, 1990). This species is used as a tracer of high runoff, environmentally unstable and upwelling conditions, similar to the well-established biological proxy *Globigerina bulloides*. Their growth is accelerated by eutrophic environments (Kimoto *et al.*, 2009).

The number of extinctions and speciations of the planktonic foraminifera at the end of the Pleistocene was extremely few. Modern species have originated mainly in the Pliocene and Pleistocene. Although 60% of modern planktonic foraminifera thrive in the tropics and subtropics, 30% are cosmopolitan and only 10% are restricted to temperate condition. Others, such as the small, living spinose *Turborotalita* (e.g., *T. quinqueloba*) are restricted to the cold to temperate waters of the South Atlantic, but they are especially abundant during spring bloom conditions at low latitudes, feeding mostly on algae and small zooplankton (Asano *et al.*1968; Kroon *et al.*, 1988). Common forms in the thermocline waters of the tropics are *Candeina*, and in the tropics and subtropics are *Sphaeroidinella*. Quadrate forms, such as *Dentoglobigerina*, can inhabit the upper and deep thermocline (Pearson and Wade, 2009). Modern *Streptochilus globulosum* have intermediate oxygen and carbon stable isotope values, suggestive of a thermocline-dwelling habitat (Resig and Kroopnick, 1983).

6.5 Conclusion

Planktonic foraminifera have played and continue to play a vital role in the marine ecosystem. Over geological time, since the Jurassic, they appear to have evolved on a number of discrete occasions from benthic foraminiferal stock. However, once established they have shown themselves to be highly versatile and adaptive. They have evolved to fill most surface water niches in the oceans and have adapted to exploit various climatic zones and supplies of nutrients. Sexton and Norris (2008) in studying the dispersal of the foraminifera *Truncorotalia truncatulinoides* (Plate 6.7, Figs. 18–20) inferred that tectonic and water-mass barriers to dispersal are, in most cases, very weak. However, the seawater temperature within the climate zones influences the latitudinal distribution of planktonic foraminifera, while supply of nutrients affects their vertical distribution. Today, high-temperature gradients between the poles and the equator mean that the temperate northern hemisphere forms are isolated from the southern temperate forms by warm tropical waters. Therefore, today, climatic zones and latitudinal temperature gradients determine the distribution of planktonic foraminiferal. This contrasts with the paleogeographic provincialism which is exhibited by the more static larger benthic foraminifera (e.g., their American, Tethyan, and Indo-Pacific provinces, see BouDagher-Fadel, 2008; BouDagher-Fadel and Price, 2010).

The rate of evolution of the planktonic foraminifera, their morphological diversity, and their constrained temporal and latitudinal extent means that fossil forms have become a hugely valuable biostratigraphic and paleoenvironmental tool. They allow, within climate zones, global correlations to be made that are much more difficult to do with other foraminiferal groups, which are either long ranging (e.g., the smaller benthic foraminifera) or show provincialism (e.g., the larger benthic foraminifera). Their role as markers for biostratigraphical zonation and correlation underpins most drilling of marine sedimentary sequences and establishes, therefore, their importance to hydrocarbon exploration and hence to the modern global economy.

We have presented a revised and self-consistent taxonomy and phylogeny for the numerous orders and superfamilies, families, and genera that can be identified for the planktonic foraminifera. We have also presented revised stratigraphic range charts (http://dx.doi.org/10.14324/111.9781910634257) for the key planktonic foraminiferal species, which we hope will aid practical biostratigraphic studies in the future. Although we have endeavoured to capture the great diversity of the research presented in literature, there are still considerable areas of uncertainty and ignorance, both about extinct and extant forms. These unanswered questions make the study of planktonic foraminifera still a challenging and exciting field. We hope that this book makes a helpful and useful contribution to what will be a continuing, vibrant, and globally significant research endeavour.

Plate 6.1. Figure 1. *Candeina praenitida* (Blow). Holotype, figured by Blow (1969), Somalia, East Africa. Surface. Smooth, few, or no pustules, as in descendant *C. nitida*, usually more high spired than in ancestral *C. somaliensis*. Miocene, N16a–N17a, x100. Figures 2 and 3. *Candeina somaliensis* new species. Holotype, figured as paratype of *Candeina praenitida* by Blow (1969), Surface is postulate-covered, as in ancestral *Tenuitellinata juvenilis* (Bolli), usually not so high spired as *C. praenitida*, and surface not smooth. Jamaica, Miocene, N15–N16b, x100. Figure 4. *Candeina nitida* d'Orbigny. Reproduced from Banner's Collection UCL, France, Miocene, N16, x100. Figure 5. *Polyperibola christiani* Liska. Paratype, figured by Liska (1980), Trinidad, Miocene, N16, x133. Figures 6-7. *Globigerinita flparkerae* (Brönnimann and Resig). Figured by Blow (1979) as *Globigerinoides parkerae* Bermudez, Jamaica, Holocene, N23b, x133. Figures 8–13. *Tenuitellinata bradyi* (Wiesner). Figured by BouDagher-Fadel *et al.* (1997). A microperforate species with scattered pustules concentrated near the umbilicus, NHM collection, Holocene, N23b, x500. Figure 14. *Tinophodella glutinata* (Egger). Italy, UCL Collection, Pliocene, N19, x250. Figure 15. *Globigerinita uvula* (Ehrenberg). Trinidad, UCL Collection, Burdigalian, N5, x200. Figures 16–18. *Tenuitella anfracta* (Parker).Figured by BouDagher-Fadel *et al.* (1997), Indian Ocean, NHM collection, Holocene, N23b, x250. Figure 19. *Catapsydrax dissimilis* (Cushman and Bermudez). Trinidad, UCL Collection, Miocene, N4, x133. Figure 20. *Catapsydrax unicavus* Bolli, Loeblich and Tappan. Barbados, UCL Collection, Miocene, N4, x70.

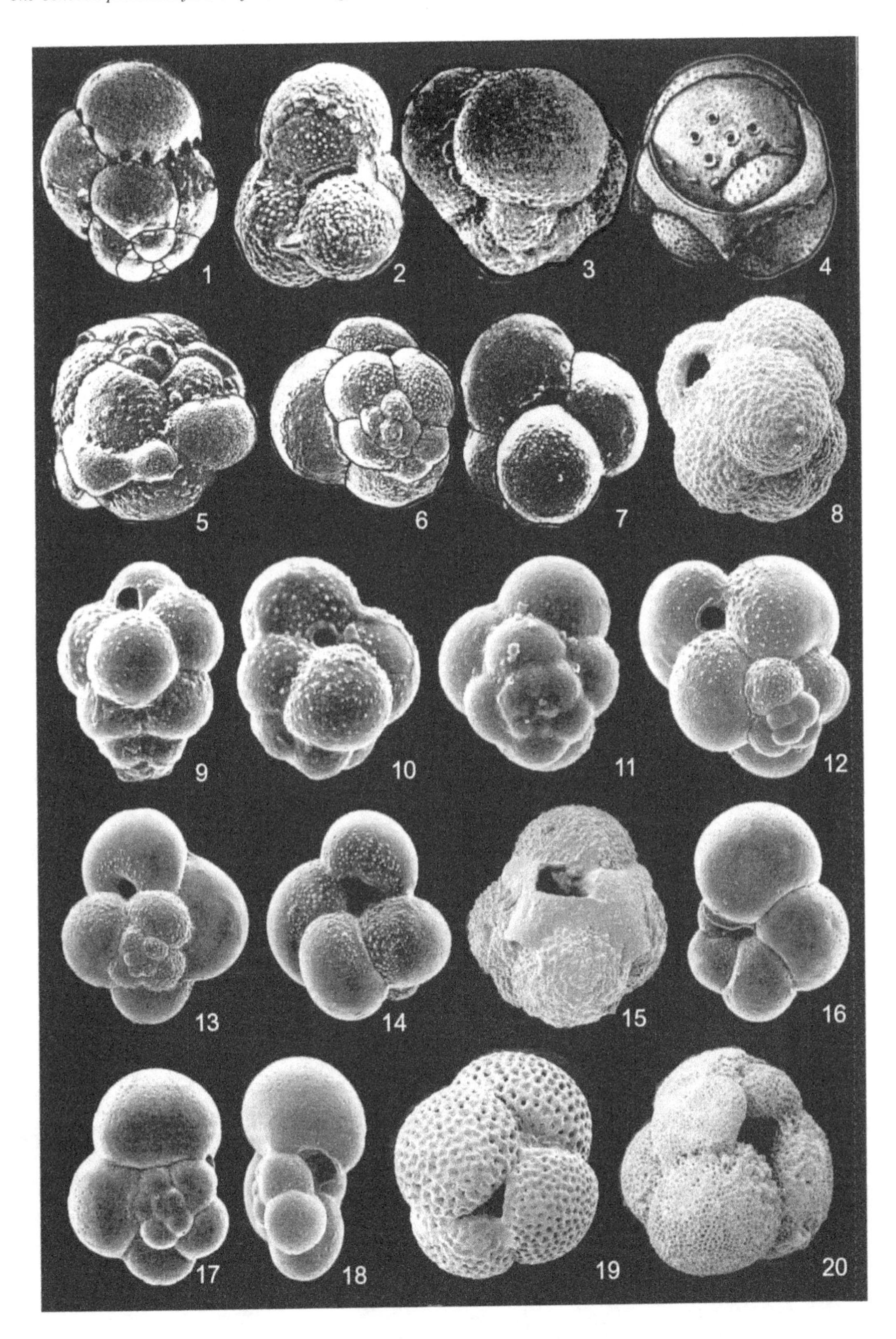
1
2
3
4
5
6
7
8
9
10
11
12
13
14
15
16
17
18
19
20

Plate 6.2. Figure 1. *Dentoglobigerina yeguaensis* (Weinzierl and Applin). Mississippi, UCL Collection, Eocene, P13, x80. Figures 2–6. *Dentoglobigerina baroemoenensis* (Le Roy). Trinidad, UCL Collection, Miocene, N13, x100. Figure 7. *Dentoglobigerina venezuelana* (Hedberg). Trinidad, UCL Collection, Miocene, N4, x80. Figure 8. *Globigerina praebulloides* Blow. Venezuela, UCL Collection, Miocene, N12, x133. Figure 9. *Globigerina bulloides* d'Orbigny. Venezuela, UCL Collection, Miocene, N17, x80. Figure 10. *Globigerina bulbosa* Le Roy. Figured by Blow (1979), Venezuela, N12, x100. Figures 11-12. *Globigerina praecalida* Blow. Holotype, figured by Blow (1979), Pacific Ocean, Holocene, x100. Figures 13–15. *Globigerina eamesi* Blow. Figured by Kennett and Srinivasan (1983), DSDP site 281-13-2, Miocene, N4, x130. Figure 16. *Globigerina umblicata* Orr and Zaizteff. Figured by Kennett and Srinivasan (1983), DSDP Site 173-33, Pleistocene, N22, x80. Figure 17. *Globoturborotalita* sp. Banner Collection UCL, Miocene, N15, x100. Figures 18-19. *Globigerinoides altiaperturus* Bolli. DSDP site 208, UCL Collection, Miocene, N6, x130. Figure 20. *Globigerinoides amplus* Peronig. Mexico, UCL Collection, Miocene, N12, x100. Figure 21. *Globigerinoides conglobatus* (Brady). Bermuda, UCL Collection, Holocene, N23b, x58.

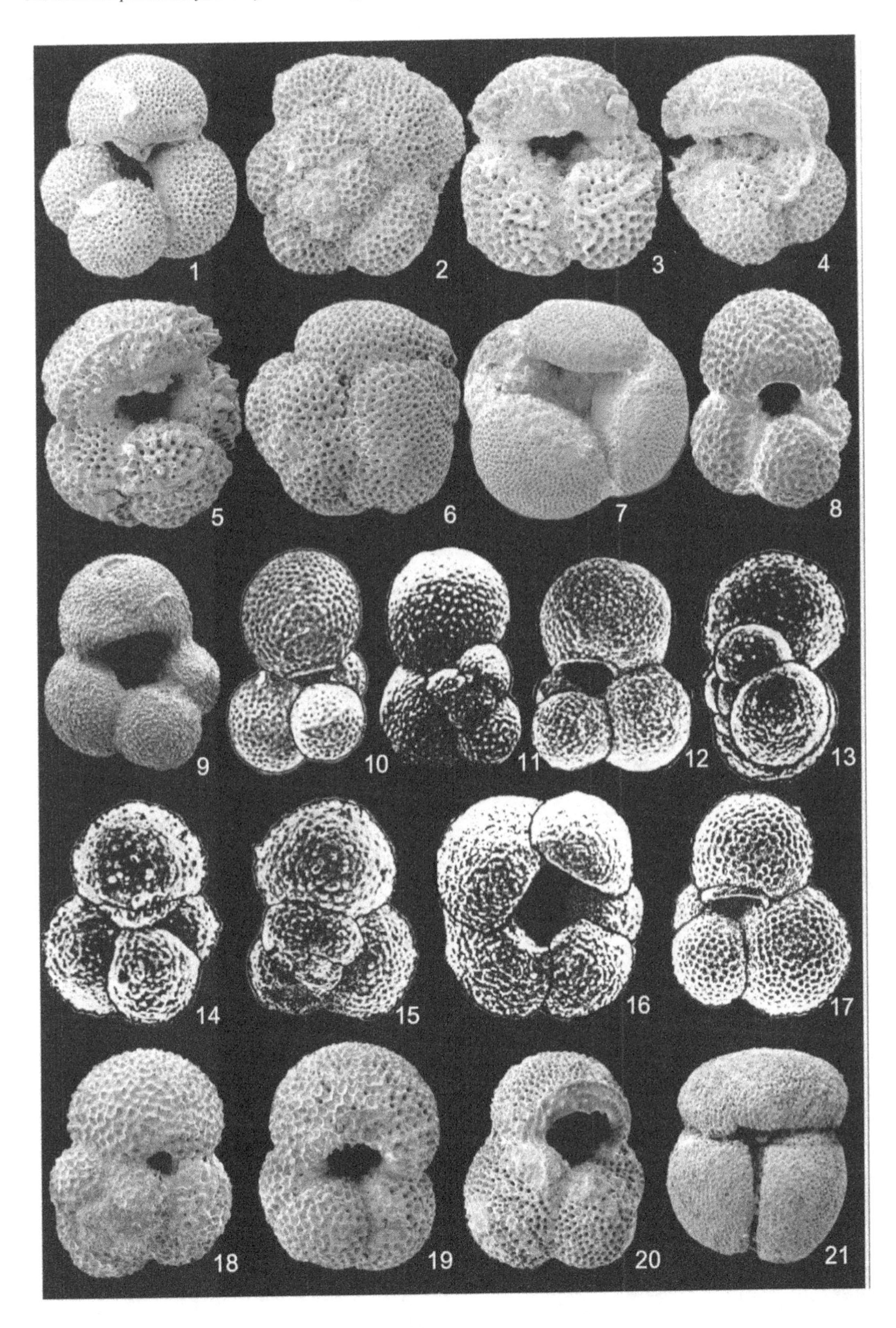
1
2
3
4
5
6
7
8
9
10
11
12
13
14
15
16
17
18
19
20
21

Plate 6.3. Figures 1-2. *Globigerinoides obliquus* Bolli. DSDP site 289, UCL Collection, Miocene, N8, x58. Figures 3-4. *Globigerinoides primordius* Blow and Banner. Trinidad, Cipero Formation, UCL Collection, Miocene, N4, x100. Figures 5-6. *Globigerinoides quadrilobatus* (d'Orbigny). Venezuela, UCL Collection, Miocene, N17, x100. Figures 7-8. *Globigerinoides ruber* (d'Orbigny). Venezuela, UCL Collection, Miocene, N17, x100. Figures 9-10. *Globigerinoides sacculifer* (Brady). Cyprus, UCL Collection, Pliocene, N19, x130. Figure 11–13. *Globigerinoides fistulosus* (Schubert). (11-12) Figured by Kennett and Srinivasan (1983), DSDP site 208-3, Pliocene, N21, (13) Banner's Collection UCL, x33. Figures 14–17. *Globoquadrina dehiscens* (Chapman, Parr, and Collins). (14-16) figures reproduced from Banner's collection UCL, (17) Figured by Kennett and Srinivasan, DSDP site 289, Miocene, N17, x130. Figures 18–20. *Globorotaloides suteri* Bolli. Figured by Kennett and Srinivasan (1983), DSDP site 208, Miocene, N4, x160.

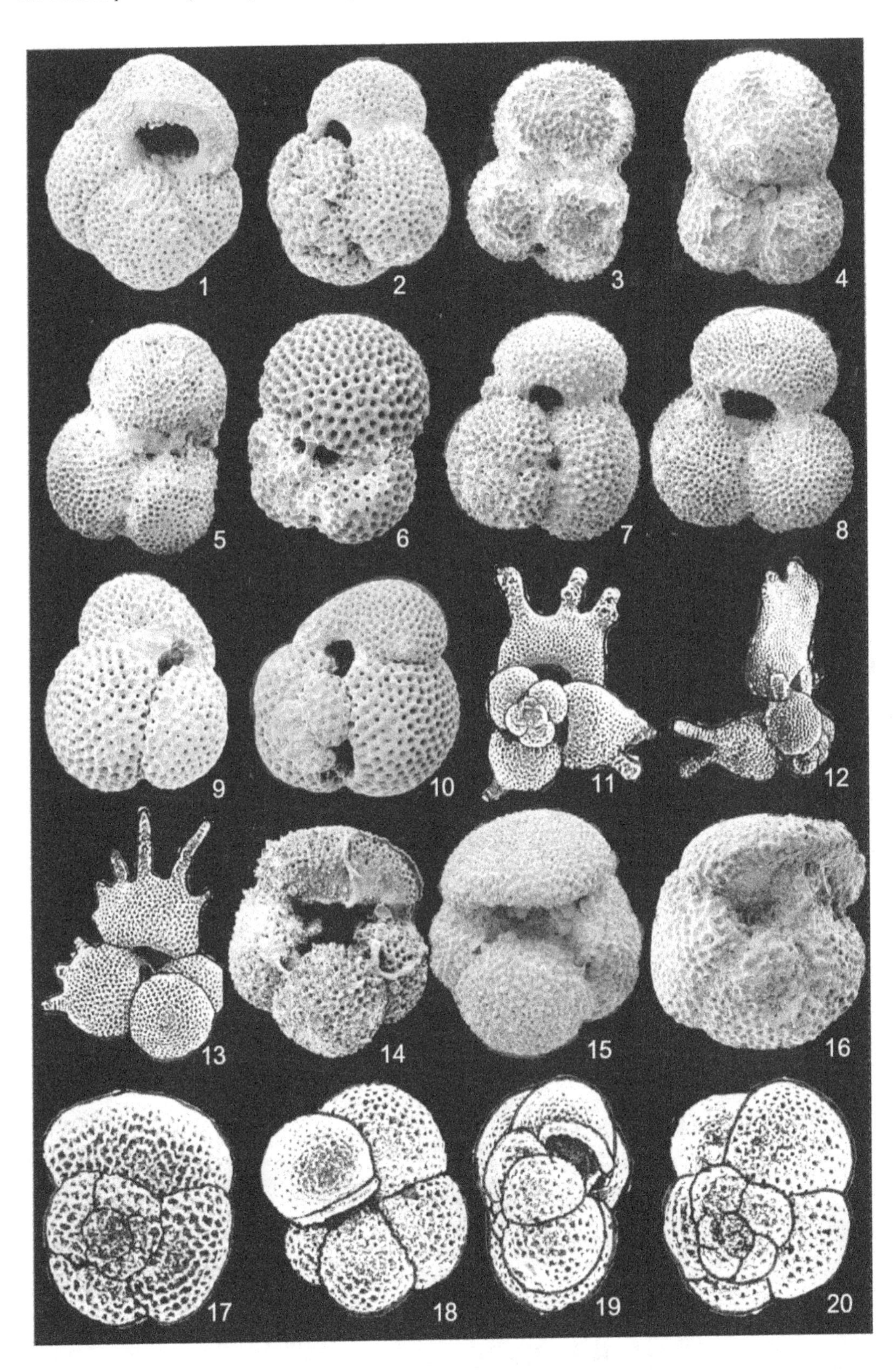
1
2
3
4
5
6
7
8
9
10
11
12
13
14
15
16
17
18
19
20

Plate 6.4. Figures 1-2. *Globoturborotalita brazieri* (Jenkins). Figured by Kennett and Srinivasan (1983), DSDP Site 208, Miocene, N5, x140. Figures 2–4. *Globoturborotalita woodi* (Jenkins). Figured by Kennett and Srinivasan (1983), DSDP Site 207, Miocene, N16, x150. Figures 5-6. *Globoturborotalita decoraperta* (Takayanagi and Saito). Figured by Kennett and Srinivasan (1983), DSDP Site 207, N17, x145. Figures 7-8. *Globoturborotalita druryi* (Akers). Figured by Kennett and Srinivasan (1983), DSDP Site 289, Miocene, N7, x185. Figures 9-10. *Globoturborotalita nepenthoides* (Brönnimann and Resig). Paratype, figured by Brönnimann and Reisig (1971), SW Pacific, Miocene, N12, x200. Figures 11-12. *Globoturborotalita nepenthes* (Todd). Figured by Kennett and Srinivasan (1983), DSDP Site 289, Miocene, N16, x108. Figures 13–15. *Bolliella adamsi* Banner and Blow. South Pacific, Banner's Collection UCL, Holocene, N23b, x50. Figure 16. *Globigerinella aequilateralis* (Brady). Venezuela, UCL Collection, Pleistocene, N23a, x50. Figures 17-18. *Clavatorella nicobarensis* Srinivasan and Kennett. Figured by Kennett and Srinivasan (1983), Indian Ocean, Pliocene, N19, x80. Figures 19–20. *Globorotalia archeomenardii* Bolli. Figured by Kennett and Srinivasan (1983), DSDP Site 289, Miocene, N9, x100.

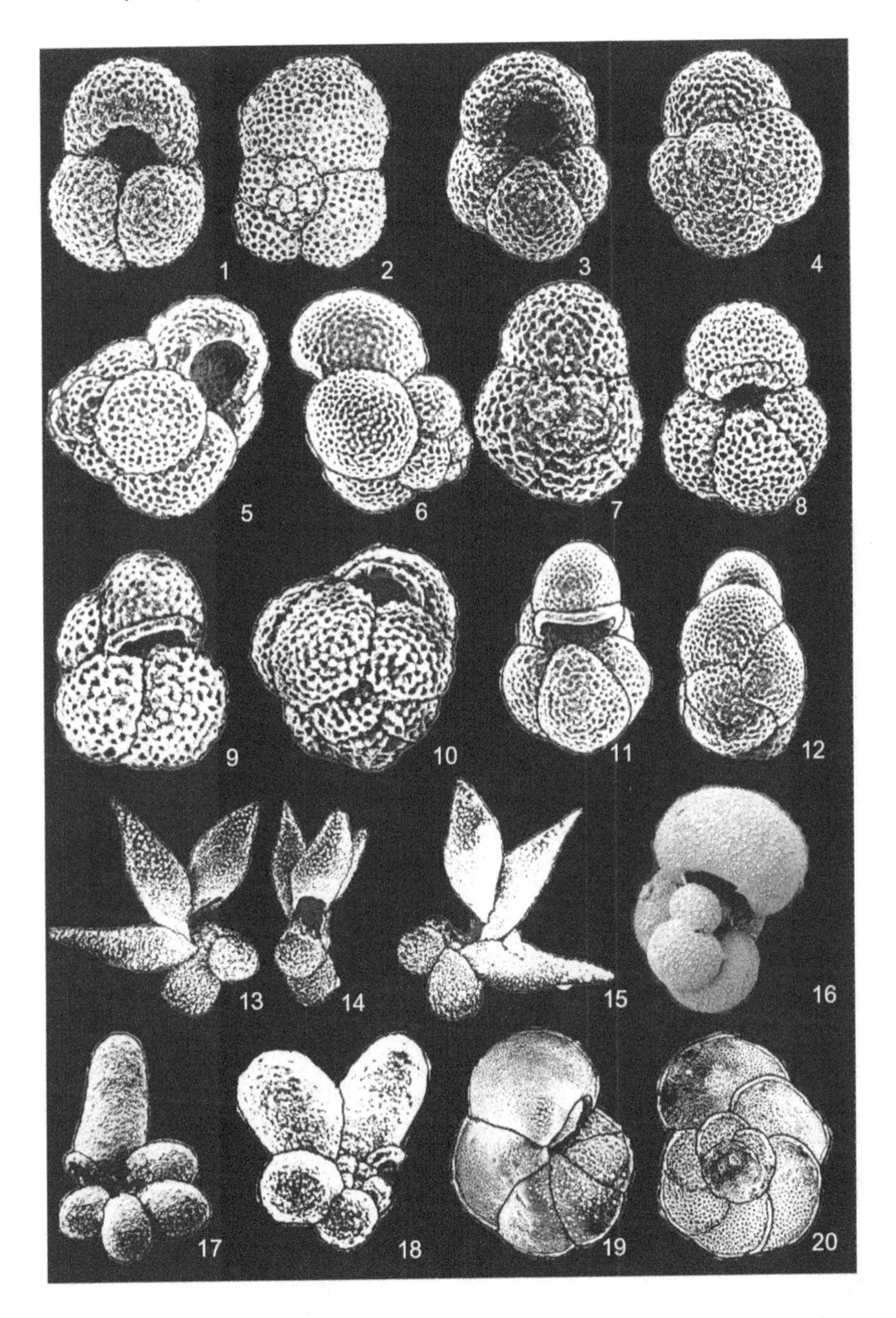
1
2
3
4
5
6
7
8
9
10
11
12
13
14
15
16
17
18
19
20

Plate 6.5. Figure 1. *Globorotalia praemenardii* Cushman and Stainforth. Trinidad, Miocene, N12, UCL Collection, x100. Figures 2–6. *Globorotalia menardii* (Parker, Jones, and Brady) = *Globorotalia cultrata* (d'Orbigny). *Globorotalia menardii* and *cultrata* are considered to be synonyms, and *menardii* is junior to *cultrata* but it is given seniority because of its wide usage. Near-topotype, Bermuda, (2, 4-6) from UCL Collection, (3) from Banner's Collection UCL, Holocene, N23b, x30. Figures 7-8. *Globorotalia multicamerata* Cushman and Jarvis. (7) Figured by Kennett and Srinivasan (1983), DSDP Site 289, Miocene, N18, x30, (8) Figured by Blow (1979), Jamaica, Miocene, N19, x30. Figures 9-10. *Globorotalia fimbriata* (Brady). Trinidad, UCL Collection, Pliocene, N21, x57. Figures 11-12. *Globorotalia neoflexuosa* Srinivasan, Kennett, and Bé. Indian Ocean, Banner's Collection UCL, Holocene, N23b, x29. Figures 13-14. *Globorotalia miocenica* Palmer. Trinidad, (13) Banner's Collection UCL, x100, (14) UCL Collection, x113, Pliocene, N21. Figures 15-16. *Globorotalia limbata* (Fornasini). (15) Figured by Kennett and Srinivasan (1983), (16) Banner's Collection UCL, DSDP Site 289, Late Miocene, N17, x57. Figures 17-18. *Globorotalia exilis* Blow. Holotype, figured by Blow (1969), West Indies, Pliocene, N19, x57. Figures 19-20. *Globorotalia inflata* (d'Orbigny). Tasman Sea, UCL Collection, Holocene, N23b, x100.

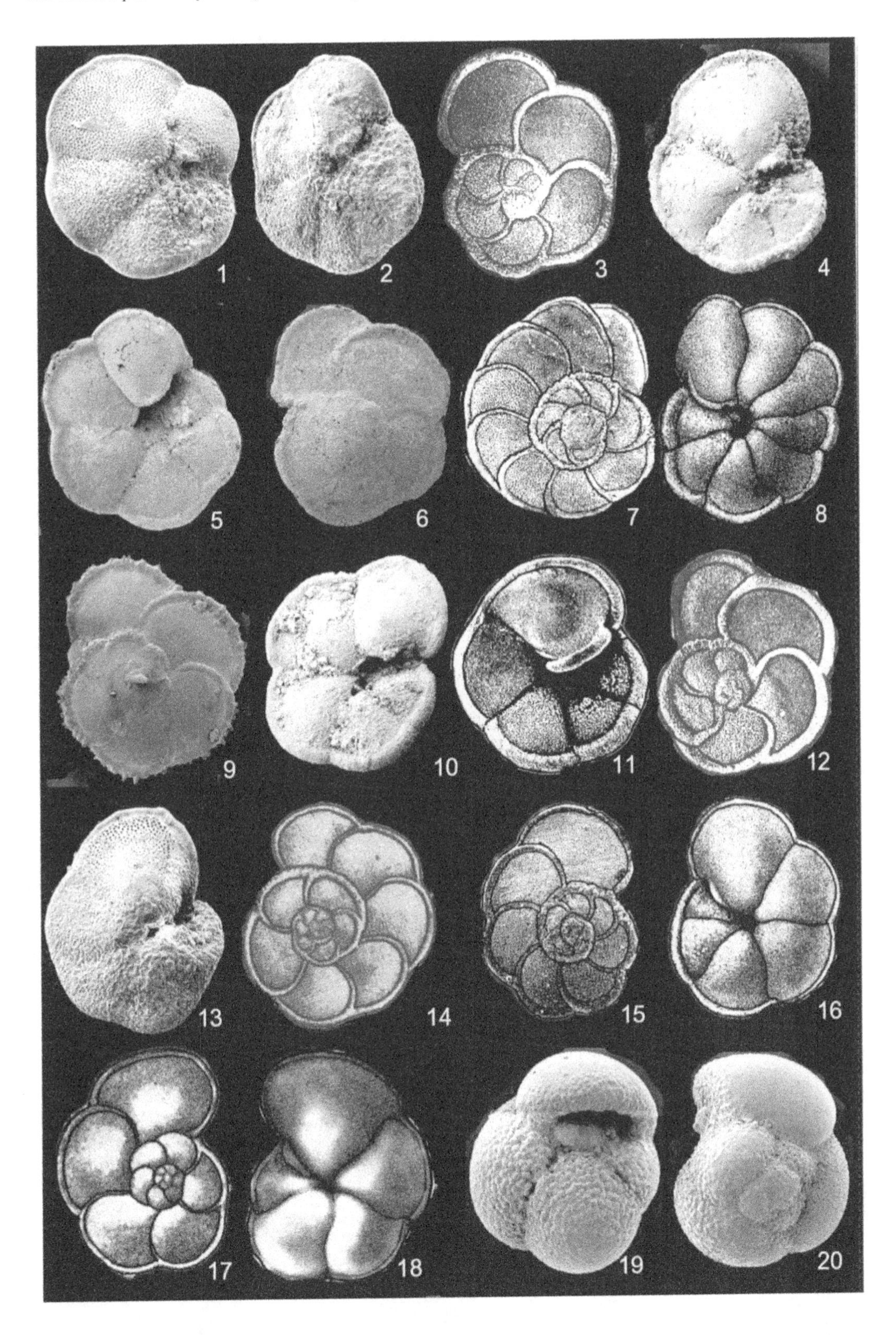
1
2
3
4
5
6
7
8
9
10
11
12
13
14
15
16
17
18
19
20

Plate 6.6. Figures 1–3. *Paragloborotalia peripheroronda* (Blow and Banner). (1–2) Figured by Kennett and Srinivasan (1983), DSDP Site 289, Miocene, N4, x94, (3) UCL Collection, Trinidad, Miocene, N8, x100. Figures 4-5. *Paragloborotalia peripheroacuta* (Blow and Banner). Figured by Kennett and Srinivasan (1983), DSDP Site 289, Miocene, N10, x200. Figures 6-7. *Globorotalia praefohsi* (Blow and Banner). Trinidad, UCL Collection, Miocene, N12b, x133. Figures 8-9. *Globorotalia fohsi* (Cushman and Ellisor). Banner's Collection UCL, Miocene, N13a, x133. Figures 10-11. *Globorotalia lobata* (Bermudez). Figured by Kennett and Srinivasan (1983), DSDP Site 289, Miocene, N1, x100. Figures 12-13. *Globorotalia robusta* (Bolli). Figured by Kennett and Srinivasan (1983), DSDP Site 289, Miocene, N12b, x100. Figures 14–16. *Globorotalia hirsuta* (d'Orbigny). Trinidad, UCL Collection, Pliocene, N21b, x100. Figures 17-18. *Globorotalia margaritae* (Bolli and Bermudez). Cyprus, UCL Collection, Pliocene, N19, x80. Figures 19-20. *Globorotalia praehirsuta* (Blow). Holotype, figured by Bolli (1969), Italy, Pleistocene, N22, x100. Figures 21-22. *Globorotalia primitiva* (Cita). Holotype, figured by Cita (1973), Tyrrhenian Basin, Western Mediterranean, Pliocene, N19, x133.

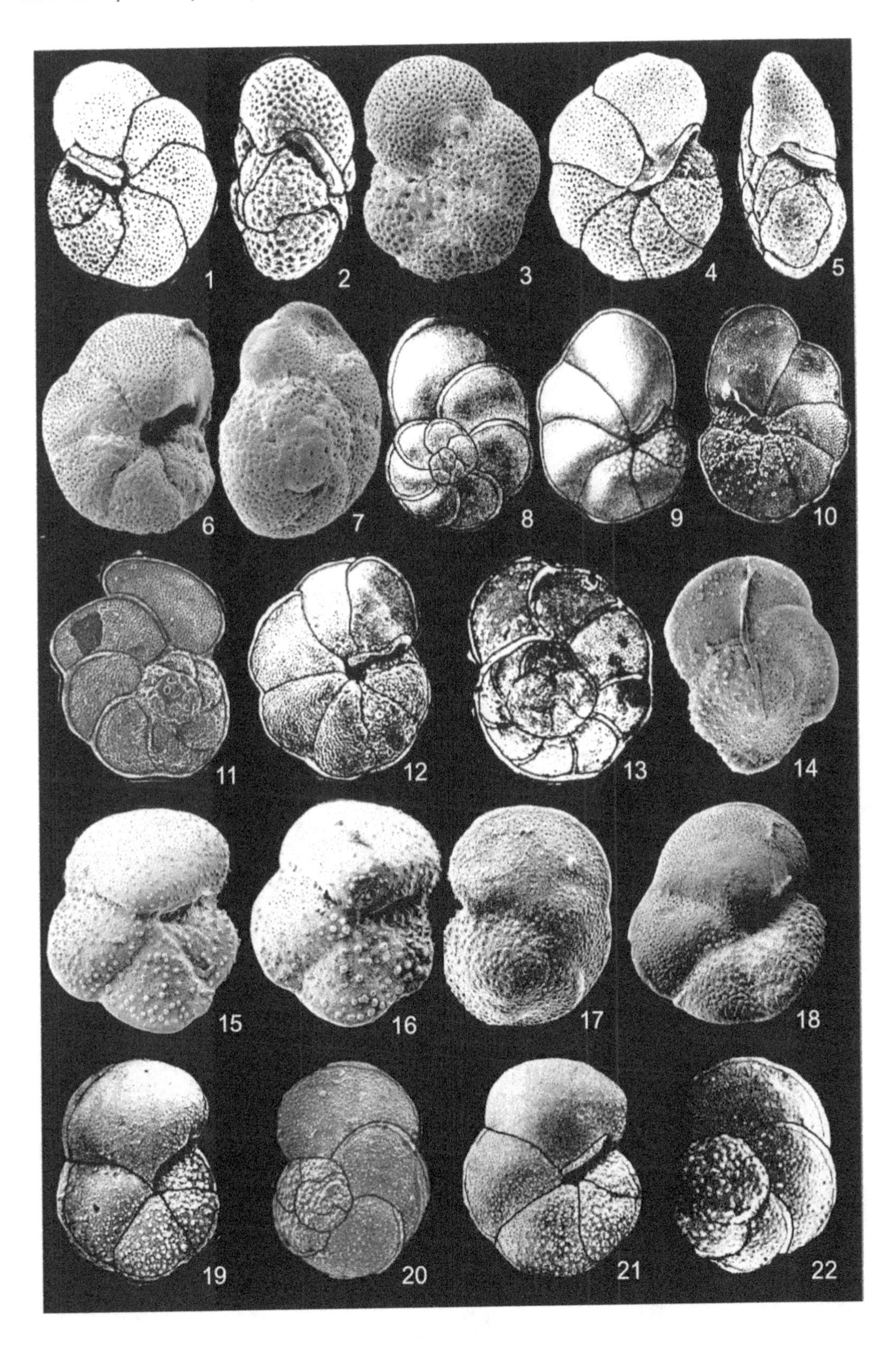
1
2
3
4
5
6
7
8
9
10
11
12
13
14
15
16
17
18
19
20
21
22

Plate 6.7. Figure 1. *Globorotalia praescitula* (Blow). Trinidad, UCL Collection, Miocene, N7, x67. Figures 2-3. *Globorotalia scitula* (Brady). Trinidad, UCL Collection, Miocene, N12, x100. Figure 4. *Globorotalia juanai* (Bermudez and Bolli). Figured by Kennett and Srinivasan (1983), DSDP 289, Miocene, N17, x99. Figures 5–8. *Globorotalia tumida* (Brady). (5, 6) Trinidad, UCL Collection, x30; (7, 8) figured by Kennett and Srinivasan (1983), DSDP Site 206-4, x35, Pliocene, N19. Figures 9–11. *Globorotalia ungulata* (Bermudez). (9) Topotype, UCL Collection, Venezuela, x67; (10, 11) figured by Kennett and Srinivasan (1983), DSDP Site 502, x58, Pliocene, N21. Figures 12-13. *Neogloboquadrina humerosa* (Takayanagi and Saito). Topotypes, Nobori Formation, Shikoku, Japan, UCL Collection, Pliocene, N21, x80. Figures 14-15. *Neogloboquadrina dutertrei* (d'Orbigny). Cuba, UCL Collection, Pliocene, N21, x60. Figures 16-17. *Truncorotalia crassaformis* (Galloway and Wissler). Jamaica, UCL Collection, Pleistocene, N22, x66. Figures 18–20. *Truncorotalia truncatulinoides* (d'Orbigny). Tasman Sea, UCL Collection, Holocene, N23b, x40.

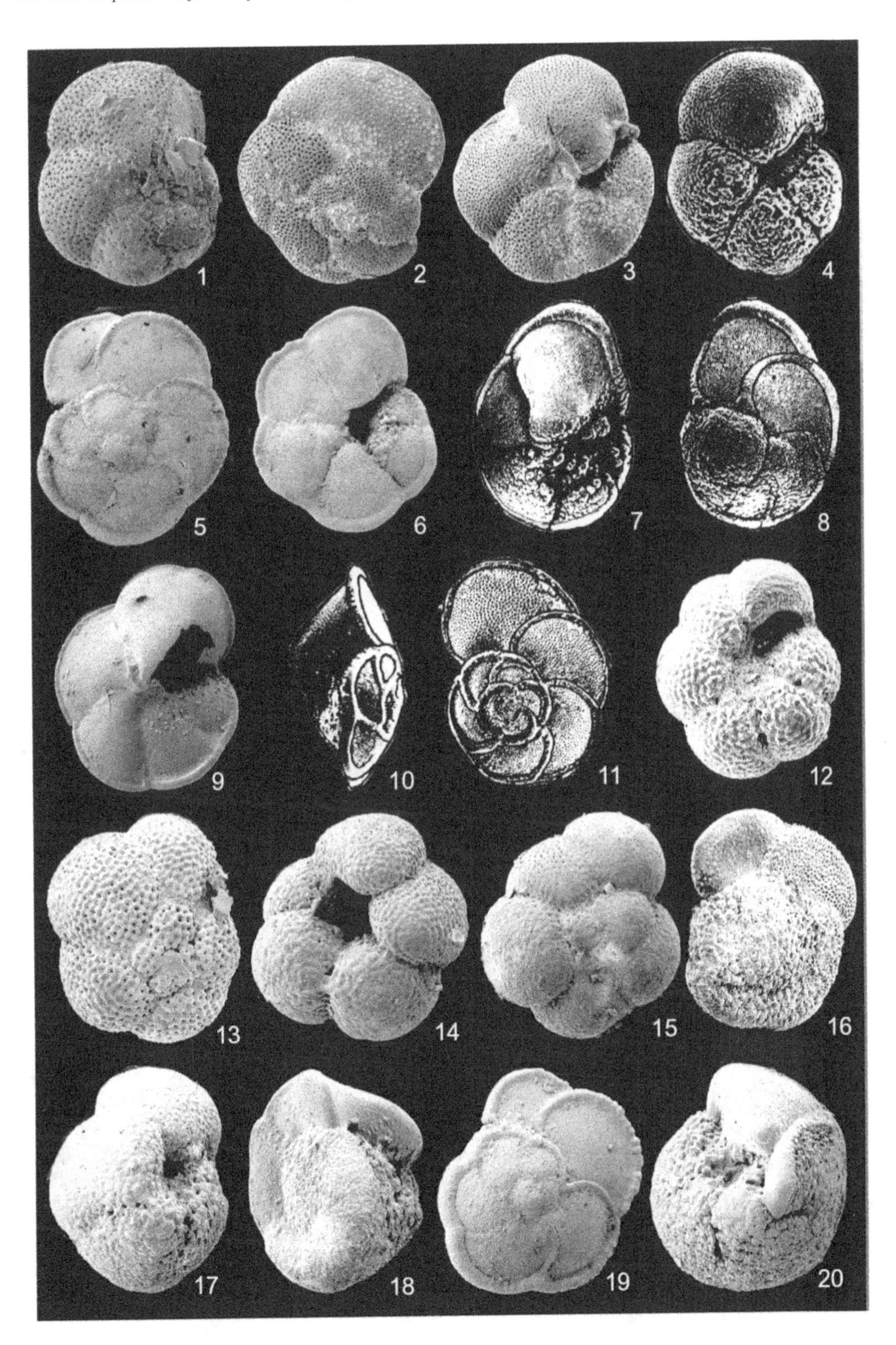
1
2
3
4
5
6
7
8
9
10
11
12
13
14
15
16
17
18
19
20

Plate 6.8. Figures 1–5. *Paragloborotalia obesa* (Bolli). Trinidad, UCL Collection, Miocene, N9, x133. Figures 6–8. *Paragloborotalia pseudopima* (Blow). Paratypes, figured by Blow (1969), New Guinea, Pliocene, N21, x80. Figures 9-10. *Protentella prolixa* (Lipps). Figured by Kennett and Srinivasan (1983), Newport Bay, California, Middle Miocene, x175. Figure 11. *Hastigerina pelagica* (d'Orbigny). Venezuela, UCL Collection, Holocene, N23b, x57. Figure 12. *Orcadia riedeli* (Rögl and Bolli). Figured by BouDagher-Fadel *et al.* (1997), NHM collection, Holocene, N23b, x227. Figures 13–15. *Neoacarinina blowi* (Thompson). Banner's Collection UCL, Pleistocene, N23a, x80. Figures 16–19. *Orbulina universa* (d'Orbigny). Bermuda,UCL Collection, Holocene, N23b,x 40.

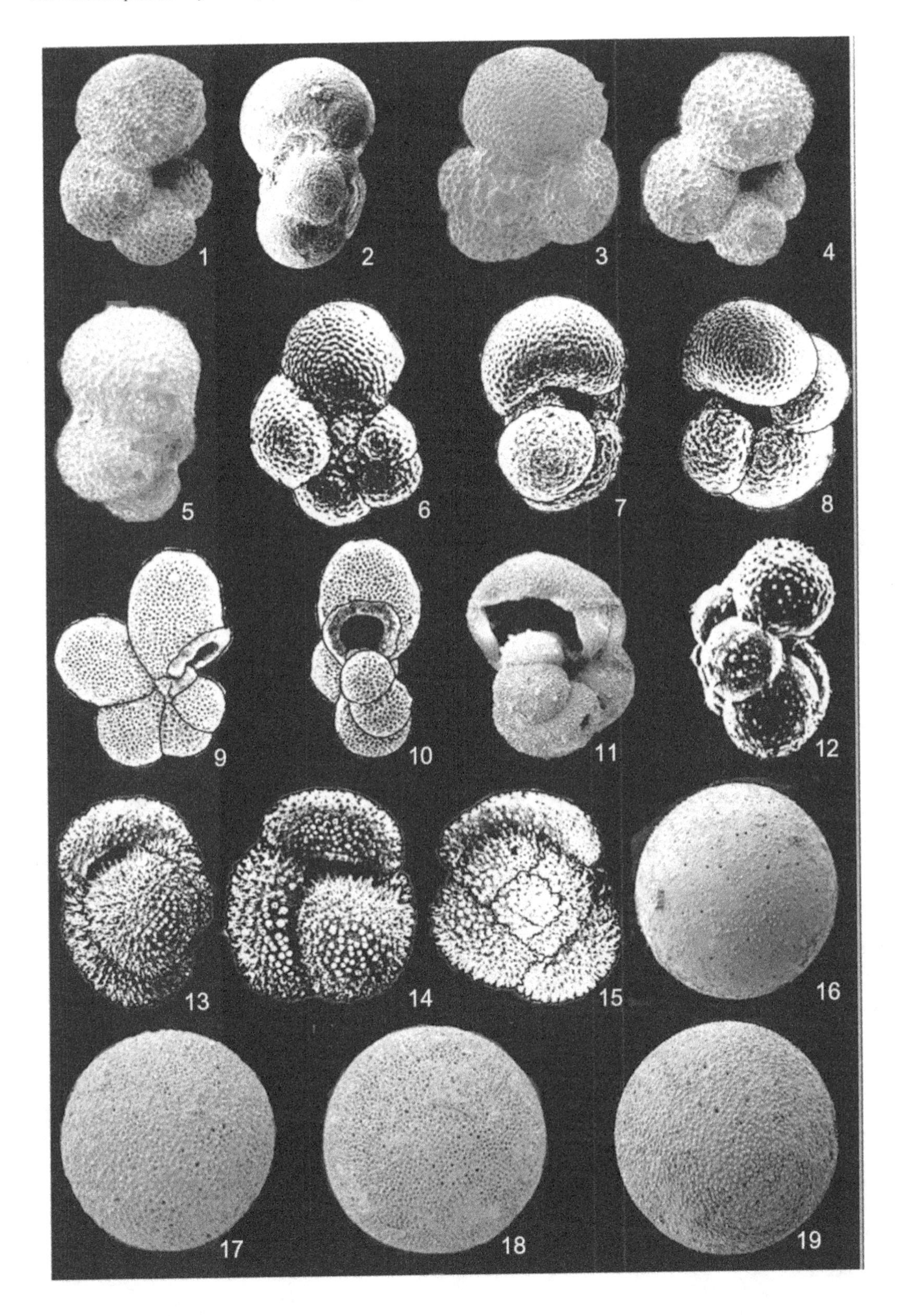
1
2
3
4
5
6
7
8
9
10
11
12
13
14
15
16
17
18
19

Plate 6.9. Figure 1. *Orbulina universa* (d'Orbigny). Bermuda, UCL Collection, Holocene, N23b, x100. Figure 2. *Orbulina suturalis* (Brönnimann). Trinidad, UCL Collection, Miocene, N12, x40. Figure 3. *Orbulina bilobata* (d'Orbigny). Cyprus, UCL Collection, Miocene, N13, x40. Figure 4. *Praeorbulina sicana* (de Stefani). Banner's Collection UCL, Miocene, N8, x57. Figure 5. *Praeorbulina curva* (Blow). Figured by Kennett and Srinivasan (1983), DSDP Site 289, Miocene, N8, x55. Figure 6. *Praeorbulina glomerosa* (Blow). Figured by Kennett and Srinivasan (1983), DSDP Site 289, Miocene, N8, x67. Figure 7. *Praeorbulina circularis* (Blow). Figured by Kennett and Srinivasan (1983), DSDP Site 207, *O. suturalis* Zone. Early Middle Miocene, x55. Figures 8–10. *Pulleniatina obliquiloculata* (Parker and Jones). Challenger Expedition, Station 224, UCL Collection, Holocene, N23b, x80. Figures 11-12. *Pulleniatina primalis* (Banner and Blow). (11) Banner's Collection UCL, (12) Figured by Kennett and Srinivasan (1983), DSDP Site 289-17-4, Early Pliocene, N19, x100. Figure 13. *Pulleniatina praespectabilis* (Brönnimann and Resig). Holotype, figured by Brönnimann and Resig (1971), New Guinea, Pliocene, N20, x67. Figure 14. *Prosphaeroidinella disjuncta* (Finlay). Venezuela, UCL Collection, Miocene, N7, x80. Figure 15. *Prosphaeroidinella parkerae* (Ujiié). Banner Collection UCL, Miocene, N17, x57.

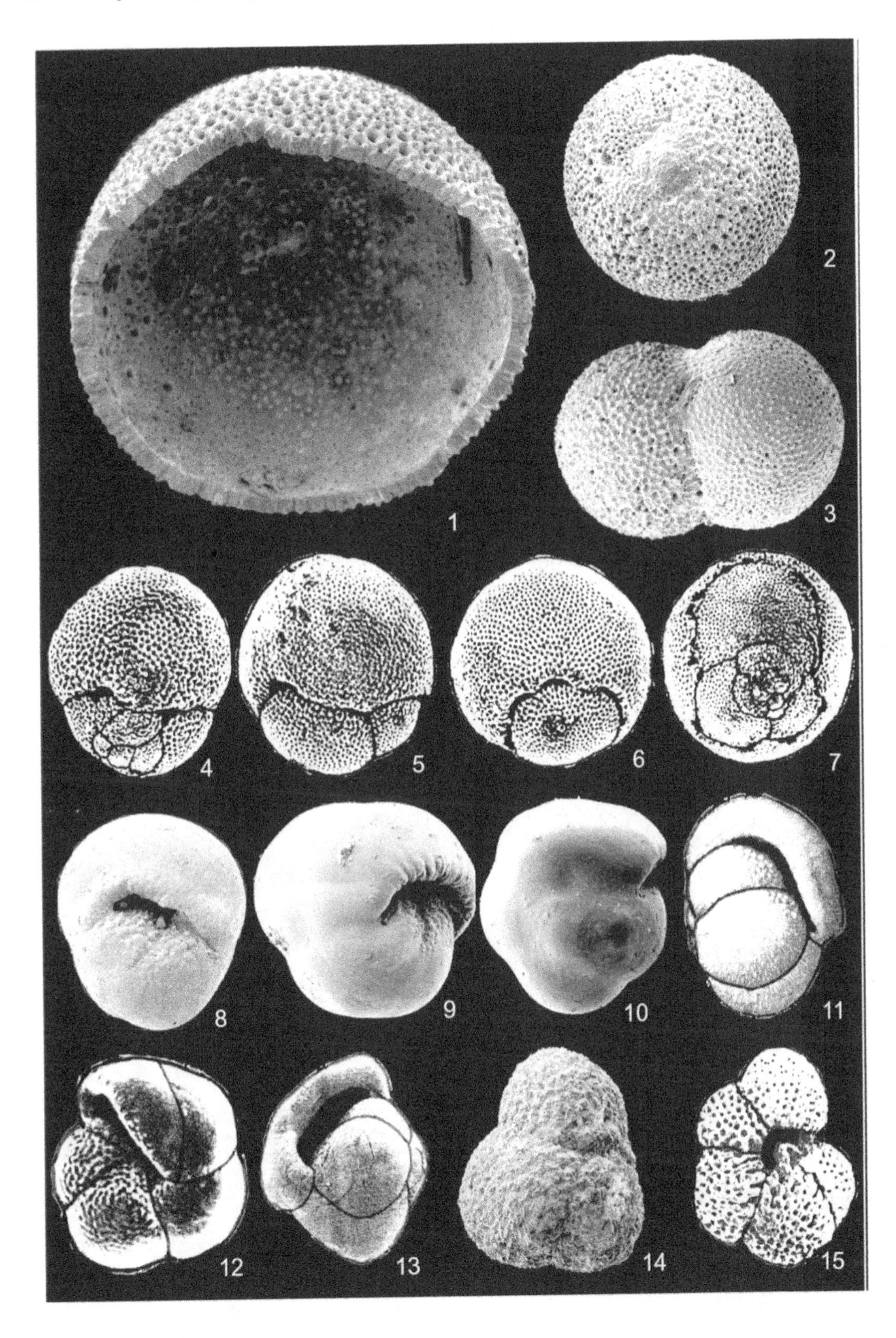
1
2
3
4
5
6
7
8
9
10
11
12
13
14
15

Plate 6.10. Figures 1–3. *Prosphaeroidinella disjuncta* (Finlay). Venezuela, UCL Collection, Miocene, N7, x80. Figures 4-5. *Sphaeroidinellopsis multiloba* (Le Roy). Venezuela, UCL Collection, Miocene, N7, x100. Figures 6-7. *Sphaeroidinellopsis seminulina* (Schwager). (6) Venezuela, UCL Collection, Miocene, N7, x100; (7) Figured by Kennett and Srinivasan (1983), DSDP Site 289, Pliocene, N19, x120. Figure 8. *Sphaeroidinellopsis paenedehiscens* (Blow). Trinidad, UCL Collection, Miocene, N17b, x57. Figure 9. *Sphaeroidinella dehiscens* (Parker and Jones). Banner's Collection UCL, Holocene, N23b, x57. Figure 10. *Sphaeroidinella excavata* (Banner and Blow). Banner's Collection UCL, Holocene, N23b, x27. Figures 11-12. *Turborotalita humilis* (Brady). Challenger Expedition, UCL Collection, Holocene, N23b, x266. Figure 13. *Cassigerinella boudecensis* Pokorný = *Cassigerinella chipolensis* (Cushman and Ponton). Figured by Li (1986), Oligocene, Trinidad, x285. Figures 14-15. *Riveroinella martinezpicoi* (Bermudez and Seiglie). Banner's Collection UCL, Miocene, N6, x330. Figures 16-18. *Gallitellia vivans* (Cushman). Banner's Collection UCL, Holocene, N23b, x200. Figures 19-20. *Streptochilus martini* (Pijpers). Banner's Collection UCL, Oligocene, P19, x200. Figures 21, 23. *Streptochilus globigerus* (Schwager). Banner's Collection UCL, Pliocene, N19, x100. Figure 22. *Streptochilus globulosus* (Cushman). Banner's Collection UCL, Holocene, N23b, x200. Figure 24. *Streptochilus pristinus* (Brönnimann and Resig). Banner's Collection UCL, Miocene, N4, x200.

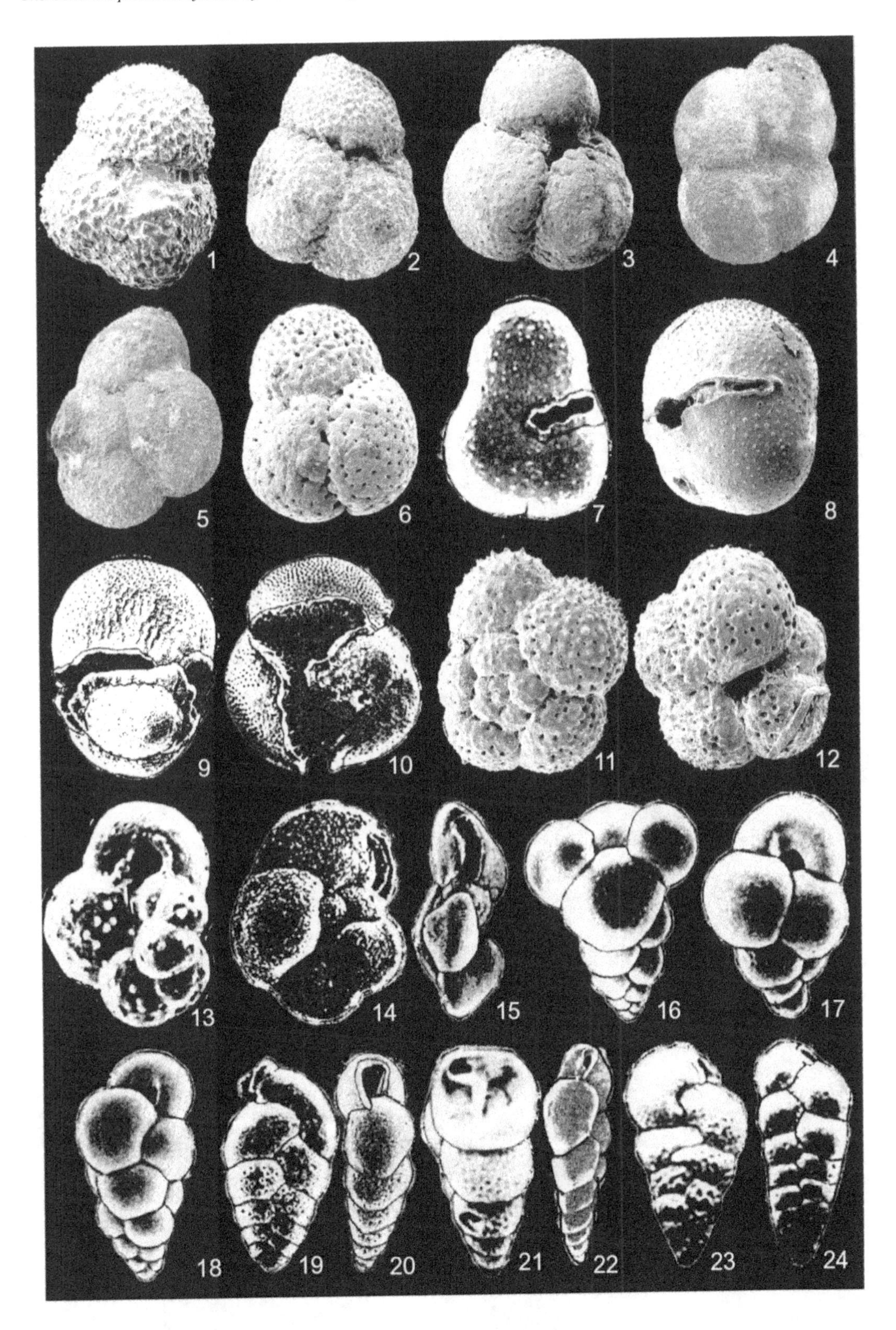
1
2
3
4
5
6
7
8
9
10
11
12
13
14
15
16
17
18
19
20
21
22
23
24

Plate 6.11. All figures are figured by Postuma (1971). Figure 1. *Globigerinatella insueta* (Cushman and Stainforth). Trinidad, Miocene, N7, x86. Figure 2. *Globigerinita naparimaensis* (Brönnimann). East Gulf, Miocene, N16, x160. Figure 3. *Globigerinoides altiaperturus* (Bolli). Trinidad, Miocene, N6, x73. Figure 4. *Globigerinoides conglobatus* (Brady). Pacific Ocean, Pliocene, N19, x47. Figure 5. *Globigerinoides extremus* (Bolli and Bermudez). Venezuela, Miocene, N18, x65. Figure 6. *Globigerinoides sacculifer* (Brady). Italy, Miocene, N17, x70. Figure 7. *Globigerinoides fistulosus* (Schubert). A 3344, Pliocene, N19, x33. Figure 8. *Globigerinoides obliquus* Bollii. Trinidad, Miocene, N17, x87. Figure 9. *Praeorbulina sicana* (de Stefani). Trinidad, Miocene, N8, x93. Figure 10. *Globigerinoides trilobus* (Reuss). Venezuela, Miocene, N15, x90. Figure 11. *Dentoglobigerina altispira* (Reuss). Trinidad, Miocene, N15, x90. Figure 12. *Globoquadrina dehiscens* (Cushman, Parr, and Collins). Trinidad, Miocene, N17, x83. Figure 13. *Truncorotalia crassaformis* (Galloway and Wissler). Indonesia, Pliocene, N19, x90. Figure 14. *Neogloboquadrina dutertrei* (d'Orbigny). Pacific Ocean, Pliocene, N19, x95. Figure 15. *Globorotalia praefohsi* (Blow and Banner). Trinidad, Miocene, N12b, x67. Figure 16. *Truncorotalia truncatulinoides* (d'Orbigny). Pacific Ocean, Pleistocene, N22, x60. Figure 17. *Globorotalia plesiotumida* (Blow and Banner). Jamaica, Miocene, N17, x60. Figure 18. *Pulleniatina obliquiloculata* (Parker and Jones). Challenger Expedition from Station 224, Holocene, N23b, x53. Figure 19. *Sphaeroidinella dehiscens* (Parker and Jones). Pacific Ocean, Pliocene, N19, x57. Figure 20. *Sphaeroidinellopsis subdehiscens* (Blow). Trinidad, Miocene, N17, x100.

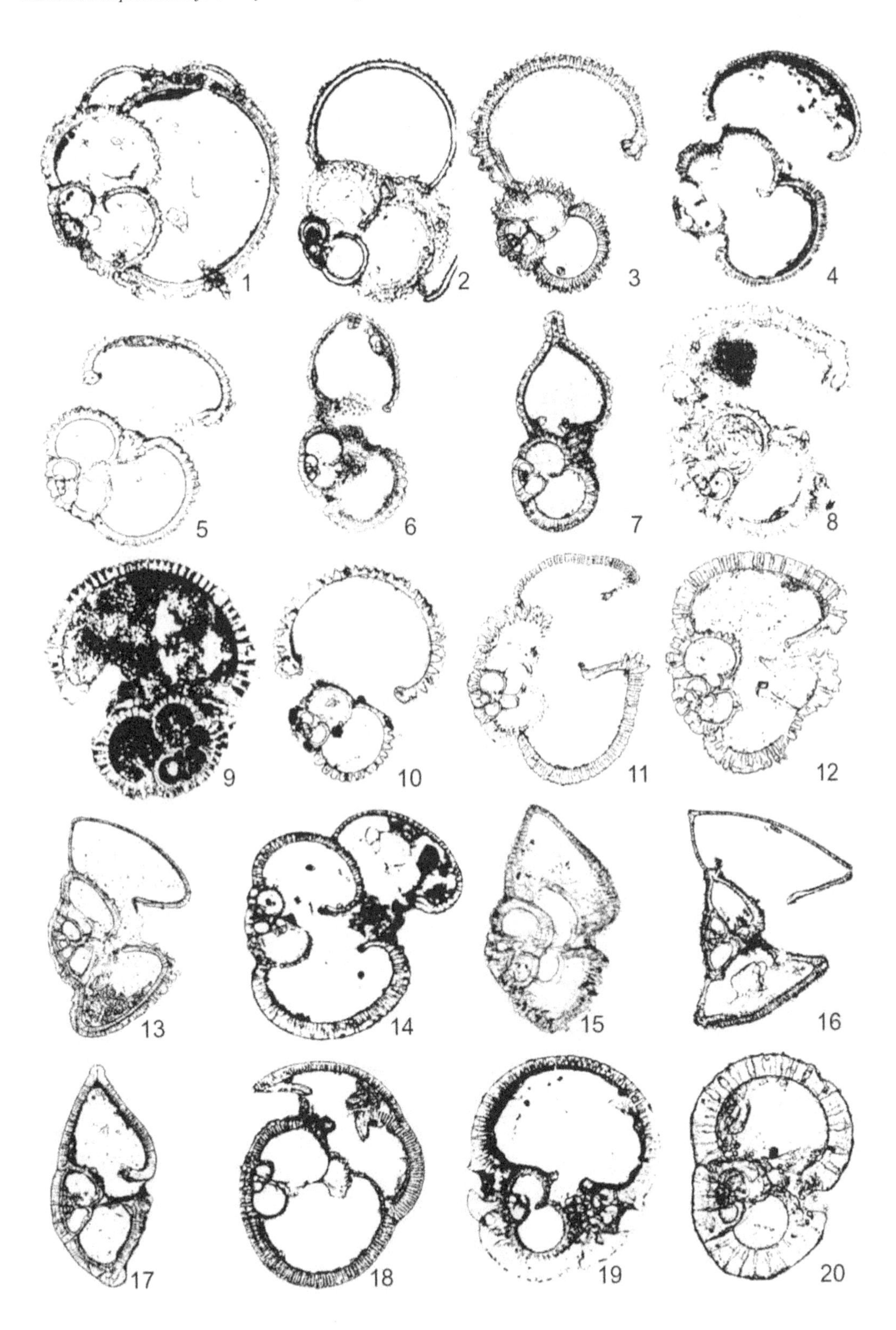
1
2
3
4
5
6
7
8
9
10
11
12
13
14
15
16
17
18
19
20

Plate 6.12. All images from the UCL Collection. Figure 1. *Globigerinoides altiaperturus* (Bolli). Lebanon, Miocene, N6, x110. Figure 2. *Globorotalia praemenardii* (Cushman and Stainforth). Indonesia, Miocene, N12, x52. Figure 3. *Dentoglobigerina altispira* (Cushman and Jarvis). Indonesia, Miocene, N7, x140. Figure 4. *Globorotalia plesiotumida* (Blow and Banner). Indonesia, Miocene, N17, x35. Figure 5. (A) *Dentoglobigerina altispira* (Cushman and Jarvis), (B) *Orbulina universa* (d'Orbigny). Indonesia, Miocene, N12, x37. Figure 6. *Sphaeroidinellopsis subdehiscens* (Blow). Indonesia, Miocene, N16, x25. Figure 7. (A) *Globigerinoides* sp., (B) *Neogloboquadrina* sp., (C) *Globorotalia margaritae* (Bolli and Bermudez). Indonesia, Pliocene, N20b, x37.

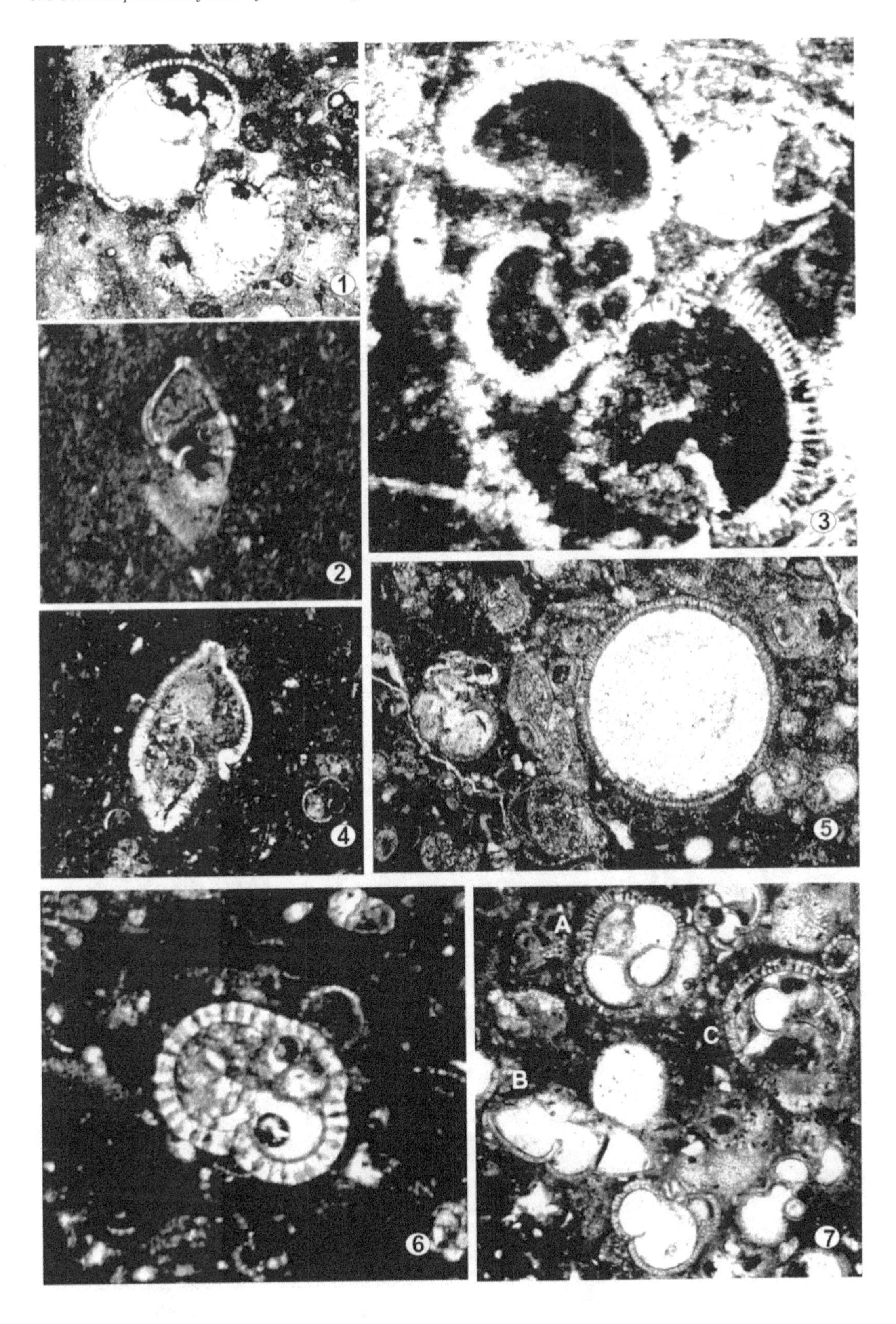
1
2
3
4
5
6
7
A
B
C

Plate 6.13. All images from the UCL Collection. Figure 1. (A) *Globoquadrina* sp., (B) *Globorotalia menardii* (d'Orbigny). Indonesia, Miocene, N17, x38. Figure 2. *Globoquadrina dehiscens* (Chapmin, Parr, and Collins). Indonesia, Miocene, N16, x35. Figure 3. (A) *Sphaeroidinellopsis subdehiscens* (Blow), (B) *Globigerinoides trilobus* (Reuss), (C) *Globorotalia margaritae* (Bolli and Bermudez). Indonesia, Pliocene, N20b, x20. Figure 4. (A) *Orbulina universa* (d'Orbigny), (B) *Globorotalia inflata* (d'Orbigny), Indonesia, Pliocene, N20b, x21. Figure 5. (A) *Globigerinoides* sp., (B) *Globorotalia margaritae* (Bolli and Bermudez). Indonesia, Pliocene, N20b, x33. Figure 6. (A) *Globorotalia fohsi* (Cushman and Ellisor), (B) *Paragloborotalia* sp. Indonesia, Miocene, N12b, x35. Figure 7. *Globigerinoides quadrilobatus* (d'Orbigny). Indonesia, Miocene, N16, x32.

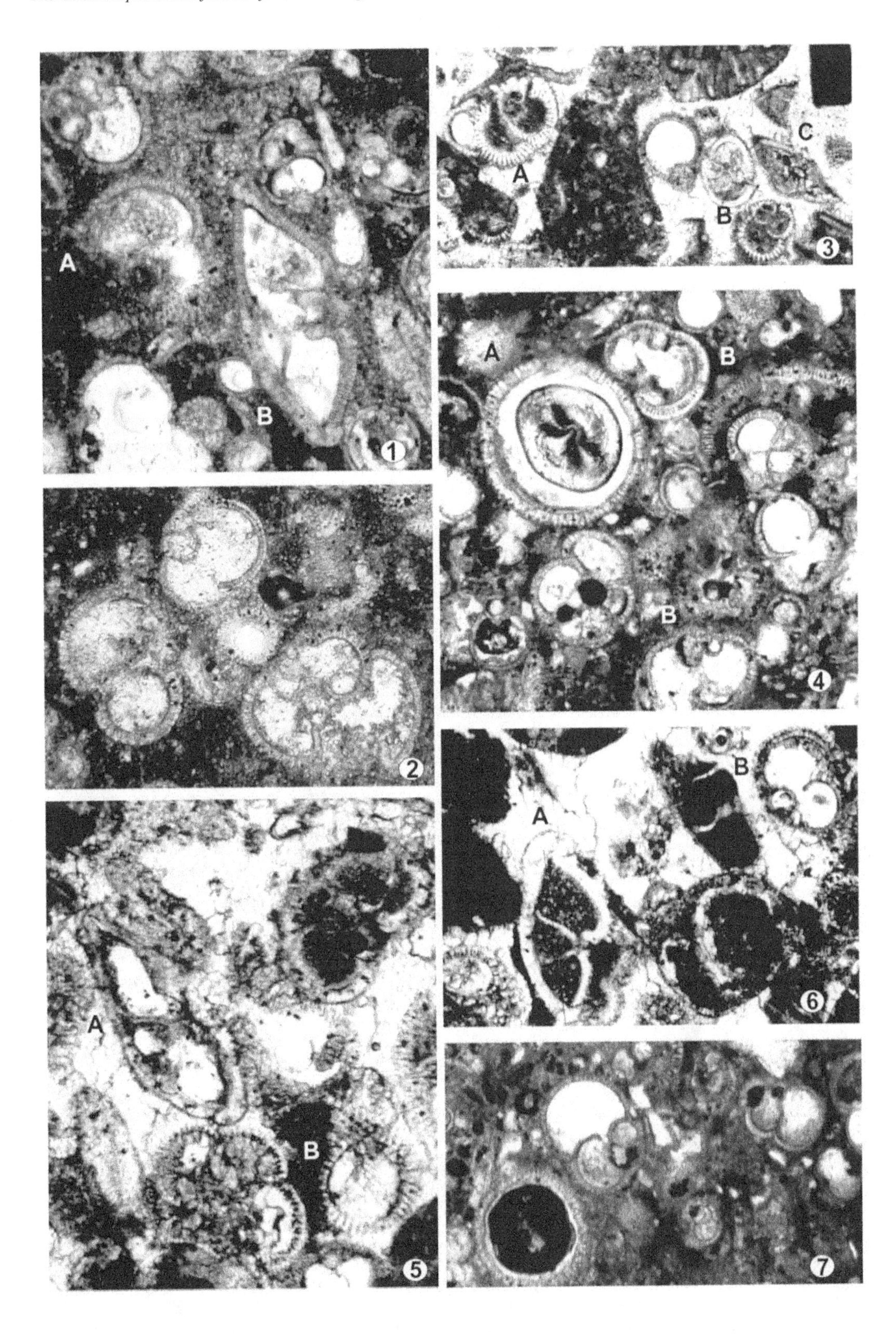
A
B
1
2
A
B
5
A
B
C
3
A
B
B
4
A
B
6
7

Plate 6.14. All images from the UCL Collection. Figure 1. (A) *Globigerinoides* sp., (B) Operculina sp., (C) *Globorotalia scitula* (Brady). Indonesia, Miocene, N12, x28. Figure 2. *Pulleniatina primalis* (Banner and Blow). Indonesia, Pliocene, N19, x23. Figure 3. (A) *Globorotalia* sp., (B) *Globorotalia margaritae* (Bolli and Bermudez), (C) *Globigerinoides* sp., (D) *Globoquadrina* sp. Indonesia, Pliocene, N20, x28. Figure 4. (A) *Globorotalia miocenica* (Palmer), (B) *Sphaeroidinella dehiscens* (Parker and Jones). Plioccene, N19, x38. Figure 5. (A) *Sphaeroidinellopsis subdehiscens* (Blow), (B) *Dentoglobigerina altispira* (Cushman and Jarvis), (C) *Globorotalia fohsi* (Cushman and Ellisor). Indonesia, Miocene, N13a, x33. Figure 6. *Globorotalia margaritae* (Bolli and Bermudez). Indonesia, Pliocene, N20, x43.

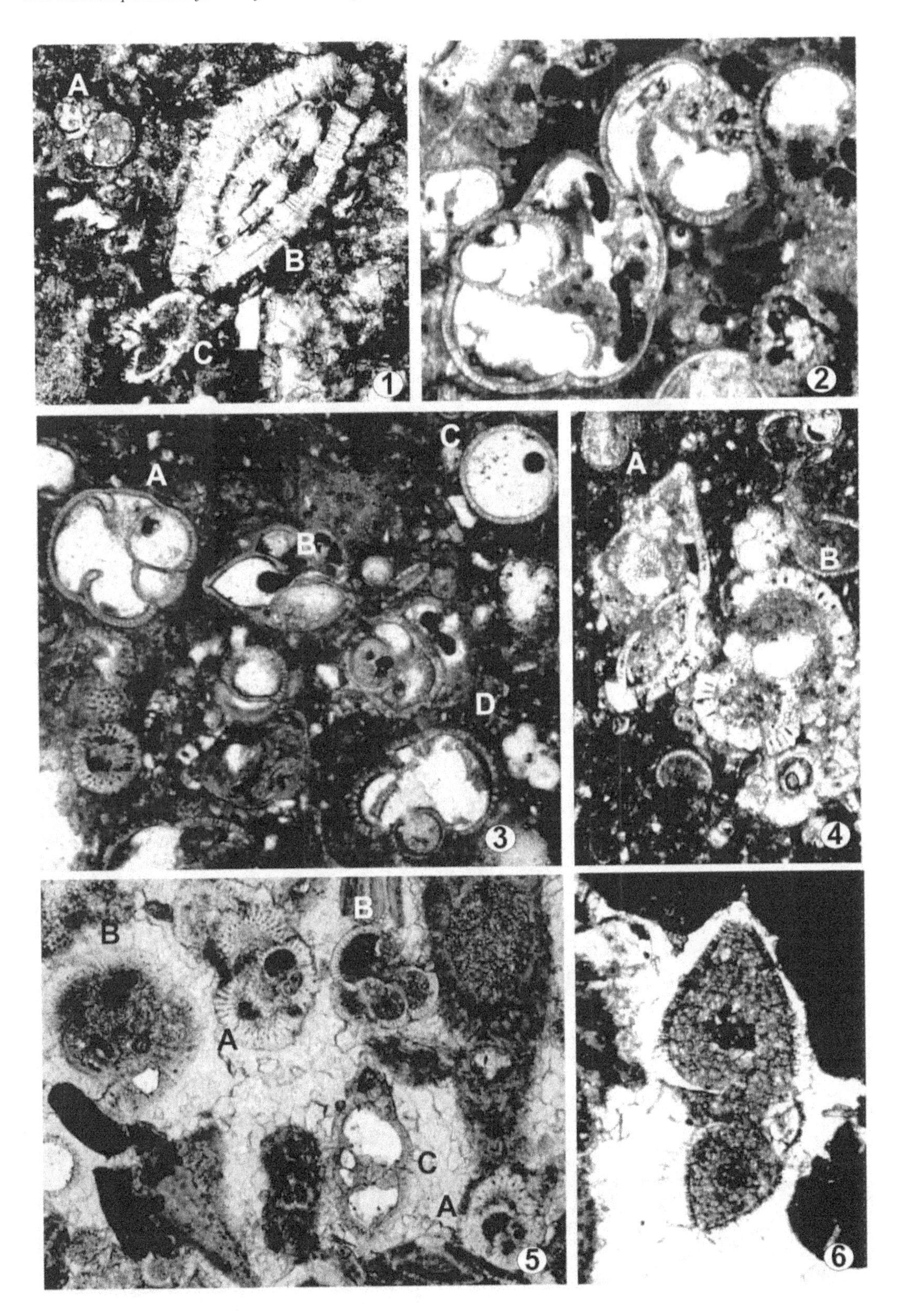
A
B
C
1
2
A
B
C
D
3
A
B
4
B
B
A
C
A
5
6

Plate 6.15. All images from the UCL Collection. Figure 1. (A) *Globigerina* sp., (B) *Orbulina universa* (d'Orbigny). Indonesia, Miocene, N12, x18. Figure 2. *Globorotalia scitula* (Brady). Indonesia, Miocene, N9, x23. Figure 3. *Dentoglobigerina altispira* (Cushman and Jarvis). Indonesia, Miocene, N9, x24. Figure 4. *Globigerinoides ruber* (d'Orbigny). Indonesia, Miocene, N16, x40. Figure 5. *Globorotalia margaritae* (Bolli and Bermudez). Indonesia, Pliocene, N20, x23. Figure 6. *Paragloborotalia peripheroronda* (Blow and Banner). Indonesia, Miocene, N9, x23. Figure 7. (A) *Paragloborotalia mayeri* (Cushman and Ellisor), (B) *Orbulina* sp., Indonesia, Miocene, N12, x22. Figure 8. *Globigerinoides trilobus* (Reuss). Indonesia, Miocene, N12, x82. Figure 9. (A) *Pulleniatina* primalis (Banner and Blow), (B) *Globoquadrina* sp., Pliocene, N19, x18. Figure 10. *Paragloborotalia mayeri* (Cushman and Ellisor). Indonesia, Miocene, N12, x22. Figure 11. *Catapsydrax parvulus* (Bolli, Loeblich, and Tappan). Indonesia, Miocene, N12, x50. Figure 12. *Globoquadrina dehiscens* (Chapmin, Parr, and Collins). Indonesia, Miocene, N16, x38.

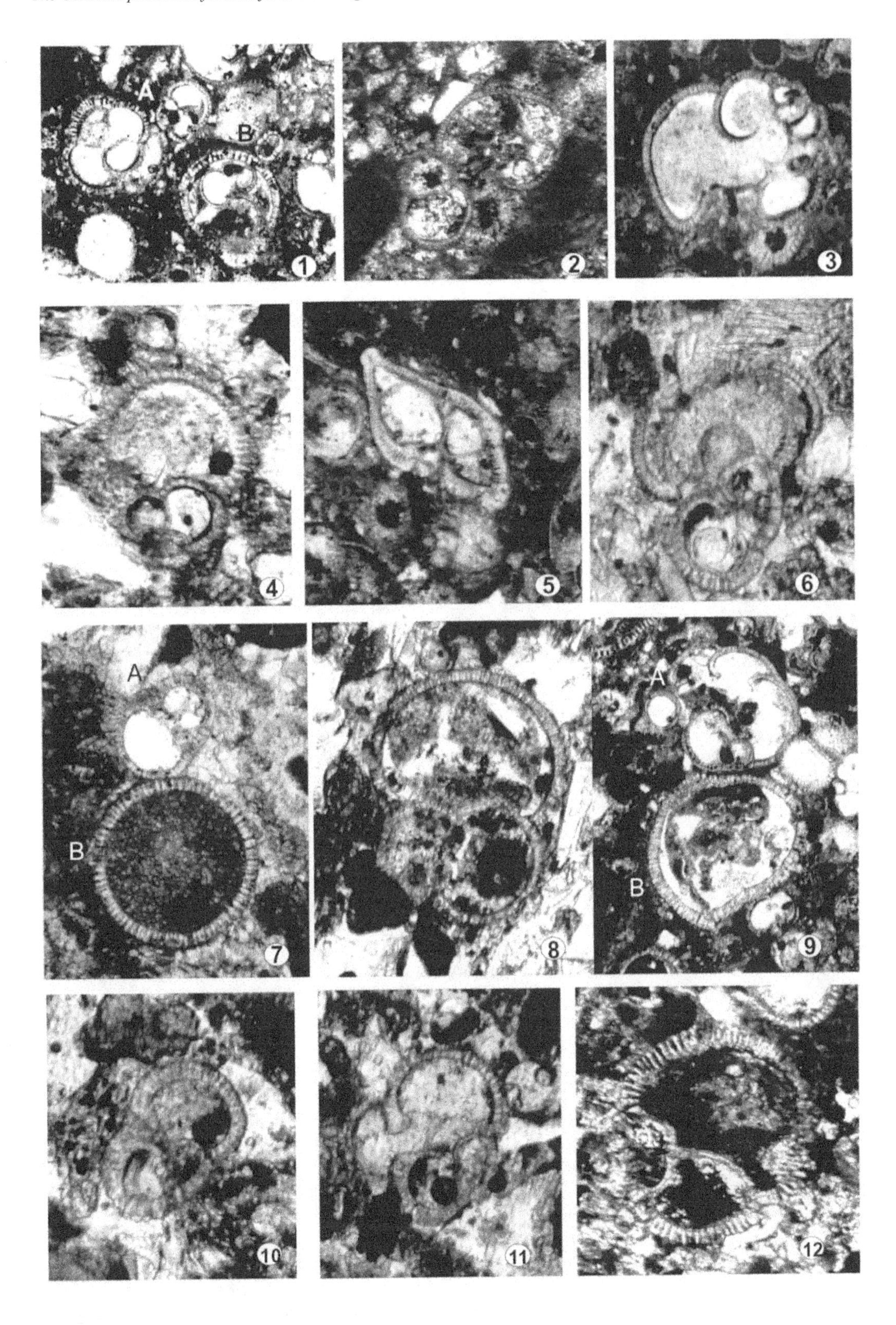
A
B
1
2
3
4
5
6
A
B
7
8
A
B
9
10
11
12

References

Abramovich, S., Keller, G., 2003. Planktic foraminiferal response to the latest Maastrichtian abrupt warm event: a case study from mid-latitude Atlantic Site525A. Mar. Micropaleontol. 48, 225–249.

Adams, C.G., Lee, D.E., Rosen, B.R., 1990. Conflicting isotopic and biotic evidence for tropical sea-surface temperatures during the Tertiary. Palaeogeogr. Palaeoclimatol. Palaeoecol. 77, 289–313.

Adshead, P.C., 1980. Pseudopodial variability and behaviour of globigerinids (foraminiferida) and other planktonic sarcodina developing in cultures. In: Sliter, W.V. (Ed.), Studies in Marine Micropaleontology and Paleoecology, a Memorial Volume to Orville L. Bandy, special publication, 19, 96–126. Allen Press.

Agnini, C., Macrı, P., Backman, J., Brinkhuis, H., Fornaciari, E., Giusberti, L., Luciani, V., Rio, D., Sluijs, A., Speranza, F., 2009. An early Eocene carbon cycle perturbation at ~52.5 Ma in the Southern Alps: chronology and biotic response. Paleoceanography 24, PA2209. http://dx.doi.org/10.1029/2008PA001649.

Aitchison, J.C., Ali, J.R., Davis, A.M., 2007. When and where did India and Asia collide? J. Geophys. Res. 112, B05423.

Alegret, L., Arenillas, I., Arz, J.A., Diaz, C., Grajales-Nishimura, J.M., Melendez, A., Molina, E., Rojas, R., Soria, A.R., 2005. Cretaceous-Paleogene boundary deposits at Loma Capiro, central Cuba: evidence for the Chicxulub impact. Geology 33, 721–724.

Alexander, S.P., 1985. The Cytology of Certain Benthonic Foraminifera in Relation to Test Structure and Function. University of Wales, PhD Thesis.

Alexander, S.P., Banner, F.T., 1984. The functional relationship between skeleton and cytoplasm in *Haynesina germanica* (Ehrenberg). J. Foraminiferal Res. 14, 159–170.

Allmon, W.D., 2001. Nutrients, temperature, disturbance, and evolution: a model for the late Cenozoic marine record of the western Atlantic. Palaeogeogr. Palaeoclimatol. Palaeoecol. 166, 9–26.

Anderson, O.R., Faber Jr., W.W., 1984. An estimation of calcium carbonate deposition rate in a planktonic foraminifer *Globigerinoides sacculifer* using ^{45}Ca as a tracer: a recommended procedure for improved accuracy. J. Foraminiferal Res. 14, 303–308.

Ando, A., Huber, B.T., MacLeod, K.G., 2010. Depth-habitat reorganization of planktonic foraminifera across the Albian/Cenomanian boundary. Paleobiology 36, 357–373.

Angell, R.B., 1967. The process of chamber formation in the foraminifer *Rosalina floridans*. J. Protozool. 14, 566–574.

Apthorpe, M., 2002. Early Bajocian planktonic foraminifera from western Australia. In: Revets, S. (Ed.), FORAMS 2002, International Symposium on Foraminifera. University of Western Australia, Perth, Australia, Volume of Abstracts, 81.

Archibald, J.M., 2008. The eocyte hypothesis and the origin of eukaryotic cells. Proc. Natl. Acad. Sci. U.S.A. 51, 20049–20050.

Archibald, J.M., Longet, D., Pawlowski, J., Keeling, P.J., 2003. A novel polyubiquitin structure in Cercozoa and Foraminifera: evidence for a new eukaryotic supergroup. Mol. Biol. Evol. 20, 62–66.

Asano, K., Ingle, J.C., Takayanagi, Y., 1968. Origin and development of *Globigerina quinqueloba* Natland in the North Pacific: sciences reports of Tohoku University. 2nd Ser. (Geol.) 39, 213–241.

Aurahs, R., Göker, M., Grimm, G., Hemleben, V., Hemleben, C., Schiebel, R., Kučera, M., 2009. Using the multiple analysis approach to reconstruct phylogenetic relationships among planktonic foraminifera from highly divergent and length-polymorphic SSU rDNA sequences. Bioinform. Biol. Insights 2009 (3), 155–177.

Aze, T., Ezard, T., Purvis, A., Coxall, H., Stewart, D., Wade, B., Pearson, P., 2011. A phylogeny of Cenozoic macroperforate planktonic foraminifera from fossil data. Biol. Rev. 86, 900–927.

Baldauf, S.L., 2008. An overview of the phylogeny and diversity of eukaryotes. J. Syst. Evol. 46, 263–273.

Bambach, R.K., 2006. Phaenerozoic biodiversity mass extinctions. Annu. Rev. Earth Planet. Sci. 34, 117–155.

Bandy, O.L., 1960. The geological significance of coiling ratios in the foraminifera *Globigerina pachyderma* (Ehrenberg). J. Paleontol. 34, 671–681.

Banner, F.T., 1978. Form and function in coiled benthic foraminifera. Br. Micropalaeonotol. 8, 11–12.

Banner, F.T., Blow, W.H., 1965. Progress in the planktonic foraminiferal biostratigraphy of the Neogene. Nature 208, 1164–1166.

Banner, F.T., Desai, D., 1988.A review and revision of the Jurassic—early Cretaceous *Globigerinina*, with especial reference to the Aptian assemblages of Speeton (North Yorkshire, England). J. Micropalaeontol. 7, 143–185.

Banner, F.T., Strank, A.R.E., 1987. On *Wondersella athersuchi*, a new stratigraphically significant hedbergellid foraminiferan from the Cretaceous Shuaiba Formation of the Middle East. J. Micropalaeontol. 6, 39–48.

Banner, F.T., Strank, A.R.E., 1988. *Wondersella athersuchi* Banner and Strank: a corrective note on its systematic position. J. Micropalaeontol. 7, 186.

Banner, F.T., Williams, E., 1973. Test structure, organic skeleton and extrathalamous cytoplasm of *Ammonia* Brunnich. J. Foraminiferal Res. 3, 49–69.

Banner, E.T., Pereira, C.P.G., Desai, D., 1985. 'Tretomphaloid' float chambers in the Discorbidae and Cymbaloporidae. J. Foraminiferal Res. 15, 159–174.

Banner, F.T., Copestake, P., White, M.R., 1993. Barremian–Aptian Praehedbergellidae of the North Sea area: a reconnaissance. Bull. Br. Mus. 49, 1–30.

Baudin, F., 2005. A late Hauterivian short-lived anoxic event in the Mediterranean Tethys: the "Faraoni event" C. R. Geosci. 337, 1532–1540.

Bé, A.W.H., 1968. Shell porosity of recent planktonic foraminifera as a climatic index. Science 161, 881–884.

Bé, A.W.H., 1969. Planktonic foraminifera. Arctic Map Folio 11, 9–12.

Bé, A.W.H., 1980. Gametogenic calcification in a spinose planktonic foraminifer, *Globigerinoides sacculifer* (Brady). Mar. Micropaleontol. 5, 283–310.

Bé, A.W.H., Anderson, O.R., 1976. Gametogenesis in planktonic foraminifera. Science 192, 890–892.

Bé, A.W.H., Hemleben, C., 1970. Calcification in a living planktonic foraminifera Globigerinoides sacculifer (Brady). Neues Jahrbuch für Geologie und Palaontologie, Monatshefte 134, 221–234.

Bé, A.W.H., 1982. Biology of planktonic foraminifera. In: Broadhead, T.W. (Ed.), Foraminifera: Notes for a Short Course. Studies in Geology, University of Tennessee, Knoxville. Vol. 6, pp. 51–92.

Bé, A.W.H., McIntyre, A., Breger, D.L., 1966. Shell microstructure of a planktonic foraminifer, *Globorotalia menardii* (d'Orbigny). Eclogae Geol. Helv. 59, 885–896.

Bé, A.W.H., Harrison, S.M., Lott, L., 1973. *Orbulina universa* d'Orbigny in the Indian Ocean. Micropaleontology 19, 150–192.

Bé, A.W.H., Hemleben, Ch., Anderson, O.R., Spindler, M., 1979. Chamber formation in planktonic foraminifera. Micropaleontology 25, 294–307.

Bé, A.W.H., Hemleben, C., Anderson, O.R., Spindler, M., Hacunda, J., Tunitivate-Choy, S., 1977. Laboratory and field observation of living planktonic foraminifera. Micropaleontology 23, 155–179.

Bé, A.W.H., Spero, H.J., Anderson, O.R., 1982. Effects of symbiont elimination and reinfection on the life processes of the planktonic foraminifer *Globigerinoides sacculifer*. Mar. Biol. 70, 73–86.

Beckmann, J.P., 1957. *Chiloguembelina* Loeblich and Tappan and related Foraminifera from the lower Tertiary of Trinidad, B.W.I. Bull. U.S. Natl. Museum 215, 83–95.

Benjamini, C., Reiss, Z., 1979. Wall-hispidity and perforation in Eocene planktonic foraminifera. Micropaleontology 25 (2), 141–150.

Berggren, W.A., 1962. Some planktonic Foraminifera from the Maestrichtian and type Danian stages of Denmark and southern Sweden. Stockholm Contrib. Geol. 9, 1–102.

Berggren, W.A., 1969. Cenozoic chronostratigraphy, planktonic foraminiferal zonation and the radiometric time scale. Nature 224, 1072–1075.

Berggren, W.A., 1971. Multiple phylogenetic zonations of the Cenozoic based on planktonic foraminifera. In: Farinacci, A. (Ed.), Proceedings of the II Planktonic Conference, Roma 1970, 41–56.

Berggren, W.A., 1977. Atlas of Paleogene planktonic foraminifera: some species of the genera *Subbotina, Planorotalites, Morozovella, Acarinina* and *Truncorotaloides*. In: Ramsey, A.T.S. (Ed.), Oceanic Micropaleontology. Academic Press, London, pp. 205–300.

Berggren, W.A., Kent, D.V., Swisher III, C.C., Berggren, W.A., Hilgen, F.J., Langereis, C.G., Kent, D.V., Obradovich, J.D., Raffi, I., Raymo, M.E., Shackleton, N.J., 1995a. Late Neogene chronology: new perspectives in high resolution stratigraphy. American Bulletin of the Geologival Society 107, 1272–1287.

Berggren, W.A., Kent, D.V., Swisher III, C.C., Aubry, M.-P., 1995b.A revised Cenozoic geochronology and chronostratigraphy. In: Berggren, W.A., Kent, D.V., Aubry, M.-P., Hardenbol, J. (Eds.), Geochronology, Time Scales and Global Stratigraphic Correlation: A Unified Temporal Framework for an Historical Geology54, SEPM Special Publication, pp. 129–212.

Berggren, W.A., Miller, K.G., 1988. Paleogene tropical planktonic foraminiferal biostratigraphy and magnetobiochronology. Micropaleontology 34, 362–380.

Berggren, W.A., Norris, N.D., 1993. Origin of the genus *Acarinina* and revisions to Paleocene biostratigraphy. In: Abstracts with Programs, Geological Society of America Annual Meeting, A359.

Berggren, W.A., Norris, R.D., 1997. Biostratigraphy, phylogeny and systematics of Paleocene trochospiral planktic foraminifera. Micropaleontology 43, 1–116.

Berggren, W.A., Pearson, P.N., 2005. A revised tropical and subtropical Paleogene planktonic foraminiferal zonation. J. Foraminiferal Res. 35, 279–298.

Berggren, W.A., Prothero, D.R., 1992. Eocene-Oligocene climatic and biotic evolution: an overview. In: Prothero, D.R., Berggren, W.A. (Eds.), Eocene-Oligocene Climatic and Biotic Evolution. Princeton University Press, Princeton, pp. 1–28.

Berggren, W.A., Phillips, J.D., Bertels, A., Wall, D., 1967. Late Pliocene-Pleistocene stratigraphy in deep sea cores from the south-central North Atlantic. Nature 216, 253–254.

Berggren, W.A., Kent, D.V., Flynn, J.J., Van Couvering, J.A., 1985. Cenozoic geochronology. Geol. Soc. Am. Bull. 96, 1407–1418.

Berggren, W.A., Hilgen, F.J., Langereis, C.G., Kent, D.V., Obradovich, J.D., Raffi, I., Raymo, M.E., Shackleton, N.J., 1995a. Late Neogene chronology: new perspectivesin high-resolution stratigraphy. Geol. Soc. Am. Bull. 107, 1272–1287.

Berggren, W.A., Kent, D.V., Swisher, C.C., Aubry, M.-P., 1995b. A revised Cenozoic geochronology and chronostratigraphy. In: Berggren, W.A., Kent, D.V., Hardenbol, J. (Eds.), Geochronology, Time Scales and Global Stratigraphic Correlations: A Unified Temporal Framework for an Historical Geology, Society of Economic Paleontologists and Mineralogists Special Publication, 54, Tulsa, OK, 129–212.

Belyea, P.R., Thunell, R.C., 1984. Fourier shape analysis and planktonic foraminiferal evolution: the *Neogloboquadrina-Pulleniatina* lineages. J. Paleonotol. 58, 1026–1040.

Berney, C., Pawlowski, J., 2003. Revised small subunit rRNA analysis provides further evidence that Foraminifera are related to Cercozoa. J. Mol. Evol. 57, S120–S127.

Bijma, J., Faber, W.W.J., Hemleben, C., 1990. Temperature and salinity limits for growth and survival of some planktonic foraminifers in laboratory cultures. J. Foraminiferal Res. 20, 95–116.

Blow, W.H., 1969.Late Middle Eocene to recent planktonic foraminiferal biostratigraphy. In: Brönnimann,P., Renz,H.H. (Eds.), Proceedings of the First International Conference on Planktonic Microfossils1, E.J. Brill, Leiden, pp. 199–422.

Blow, W.H., 1979. A study of the morphology, taxonomy, evolutionary relationship and the stratigraphical distribution of some Globigerinidae (mainly Globigerinacea). In: Brill, E.J. (Ed.), The Cainozoic Globigerinidae, Leiden, Netherlands, 3 vols. 1–1413.

Blow, W.H., Banner, F., 1962. The mid-tertiary (upper Eocene to Aquitanian) globigerinaceae. In: Eames, F.E. *et al.*, (Ed.), Fundamentals of Mid-Tertiary Stratigraphical Correlation. Cambridge University Press, Cambridge, pp. 61–163.

Boersma, A., Premoli Silva, I., 1991. Distribution of Paleogene planktonic foraminifera analogies with the Recent? Palaeogeogr. Palaeoclimatol. Palaeoecol. 83, 29–48.

Boersma, A., Premoli Silva, L., Shackleton, N.J., 1987. Atlantic Eocene planktonic foraminiferal paleohydrographic indicators and stable isotope paleoceanography. Paleoceanography 2, 287–331.

Bolli, H.M., 1957. The genera *Globigerina* and *Globorotalia* in the Paleocene-Lower Eocene Lizard Springs Formation of Trinidad, B.W.I. In: Loeblich Jr., A.R. *et al.*, (Ed.), Studies in Foraminifera, Bulletin of the United States National Museum 215, 61–82.

Bolli, H.M., 1966. Zonation of Cretaceous to Pliocene marine sediments based on Planktonic foraminifera. Boletín informativo dela Asociacion Venezolanade Geología, Mineríay Petroleo, 9, 1–34.

Bolli, H.M., Saunders, J.B., 1985. Oligocene to Holocene low latitude planktic foraminifera. In: Bolli, H.M., Saunders, J.B., Perch-Nielsen, K. (Eds.), Plankton Stratigraphy. Cambridge University Press, Cambridge, pp. 155–262.

Bolli, H.M., Loeblich, A.R., Tappan, H., 1957. The Planktonic foraminiferal families Hantkeninidae, Orbulinidae, Globorotaliidae, and Globotruncanidae. U.S. Natl. Museum Bull. 215, 3–50.

Boltovskoy, E., Wright, R., 1976. Recent Foraminifera. Dr. W. Junk Publishers, The Hague, 515p.

Bornemann, A., Norris, R.D., 2007. Size-related stable isotope changes in Late Cretaceous planktic foraminifera: implications for paleoecology and photosymbiosis. Mar. Micropaleontol. 65, 32–42.

Bornemann, A., Norris, R., Friedrich, O., Beckmann, B., Schouten, S., Sinninghe Damsté, J., Vogel, J., Hofmann, P., Wagner, T., 2008. Isotopic evidence for glaciation during the cretaceous supergreenhouse. Science 319, 189–192.

Borsetti, A.M., Cati, F., Colalongo, M.L., Sartoni, S., 1979. Biostratigraphy and absolute ages of the Italian Neogene. Annales Geologia Hellenica. In: 7th International Mediterranean Neogene Congress, Athens 183–197.

BouDagher-Fadel, M.K., 1996. Re-evaluation of the Early Cretaceous planktonic foraminifera (Praehedbergellidae). Cretaceous Res. 17, 761–771.

BouDagher-Fadel, M.K., 2008. Evolution and Geological Significance of Larger Benthic Foraminifera. Developments in Palaeontology and Stratigraphy 21, Elsevier, Amsterdam, p. 544.

BouDagher-Fadel, M.K. 2012. Globigerinitoidea, a new Cenozoic planktonic foraminiferal superfamily, with an emended family and species, Mircoapaleonotology. 58, 396.

BouDagher-Fadel, M.K., 2012. Biostratigraphic and Geological Significance of Planktonic Foraminifera. Developments in Palaeontology & Stratigraphy 22, Elsevier, Amsterdam, p. 275

BouDagher-Fadel, M.K., Price, G.D., 2010. The evolution and palaeogeographic distribution of the lepidocyclinids. J. Foraminiferal Res. 40, 79–108.

BouDagher-Fadel, M.K., Banner, F.T., Whittaker, J.E., 1997. Early Evolutionary History of Planktonic Foraminifera. British Micropalaeontological Society Publication Series, Chapman and Hall Publishers, London, p. 269.

Brown N.K., Jr., 1969. Heterohelicidae Cushman, 1927, amended, a cretaceous planktonic foraminiferal family. In: Brönnimann, P., Renz, H.H. (Eds.), Proceedings of the First International Conference on Planktonic Microfossils, Geneva 1967. E.J. Brill, Leiden, vol. 2, pp. 21–67.

Brady, H.B., 1884. Report on the foraminifera dredged by H.M.S. 'Challenger' during the years 1873–1876. Report of the scientific results of the exploring voyage of H.M.S. Challenger, Zoology, 49: 1–814, pls, 1–115.

Brinkhuish, H., Zachariasswe, J., 1988. Dinoflagellate cysts, sea level changes and planktonic foraminifers across the Cretaceous-Tertiary boundary at Al Haria. northwest Tunisia. Mar. Micropaleontol. 13, 153–191.

Brönnimann, P., 1950. The genus *Hantkenina* Cushman in Trinidad and Barbados, B. W. I. J. Paleontol. 24, 397–420.

Brummer, G.J.A., Hemleben, C., Spindler, M., 1987. Ontogeny of extant globigerinoid planktonic foraminifera; a concept exemplified by *Globigerinoides sacculifer* (Brady) and *G. ruber* (d'Orbigny). Mar. Micropaleontol. 12, 357–381.

Caron, D.A., Bé, A.W.H., Anderson, O.R., 1981. Effect of variations in light intensity on life processes of planktonic foraminifer *Globigerinoides sacculifer* in laboratory cultures. J. Mar. Biol. Assoc. U.K. 62, 435–451.

Caron, M., Homewood, P., 1983. Evolution of early foraminifers. Mar. Micropaleontol. 7, 453–462.

Carpenter, W.B., Parker, W.K., Jones, T.R., 1862. Introduction to the study of the foraminifera. Hardwicke, London, pp 319.

Carter, D.J., Hart, M.B., 1977. Aspects of mid-Cretaceous stratigraphical micropalaeontology. Bull. Br. Mus. 29, 1–135.

Cavalier-Smith, T., 2002. Chloroplast evolution: secondary symbiogenesis and multiple losses. Curr. Biol. 12, R62–R64.

Chaisson, W., Leckie, M., 1993. High-resolution Neogene planktonic foraminiferal biostratigraphy of Site 806, Ontong Java Plateau (western equatorial Pacific). In: Berger, W., Kroenke, L.W., Mayer, L.A. *et al.*, (Eds.), Proceedings of the Ocean Drilling Program, Scientific Results, Ocean Drilling Program, College Station, TX, Vol. 130, pp. 137–178.

Chaisson, W.P., Pearson, P.N., 1997. Planktonic foraminifer biostratigraphy at Site 925: Middle Miocene–Pleistocene. In: Shackleton, N.J., Curry, W.B., Richter, C., Bralower, T.J. (Eds.), Proceeding of the Ocean Drilling Program, Scientific Results, College Station TX, 154, 3–31.

Chaisson, W.P., Ravelo, A.C., 1997. Changes in upper water-column structure at Site 925, late Miocene–Pleistocene: planktonic foraminifer assemblage and isotopic evidence. In: Shackleton, N.J., Curry, W.B., Richter, C., Bralower, T.J. (Eds.), Proceedings of the Ocean Drilling Program, Scientific Results, Ocean Drilling Program, College Station, TX. Vol. 154, pp. 255–268.

Cifelli, R., 1982. Early occurrences and some phylogenetic implications of spiny, honeycomb textured planktonic foraminifers. J. Foraminiferal Res. 12, 105–115.

Coccioni, R., Baudin, F., Cecca, F., Chiari, M., Galeotti, S., Gardin, S., Salvini, G., 1998. Integrated stratigraphic, palaeontological, and geochemical analysis of the uppermost Hauterivian Faraoni Level in the Fiume Bosso section, Umbria–Marche Apennines, Italy. Cretaceous Res. 19, 1–23.

Courtillot, V.E., Renne, P.R., 2003. On the ages of flood basalt events. C. R. Geosci. 335, 113–140.

Coxall, H.K., Huber, B.T., Pearson, P.N., 2003. Origin and morphology of the Eocene planktonic foraminifera *Hantkenina*. J. Foraminiferal Res. 33, 237–261.

Coxall, H.K., Wilson, P.A., Pearson, P.N., Sexton, P.F., 2007. Iterative evolution of digitate planktonic foraminifera. Paleobiology 33, 495–516.

Cramer, B.S., Kent, D.V., 2005. Bolide summer: the Paleocene/Eocene thermal maximum as a response to an extraterrestrial trigger. Palaeogeogr. Palaeoclimatol. Palaeoecol. 224, 144–166.

Cushman, J.A., 1933. Foraminifera, their classification and economic use, Special Publications of the Cushman Laboratory of Foraminiferal Research, 2nd edition.

Cushman, J.A., Wickenden, R.T., 1930. The development of *Hantkenina* in the Cretaceous with a description of a new species. Contr. Cushman Lab. Foram. Res. 6, 39–43.

D'Hondt, S.L., 1991. Phylogenetic and stratigraphic analysis of earliest Paleocene biserial and wiserial planktonic foraminifera. J. Foraminiferal Res. 21, 168–181.

D'Hondt, S., Zachos, J.C., 1993. On stable isotopic variation and earliest paleocene planktic foraminifera. Paleoceanography 8, 527–547.

D'Hondt, S., Zachos, J.C., 1998. Cretaceous foraminifera and the evolutionary history of planktic photosymbiosis. Paleobiology 24, 512–523.

D'Hondt, S., Zachos, J.C., Schultz, G., 1994. Stable isotopic signals and photosymbiosis in late paleocene planktonic foraminifera. Paleobiology 20, 391–406.

Darling, K.F., Wade, C.M., 2008. The genetic diversity of planktic foraminifera and the global distribution of ribosomal RNA genotypes. Mar. Micropaleontol. 67, 216–238.

Darling, K.F., Pearson, D., Wade, C.M., Leigh Brown, A.J., 1996. Molecular phylogeny of the planktic foraminifera. J. Foraminiferal Res. 26, 324–330.

Darling, K.F., Wade, C.M., Kroon, D., Leigh Brown, A.J., 1997. Planktic foraminiferal molecular evolution and their polyphyletic origins from benthic taxa. Mar. Micropaleontol. 30, 251–266.

Darling, K.F., Wade, C.M., Kroon, D., Leigh Brown, A.J., Bijma, J., 1999. The diversity and distribution of modern planktonic foraminiferal small subunit ribosomal RNA genotypes and their potential as tracers of present and past ocean circulations. Paleoceanography 14, 3–12.

Darling, K.F., Wade, C.M., Steward, I.A., Kroon, D., Dingle, R., Leigh Brown, A.J., 2000. Molecular evidence for genetic mixing of Arctic and Antarctic subpolar populations of planktonic foraminifers. Nature 405, 43–47.

Darling, K.F., Kucera, M., Pudsey, C.J., Wade, C.M., 2004. Molecular evidence licks cryptic diversification in planktonic protists to Quaternary climate dynamics. Proc. Natl. Acad. Sci. U.S.A. 101, 7657–7662.

Darling, K.F., Kucera, M., Kroon, D., Wade, C.M., 2006. A resolution for the coiling direction paradox in *Neogloboquadrina pachyderma*. Paleoceanography 21, 2011.

Darling, K.F., Kucera, M., Wade, C.M., 2007. Global molecular phylogeography reveals persistent Arctic circumpolar isolation in a marine planktonic protist. Proc. Natl. Acad. Sci. U.S.A. 104, 5002–5007.

Darling, K.F., Thomas, E., Kasemann, S.A., Seears, H.A., Smart, C.W., Wade, C.M., 2009. Surviving mass extinction by bridging the benthic/planktic divide. Proc. Natl. Acad. Sci. U.S.A. 106, 12629–12633.

de Vargas, C., Zaninetti, L., Hilbrecht, H., Pawlowski, J., 1997. Phylogeny and rates of molecular evolution of planktonic foraminifera: SSU rDNA sequences compared to the fossil record. J. Mol. Evol. 45, 285–294.

de Vargas, C., Norris, R., Zaninetti, L., Gibb, S.W., Pawlowski, J., 1999. Molecular evidence of cryptic speciation in planktonic foraminifers and their relation to oceanic provinces. Proc. Natl. Acad. Sci. U.S.A. 96, 2864–2868.

de Vargas, C., Renaud, S., Hilbrecht, H., Pawlowski, J., 2001. Pleistocene adaptive radiation in *Globorotalia truncatulinoides*: genetic, morphologic, and environmental evidence. Paleobiology 27, 104–125.

de Vargas, C., Bonzon, M., Rees, N., Pawlowski, J., Zaninetti, L., 2002. A molecular approach to biodiversity and ecology in the planktonic foraminifera *Globigerinella siphonifera* (d'Orbigny). Mar. Micropaleontol. 45, 101–116.

Desmares, D., Grosheny, D., Beaudoin, B., 2008. Ontogeny and phylogeny of Upper Cenomanian rotaliporids (Foraminifera). Mar. Microplaeontol. 69, 91–105.

Dickens, G.R., 2001. Modeling the global carbon cycle with a gas hydrate capacitor: significance for the latest Paleocene Thermal maximum. In: Paul, C., Dillon, W.P. (Eds.), Natural Gas Hydrates: Occurrence, Distribution, and Detection. Geophysical Monograph Series, AGU, Washington, DC, Vol. 124, pp. 19–38.

Dolenec, T., Pavsic, J., Lojen, S., 2000. Ir anomalies and other elemental markers near the Palaeocene-Eocene boundary in a flysch sequence from the Western Tethys (Slovenia). Terra Nova 12, 199–204.

Duguay, L.E., 1983. Comparative laboratory and field studies on calcification and carbon fixation in foraminiferal-algal associations. J. Foraminiferal Res. 13, 252–261.

Eldholm, O., Coffin, M.F., 2000. Large igneous provinces and plate tectonics. In: Richards, M.A., Gordon, R.G., van der Hilst, R.D. (Eds.), The History and Dynamics of Global Plate Motions. American Geophysical Union Geophysical Monograph 121, American Geophysical Union, Washington, DC, pp. 309–326.

Embry, J.-C., Vennin, E., Van Buchem, F.S.P., Schroeder, R., Pierre, C., Aurell, M., 2010. Mesozoic and Cenozoic Carbonate Systems of the Mediterranean and the Middle East: Stratigraphic and Diagenetic Reference Models. Geological Society, Special Publications, London 329, 113–143.

Emelyanov, E.M., 2005. Barrier Zones in the Ocean. Springer, Berlin, Heidelberg, New York 636 pp.

Erba, E., 2006. The first 150 million years history of calcareous nannoplankton: biosphere–geosphere interactions. Palaeogeogr. Palaeoclimatol. Palaeoecol. 232, 237–250.

Erez, J., 1983. Calcification rates, photosynthesis and light in planktonic foraminifera. In: Westbroek, P., de Jong, E. (Eds.), Biomineralization and Biological Metal Accumulation. D Reidel Publishing Company, Dortrecht, pp. 307–312.

Erez, J., Almogi, A., Abraham, S., 1991. On the life history of planktonic foraminifera: lunar reproduction cycle in *Globigerinoides sacculifer* (Brady). Paleoceanography 6, 295–306.

Ericson, D.B., 1959. Coiling direction of *Globigerina pachyderma* as a climatic index. Science 130, 219–220.

Ericson, D.B., Wollin, G., Wollin, J., 1954. Coiling direction of *Globorotalia truncatulinoides* in deep-sea cores. Deep Sea Res. 2, 152–158.

Faber, W.W., Anderson, O.R., Lindsey, J.L., Caron, D.A., 1985. Algal foraminiferal symbiosis in the planktonic foraminifer *Globigerinella aequilateralis*, I. Occurrence and stability of two mutually exclusive chrysophytes endosymbionts and their ultrastructure. J. Foraminiferal Res. 18, 334–343.

Fairbanks, R.G., Sverdlove, M., Free, R., Wiebe, P.H., Bé, A.W., 1982. Vertical distribution and isotopic fractionation of living planktonic foraminifera from the Panama Basin. Nature 298 (5877), 841–844.

Falkowski, P.G., Dubinsky, Z., Muscatine, L., McCloskey, L., 1993. Population control in corals. Bioscience 43, 606–611.

Falkowski, P.G., Katz, M.E., Knoll, A.H., Quigg, A., Raven, J.A., Schofield, O., Taylor, F.J.R., 2004. The evolution of modern eukaryotic phytoplankton. Science 305, 354–360.

Fedorov, A.V., Dekens, P.S., McCarthy, M., Ravelo, A.C., deMenocal, P.B., Barreiro, M., Pacanowski, R.C., Philander, S.G., 2006. The Pliocene paradox (mechanisms for a permanent El Niño). Science 312, 1485–1489.

Feldman, A.H., 2003. The evolutionary origin and development of the neogene planktonic foraminiferal *globorotalia* (*truncorotalia*) subgenus: the mode and tempo of speciation and the origin of coiling direction reversals and dominance. PhD thesis, State University of Florida.

Flakowski, J., Bolivar, I., Fahrni, J., Palowski, J., 2005. Actin phylogeny of foraminifera. J. Foraminiferal Res. 35, 93–102.

Flower, B.P., Kennett, J.P., 1993. Middle Miocene ocean–climate transition: high resolution oxygen and carbon isotopic records from DSDP Site 588A, southwest Pacific. Paleoceanography 8, 811–843.

Flower, B.P., Kennett, J.P., 1995. Middle Miocene deep water paleoceanography in the southwest Pacific: relating with East Antarctic ice sheet development. Paleoceanography 10 (1095), 1112.

Fraile, I., Schulz, M., Mulitza, S., Kucera, M., 2008. Predicting the global distribution of planktonic foraminifera using a dynamic ecosystem model. Biogeosciences 5, 891–911.

Fuchs, W., 1967. U ber Ursprung und Phylogenie der Trias-"Globigerinen" und die Bedeutung dieses Formenkreises für das echte Plankton. Verh. Geol. Bundesanstalt 110, 135–176.

Fuchs, W., 1973. Ein Beitrag zur Kenntnis der Jura, Globigerinen und verwandter Formen an Hand polnischen Materials des Callovien and Oxfordien. Verh. Geol. Bundesanstalt 116, 445–487.

Fuchs, W., 1975. Zur Stammesgeschichte der Planktonforaminiferen und verwandter Formen im Mesozoikum. Jahrbuch der Geologischen Bundesanstalt 118, 193–246.

Gast, R.J., Caron, D.A., 1996. Molecular phylogeny of symbiotic dinoflagellates from planktonic foraminifera and radiolarians. Mol. Biol. Evol. 13, 1192–1197.

Georgescu, M., Huber, B.T., 2006. *Paracostellagerina* nov. gen., a meridionally costellate planktonic foraminiferal genus of the middle Cretaceous (late Albian-earliest Cenomanian). J. Foraminiferal Res. 36, 368–373.

Gibbard, P.L., Head, M.J., Walker, M.J.C., the Subcommission on Quaternary Stratigraphy, 2009. Formal ratification of the

Quaternary System/Period and the Pleistocene Series/Epoch with a base at 2.58 Ma. J. Quatern. Sci. 25, 96–102.

Goldstein, S.T., 1999. Foraminifera: a biological overview. In: Gupta, B.K.S. (Ed.), Modern Foraminifera. Kluwer Academic Publishers, Dordrecht, pp. 37–55.

Gonzàlez-Donoso, J.-M., Linares, D., Caron, M., Robazynski, F., 2007. The rotaliporids, a polyphyletic group of Albian-Cenomanian planktonic foraminifera. Emendation of Genera. J. Foraminiferal Res. 37, 175–186.

Gorbachik, T.N., Kusnetsova, K.I., 1986. Study of shells mineral composition of planktonic foraminifera. Vopr. Mikropaleontol. 28, 42–44.

Gorbatchik, T.N., Moullade, M., 1973. Caracteres microstructuraux de la paroi du test des foraminiferes planctoniques du Cretace inferieur et leur signification sur le plan taxinomique. C. R. Acad. Sci. Paris, 227, ser. D, S. 2661.

Gorbachik, T.N., Poroshina, L.A., 1979. New planktonic foraminifers from the Berriasian deposits of Azerbaijan. J. Paleonotol. 3, 22–28.

Görög, A., 1994. Early Jurassic planktonic foraminifera from Hungary. Micropaleonotology 40, 225–260.

Ghosh, P., Bhattacharaya, S.K., 2001. CO_2 levels in the late Palaeozoic and Mesozoic atmosphere from the $\delta^{13}C$ values of the pedogenic carbonate and the organic matter. Palaeogeogr. Palaeoclimatol. Palaeoecol. 170, 285–296.

Gould, S.J., 2002. The Structure of Evolutionary Theory. Harvard University Press, Cambridge, Massachusetts, p. 1433.

Gradstein, F.M., Ogg, J.G., Smith, A.G., 2004. A Geologic Time Scale. Cambridge University Press, Cambridge, p. 588.

Grigelis, A.A., 1958. *Globigerina oxfordiaiza* sp.n.—an occurrence of *Globigerina* in the Upper Jurassic strata of Lithuania. Nauchnye Doklady Vysshei Shkoly. Geologo-Geograficheskie Nauki 3, 109–111.

Gvirtzman, G., Buchbinder, B., 1978. Recent and Pleistocene Coral Reefs and Coastal Sediments of the Gulf of Elat. Sedimentology in Israel, Cyprus and Turkey, Tenth International Congress on Sedimentology, Jerusalem, Guidebook, Part II: Postcongress, Israel, 160–191.

Haarmann, T., Hathorne, C., Mohtadi, M., Groeneveld, J., Kölling, M., Bickert, T., 2011. Mg/Ca ratios of single planktonic foraminifer shells and the potential to reconstruct the thermal seasonality of the water column. Paleoceanography 26, PA3218. http://dx.doi.org/10.1029/2010PA002091.

Hallam, A., 1984. Continental humid and arid zones during the Jurassic and Cretaceous. Palaeogeogr. Palaeoclimatol. Palaeoecol. 47, 195–223.

Hallam, A., Wignall, P.B., 1997. Mass extinctions and their aftermath. Oxford University Press, Oxford, p. 307.

Hallam, A., Wignall, P.B., 2000. Facies change across the Triassic-Jurassic boundary in Nevada, USA. J. Geol. Soc. 156, 453–456.

Hallock, P., 1981. Production of carbonate sediments by selected large benthic foraminifera on two Pacific coral reefs. J. Sediment. Petrol. 51, 467–474.

Hallock, P., 1999. Symbiont-bearing foraminifera. In: Gupta, B.K.S. (Ed.), Modern Foraminifera. Kluwer Academic Publishers, Dordrecht, pp. 123–139.

Hallock, P., Röttger, R., Wetmore, K., 1991. Hypotheses on form and function in foraminifera. In: Lee, J.J., Anderson, O.R. (Eds.), Biology of Foraminifera. Academic Press, New York, pp. 41–72.

Haq, B.U., Hardenbol, J., Vail, P.R., 1988. Mesozoic and Cenozoic chronostratigraphy and cycles of sea-level change. In: Wilgus, C.K., Ross, C.A., Posamentier, H., Kendall, C.G.St.C. (Eds.), Sea-Level Changes: An Integrated Approach, vol. 42, pp. 71–108. SEPM Special Publication.

Hart, M.B., 2006. The origin and evolution of planktic foraminifera. In: de Souza Carvalho, I. (Ed.), Forams 2006.Anuario do Instituto de Geociencias. Rio de Janeiro, vol. 29, p. 167.

Hart, M.B., Oxford, M.J., Hudson, W., 2002. The early evolution and palaeobiogeography of Mesozoic planktonic foraminifera. Geol. Soc. Spec. Publ. Lond. 194, 115–125.

Hemleben, C., 1975. Spine and pustule relationships in some recent planktonic foraminifera. Micropaleontology 21, 334–341.

Hemleben, C., Spindler, M., 1983. Recent advances in research on living planktonic foraminifera. Utrecht Micropaleontol. Bull. 30, 141–170.

Hemleben, C.H., Anderson, O.R., Berthold, W., Spindler, M., 1986. Calcification and chamber formation in Foraminifera-a brief overview. In: Leadbeater, B.S.C., Riding, R. (Eds.), Biomineralization in lower plants and animals, Systematics Association Special, vol. 30, pp. 237–249. Clarendon.

Hemleben, C., Spindler, M., Anderson, O., 1989. Modern planktonic foraminifera. Springer-Verlag, Heidelberg, Tokyo,

New York.

Hemleben, C., Mühlen, D., Olsson, R.K., Berggren, W.A., 1991. Surface texture and the first occurrence of spines in planktonic foraminifera from the early Tertiary. Geologische Jahrbuch 128, 117–146.

Hesselbo, S.P., Robinson, S.A., Surlyk, F., Piasecki, S., 2002. Terrestrial and marine extinction at the Triassic-Jurassic boundary synchronized with major carbon-cycle perturbations, a link to initiation of massive volcanism? Geology 30, 251–254.

Heuser, A., Eisenhauer, A., Böhm, F., Wallmann, K., Gussone, N., Pearson, P.N., Nägler, T.F., Dullo, W.C., 2005. Calcium isotope ($\delta^{44/40}$Ca) variations of Neogene planktonic foraminifera. Paleoceanography 20, PA2013. http://dx.doi.org/ 10.1029/2004PA001048.

Higgins, J.A., Schrag, D.P., 2006. Beyond methane, towards a theory for Paleocene-Eocene thermal maximum. Earth Planet. Sci. Lett. 245, 523–537.

Hilbrecht, H., Thierstein, H.R., 1996. Benthic behavior of planktic foraminifera. Geology 24, 200–202.

Hochuli, P.A., Menegatti, A.P., Weissert, H., Riva, A., Erba, E., Premoli Silva, I., 1999. Episodes of high productivity and cooling in the early Aptian Alpine Tethys. Geology 27, 657–660.

Hofker, J., 1960. Planktonic foraminifera in the Danian of Denmark. Contributions of the Cushman. Found. Foraminiferal Res. 11, 73–86.

Holbourn, A., Kuhnt, W., Kawamura, H., Jian, Z., Grootes, G., Erlenkeuser, H., Xu, J., 2005. Orbitally paced paleoproductivity variations in the Timor Sea and Indonesian Throughflow variability during the last 460 kyr. Paleoceanography 20, PA3002. http://dx.doi.org/10.1029/2004PA001094.

Holmes, N.A., 1984. An emendation of the genera *Beella* Banner and Blow 1959, and *Turborotalita* Banner and Blow, 1962, with notes on *Orcadia* Boltovosky and Watanabe, 1982. J. Foraminiferal Res. 14, 101–110.

Hottinger, L., 2006. Illustrated glossary of terms used in foraminiferal research. Carnets de Géologie/Notebooks on Geology. Memoir 2006/02.

Huber, B.T., 1994. Ontogenetic morphometrics of some Late Cretaceous trochospiral planktonic foraminifera from the Austral Realm. Smithsonian Contributions to Paleobiology 77, 1–85.

Huber, B.T., Bijma, J., Darling, K., 1997. Cryptic speciation in the living planktonic foraminifer *Globigerinella siphonifera* (d'Orbigny). Paleobiology 23, 33–62.

Huber, R., Meggers, H., Baumann, K.H., Raymo, M.E., Henrich, R., 2000. Shell size variation of the planktic foraminifer *Neogloboquadrina pachyderma* sin. in the Norwegian-Greenland Sea during the last 1.3 Ma—implications for pale-oceanographic reconstructions. Palaeogeogr. Palaeoclimatol. Palaeoecol. 160, 193–212.

Huber, B.T., Olsson, R.K., Pearson, P.N., 2006. Taxonomy of Eocene microperforate planktonic foraminifera (*Jenkinsina, Cassigerinelloita, Chiloguembelina, Zeauvigerina, Tenuitella,* and *Cassigerinella*) and Problematica (*Dipsidripella* and *Tenuitella*?). In: Pearson, P.N., Olsson, R.K., Hemleben, C., Huber, B.T., Berggren, W.A. (Eds.), Atlas of Eocene Planktonic Foraminifera. Allen Press, Lawrence, Kansas, pp. Cushman Foundation Foraminiferal Research, Special Publication 41, 461–508.

Hudson,W., Hart,M.B., Smart,C.H., 2009. Palaeobiogeography of early planktonic foraminifera. Bull. Soc. Géol. France. 180, 27–38.

Hull, P.M., Norris, R.D., Bralower, T.J., Schueth, J.D., 2011. A role for chance in marine recovery from the end-Cretaceous extinction. Nat. Geosci. 4, 856–860. http://dx.doi.org/10.1038/NGEO1302.

Hutchinson, G.E., 1967. A Treatise on Limnology, Wiley, New York. Vol. 2. 1115p.

Iaccarino, S., 1985. Mediterranean Miocene and Pliocene planktic foraminifera. In: Bolli, H.M., Saunders, J.B., Perch-Nielsen, K. (Eds.), Plankton Stratigraphy. Cambridge University Press, Cambridge, pp. 283–314.

Iaccarino, S., Salvatorini, G., 1982. A framework of planktonic foraminiferal biostratigraphy for early Miocene to late Pliocene Mediterranean area. Paleontol. Stratigr. Evol. 2, 115–125.

Ivočeva, P., Trifonova, E., 1961. Tithonian *Globigerina* from northwest Bulgaria. Trudove Varkhu Geologiyata na Bulgariya, (Seriya Paleontologiya) 3, 343–347.

Jain, S., Collins, L.S., 1978. Trends in Caribbean paleoproductivity related to the Neogene closure of the Central American Seaway. Mar. Micropaleontol. 63, 57–74.

Jakobsson, M., Backman, J., Rudels, B., Nycander, J., Frank, M., Mayer, L., Jokat, W., Sangiorgi, F., O'Regan, M., Brinkhuis, H., King, J., Moran, K., 2007. The early Miocene onset of a ventilated circulation regime in the Arctic Ocean. Nature 447, 986–990.

Jenkins, D.G., 1966. Planktonic foraminiferal zones and new taxa from the Danian to lower Miocene of New Zealand. N.

Z. J. Geol. Geophys. 8, 1088–1126.

Jenkins, D.G., 1967. Cenozoic planktonic foraminifera of New Zealand. N. Z. Geol. Surv. Paleontol. Bull. 42, 1–278.

Jenkins, D.G., 1971. Stratigraphic position of New Zealand Pliocene-Pleistocene boundary. N. Z. J. Geol. Geophys. 14, 418–432.

Jenkins, D.G., 1978. Neogene planktonic foraminifers from Deep Sea Drilling Project Leg 40 Sites 360 and 362 in the southeastern Atlantic. In: Bolli, H.M., Ryan, W.B.F. *et al.*, (Eds.), Initital Reports Deep Sea Drilling Project40, U.S. Govt. Printing Office, Washington, DC, pp. 723–741.

Jenkins, D.G., Saunders, J.B., Cifelli, R., 1981. The relationship of *Globigerinoides bisphericus* Todd 1954 to *Praeorbulina sicana* (de Stefani) 1952. J. Foraminiferal Res. 11, 262–267.

Jenkyns, H.C., 1988. The Early Toarcian (Jurassic) anoxic event, stratigraphic, sedimentary, and geochemical evidence. Am. J. Sci. 288, 101–151.

Jenkyns, H.C., 2003. Evidence for rapid climate change in the Mesozoic–Palaeogene greenhouse world. Philos. Trans. R. Soc.A 361, 1885–1916.

Jenkyns, H.C., Clayton, C.J., 1997. Lower Jurassic epicontinental carbonates and mudstones from England and Wales, chemostratigraphic signals and the Early Toarcian anoxic event. Sedimentology 44, 687–706.

Jenkyns, H.C., Wilson, P.A., 1990. Stratigraphy, paleoceanography, and evolution of cretaceous pacific guyots: relics from a greenhouse earth. Am. J. Sci. 299, 341–392.

Jiang, S., Bralower, T.J., Patzkowsky, M.E., Kump, L.R., Schueth, J.D., 2010. Geographic controls on nannoplankton extinction across the Cretaceous/Palaeogene boundary. Nat. Geosci. 3, 280–285.

Johnson, C.C., Kauffman, E.G., 1990. Originations, radiations and extinctions of Cretaceous rudistid bivalve species in the Caribbean Province. In: Kauffman, E.G., Walliser, O.H. (Eds.), Extinction events in earth history. Springer-Verlag, Berlin-Heidelberg, New York, pp. 305–324.

Johnson, C.C., Barron, E., Kauffman, E., Arthur, M., Fawcett, P., Yasuda, M., 1996. Middle Cretaceous reef collapse linked to ocean heat transport. Geology 24, 376–380.

Johnson, K.G., Todd, J.A., Jackson, J.B.C., 2007. Coral reef development drives molluscan diversity increase at local and regional scales in the late Neogene and Quaternary of the southwestern Caribbean. Paleobiology 33, 24–52.

Jolivet, L., Augier, R., Robin, C., Suc, J.-P., Rouchy, J.M., 2006. Lithospheric-scale geodynamic context of the Messinian salinity crisis. Sediment. Geol. 188–189, 9–33.

Kaneps, A.G., 1975. Cenozoic planktonic foraminifera from Antarctic Deep Sea sediments, Leg 28, DSDP. In: Hays, D.E. *et al.*, (Ed.), Initial Reports of the Deep Sea Drilling Project, 128. U.S. Government Printing Office, Washington, DC, pp. 573–583.

Kasimova, G.D., Aliyeva, D.G., 1984. Planktonic foraminifera of the Middle Jurassic beds of Azerbaijan. In: Alizade, K.A. (Ed.), Voprosy Paleontologii i Stratigrafi Azerbaijana, 479, 8–19.

Keeling, P.J., 2001. Foraminifera and Cercozoa are related in actin phylogeny: two orphans find a home? Mol. Biol. Evol. 18, 1551–1557.

Keller, G., 1981. Miocene biochronology and paleoceanography of the North Pacific: CENOP Symposium Volume. Mar. Micropaleontol. 6, 535–551.

Keller, G., Stinnesbeck, W., Adatte, T., Stueben, D., 2003. Multiple impacts across the Cretaceous-Tertiary boundary. Earth Sci. Rev. 62, 327–363.

Kelley, S.P., Gurov, E., 2002. Boltysh, another end-Cretaceous impact. Meteorit. Planet. Sci. 37, 1031–1043.

Kennett, J.P., Srinivasan, M.S., 1983. Neogene planktonic foraminifera: a phylogenetic atlas. Hutchinson Ross, Stroudsburg, Pennsylvania, 265p.

Kerr, R.A., 2006. Creatures great and small are stirring the ocean. Science 313, 1717.

Kiessling, W., 2002. Secular variations in the Phanerozoic reef ecosystem. In: Kiessling, W. *et al.*, (Ed.), Phanerozoic reef patterns: SEPM (Society for Sedimentary Geology). Special Publication, Berlin, Germany, 72, pp. 625–690.

Kimoto, K., Ishimura, T., Tsunogai, U., Itaki, T., Ujiié, Y., 2009. The living triserial planktic foraminifer *Gallitellia vivans* (Cushman): distribution, stable isotopes, and paleoecological implications. Mar. Micropaleontol. 71, 71–79.

Korchagin, O.A., Kuznetsova, K.I., Bragin, N.Yu., 2003. Find of Early Planktonic Foraminifers in the Triassic of the Crimea. Dokl. Earth Sci. 390A, 482–486.

Koutsoukos, C.A., Leafy, P.N., Hart, M.B., 1989. *Favusella washitensis* Michael (1972): evidence of ecophenotypic adaptation of a planktonic foraminifer to shallow-water carbonate environments during tile mid-Cretaceous. J.

Foraminiferal Res. 19, 324–336.

Kroon, D., Nederbragt, A.J., 1990. Ecology and paleoecology of triserial planktic foraminifera. Mar. Micropaleontol. 16, 25–38.

Kroon, D., Wouters, P.F., Moodley, L., Ganssen, G., Troelstra, S.R., 1988. Phenotypic variation of *Turborotalita quinqueloba* (Natland) tests in living populations and in the Pleistocene of an eastern Mediterranean piston core. In: Brummer, G.J.A., Kroon, D. (Eds.), Planktonic Foraminifers as Tracers of Ocean-Climate History. Free University Press, Amsterdam, pp. 131–147.

Kucera, M., Kennett, J.P., 2002. Causes and consequences of a middle Pleistocene origin of the modern planktonic foraminifer *Neogloboquadrina pachyderma* sinistral. Geology 30, 539–542.

Kuroda, J., Ogawa, N., Tanimizu, M., Coffin, M., Tokuyama, H., Kitazato, H., Ohkouchi, N., 2007. Contemporaneous massive subaerial volcanism and late cretaceous Oceanic Anoxic Event 2. Earth Planet. Sci. Lett. 256, 211–223.

Kuroyanagi, A., Kawahata, H., 2004. Vertical distribution of living planktonic foraminifera in the seas around Japan. Mar. Micropaleontol. 53, 173–196.

Kuznetsova, K.I., 2002. The earliest (Early Jurassic) stage and peculiarities of the evolution of planktonic foraminifers. Dokl. Earth Sci. 383A, 262–266.

Kuznetsova, K.I., Gorbachik, T.N., 1980. Novienakhodki planktonnikh foraminifer, v verkhyurskikh otlzhenii Krime. Dokl. Akad. Nauk SSSR 254, 748–751.

Langer, M.R., 2008. Assessing the contribution of foraminiferan protists to global ocean carbonate production. J. Eukaryot. Microbiol. 55, 163–169.

Latimer, J.C., Filippelli, G.M., 2002. Eocene to Miocene terrigenous inputs and export production, geochemical evidence from ODP Leg 177, Site 1090. Palaeogeogr. Palaeoclimatol. Palaeoecol. 182, 151–164.

Lawver, L.A., Gahagan, L.M., Coffin, M.F., 1992. The development of paleoseaways around Antarctica. In: Kennett, J.P., Warnke, D.A. (Eds.), The Antarctic Paleoenvironment, A Perspective on Global Change, American Geophysical Union Monograph 56, 7–30.

Leary, P.N., Hart, M.B., 1988. Comparison of the Cenomanian Foraminiferida from Goban Spur, Site 551, DSDP Leg 80 (Western Approaches) and Dover (SE England). Proc. Ussher Soc. 7, 81–85.

Leckie, R.M., 2009. Seeking a better life in the plankton. Proc. Natl. Acad. Sci. U.S.A. 106, 14183–14184.

Leckie, R.M., Bralower, T.J., Cashman, R., 2002. Oceanic anoxic events and plankton evolution: biotic Response to tectonic forcing during the mid-Cretaceous. Paleoceanography 17, 13–29.

Lee, J.J., 1990. Phylum Granuloreticulosa (Foraminifera). In: Margulis, L., Corliss, J.O., Melkonian, M., Chapman, D.J. (Eds.), Handbook of Protoctista. Jones and Bartlett, Boston, pp. 524–548.

Lee, J.J., Anderson, O.R., 1991. Symbiosis in foraminifera. In: Lee, J.J., Anderson, O.R. (Eds.), Biology of Foraminifera. Academic Press, New York, pp. 157–220.

Lee, J.J., Freudenthal, H., Kossoy, V., Bs, A., 1965. Cytological observations on two planktonic foraminifera, *Globigerina bulloides* d'Orbigny 1826, and *Globigerinoides ruber* (d'Orbigny 1839) Cushman, 1927. J. Protozool. 12, 531–542.

Li, Q., 1987. Origin, phylogenetic development and systematic taxonomy of the *Tenuitella plexus* (Globigerinitidae, Globigerininina). J. Foraminiferal Res. 14, 298–320.

Li, Q., McGowran, B., Boersma, A., 1995. Early Paleocene *Parvularugoglobigerina* and late Eocene *Praetenuitella*: does evolutionary convergence imply similar habitat? J. Micropalaeontol. 14, 119–134.

Lipps, J.H., 1966. Wall structure, systematics, and phylogeny studies of Cenozic planktonic foraminifera. J. Paleontol. 40, 1257.

Lipps, J.H., 1979. Ecology and paleoecology of planktic foraminifera. In: Lipps, J.H., Berger, W.H., Buzas, M.A., Douglas, R.G., Ross, C.A. (Eds.), Foraminiferal Ecology and Paleoecology, 62–104. SEPM Short Course Notes 6. Society of Economic Paleontologists and Mineralogists.

Liu, C., Olsson, R.K., 1992. Evolutionary radiation of microperforate planktonic foraminifera following the K/T mass extinction event. J. Foraminiferal res. 22, 328–346.

Liu, C., Olsson, R.K., 1994. On the origin of Danian normal perforate planktonic foraminifera from *Hedbergella*. J. Foraminiferal Res. 24, 61–74.

Liu, C., Olsson, R.K., Huber, B.T., 1998. A benthic paleohabitat for *Praepararotalia* gen. nov. and *Antarcticella* Loeblich and Tappan. J. Foraminiferal Res. 28, 3–8.

Loeblich Jr., A.R., Tappan, H., 1964. Protista 2, Sarcodina, chiefly "Thecamoebians" and Foraminiferida. In: Moore, R.C.

(Ed.), Treatise on Invertebrate Paleontology, Part C, Vols. 1 and 2. University Kansas Press, Kansas.

Loeblich Jr., A.R., Tappan, H., 1988. Foraminiferal Genera and their Classification,2 Vols. Van Nostrand Reinhold, New York 2047pp.

Lohmann, G.P., 1992. Increasing seasonal upwelling in the subtropical south Atlantic over the past 700,000 yrs: evidence from deep living planktonic foraminifera. Mar. Micropaleontol. 19, 1–12.

Longet, D., Archibald, J.M., Keeling, P.J., Pawlowski, J., 2003. Foraminifera and Cercozoa share a common origin according to RNA polymerase II phylogemies. Int. J. Syst. Evol. Microbiol. 53, 1735–1739.

Longoria, J.F., 1974. Stratigraphic, morphologic and taxonomic studies of Aptian planktonic foraminifers. Revista española de micropaleontología, Spec. Issue, 1–107.

Louis-Schmid, B., Rais, P., Bernasconi, S.M., 2007. Detailed record of the mid-Oxfordian (Late Jurassic) positive carbon-isotope excursion in two hemipelagic sections (France and Switzerland), A plate tectonic trigger? Palaeogeogr. Palaeoclimatol. Palaeoecol. 248, 459–472.

Lu, G., Keller, G., 1995. Ecological stasis and saltation: species richness change in planktic foraminifera during the late Paleocene to early Eocene, DSDP Site 577. Paleogeogr. Paleoclimatol. Paleoecol. 117, 211–227.

Lunt, D.J., Ridgwell, A., Sluijs, A., Zachos, J., Hunter, S., Haywood, A., 2011. A model for orbital pacing of methane hydrate destabilization during the Palaeogene. Nat. Geosci. http://dx.doi.org/10.1038/NGEO126.

Majewski, W., 2003. Water-depth distribution of Miocene planktonic foraminifera from ODP site 744, Southern Indian Ocean. J. Foraminiferal Res. 33 (2), 144–154.

Mann, S., 2001. Biomineralization: Principles and Concepts in Bioinorganic Materials Chemistry. Oxford University Press, Oxford, United Kingdom, 198p.

Margulis, L., Fester, R. (Eds.), 1991. Symbiosis as a Source of Evolutionary Innovation. MIT Press, Cambridge, MA.

MacLeod, N., 2001. The role of phylogeny in quantitative paleobiological analysis. Paleobiology 27, 226–241.

McGowran, B., Berggren, W.A., Hilgen, F., Steininger, F., Aubry, M.-P., Lourens, L., Van Couvering, J., 2009. Neogene and Quaternary coexisting in the geological time scale: the inclusive compromise. Earth Sci. Rev. 96, 249–262.

Miller, Kenneth G., Browning, James V., Aubry, Marie-Pierre, Wade, Bridget S., Katz, Miriam E., Kulpecz, Andrew A., Wright, James D., 2008. Eocene-Oligocene global climate and sea-level changes: St. Stephens Quarry, Alabama. Geol. Soc. Am. Bull. 120, 34–53.

Miller, K.G., Mountain, G.S., Wright, J., Browning, J.V., 2011.A 180-million-year record of sea level and ice volume variations from continental margin and deep-sea isotopic records. Oceanography 24, 40–53.

Morozova, V.G., 1959. Stratigraphy of the Danian-Montian deposits of Crimea by means of Foraminifera (in Russian). Dokl. Akad. Nauk SSSR 124, 1113–1116.

Morozova, V.G., 1960. Stratigraphical zonation of Danian-Montian deposits in the U.S.S.R. and the Cretaceous-Paleogene boundary—XXI International Geology Congress. Dokl. Soviet Geol. 5, 97–102.

Moullade, M., Bellier, J.-P., Tronchetti, G., 2002. Hierarchy of criteria, evolutionary processes and taxonomic simplification in the classification of Lower Cretaceous planktonic foraminifera. Cretaceous Res. 23, 111–148.

Murray, J.W., 1991. Ecology and palaeoecology of benthic foraminifera. Longman, Harlow. Mutti, M., Piller, W.E., Betzler, C. (Eds.), 2011. Carbonate Systems During the Oligocene-Miocene Climatic Transition. Special Publication 42 of the International Associations of the Sedimentologists, 312.

Nederbragt, A.J., 1990. Biostratigraphy and Paleoceanographic Potential of the Cretaceous Planktic Foraminifera Heterohelicidae. Ph.D. Centrale Huisdrukkerij Vrije University, Amsterdam.

Nikolaev, S.I., Berney, C., Fahrni, J., Bolivar, I., Polet, S., Mylnikov, A.P., Aleshin, V.V., Petrov, N.B., Pawlowski, J., 2004. The twilight of Heliozoa and rise of Rhizaria: an emerging supergroup of amoeboid eukaryotes. Proc. Natl. Acad. Sci. U.S.A. 101, 8066–8071.

Norris, R.D., 1991. Parallel evolution in the keel structure of planktonic foraminifera. J. Foraminiferal Res. 21, 319–331.

Norris, R.D., 1996. Symbiosis as an evolutionary innovation in the radiation of Paleocene planktic foraminifera. Paleobiology 22, 461–480.

Norris, R.D., 1998. Planktonic foraminifer biostratigraphy: Eastern Equatorial Atlantic. In: Mascle, J., Lohmannn, G.P., Moullade, M. (Eds.), Proceedings of the Ocean Drilling Program, Scientific Results 159, 445–479.

Norris, R.D., Nishi, H., 2001. Evolution of iterative trends in coiling of Paleocene-Eocene tropical planktic foraminifera. Paleobiology 27 (2), 327–347.

Norris, R.D., Wilson, P.A., 1998. Low-latitude sea-surface temperatures for the mid-Cretaceous and the evolution of planktic foraminifera. Geology 26, 823–826.

O'Dogherty, L., Sandoval, J., Vera, J.A., 2000. Ammonite faunal turnover tracing sea level. J. Geol. Soc. Lond. 157, 723–736.

Oberhauser, R., 1960. Foraminiferen und Mikrofossilien "incertae sedis" der ladinischen und karnischen Stude der Trias aus den Ostalpen und Persien. Jahrbuch der Geologischen Bundesanstalt Sonderbdand, Wien 5, 5–46.

Olsson, R.K., 1963. Latest Cretaceous and earliest Tertiary stratigraphy of New Jersey Coastal Plain. Bull. Am. Assoc. Petrol. Geol. 47, 643–665.

Olsson, R.K., 1970. Planktonic foraminifera from base of Tertiary, Millers Ferry, Alabama. J. Paleontol. 44, 598–604.

Olsson, R.K., 1982. Cenozoic planktonic foraminifera: a paleobiogeographic summary. Notes for a Short Course organised by Buzas, M.A. and Sen Gupta, B.K, University of Tennessee, 1–26.

Olsson, R.K., Hemleben, C., Berggren, W.A., Liu, C., 1992. Wall texture classification of planktonic foraminifera genera in the lower Danian. J. Foraminiferal Res. 22, 195–213.

Olsson, R.K., Hemleben, C., Berggren, W.A., Huber, B.T., 1999. Atlas of Paleocene planktonic foraminifera. Smithsonian Contrib. Paleobiol. 85, 252.

Ortiz, J.D., Mix, A.C., Collier, R.W., 1995. Environmental control of living symbiotic and asymbiotic foraminifera of the California Current. Paleoceanography 10, 987–1009.

Orue-Etxebarria, X., 1985. Descripcion de dos nuevas especies de foraminíferos planctonicos en el Eoceno costero de la provincia de Bizkaia. Revista Española de Micropaleontología 17, 467–477.

Ottens,J.J.,1992.Plankticforaminifera as indicators of ocean environments in the northeast Atlantic. Enschede:FEBOB.V. 189pp.

Oxford, M.J., Gregory, F.J., Hart, M.B., Henderson, A.S., Simmons, M.D., Watkinson, M.P., 2002. Jurassic planktonic foraminifera from the United Kingdom. Terra Nova 14, 205–209.

Pawlowski, J., Bolivar, I., Guiard-Maffia, J., Gouy, M., 1994. Phylogenetic position of the foraminifera inferred from LSU rRNA gene sequences. Mol. Biol. Evol. 11, 929–938.

Pawlowski, J., Bolivar, I., Fahrni, J.F., Cavalier-Smith, T., Gouy, M., 1996. Early origin of foraminifera suggested by SSU rRNA gene sequences. Mol. Biol. Evol. 13, 445–450.

Pearson, P.N., 1993. A lineage phylogeny for the Paleogene planktonic foraminifera. Micropaleontology 39, 193–232.

Pearson, P.N., 1998. Speciation and extinction asymmetries in paleontological phylogenies; evidence for evolutionary progress?: Paleobiology 23, 305–335.

Pearson, P.N., Berggren, W.A., 2006. Taxonomy, biostratigraphy and phylogeny of *Morozovelloides* n. gen. In: Pearson, P.N., Olsson, R.K., Huber, B.T., Hemleben, C., Berggren, W.A. (Eds.), Atlas of Eocene Planktonic Foraminifera, Cushman Foundation Special Publication, 41, 327–342.

Pearson, P.N., Chaisson, W.P., 1997. Late Paleocene to middle Miocene planktonic foraminifer biostratigraphy of the Ceara Rise. In: Shackleton, N.J., Curry, W.B., Richter, C., Bralower, T.J. (Eds.), Proceeding of the Ocean Drilling Program, Scientific Results154, 33–68.

Pearson, P.N., Wade, B.S., 2009a. Taxonomy and stable isotope paleoecology of well preserved planktonic foraminifera from the uppermost Oligocene of Trinidad. J. Foraminiferal Res. 39, 191–217.

Pearson, P.N., Shackleton, N.J., Hall, M.A., 1993. Stable isotope paleoecology of middle Eocene planktonic foraminifera and multi-species isotope stratigraphy, Deep Sea Drilling Project Site 523, South Atlantic. J. Foraminiferal Res. 23, 123–140.

Pearson, P.N., Shackleton, N.J., Weedon, G.P., Hall, M.A., 1997. Multispecies planktonic foraminifer stable isotope stratigraphy through Oligocene/Miocene boundary climatic cycles, Site 926. Proc Ocean Drill. Prog. Sci. Results 154, 441–449.

Pearson, P.N., Ditchfield, P.W., Singano, J., Harcourt-Brown, K.G., Nicholas, C.J., Olsson, R.K., Shackleton, N.J., Hall, M.A., 2001. Warm tropical sea surface temperatures in the Late Cretaceous and Eocene epochs. Nature 413, 481–487.

Pearson, P.N., Olsson, R.K., Hemleben, C., Huber, B.T., Berggren, W.A. (Eds.), 2006a. In: Atlas of Eocene Planktonic Foraminifera 41, Spec. Publ. Cushman Found., Washington, DCJ. Foraminiferal Res, p. 513.

Pearson, P.N., Premec-Fucek, V., Premoli Silva, I., 2006b. Taxonomy, biostratigraphy, and phylogeny of Eocene *Turborotalia*. In: Special Publication Cushman Foundation for Foraminiferal Research, 433–460. ISSN 0070-2242. 41.

Pearson, P.N., van Dongen, B.E., Nicholas, C.J., Pancost, R.D., Schouten, S., Singano, J.M., Wade, B.S., 2007. Stable warm tropical climate through the Eocene Epoch. Geology 35, 211–214.

Peck, V.L., Hall, I.R., Zahn, R., Elderfield, H., 2008. Millennial-scale surface and subsurface paleothermometry from the northeast Atlantic, 55–8 ka BP. Paleoceanography 23, PA3221. http://dx.doi.org/10.1029/2008PA001631.

Pessagno Jr., E.A., 1967. Upper Cretaceous stratigraphy of the western Gulf Coastal Plain. Palaeontogr. Am. 5, 245–445.

Petrizzo, M.R., Huber, B.T., Wilson, P.A., MacLeod, K.G., 2008. Late Albian paleoceanography of the western subtropical North Atlantic. Paleoceanography PA1213. http://dx.doi.org/10.1029/2007PA001517.

Poore, R.Z., Berggren, W.A., 1975. Late Cenozoic planktonic foraminiferal biostratigraphy and paleoclimatology of Hatton-Rockall Basin DSDP Site 116. J. Foraminiferal Res. 5, 270–293.

Postuma, J.A., 1971. Manual of Planktonic Foraminifera. Elsevier, Amsterdam, 420pp.

Premoli Silva, I., Sliter, W.V., 1995. Cretaceous planktonic foraminiferal biostratigraphy and evolutionary trends from the Bottaccione section, Gubbio, Italy. Paleontogr. Ital. 82, 1–89.

Premoli Silva, I., Sliter, W.V., 1999. Cretaceous paleoceanography: evidence from planktonic foraminiferal evolution. In: Barrera, E., Johnson, C.C. (Eds.), Evolution of the Cretaceous Ocean-Climate System. Geological Society of America. Special Paper 332, 301–328.

Premoli-Silva, I., Bolli, H.M., 1973. Late Cretaceous to Eocene planktonic foraminifera and stratigraphy of Leg 15 sites in the Caribbean Sea. In: Edgar, N.T., Saunders, J.B. *et al.*, (Eds.), Initial Reports of the Deep Sea Drilling Project 15, U.S. Government Printing Office, Washington, DC, pp. 449–547.

Premoli Silva, I., Wade, B.S., Pearson, P.N., 2006. Chapter 7: Taxonomy of *Globigerinatheka* and *Orbulinoides*. In: Pearson, P.N., Olsson, R.K., Huber, B.T., Hemleben, C., Berggren, W.A. (Eds.), Atlas of Eocene Planktonic Foraminifera, Cushman Foundation Special Publication, 41, 169–212.

Quillévéré, F., Norris, R.D., Moussa, I., Berggren, W.A., 2001. Role of photosymbiosis and biogeography in the diversification of early Paleogene acarininids (planktonic foraminifera). Paleobiology 27, 311–326.

Resig, J., Kroopnick, P., 1983. Isotopic and distributional evidence of a planktonic habit for the foraminiferal genus Streptochilus Brönnimann and Resig, 1971. Mar. Micropaleontol. 8, 235–248.

Rey, M., 1939. Distribution stratrigraphique des *Hantkenina* dans le Nummulitique du Rharb (Maroc). Bulletin de la Societe Geologique de France 5, 321–341.

Robaszynski, F., Caron, M., 1979. Atlas de Foraminife`res planctoniqes du Crétacé moyen (mer Boréal et Tethys), I et II. (Coordinators) Cahiers de Paléontologie, I, 1–185; II, 1–181.

Robaszynski, F., Caron, M., Gonzales, J.M., Wonders, A.H. (Eds.), 1984. Atlas of Late Cretaceous globotruncanids. Rev. Micropaléontol. 26, 145–305.

Rögl, F., Egger, H., 2010. The missing link in the evolutionary origin of the foraminiferal genus *Hantkenina* and the problem of the Lower/Middle Eocene boundary. Geology 38, 23–26.

Rögl, F., Steininger, F.F., 1983. Vom Zerfall der Tethys zu Mediterran und Paratethys. Annalen des Naturhistorischen Museums in Wien 85A, 135–163.

Rohling, E.J., Cooke, S., 1999. Stable oxygen and carbon isotopes in foraminiferal carbonate shells. In: Barun, K. Sen Gupta (Ed.), Modern Foraminifera, Kluwer Academic Publishers, Great Britain, pp. 239–258.

Roth, J.M., Droxler, A.W., Kameo, K., 2000. The Caribbean carbonate crash at the middle to late Miocene transition, linkage to the establishment of the modern global ocean conveyor. In: Leckie, R.M., Sigurdsson, H., Acton, G.D., Draper, G. (Eds.), Proceedings of the Ocean Drilling Program, Scientific Results165, 249–273.

Said, R., Kenawy, A., 1956. Upper Cretaceous and Lower Tertiary foraminifera from northern Sinai, Egypt. Micropaleontology 2, 105–173.

Saito, T., 1976. Geologic significance of coiling direction in the planktonic foraminifera *Pulleniatina*. Geology 4, 305–309.

Saito, T., Thompson, P.R., Breger, D., 1976. Skeletal ultramicrostructure of some elongate-chambered planktonic foraminifera and related species. In: Takayanagi, Y., Saito, T. (Eds.), Progress in micropaleontology; selected papers in honor of Prof. Kiyoshi Asano. Micropaleontology Press, New York, pp. 278–304.

Saito, T., Thompson, P.R., Breger, D., 1981. Systematic Index of Recent and Pleistocene Planktonic Foraminifera. University of Tokyo Press, Tokyo, pp. 1–190.

Samuel, O., Salaj, J., 1968. Microbiostratigraphy and Foraminifera of the Slovak Carpathian Paleogene. Geologicky Ustav Dionyza Stura, Bratislava 1-232.

Samson, Y., Janin, M.-C., Bignot, G., Guyader, J., Breton, G., 1992. Les Globigerines (Foraminiferes planctoniques) de l'Oxfordien inferieur de Villers-sur-Mer (Calvados, France) dans leur gisement. Rev. Paléobiol. 11, 409–431.

Schmitz, B., Speijer, R.P., Aubry, M.-P., 1996. Latest Paleocene benthic extinction event on the southern Tethyan shelf (Egypt); foraminiferal stable isotopic(^{13}C, ^{18}O) records. Geology 24, 347–350.

Schneider, C.E., Kennett, J.P., 1999. Segregation and speciation in the Neogene planktonic foraminiferal clade, *Globoconella*. Paleobiology 25, 383–395.

Schulte, P., and 40 others, 2010. The Chicxulub asteroid impact and mass extinction at the Cretaceous-Paleogene boundary. Science 327 (5970), 1214–1218. http://dx.doi.org/10.1126/science.1177265.

Scott, G.H., 1974. Biometry of the foraminiferal shell. In: Hedley, R.H., Adams, C.G. (Eds.), Foraminifera, Academic Press, New York, 1, pp. 55–151.

Seears, H.A., Darling, K.F., Wade, C.M., 2012. Ecological partitioning and diversity in tropical planktonic foraminifera. BMC Evol. Biol. 12, 54. http://dx.doi.org/10.1186/1471-2148-12-54.

Sen Gupta, B.K., 1999. Introduction to modern foraminifera. In: Sen Gupta, B.K. (Ed.), Modern Foraminifera. Kluwer Academic Publishers, Dordrecht, p. 384.

Sen Gupta, B.K., Platon, E., Bernhard, J.M., Aharon, P., 1997. Foraminiferal colonization of hydrocarbon-seep bacterial mats and underlying sediment, Gulf of Mexico Slope. J. Foraminiferal Res. 27, 292–300.

Sexton, P.F., Norris, R.D., 2008. Dispersal and biogeography of marine plankton: long-distance dispersal of the foraminifer *Truncorotalia truncatulinoides*. Geology 36, 899–902.

Shackleton, N.J., 1967. Oxygen isotope analyses and Pleistocene temperatures reassessed. Nature 215, 15–17.

Shackleton, N.J., Opdyke, N.D., 1973. Oxygen isotope and palaeomagnetic stratigraphy of Equatorial Pacific core V28-238: oxygen isotope temperatures and ice volumes on a 105 year and a 106 year scale. Quater. Res. 3, 39–55.

Shackleton, N.J., Vincent, E., 1978. Oxygen and carbon isotope studies in recent foraminifera from the southwest Indian Ocean. Mar. Micropaleontol. 3, 1–13.

Shackleton, N.J., Hall, M.A., Bleil, U., 1985. Carbon isotope stratigraphy, Site 577. In: Heath, G.R., Burckle, L.D. *et al.*, (Eds.), Initial Reports DSDP, 86: (U.S. Govt. Printing Office), Washington, pp. 503–511.

Shackleton, N.J., Hall, M.A., Pate, D., 1995. Pliocene stable isotope stratigraphy of Site 864. In: Pisias, N.G., Mayer, L.A., Janecek, T.R., Palmer-Julson, A., van Andel, T.H. (Eds.), Proceedings of the Ocean Drilling Program, Scientific Result 138, Ocean Drilling Program, College Station, TX, pp. 337–355.

Sheehan, R., Banner, F.T., 1972. The pseudopodia of *Elphidircm incertum*. Rev. Esp. Micropaleontol. 4, 31–63.

Shokhina, V.A., 1937. The Genus *Hantkenina* and Its Stratigraphical Distribution in the North Caucasus: Problems of Paleontology. Publication of the Laboratory of Palaeontology, Moscow University, 2–3, pp. 425–452.

Sinha, D.K., Singh, A.K., 2008. Late Neogene planktic foraminiferal biochronology of the ODP Site 763a, Exmouth Plateau, southeast Indian Ocean. J. Foraminferal Res. 38, 251–270.

Smit, J., 1977. Discovery of a planktonic Foraminiferal association between the *Abathomphalus mayaroensis* Zone and the *'Globigerina' eugubina* Zone at the Cretaceous/Tertiary Boundary in the Barranco del Gredero (Caravaca, S.E. Spain). Koninklijke Nederlandse Akademie van Wetenschappen Proc., Ser. B 80, 280–301.

Smit, J., 1982. Extinction and evolution of planktonic Foraminifera after a major impact at the Cretaceous/Tertiary Boundary: GSA Special Paper. 190, 329–352.

Smith, A.G., Pickering,K.T., 2003. Oceanic gateways as a critical factor to initiate ice-house Earth. J. Geol. Soc. Lond. 160, 337–340.

Soldan, D.M., Petrizzo, M.R., Premoli Silva, I., Cau, A., 2011. Phylogenetic relationships and evolutionary history of the Paleogene genus *Igorina* through parsimony analysis. J. Foraminiferal Res. 41 (3), 260–284.

Sole, R.V., Montoya, J.M., Erwin, D.H., 2002. Recovery after mass extinction: evolutionary assembly in large-scale biosphere dynamics. Philos. Trans. R. Soc. Lond. B 357, 697–707.

Spero, H.J., DeNiro, M.J., 1987. The influence of symbionts photosynthesis on the $\delta^{18}O$ and $\delta^{13}C$ values of planktonic foraminiferal shell calcite. Symbiosis 4, 213–228.

Spezzaferri, S., Pearson, P., 2009. Distribution and ecology of *Catapsydrax indianus*, a new planktonic foraminifer index species for the late oligocene-early miocene. J. Foraminiferal Res. 39, 112–119.

Spindler, M., Hemleben, C., 1980. Endocytobiology Endosymbiosis and Cell Biology. Walter de Gruyter & Co., Berlin.

Spindler, M., Dieckmann, G.S., 1986. Distribution and abundance of the planktic foraminifer *Neogloboquadrina pachyderma* in sea ice of the Weddell Sea (Antarctica). Polar Biol. 5, 185–191.

Spindler, M., Hemleben, C., Bayer, U., Bé, A.W.H., Anderson, O.R., 1979. Lunar periodicity of reproduction in the planktonic foraminifer *Hastigerina pelagica*. Mar. Ecol. Prog. Ser. 1, 61–64. http://dx.doi.org/10.3354/meps001061.

Spindler, M., Hemleben, Ch., Salomow, J., Smit, L., 1984. Feeding behavior of some planktonic foraminifera in laboratory cultures. J. Foraminiferal Res. 14, 231–249.

Sprovieri, R., Sprovieri, M., Caruso, A., Pelosi, N., Bonomo, S., Ferraro, L., 2006. Astronomic forcing on the planktonic foraminifera assemblage in the Piacenzian Punta Piccola section (southern Italy). Paleoceanography 21, PA4204. http://dx. doi.org/10.1029/2006PA001268.

Stainforth, R.M., Lamb, J.L., Luterbacher, H., Beard, J.H., Jeffords, R.M., 1975. Cenozoic planktonic foraminiferal zonation and characteristics of index forms. Univ. Kansas Paleontol. Contrib. 62, 1–425.

Stam, B., 1986. Quantitative analysis of Middle and Late Jurassic foraminifera from Portugal and its implications for the Grand Banks of Newfoundland. Utrecht Micropaleontol. Bull. 34, 1–168.

Stanley, S.M., Wetmore, K.L., Kennett, J.P., 1988. Macroevolutionary differences between the two major clades of Neogene planktonic foraminifera. Paleobiology 14, 235–249.

Steuber, T., Rauch, M., Masse, J.-P., Graaf, J., Malkoc, M., 2005. Low-latitude seasonality of Cretaceous temperatures in warm and cold episodes. Nature 437, 1341–1344.

Stewart, I.A., Darling, K.F., Kroon, D., Wade, C.M., Troelstra, S.R., 2001. Genotypic variability in subarctic Atlantic planktic foraminifera. Mar. Micropaleontol. 43, 143–153.

Stewart, J.A., Wilson, P.A., Edgar, K.M., Anand, P., James, R.H., 2012. Geochemical assessment of the palaeoecology, ontogeny, morphotypic variability and palaeoceanographic utility of "*Dentoglobigerina*" *venezuelana*". Marine Micropaleontology 84, 74–86.

Storey, M.R., Duncan, A., Swisher, C.C., 2007. Paleocene-Eocene thermal maximum and the opening of the northeast Atlantic. Science 316, 587–589.

Stott, L.D., Kennett, J.P., 1990. Antarctic Palaeogene planktonic foraminifera biostratigraphy: ODP Leg 113 Sites 689 and 690. Proc. Ocean Drill. Prog. Sci. Results 113, 549–569.

Stüben, D., Kramar, U., Harting, M., Stinnesbeck, W., Keller, G., 2005. High-resolution geochemical record of Cretaceous-Tertiary boundary sections in Mexico: new constraints on the K/T and Chicxulub events. Geochim. Cosmochim. Acta 69, 2559–2579.

Subbotina, N.N., 1953. Fossil foraminifera of the USSR. Globigerinidae, Hantkeninidae and Globorotaliidae [in Russian]. Trudy Vsesoyuznogo Neftyanogo Nauchno-Isledovatelskogo Geologo-Razvedochnogo Instituta (VNIGRI). Novaya Seriya 76, 1–296.

Takayanagi, Y., Saito, T., 1962. Planktonic foraminifers from the Nobori Formation, Shikoku, Japan. Sci. Rep. Tohoku Univ. Ser. 2 (5), 67–106.

Taniguchit, A., Bé, A.W.H., 1985. Variation with depth in the number of chambers in planktonic foraminiferal shells. J. Oceanogr. 1573-868X41 (1. 1985), 56–58.

ter Kuile, B., 1991. Mechanisms for calcification and carbon cycling in algal symbiont-bearing foraminifera. In: Lee, J.J., Anderson, O.R. (Eds.), Biology of Foraminifera. Academic Press, New York, pp. 73–89.

Thalmann, H.E., 1932. Die Foraminiferen-gattung *Hantkenina* Cushman, 1924, und ihre regional-stratigraphische Verbreitung. Eclogae Geol. Helv. 25, 287–292.

Thalmann, H.E., 1942. Foraminiferal genus *Hantkenina* and its subgenera. Am. J. Sci. 240, 809–823.

Thomas, E., Brinkhuis, H., Huber, M., Roehl, U., 2006. An ocean view of the early Cenozoic Greenhouse World. Oceanography 19, (Special Volume on Ocean Drilling), 63–72.

Toumarkine, M., Bolli, H.M., 1970. Evolution de *Globorotalia cerroazulensis* (Cole) dans l'Eocene Moyen et Superieur de Possagno (Italie). Revue de Micropaleonotologie 13, 131–145.

Turco, E., Bambini, A.M., Foresi, L., Iaccarino, S., Lirer, F., Mazzei, R., Salvatorini, G., 2002. Middle Miocene high-resolution calcareous plankton biostratigraphy at Site 926 (Leg 154, equatorial Atlantic Ocean): palaeoecological and palaeobiogeographical implications. Geobios Memoire special, no. 24, 257–276.

Twitchett, R.J., 2006. The palaeoclimatology, palaeoecology and palaeoenvironmental analysis of mass extinction events. Palaeogeogr. Palaeoclimatol. Palaeoecol. 232, 190–213.

Ujiié, Y., Lipps, J.H., 2009. Cryptic diversity in planktonic foraminifera in the Northwest Pacific Ocean. J. Foraminiferal Res. 39, 145–154.

Ujiié, Y., Kimoto, K., Pawlowski, J., 2008. Molecular evidence for an independent origin of modern triserial planktonic foraminifera from benthic ancestors. Mar. Micropaleontol. 69, 263–340.

Van Hinte, J.E., 1965. An approach to *Orbitoides*. Proceedings of the Koninklijke Nederlandse Akademie van Wetenschappen, series B 68, 57–71.

Van Hinte, J.E., 1968. The Late Cretaceous larger foraminifer *Orbitoides douvillei* (Silvestri) at its type locality Belvés, SW France. Proc. Koninklijke Nederlandse Akademie van Wetenschappen ser. B 71, 359–372.

Van Couvering, J.A., Aubry, M.-P., Berggren, W.A., Gradstein, F.M., Hilgen, F.J., Kent, D.V., Lourens, L.J., McGowran, B., 2009. What, if anything, is Quaternary? Episodes 32 (2), 125–126.

Varol, O., Houghton, S., 1996. Didemnid ascidian spicules. J. Micropaleotol. 15, 135–149.

Verga, D., Premoli Silva, I., 2003. Early Cretaceous planktonic foraminifera from the Tethys. The small-sized representatives of the genus *Globigerinelloides*. Cretaceous Res. 24, 305–334.

Vilks, G., Walker, D.A., 1974. Morphology of *Orbulina universa* d'Orbigny, in relation to other spinose planktonic foraminifera. J. Foraminiferal Res. 4, 1.

von der Heydt, A., Dijkstra, H.A., 2006. Effect of ocean gateways on the global ocean circulation in the late Oligocene and early Miocene. Paleoceanography 21, PA1011. http://dx.doi.org/10.1029/2005PA001149.

Wade, C.M., Darling, K.F., 2002. Fossilized records of past seas. Microbiol. Today 29, 183–185.

Wade, B.S., Kroon, D., 2002. Middle Eocene regional climate instability: evidence from the western North Atlantic. Geology 30, 1011–1014.

Wade, C.M., Darling, K.F., Kroon, D., Leigh Brown, A.J., 1996. Early evolutionary origin of the planktic foraminifera inferred from small subunit rDNA sequence comparisons. J. Mol. Evol. 43, 672–677.

Wade, B.S., Al-Sabouni, N., Hemleben, C., Kroon, D., 2008. Symbiont bleaching in fossil planktonic foraminifera. Evol. Ecol. 22, 253–265.

Wade, B.S., Pearson, P.N., Berggren, W.A., Pälike, H., 2011. Review and revision of Cenozoic tropical planktonic foraminiferal biostratigraphy and calibration to the geomagnetic polarity and astronomical time scale. Earth-Science Reviews, 104, 111–142.

Walliser, O.H., 1996. Global events and event stratigraphy in the Phanerozoic. In: Walliser, O.H. (Ed.), Global Events and Event Stratigraphy. Springer, Berlin, pp. 7–19.

Walliser, O.H., 2003. Sterben und Neubeginn im Spiegel der Palaeofauna. Die Bedeutung der globalen Faunenschnitte für die Stammesgeschichte. In: Hansch, W. (Ed.), Katastrophen in der Erdgeschichte. Wendezeiten des Lebens, museo Heilbronn (Sta¨dtische Museen), Vol. 19, pp. 60–69.

Walter, B., 1989. Au Valanginien supérieur, une crise de la faune des bryozoaires, indication d'un important refroidissement dans le Jura. Palaeogeogr. Palaeoclimatol. Palaeoecol. 74, 255–263.

Ward, P.D., 2006. Out of Thin Air. Joseph Henry Press, Washington, DC, pp. 282.

Weissert, H., Erba, E., 2004. Volcanism, CO_2 and palaeoclimate: a Late Jurassic–Early Cretaceous carbon and oxygen isotope record. J. Geol. Soc. Lond. 161, 695–702.

Wernli, R., 1995. Les foraminife`res globigériniformes (Oberhauserellidae) du Toarcien inférieur de Teysachaux (Préalpes médianes, Fribourg, Suisse). Rev. Paléobiol. 14, 257–269.

Wernli, R., Görög, A., 2000. Détermination of Bajocian protoglobigerinids (Foraminifera) in thin sections. Rev. Paléobiol. 19, 399–407.

Wernli, R., Görög, A., 2007. Protoglobigerinids and Oberhauserellidae (Foraminifera) of the Bajocian-Bathonian of the Southern Jura Mts, France. Rev. Micropaléontol. 50, 185–205.

West, O.L.O., 1995. A hypothesis for the origin of fibrillar bodies in planktic foraminifera by bacterial endosymbiosis. Mar. Micropaleontol. 26, 131–135.

White, T.J., Burns, T., Lee, S., Taylor, J., 1990. Amplification and direct sequencing of fungal ribosomal RNA genes for phylogenetics. In: Innis, M.A., Gelfland, D.H., Sninsky, J.J., White, T.J. (Eds.), PCR Protocols: A Guide to Methods and Applications. Harcourt Brace Jovanovich, San Diego, pp. 315–322.

Whiteside, J., Olsen, P., Eglinton, T., Brookfield, M., Sambrotto, R., 2010. Compound-specific carbon isotopes from Earth's largest flood basalt eruptions directly linked to the end-Triassic mass extinction. Proc. Natl. Acad. Sci. U.S.A. 107, 4872–4877.

Wignall, P.B., 2010. Safer in the south. Nat. Geosci. 3, 228–229.

Wilson, P.A., Norris, R.D., 2001. Warm tropical ocean surface and global anoxia during the mid-Cretaceous period. Nature 412, 425–429.

Whitfield, J.B., Lockhart, P.J., 2007. Deciphering ancient rapid radiations. Trends Ecol. Evol. 22, 258–265.

Woods, A.D., 2005. Paleoceanographic and paleoclimatic context of Early Triassic time. C. R. Acad. Sci. Pale, Vol 4. 395–404.

Wray, C.G., Langer, M.R., DeSalle, R., Lee, J.J., Lipps, J.H., 1995. Origin of the foraminifera. Proc. Natl. Acad. Sci. U.S.A. 92, 141–145.

Xu, X., Kimoto, K., Oda, M., 1995. Predominance of left-coiling *Globorotalia truncatulinoides* (d'Orbigny) between 115,000 and 50,000 years BP: a latest foraminiferal biostratigraphic event in the western North Pacific. Quatern. Res. 34, 39–47.

Yedid, G., Ofria, C.A., Lenski, R.E., 2009. Selective press extinctions, but not random pulse extinction cause delayed ecological recovery in communities of digital organisms. Am. Nat. 173, 139154.

Zachos, J.C., Pagani, M., Sloan, L., Thomas, E., Billups, K., 2001. Trends, rhythms, and aberrations in global climate 65 Ma to present. Science 292, 686–693.

Zachos, J.C., Dickens, G.R., Zeebe, R.E., 2008. An early Cenozoic perspective on greenhouse warming and carbon-cycle dynamics. Nature 451, 279.

Zhang, J., Scott, D.B., 1995. New planktonic foraminiferal genus and species from the upper Oligocene, DSDP Hole 366A, Leg 41. Micropaleonotology 41, 77–83.

Subject Index

Note: Page numbers followed by "*f*" indicate figures.

CPSIA information can be obtained
at www.ICGtesting.com
Printed in the USA
FSHW010309250620
71522FS

9 781910 634257